Petrology: Composition and Structure of Rocks

Petrology: Composition and Structure of Rocks

Edited by Seymor White

SYRAWOOD
PUBLISHING HOUSE

New York

Published by Syrawood Publishing House,
750 Third Avenue, 9th Floor,
New York, NY 10017, USA
www.syrawoodpublishinghouse.com

Petrology: Composition and Structure of Rocks
Edited by Seymor White

© 2018 Syrawood Publishing House

International Standard Book Number: 978-1-68286-557-6 (Hardback)

Cataloging-in-Publication Data

Petrology : composition and structure of rocks / edited by Seymor White.
   p. cm.
Includes bibliographical references and index.
ISBN 978-1-68286-557-6
1. Petrology. 2. Rocks. 3. Physical geology. I. White, Seymor.
QE431.2 .P48 2018
552--dc23

TABLE OF CONTENTS

PREFACE

The world is advancing at a fast pace like never before. Therefore, the need is to keep up with the latest developments. This book was an idea that came to fruition when the specialists in the area realized the need to coordinate together and document essential themes in the subject. That's when I was requested to be the editor. Editing this book has been an honour as it brings together diverse authors researching on different streams of the field. The book collates essential materials contributed by veterans in the area which can be utilized by students and researchers alike.

Petrology is a branch of geology which is mainly concerned with the study and classification of rocks. It can be divided into a number of branches such as igneous petrology, sedimentary petrology, metamorphic petrology and experimental petrology. As this field is emerging at a fast pace, this book will help the readers to better understand the concepts of petrology. This covers in detail some existent theories and innovative models revolving around petrology. It attempts to understand the multiple branches that fall under this discipline.

Each chapter is a sole-standing publication that reflects each author's interpretation. Thus, the book displays a multi-facetted picture of our current understanding of application, resources and aspects of the field. I would like to thank the contributors of this book and my family for their endless support.

Editor

Soil Organic Carbon Loss and Selective Transportation under Field Simulated Rainfall Events

Xiaodong Nie[1,2], Zhongwu Li[1,2]*, Jinquan Huang[3], Bin Huang[1,2], Yan Zhang[1,2], Wenming Ma[1,2], Yanbiao Hu[1,2], Guangming Zeng[1,2]

1 College of Environmental Science and Engineering, Hunan University, Changsha, PR China, 2 Key Laboratory of Environmental Biology and Pollution Control (Hunan University), Ministry of Education, Changsha, PR China, 3 Department of Soil and Water Conservation, Yangtze River Scientific Research Institute, Wuhan, PR China

Abstract

The study on the lateral movement of soil organic carbon (SOC) during soil erosion can improve the understanding of global carbon budget. Simulated rainfall experiments on small field plots were conducted to investigate the SOC lateral movement under different rainfall intensities and tillage practices. Two rainfall intensities (High intensity (HI) and Low intensity (LI)) and two tillage practices (No tillage (NT) and Conventional tillage (CT)) were maintained on three plots (2 m width × 5 m length): HI-NT, LI-NT and LI-CT. The rainfall lasted 60 minutes after the runoff generated, the sediment yield and runoff volume were measured and sampled at 6-min intervals. SOC concentration of sediment and runoff as well as the sediment particle size distribution were measured. The results showed that most of the eroded organic carbon (OC) was lost in form of sediment-bound organic carbon in all events. The amount of lost SOC in LI-NT event was 12.76 times greater than that in LI-CT event, whereas this measure in HI-NT event was 3.25 times greater than that in LI-NT event. These results suggest that conventional tillage as well as lower rainfall intensity can reduce the amount of lost SOC during short-term soil erosion. Meanwhile, the eroded sediment in all events was enriched in OC, and higher enrichment ratio of OC (ERoc) in sediment was observed in LI events than that in HI event, whereas similar ERoc curves were found in LI-CT and LI-NT events. Furthermore, significant correlations between ERoc and different size sediment particles were only observed in HI-NT event. This indicates that the enrichment of OC is dependent on the erosion process, and the specific enrichment mechanisms with respect to different erosion processes should be studied in future.

Editor: Vanesa Magar, Centro de Investigacion Cientifica y Educacion Superior de Ensenada, Mexico

Funding: The study was funded by the National Natural Science Foundation of China (40971179, 41271294), the Program for New Century Excellent Talents in University (NCET-09-330), and the Natural Science Foundation of Hunan Province of China (11JJ3041). The funders had no role in study design, data collection and analysis, decision to publish, or preparation of the manuscript.

Competing Interests: The authors have declared that no competing interests exist.

* Email: lzw@hnu.edu.cn

Introduction

Soil erosion has attracted more and more attention from all over the world for its impact on carbon geochemical cycles between soils and the atmosphere [1,2]. However, it is still a controversial issue on the role of soil erosion on carbon cycles, with the most famous debate is the carbon source or sink [3–7]. The substance of the issue is the poor understanding of soil erosion process and the included carbon dynamics. The Intergovernmental Panel on Climate Change (IPCC) [8] suggested that lateral carbon movement was the source of the greatest uncertainty in the global carbon balance. Furthermore, Kuhn et al. [1] indicated that the movement of soil organic carbon (SOC), both its particulate and dissolved forms, through agricultural landscapes is not fully understood.

Loss of SOC from the ecosystem occurs as a result of three processes: (i) physical removal by water (erosion); (ii) release of carbon into the atmosphere; and (iii) leaching [9,10]. While in water erosion, most of the SOC is lost through the procedure (i). Researchers considered that the physical removal of SOC undergo four stages during the erosion processes [11]. Firstly, the macroaggregates are detached and dispersed into microaggregates by raindrop impact, and release organic carbon (OC) at the same time. Secondly, SOC is transported by runoff in form of either dissolved organic carbon (DOC) or sediment-bound organic carbon (SBOC). Thirdly, the coarse or heavy particles were deposited in micro-depression during the migration path. Eventually, the SOC with the transportable particles or runoff are transported to outlet and deposited in concave slopes and floodplains. However, these processes are related to a number of factors, namely, rainfall intensity and kinetic energy, infiltration and runoff rates, soil properties and soil surface conditions such as soil moisture, roughness, crop residues, slope length and steepness [12,13]. Among them, rainfall intensity and tillage practice have become the focus of the erosion study. Lots of experiments were conducted to study the impact of rainfall intensity and tillage practice on soil delivery and nutrient loss [13–15].

However, most of the researches, under different rainfall intensities and tillage practices, focused on the SOC dynamics during erosion, were conducted in watershed [16–18] or laboratory [10,13,19]. Different points of view were observed between them for the variety of research conditions. For example, Lal et al. [20] suggested that no-till would decrease silt in rivers and lakes, which would lower transport of SOC and pollutant-laden sediments to aquatic ecosystems and reduce hypoxia. Also, some researchers indicated that conservation tillage practice reduce losses in soil and SOC [21,22]. However, Cogle et al. [23] found that the lost carbon from 20 cm deep tillage was consistently less from zero tillage. In addition, the study scale is also considered to be an important factor impact on the movement of SOC [24]. Schiettecatte et al. [19] indicated that at a largerscale, due to the increased probability of sediment deposition by topography and vegetation, the sediment became more enriched in OC. The problem obtained within watershed scale is the representativeness of field conditions and the extent to which the data obtained with these microcosms can be extrapolated [25]. And for laboratory experiments, the experiment condition is too ideal to simulate the natural state. While simulated rainfalls at small plot scale had been applied to investigate the detachment and sediment transport capacity of runoff [26] and the effects of water erosion on soil properties and productivity [27], and important and meaningful results were obtained. So the study at plot scale in field condition is essential for improving the understanding of SOC dynamics under different erosion processes.

Therefore, field simulated rainfall events at small plot scale were performed in this study, and the objectives of present study were to: (i) examine the carbon lateral movement at plot scale in field runoff area, (ii) investigate the selective migration processes of SOC. Such information will be useful for the study of SOC transportation, also, provide basic data for SOC migration model.

Materials and Methods

2.1. Ethics Statement

In this study, soil sampling and sample determinations conducted were permitted by the local authorities (i.e. Soil and Water Conservation Monitoring Station). We also obtained a permission from the local authorities for reporting research results to the public. In addition, the field studies did not involve endangered or protected species.

2.2. Study site

The simulated rainfall experiments were conducted at the Soil and Water Conservation Monitoring Station (111°22′ E, 27°03′ N), Hunan province, China (Fig. 1). The study area is located in subtropics humid monsoon climate zone, with an annual precipitation of 1 218.5 ~1 473.5 mm and average annual temperature of 17.1°C. The record rainfall intensity of the last five years varied between 0.10 and 2.11 mm min^{-1}, with 90% between 0.50 and 1.44 mm min^{-1}. The area is characterized by hills and sloping lands with a gradient of 3 to 30%. The soil was developed from Quaternary red soil with sandy and clay loam texture. The area is a typical red soil hilly region. Due to the dense population and unreasonable land use, this region has suffered serious soil erosion, and more than 4 195.05 km^2 cropland suffered water erosion with different degree. This area is representative of the agricultural, socio-economic and environmental situation of many slope farming areas in the region. The study carried out on this area is typical and representative of general situation in red soil hilly region.

2.3. Plot set-up and rainfall simulation

A 7 m×5 m block was taken from a typical sloping land with a 17% gradient to conduct the rainfall simulation. The block was previously planted with slope cultivated *Polygonatum odoratum*

(Mill.) Druce. After harvesting the crops, this block was abandoned for almost one year, and it was almost bare before the experiments. The soil had a bulk density of 1.65±0.15 g cm^{-3}, meanwhile with a water content of 0.15±0.02%. The soil pH was 4.47±0.10 (acidic soil), and the soil carbon is considered to be SOC. The mean SOC concentration of the surface layer was 7.47±2.48 g kg^{-1} dry soil (the mean value of 45 replicate samples ± standard error), the mean DOC concentration was 29.26±0.21 mg kg^{-1}. The soil had a clay-loamy texture with 33.44±1.27% clay particle size distribution, 27.82±1.63% silt, and 38.73±1.74% sand. Before the experiments, the block was divided into three equal plots, and each plot was designed with 2 m (width) × 5 m (length), and named Plot I, II, III respectively. Two tillage practices were maintained on the three plots: plot I and II were applied no tillage (NT) and plot III was maintained conventional tillage (CT). The plot III was disk ploughed (10 cm), while others kept natural state. The three plots were separated with a 0.5 m wide space. Each plot was bound with a metal frame inserted into the ground 15 cm in order to prevent runoff from adjacent areas. To determine ERoc, plot soil was sampled in all plots at depth of 10 cm. The boreholes were later filled and carefully leveled in order to reduce the effects of soil sampling.

For the erosion experiments, a rainfall simulator with a SPRACO cone jet nozzle mounted on the top of fixed 4.57 m long stand pipes was built. The nozzles were placed on the boundary of the plots. The median drop size was 2.4 mm with a uniformity of 89.7%. According to the local rainfall intensity variation for the past five years, rainfall intensities of 0.4~0.6 and 1.3~1.5 mm min^{-1} were used, representing the low intensity (LI) and high intensity (HI) storms of this region. Plot I, II, III were treated with HI-NT, LI-NT and LI-CT, respectively. Calibrations of rainfall intensities were conducted prior to the experiments. Four simulators were used in HI event, and two were used in LI events. For the HI event, each simulator was located on the longer side and closed to the conner of the runoff plot, the two simulators in LI events were located on the similar positions and distributed in diagonal direction. Meanwhile, five rain gauges were used to measure the actual rainfall intensity, one was placed on the top of the plot, and four were distributed around the two sides. For each rainfall event, rainfall lasted 60 min after the overland surface runoff began. Once overland surface runoff began, random runoff samples were manually and intermittently collected at 6 min intervals using a 1000 mL kettle. Each collected sample were deposited, separated from the water, dried in a forced-air oven at 105°C until constant mass was achieved and weighed for the determination of sediment concentration. All other runoff and sediment samples were collected in a marked pail and the total runoff volume in 6 min was recorded. Another sample was taken from the thoroughly mixed pail, and this sample was splitted into two portions. One portion was dried in an oven at 105°C until constant mass and then weighed for the determination of physicochemical properties, the other portion was sieved with1, 0.5 and 0.25 mm pore openings for the separation of aggregates. The separated aggregates were dried and weighed separately. The actual rainfall intensity was determined after the simulated rainfall event through the rainfall gauges. The mean rainfall intensity was found to be 1.38 mm min^{-1} for the HI–NT event plot, 0.53 mm min^{-1} for the LI-NT and LI-CT event plot.

2.4. Sample treatment and data analysis

Soil bulk density was determined by cutting ring method. SOC concentrations of soil and sediment were determined with the dichromate oxidation method of Walkley and Black [28]. Soil particle sizes were analyzed using the pipette method [29]. Total

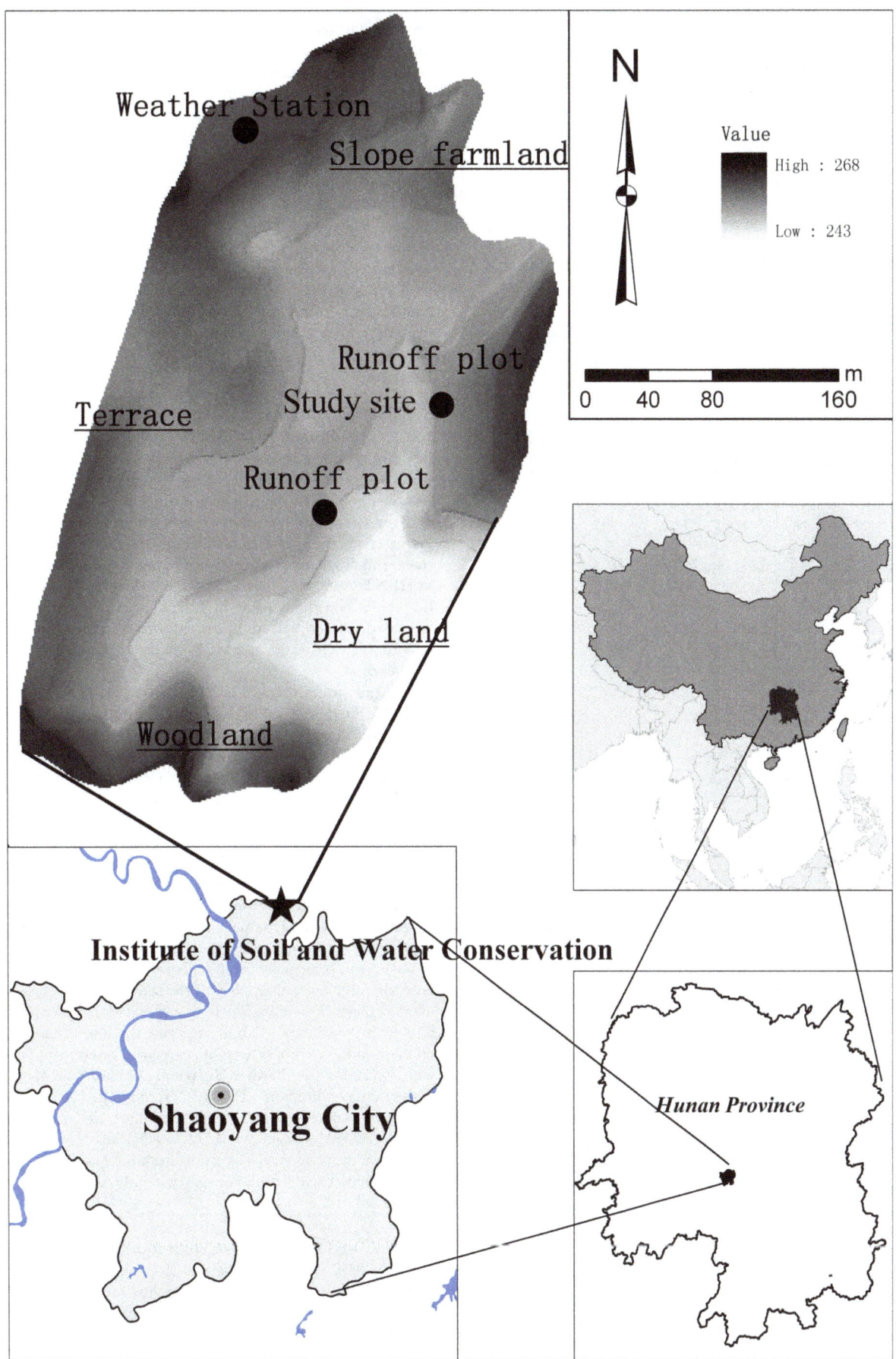

Figure 1. Location of the study area.

Table 1. The regular patterns of sediment and runoff transportation during water erosion.

Sample number	HI-NT (1.33 mm min⁻¹), TSR: 1'31"				LI-NT (0.53 mm min⁻¹), TSR: 2'31"				LI-CT (0.53 mm min⁻¹), TSR:48'0"			
	SYR (g min⁻¹)	CS (g)	RR (L min⁻¹)	CR (L)	SYR (g min⁻¹)	CS (g)	RR (L min⁻¹)	CR (L)	SYR (g min⁻¹)	CS (g)	RR (L min⁻¹)	CR (L)
1	248.93	1493.58	5.62	33.70	45.33	271.99	1.62	9.70	1.25	7.48	0.42	2.50
2	815.48	6386.479	12.25	107.20	111.35	940.08	4.77	38.30	2.74	23.91	0.55	5.80
3	659.31	10342.35	11.87	178.40	92.08	1492.56	4.78	67.00	4.37	50.12	0.80	10.60
4	514.51	13429.40	12.08	250.90	99.52	2089.68	5.18	98.10	4.93	79.70	1.00	16.60
5	680.28	17511.10	11.67	320.90	94.67	2657.68	4.75	126.60	4.88	108.95	1.08	23.10
6	524.83	20660.10	13.22	400.20	92.05	3209.98	5.00	156.60	5.80	143.77	1.22	30.40
7	586.71	24180.35	12.88	477.50	119.98	3929.86	4.93	186.20	11.53	212.95	1.57	39.80
8	508.14	27229.17	11.48	546.40	161.07	4896.29	4.95	215.90	15.77	307.60	1.65	49.70
9	564.65	30617.05	11.18	613.50	212.03	6168.45	5.08	246.40	16.69	407.76	1.70	59.90
10	665.98	34612.93	12.37	687.70	171.08	7194.95	6.03	282.60	15.95	503.46	2.18	73.00

The first sample was collected at the time of runoff began, and samples were collected at 6 minutes intervals.
TSR, time to start runoff; SYR, sediment yield rate; CS, cumulative sediment yield; RR, runoff rate; CR, cumulative runoff volume.

organic carbon concentrations for the runoff samples were measured with a Shimadzu TOC-TN analyzer.

The ERoc of sediment was calculated by dividing the SOC content of the sediment by its content in the original soil material. In this study, the ERoc of sediment was the ratio between the SOC concentration of sediment and the value of the source soil for each plot.

2.5. Statistical analysis

Statistical data analysis was performed using SPSS 20.0 for Windows. Pearson correlation was used to test the significance of correlations among ERoc and sediment size as well as the sediment particle size distribution. Differences in correlation analysis were detected using the least significant difference procedure for a multiple range test at the 0.05 significance level.

Results

3.1. Rainfall features

3.1.1. Sediment and runoff loss. Through the trials, different sediment and runoff yield rates were distinguished. For the events of HI-NT and LI-NT, two stages of sediment and runoff loss were found. In the first stage, plot soil underwent infiltration and runoff starting, the rates of sediment and runoff loss rapidly increased in the initial rainfall time (0~12 min). In the second stage, runoff loss rates reached steady state. The sediment loss rate in HI-NT event was consistent with the runoff loss rate. However, the values of sediment loss in LI-NT were first stable and then increased. In addition, different from these trends, the sediment and runoff yield rates in LI-CT increased consistently with the rainfall time.

The time to start runoff in LI-NT event was 1.66 times longer than that in HI-NT event, while, the figure in LI-CT event was 19.05 times longer than that in LI-NT event (Table 1). Despite longer time (additional 48 min) spent in runoff starting, the LI-CT event generated less runoff than LI-NT event did. The total runoff in HI-NT event was 2.43 times greater than that in LI-NT event which was 3.87 times higher than that in LI-CT event. Moreover, the sediments yield in HI-NT and LI-NT event was 68.34 and 14.38 times than that in LI-CT event, and which was correspond to the regular pattern of the lost runoff.

3.1.2. Sediment sorting. For the soil aggregates, microaggregates can be preferentially transported and macroaggregates deposit easily during water erosion. In this experiment, the sediment was principally composed of <0.25 mm aggregate which accounted for more than 58% of the sediment yield (Fig. 2). The average proportion of 0.25–1 mm aggregate was lower than 20%, and the proportion of >1 mm aggregate was lower than 10%. For all the events, the proportions of aggregates in sediment decreased with increasing size. Further, the composition of aggregates varied with rainfall duration. For LI-NT and LI-CT events, the proportions of microaggregate (<0.25 mm) first increased and then decreased, eventually reached steady state. However, for HI-NT event, the proportions of aggregates were in stable state except in 6–18 min. The dynamics of aggregates also depend with rainfall events.

3.2. SOC loss and selective migration

3.2.1. SOC loss. During water erosion, SOC loss in two forms: SBOC and DOC. Changes in loss rates of the SOC with respect to time were found to be very different among different rainfall events (Table 2). During the HI-NT event the loss rate of SBOC increased rapidly at the initial rainfall time, and then entered into a stable state. While For the LI-NT and LI-CT

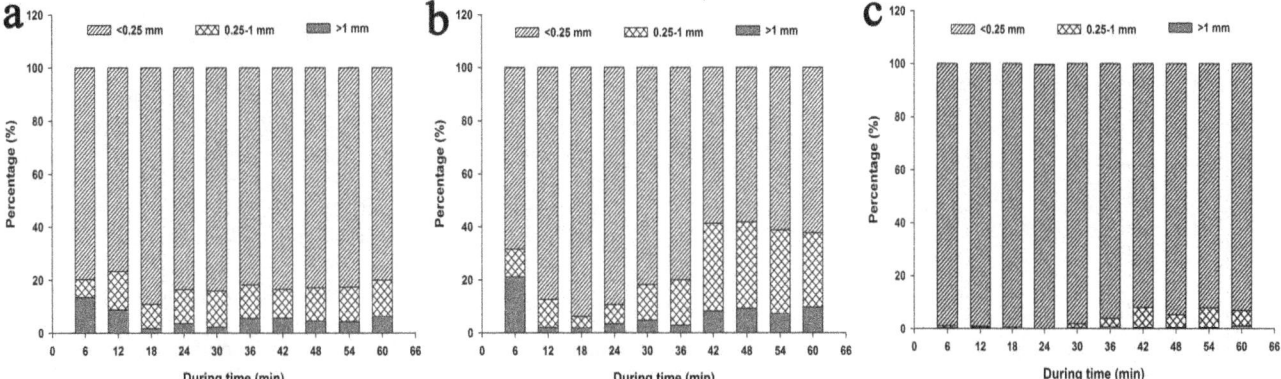

Figure 2. The distribution of different size aggregates in sediment in (a) HI-NT event, (b) LI-NT event, and (c) LI-CT event. HI-NT, high rainfall intensity-no tillage event; LI-NT, low rainfall intensity-no tillage event; LI-CT, low rainfall intensity– conventional tillage event.

events, the loss rates of SBOC increased with the whole rainfall time. In comparison to SBOC, the loss rates of DOC presented various trends. For the HI-NT event, the DOC loss rate was relatively stable during the first 30 minutes and then decreased to the end. Meanwhile, the trends of fluctuation and increasing in DOC loss rates were found in LI-NT and LI-CT events, respectively.

The lost SOC (SBOC+DOC), SBOC and DOC in different events decreased in the order: HI-NT>LI-NT>LI-CT (Table 2). For NT plots, HI rainfall event had 3.25 times higher lost SOC than the LI events did. While for the LI events, NT plot had 12.76 times higher lost SOC than CT plot did, despite the CT plot had longer rainfall duration. Further, compared to the lost SOC, it was little and approximate for the DOC loss in each event. However, a considerable of lost SBOC under different rainfall events was observed. The values of SBOC/SOC reached 94% in HI-NT and LI-NT plot, the least was 67.02% in LI-CT. SBOC was the main form of the lost SOC.

3.2.2. ERoc in sediment. In addition to the amount of the lost SOC, the selective migration processes were also studied. OC enrichment ratio (ERoc) in sediment for each event is presented in Fig. 3. ERoc curves for all the events were >1 except one value (0.96) in HI-NT event (48 min). For the events of LI-NT and LI-CT, ERoc curves had similar shapes, for example, the increasing stage (18–36 min), decline stage (36–54 min) and the peak value (36 min) were occurred at the same time. However, for HI-NT event, the ERoc curves decreased rapidly in fluctuation, and an exponential relationship (ER = 1.01+1.11 exp(−0.058t), $R^2 = 0.73$) between ERoc and duration time was found. Under LI events, the tillage practices (NT and CT) had a moderate influence on ERoc, while in NT plots, the rainfall intensity (HI and LI) had a great impact on ERoc.

3.3. Correlation analysis

3.3.1. The correlations of SOC and sediment and runoff. As the direct or indirect carrier of the lost SOC, runoff and sediment are important factors impact on the transportation of SOC. Fig. 4 displays the correlations between sediment yield, runoff volume and lost SOC for different rainfall events. Significant positive correlations (*P*<0.05) between sediment yield and lost SOC were observed in all rainfall events. First of all, the amount of the lost SOC increased with the increasing of the eroded sediment. Secondly, the linear correlations decreased with the increasing of the sediment yield (the correlation coefficient r decrease in the order: LI-CT>LI-NT>HI-NT). However, there

was not significant linear relationship (*P*>0.05) between lost SOC and runoff volume for all the rainfall events except in LI-CT event. In LI-CT event, the runoff volume was very low, the lost SOC increased with runoff volume. While in LI-NT and HI-NT events, runoff increased rapidly, and finally got into a stable state.

3.3.2. The correlations between ERoc and different size particles. Sediment particles transportation is often considered to be a cause of the selective migration of SOC. The correlations of sediment particles (different size aggregates (non-disperd particles) and sediment particle size distribution) and ERoc were analyzed (Table 3). The result shows that the correlations varied with rainfall events. In HI-NT event, ERoc had significant positive correlations with the content of clay (*P* = 0.011) and >1 mm aggregate (*P* = 0.042). Meanwhile, a significant negative correlation between ERoc and sand (*P* = 0.041) was observed. However, for the events of LI-NT and LI-CT, there were not significant correlations between ERoc and soil particles, neither sediment aggregates nor sediment particle size distribution.

Discussion

4.1. SOC loss

The eroded carbon in all events was found to be mainly in form of SBOC (more than 67%, and as high as 90% in NT events). Similar results were obtained by Lowrance and Williams [30], who found that up to 90% of SOC in runoff may be in particulate phase. This result could be mainly explained by the distribution of different forms carbon in plot soil. In this study, the content of the DOC in original soil was 0.4% of the SOC, this means that most of the SOC is insoluble in water. Thus the original source of SBOC would be greatly guaranteed. Furthermore, the erosion intensity was also an important factor impact on the composition of the lost SOC. As showed in table 2, both the lost DOC and SBOC increased with erosion intensity, but the SBOC growth rate was much higher than DOC. This indicates that the more SOC lost, the higher proportion of SBOC in lost SOC.

The correlation analysis indicated that the lost SOC was significantly correlated to the eroded sediment (P<0.05), while not always correlated to runoff. This result showed that the lost SOC was more close to sediment, as well as the lost SOC was mainly in form of SBOC (Table 2), which indicated that sediment was the direct and main carrier of the lost SOC. While, runoff, as the limiting condition of sediment transport and detachment [26], did not affect the SOC transportation directly. This is consistent with the study result that nutrient loss during soil erosion was not

Table 2. The regular patterns of SOC transportation during water erosion.

sample number	HI-NT (1.33 mm min⁻¹), TSR: 1'31"				LI-NT (0.53 mm min⁻¹), TSR: 2'31"				LI-CT (0.53 mm min⁻¹), TSR-48'0"			
	SBOCR (g min⁻¹)	CSBOC (g)	DOCR (g min⁻¹)	CDOC (g)	SBOCR (g min⁻¹)	CSBOC (g)	DOCR (g min⁻¹)	CDOC (g)	SBOCR (g min⁻¹)	CSBOC (g)	DOCR (g min⁻¹)	CDOC (g)
1	3.41	20.46	0.08	0.47	0.61	3.64	0.01	0.03	0.01	0.09	0.02	0.11
2	11.34	88.47	0.08	0.97	1.47	12.46	0.02	0.13	0.03	0.27	0.02	0.22
3	5.88	123.76	0.08	1.45	1.30	20.25	0.05	0.44	0.05	0.58	0.03	0.41
4	6.02	159.88	0.09	1.98	1.40	28.67	0.03	0.60	0.06	0.94	0.04	0.65
5	6.60	199.47	0.07	2.38	1.51	37.76	0.01	0.65	0.06	1.31	0.04	0.91
6	4.29	225.20	0.05	2.70	1.55	47.04	0.01	0.70	0.08	1.78	0.04	1.17
7	5.20	256.42	0.20	3.91	1.90	58.41	0.29	2.42	0.14	2.59	0.06	1.51
8	3.86	279.56	0.32	5.82	2.40	72.81	0.29	4.15	0.18	3.66	0.07	1.90
9	5.35	311.65	0.25	7.31	3.05	91.13	0.01	4.22	0.17	4.66	0.08	2.36
10	5.47	344.45	0.06	7.70	1.93	102.73	0.24	5.62	0.17	5.69	0.07	2.80

The first sample was collected at the time of runoff began, and samples were collected at 6 minutes intervals.
SBOCR, soil-bound organic carbon loss rate; CSBOC, cumulative soil-bound organic carbon; DOCR, dissolved organic carbon loss rate; CDOC, cumulative dissolved organic carbon.

directly related to runoff volume [23]. Consequently, it is appropriate to study the lost SOC through the study of sediment and sediment bound organic carbon.

Arnaez et al. [31] found that runoff increased linearly with rainfall intensity resulting in soil losses that also increased with rainfall intensity. Further, Zhang et al. [32] suggested that the amounts of eroded SOC were found to be strongly influenced by rainfall intensity. This study showed that more SOC was lost in HI-NT event in comparison to LI-NT and the amount of the lost SOC was significantly associated with sediment ($P<0.05$) (Fig. 4). In fact, the high rainfall intensity made seal formation and ponding time become shorter (Table 2), therefore, the infiltrated rainfall decreased quickly, meanwhile, runoff volume and velocity increased. In this way, rill erosion could become more intense and more soil and nutrients will be lost under high rainfall intensity event [13]. Consequently, higher rainfall intensity leads to great amount sediment eroded which result in more SOC loss.

While the amount of the lost SOC in LI-CT event can be ignored in comparison to that in LI-NT event. There are big difference between this result with the view that CT degrades soil structure and loosens soil surface which can accelerate soil erosion and SOC loss [21,22,33,34]. Low amount of runoff and sediment yield were considered to be the reason of the negligible lost SOC in LI-CT event. Researchers suggested that under CT condition, the amount of macropores increases, infiltration is improved and the water storage capacity of soil becomes larger, also roughness of the field surface is reported to decrease runoff velocity [22,35–37]. As a result, most of the rain water infiltrated to ground, and erosive power became very low. In general, CT changed the underlying surface properties and prevented soil loss. This result is consistent with that in the study of Cogle et al. [23], in which SOC loss in tillage areas was consistently less than that in NT areas. However, they considered that this benefit is of limited temporal value and not persistent. It is considered that the benefit occurred immediately post tillage, and before the soil had crusted [38]. Therefore, through the study, we believe that CT can make the underlying surface becomes rough and then reduce the loss of SOC, but this is limited by rainfall conditions. And we tend to attribute this result to the short rainfall time and low rainfall intensity used in our experiments. Therefore, more studies which were conducted with higher rainfall intensity and long-term monitor on CT plot should be taken in future. Nevertheless, the result is important for soil and water conservation in the areas which suffer from frequent short rains, for example the central southern China.

4.2. The selective transportation of SOC

Massey and Jackson [39] suggested that the ERoc often reflect the selectivity of OC. The study result showed that the ERoc values were >1 and indicated that the SOC could be transported preferentially. As the main carrier of the lost SOC, the eroded sediment also showed selectivity in migration process during erosion. The study result, the proportion of aggregates decreased with the increasing of aggregate size, indicated that the finer particles were preferential transported. Red soils in subtropical China are low in exchangeable sodium potential [40], and the main mechanisms of aggregate breakdown were by slaking due to fast wetting, mechanical breakdown due to raindrop impact and by runoff shear stress [41,42]. With the impact of rainfall kinetic energy [43], stream power [42] on the aggregate breakdown and transportation, a sorting process in sediment (both sediment size and density) transport was found [42]. Due to nonhomogeneous distribution of nutrients among particles of various sizes and density [10], the selective migration of particles was considered to

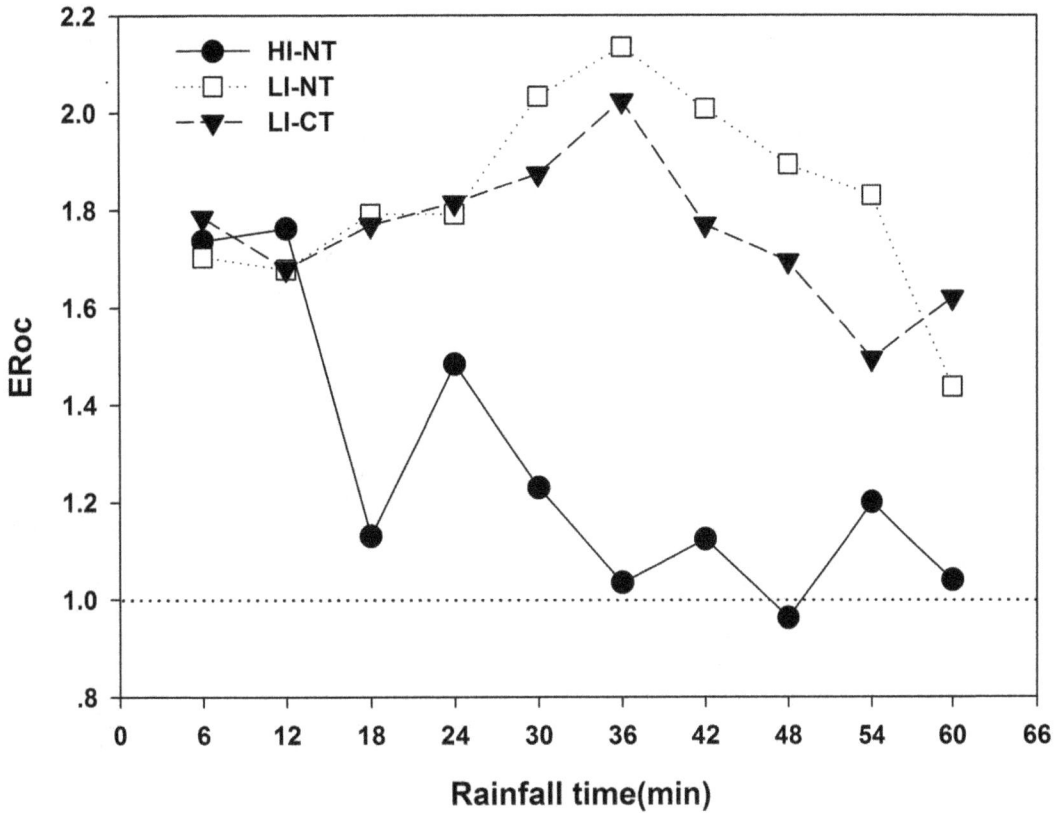

Figure 3. Dynamics of ERoc in sediment for different rainfall events. ERoc, enrichment ratio of organic carbon in sediment. HI-NT, high rainfall intensity-no tillage event; LI-NT, low rainfall intensity-no tillage event; LI-CT, low rainfall intensity-conventional tillage event.

be one of the reasons of the enrichment of OC in sediment [44]. Martinez-Mena et al. [25] suggested that the selectivity has been partly attributed to the transported of fine-sized sediments which are richer in silt and clay particles. However, this study found that the correlations between ERoc and different size particles depend on rainfall events. Significant correlations between ERoc and

sediment particles can be only found in HI-NT event. The large amount of the lost SOC (most of C was associate with the sediment particles) may help explain those significant correlations. While in LI-CT and LI-NT events, the ERoc had not any significant correlation with soil particles or sediment particle size distribution (Table 3). Jacinthe et al. [45] showed sediment collected during

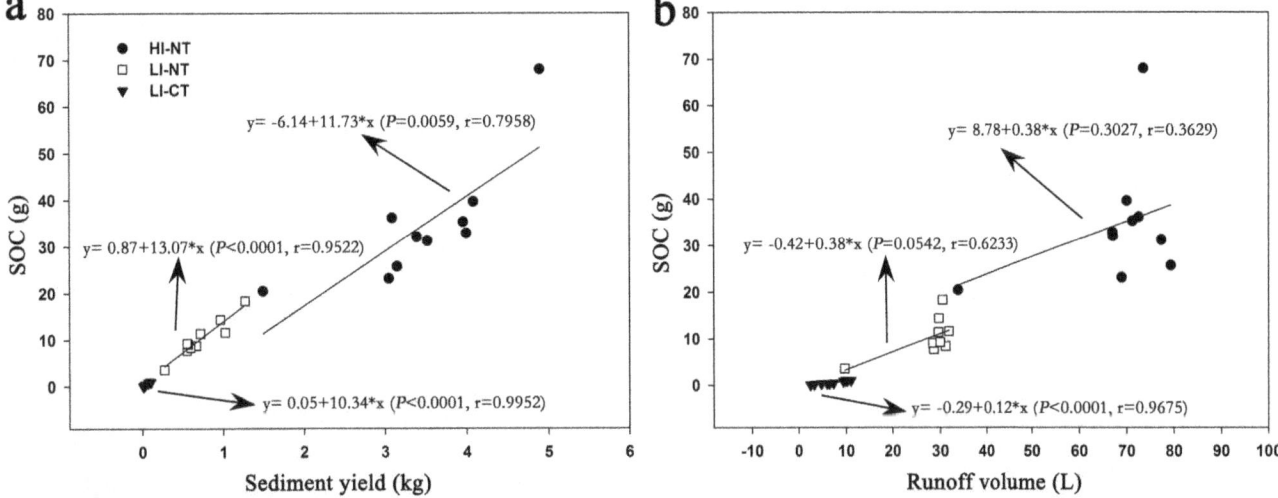

Figure 4. Correlations between the lost SOC and (a) sediment yield, (b) runoff volume. HI-NT, high rainfall intensity-no tillage event; LI-NT, low rainfall intensity-no tillage event; LI-CT, low rainfall intensity-conventional tillage event.

Table 3. Correlation analysis of ERoc and different size particles in sediment.

ERoc	Non-disperd particles			particle-size distribution		
	>1 mm	0.25–1 mm	<0.25 mm	sand	silt	clay
HI-NT	0.649*	−0.224	−0.523	−0.652*	0.236	0.760*
LI-NT	−0.343	0.105	−0.064	0.019	0.213	−0.088
LI-CT	−0.191	−0.468	0.462	–	–	–

Due to the small amount of sediment in LI-CT event, there were not enough samples to measure the particle-size distribution, and then default values were produced. HI-NT, high rainfall intensity-no tillage event; LI-NT, low rainfall intensity-no tillage event; LI-CT, low rainfall intensity-conventional tillage event. *Significant at 0.05 level.

the low-intensity storms contained more mineralized carbon (30–40% of sediment carbon) than materials displaced during the high-intensity summer storms. And the preferentially transportation of poorly decomposed non-cohesive plant fragments are often attributed to the higher ERoc in sediment in low rainfall events [46].

Researchers suggested that sediment will become less enriched in carbon as time passes during an event since the more carbon-rich fine aggregates are depleted early in the event [25]. Whereas the increasing transport capacity of runoff is also considered to be related to the decreasing of ERoc [47,48]. As a result, an exponential relationship in character (ERoc = 1.18+0.76exp(−0.046t), $R^2 = 0.81$) between ERoc and rainfall duration (t) was found by Polyakov and Lal [10]. The same result of ERoc dynamics was observed in HI-NT event. However, for LI events, despite with increasing erosive power and sediment yield, an increasing trend of the ERoc was found during 12~36 min. This suggests that the easiest transport substance were not the particles which have the highest OC concentration in LI events. Therefore, rainfall intensity had great impact on the ERoc in sediment and different enrichment mechanisms accounted for different ERoc dynamics. However, in contrast to rainfall intensity, tillage practices had a smaller impact on ERoc. For LI-NT and LI-CT events, ERoc curves had similar shapes. And the ERoc was almost the same in LI-CT and LI-NT plot when the erosive power of runoff was limited, while the difference could be observed when the erosive power growth at different degrees. Our results support the view that tillage affected the sediment yields but did not directly influence the ERoc [23]. Nevertheless, the impact of tillage on the ERoc mechanism is still unclear and further study is required.

Conclusions

The regular patterns of SOC transportation during soil erosion processes were studied through field simulated rainfall experiments. These experiments showed that the LI-NT event had 12.76 times higher lost SOC than LI-CT event did. Conventional tillage can increase rainwater infiltration and reduce soil erosion and SOC loss. However, these results were obtained under short-term and low rainfall intensity event, and more experiments at long-term and storm events are needed to confirm these results. It was also found that SOC as well as sediment particles presented selective in erosion processes, and these processes were affected by rainfall intensity but not tillage practice. The selective transportation of finer particles was not always the reason for the enrichment of OC in sediment. The specific enrichment mechanisms of OC in sediment in relation to different erosion processes have to be studied in future. In addition, due to the complexity and uncertainties of field rainfall experiments, more similar field experiments should be carried out in the future.

Acknowledgments

We would like to thank Shuguang Wang, Guiping Liu and Kunjun Li of Soil and Water Conservation Monitoring Station of Shaoyang for the help of providing the study area and facilities for field work.

Author Contributions

Conceived and designed the experiments: ZWL XDN GMZ. Performed the experiments: ZWL XDN JQH WMM BH. Analyzed the data: XDN YZ YBH. Contributed reagents/materials/analysis tools: XDN. Contributed to the writing of the manuscript: XDN ZWL.

References

1. Kuhn NJ, van Oost K, Cammeraat E (2012) Soil erosion, sedimentation and the carbon cycle Preface. Catena 94: 1–2.
2. Doetterl S, van Oost K, Six J (2012) Towards constraining the magnitude of global agricultural sediment and soil organic carbon fluxes. Earth Surf Proc Land 37: 642–655.
3. Van Oost K, Six J, Govers G, Quine T, De Gryze S (2008) Soil erosion: A carbon sink or source? Response. Science 319: 1042–1042.
4. Lal R, Pimentel D (2008) Soil erosion: A carbon sink or source? Science 319: 1040–1042.
5. Harden JW, Berhe AA, Torn M, Harte J, Liu S, et al. (2008) Soil erosion: Data say C sink. Science 320: 178–179.
6. Fullen MA, Booth CA (2006) Grass ley set-aside and soil organic matter dynamics on sandy soils in Shropshire, UK. Earth Surf Proc Land 31: 570–578.
7. Dymond JR (2010) Soil erosion in New Zealand is a net sink of CO(2). Earth Surf Proc Land 35: 1763–1772.
8. Denman KL, Brasseur G, Chidthaisong A, Ciais P, Cox PM, et al. (2007) Couplings between changes in the climate system and biogeochemistry. In: Solomon S, Qin D, Manning M, Chen Z, Marquis M, Averyt KB, Tignor M, Miller HL, editors, Climate Change 2007: The Physical Science Basis. Contribution of Working Group I to the Fourth Assessment Report of the Inter-governmental Panel on Climate Change. Cambridge, United Kingdom and New York, NY, USA.: Cambridge University Press.
9. Schreiber J (1999) Nutrient leaching from corn residues under simulated rainfall. J Environ Qual 28: 1864–1870.
10. Polyakov VO, Lal R (2004) Soil erosion and carbon dynamics under simulated rainfall. Soil Sci 169: 590–599.
11. Lal R (2005) Soil erosion and carbon dynamics. Soil Till Res 81: 137–142.
12. Chaplot VA, Le Bissonnais Y (2003) Runoff features for interrill erosion at different rainfall intensities, slope lengths, and gradients in an agricultural loessial hillslope. Soil Sci Soc Am J 67: 844–851.
13. Assouline S, Ben-Hur A (2006) Effects of rainfall intensity and slope gradient on the dynamics of interrill erosion during soil surface sealing. Catena 66: 211–220.
14. Jin K, Cornelis WM, Gabriels D, Baert M, Wu HJ, et al. (2009) Residue cover and rainfall intensity effects on runoff soil organic carbon losses. Catena 78: 81–86.
15. Girmay G, Singh B, Nyssen J, Borrosen T (2009) Runoff and sediment-associated nutrient losses under different land uses in Tigray, Northern Ethiopia. J Hydrol 376: 70–80.
16. Zhou H, Li BG, Lu YH (2009) Micromorphological analysis of soil structure under no tillage management in the black soil zone of Northeast China. J Mt Sci-Engl 6: 173–180.

17. Moebius-Clune BN, van Es HM, Idowu OJ, Schindelbeck RR, Moebius-Clune DJ, et al. (2008) Long-term effects of harvesting maize stover and tillage on soil quality. Soil Sci Soc Am J 72: 960–969.

18. Pinheiro EFM, Pereira MG, Anjos LHC (2004) Aggregate distribution and soil organic matter under different tillage systems for vegetable crops in a Red Latosol from Brazil. Soil Till Res 77: 79–84.

19. Schiettecatte W, Gabriels D, Cornelis WM, Hofman G (2008) Impact of deposition on the enrichment of organic carbon in eroded sediment. Catena 72: 340–347.

20. Lal R, Griffin M, Apt J, Lave L, Morgan MG (2004) Ecology - Managing soil carbon. Science 304: 393–393.

21. Kisic I, Basic F, Nestroy O, Mesic M, Butorac A (2002) Chemical properties of eroded soil material. J Agron Crop Sci 188: 323–334.

22. Puustinen M, Koskiaho J, Peltonen K (2005) Influence of cultivation methods on suspended solids and phosphorus concentrations in surface runoff on clayey sloped fields in boreal climate. Agr Ecosyst Environ 105: 565–579.

23. Cogle A, Rao K, Yule D, Smith G, George P, et al. (2002) Soil management for Alfisols in the semiarid tropics: erosion, enrichment ratios and runoff. Soil Use Manage 18: 10–17.

24. van Noordwijk M, Cerri C, Woomer PL, Nugroho K, Bernoux M (1997) Soil carbon dynamics in the humid tropical forest zone. Geoderma 79: 187–225.

25. Martinez-Mena M, Lopez J, Almagro M, Albaladejo J, Castillo V, et al. (2012) Organic carbon enrichment in sediments: Effects of rainfall characteristics under different land uses in a Mediterranean area. Catena 94: 36–42.

26. Schiettecatte W, Verbist K, Gabriels D (2008) Assessment of detachment and sediment transport capacity of runoff by field experiments on a silt loam soil. Earth Surf Proc Land 33: 1302–1314.

27. Li Z, Huang J, Zeng G, Nie X, Ma W, et al. (2013) Effect of Erosion on Productivity in Subtropical Red Soil Hilly Region: A Multi-Scale Spatio-Temporal Study by Simulated Rainfall. PloS one 8: e77838.

28. Walkley A, Black IA (1934) An examination of the Degtjareff method for determining soil organic matter, and a proposed modification of the chromic acid titration method. Soil Sci 37: 29–38.

29. Gee GW, Bauder JW (1986) Particle size analysis. In: Klute A, editor, Methods of Soil Analysis (2nd ed.). Am. Soc. Agron., Madison, 383–411.

30. Lowrance R, Williams RG (1988) Carbon movement in runoff and erosion under simulated rainfall conditions. Soil Sci Soc Am J 52: 1445–1448.

31. Arnaez J, Lasanta T, Ruiz-Flano P, Ortigosa L (2007) Factors affecting runoff and erosion under simulated rainfall in Mediterranean vineyards. Soil Till Res 93: 324–334.

32. Zhang X, Li Z, Tang Z, Zeng G, Huang J, et al. (2013) Effects of water erosion on the redistribution of soil organic carbon in the hilly red soil region of southern China. Geomorphology 197: 137–144.

33. West TO, Post WM (2002) Soil organic carbon sequestration rates by tillage and crop rotation: A global data analysis. Soil Sci Soc Am J 66: 1930–1946.

34. Sa JCD, Cerri CC, Dick WA, Lal R, Venske SP, et al. (2001) Organic matter dynamics and carbon sequestration rates for a tillage chronosequence in a Brazilian Oxisol. Soil Sci Soc Am J 65: 1486–1499.

35. Alakukku L (1998) Properties of compacted fine-textured soils as affected by crop rotation and reduced tillage. Soil Till Res 47: 83–89.

36. Schnug E, Haneklaus S (2002) Agricultural production technique and infiltration significance of organic farming for preventive flood protection. Landbauforsch Volk 52: 197–203.

37. Gowda P, Mulla D, Dalzell B (2003) Examining the targeting of conservation tillage practices to steep vs. flat landscapes in the Minnesota River Basin. J Soil Water Conserv 58: 53–57.

38. Yule D, Cogle A, Smith G, Rao K, George P (1991) Soil management of Alfisols for water conservation and utilization. J Indian Water Resour Soc 11: 10–13.

39. Massey H, Jackson M (1952) Selective erosion of soil fertility constituents. Soil Sci Soc Am J 16: 353–356.

40. Shi ZH, Yan FL, Li L, Li ZX, Cai CF (2010) Interrill erosion from disturbed and undisturbed samples in relation to topsoil aggregate stability in red soils from subtropical China. Catena 81: 240–248.

41. Li ZX, Cai CF, Shi ZH, Wang TW (2005) Aggregate stability and its relationship with some chemical properties of red soils in subtropical China. Pedosphere 15: 129–136.

42. Shi ZH, Fang NF, Wu FZ, Wang L, Yue BJ, et al. (2012) Soil erosion processes and sediment sorting associated with transport mechanisms on steep slopes. J Hydrol 454: 123–130.

43. Wang L, Shi ZH, Wang J, Fang NF, Wu GL, et al. (2014) Rainfall kinetic energy controlling erosion processes and sediment sorting on steep hillslopes: A case study of clay loam soil from the Loess Plateau, China. J Hydrol 512: 168–176.

44. Palis R, Ghadiri H, Rose C, Saffigna P (1997) Soil erosion and nutrient loss. III. Changes in the enrichment ratio of total nitrogen and organic carbon under rainfall detachment and entrainment. Aust J Soil Res 35: 891–905.

45. Jacinthe PA, Lal R, Owens LB, Hothem DL (2004) Transport of labile carbon in runoff as affected by land use and rainfall characteristics. Soil Till Res 77: 111–123.

46. Ghadiri H, Rose C (1991) Sorbed chemical transport in overland flow: II. Enrichment ratio variation with erosion processes. J Environ Qual 20: 634–641.

47. Sharpley A (1985) The Selection Erosion of Plant Nutrients in Runoff. Soil Sci Soc Am J 49: 1527–1534.

48. Weigand S, Schimmack W, Auerswald K (1998) The enrichment of 137Cs in the soil loss from small agricultural watersheds. Zeitschrift für Pflanzenernährung und Bodenkunde 161: 479–484.

2

Removal of Fast Flowing Nitrogen from Marshes Restored in Sandy Soils

Eric L. Sparks[1,2]*, **Just Cebrian**[1,2], **Sara M. Smith**[1]

1 Dauphin Island Sea Lab, Dauphin Island, Alabama, United States of America, **2** Marine Sciences, University of South Alabama, Mobile, Alabama, United States of America

Abstract

Groundwater flow rates and nitrate removal capacity from an introduced solution were examined for five marsh restoration designs and unvegetated plots shortly after planting and 1 year post-planting. The restoration site was a sandy beach with a wave-dampening fence 10 m offshore. Simulated groundwater flow into the marsh was introduced at a rate to mimic intense rainfall events. Restoration designs varied in initial planting density and corresponded to 25%, 50%, 75% and 100% of the plot area planted. In general, groundwater flow was slower with increasing planting density and decreased from year 0 to year 1 across all treatments. Nevertheless, removal of nitrate from the introduced solution was similar and low for all restoration designs (3–7%) and similar to the unvegetated plots. We suggest that the low NO_3^- removal was due to sandy sediments allowing rapid flow of groundwater through the marsh rhizosphere, thereby decreasing the contact time of the NO_3^- with the marsh biota. Our findings demonstrate that knowledge of the groundwater flow regime for restoration projects is essential when nutrient filtration is a target goal of the project.

Editor: Fei-Hai Yu, Beijing Forestry University, China

Funding: Funding for this project provided by the Alabama Department of Conservation and Natural Resources, State Lands Division, Coastal Section, in part, by a grant from the National Oceanic and Atmospheric Administration, Office of Ocean and Coastal Resource Management, Award #10NOS4190206. The funders had no role in study design, data collection and analysis, decision to publish, or preparation of the manuscript.

* Email: esparks4040@gmail.com

Introduction

Marsh restoration is a ubiquitous practice for mitigation of global marshland loss [1]. However, marsh restoration is expensive and labor intensive [2,3]. Compounded with the costly nature of marsh restoration, there is often inconsistency and discrepant outcomes among different techniques and designs [3]. Some studies have been conducted to evaluate cost-effectiveness of vegetative growth for restored marshes [2]; however, evaluations of the ecosystem services provided by different marsh restoration designs is scant, but should be evaluated to inform managers interested in maximizing the effectiveness of restoration projects [4–6].

Marshes provide important ecosystem services [7–10], and it has been suggested that nutrient filtration is the most economically valuable ecosystem service [4]. Processes such as denitrification and plant uptake can remove a large portion of nutrient inputs into marshes as groundwater percolates through the marsh rhizosphere [11,12].

Most nutrient filtration studies for marshes are conducted in mature natural marshes that are subjected to low to moderate flows of groundwater [12–14]. These studies have demonstrated that the presence of marsh plants increases nutrient removal through direct plant uptake as well as facilitating bacterial processes responsible for outgassing nitrogen (e.g., denitrification and anammox; [15]). However, marshes are subjected to varying groundwater flow rates from upland sources [16–18] and are dependent on factors such as rainfall intensity and soil permeability. In general, when areas are subjected to intense flow events (e.g., heavy rain), it is likely a smaller portion of the nutrients carried in these events can be removed than when the site is subjected to lower flow rates [19]. Along the northern Gulf of Mexico (nGOM) coast, there are frequent and intense rain events [20], thereby subjecting these marshes to a mixture of fast flow events, during and immediately after these rain events, and lower flow between events [16]. Assessment of nutrient removal by restored marshes under different scenarios of groundwater flow is important to improve the effectiveness of marsh restoration efforts targeting nutrient filtration as a primary goal.

In this study, we use black needlerush (*Juncus roemerinaus*) as our restored marsh plant. Black needlerush marshes are dominant on the nGOM coast [21] and have suffered significant loss over past decades primarily attributed to coastal development [22]. Due to the losses of marshes along the nGOM coast and prevalence of black needlerush, this marsh plant is the target for many restoration projects [2,23–25].

In this study, we compare groundwater flow rate and nitrate (NO_3^-) removal from an introduced groundwater solution in five black needlerush marsh restoration designs, varying in initial plant density, with unvegetated controls immediately after planting and one year after planting. Utilization of these marsh planting designs allows for comparisons of NO_3^- removal from fast flowing groundwater across designs that vary in the effort required to plant (i.e., time and cost). The groundwater plume introduced into the marsh mimics a pulse of groundwater derived from an intense rainfall event percolating through porous sediments. Expectations were groundwater flow rates, through the marsh rhizosphere, would decrease and NO_3^- removal would increase with increasing planting density over time. Results from this study can inform managers interested in maximizing restoration efficiency with the goal of reducing nutrient pollution into water bodies.

Materials and Methods

1. Site Construction

On June 11, 2010, we planted a black needlerush marsh on the outskirts of Camp Beckwith (30°23′16″ N, 87°50′31″ W) located on the eastern coast of Weeks Bay in Fairhope, Alabama, USA. The staff of this privately owned camp gave us permission to conduct this work on their property and they should be contacted for future permissions. The planting site was situated on a stretch of sandy beach with natural marsh nearby. This sandy beach was subjected to high wave energy from boat wakes. To reduce wave action at this site, a fence was constructed, prior to marsh restoration, ten meters offshore of the restoration site to reduce shoreline wave energy and erosion. The fence consisted of a wooden frame filled up with dead tree branches and trunks along with other natural debris. Black needlerush sods (approximately 20 cm long, 20 cm wide and 20 cm deep) were harvested from an adjacent marsh and planted at the restoration site. Individual sods had a black needlerush shoot density typical of nGOM salt marshes, with ranges from 1400–1800 shoots m^{-2} [2]. For experimental setup, we used a randomized block design with 3 blocks consisting of 6 plots each, yielding a total of 18 experimental plots (Fig. 1). Blocks were separated by 2 m and each plot had dimensions of 40 cm wide and 170 cm long (Fig. 1). Plots represented different restoration designs in terms of initial plant density (25%, 50%, 50%A, 75% and 100%) plus non-vegetated controls (0%) rendering 6 designs with 3 replicated plots each. Each plot contained 16 sod-sized units and the number of sods planted in the plot corresponded to the planting density treatment (e.g., 4 sods for the 25% planting density; Fig. 1). Plots planted in the 50%A design were arranged in an alternating "checkerboard" pattern (Fig. 1). To contain the introduced groundwater solution, each plot was enclosed on the top (i.e., upland) and the two lateral sides with vertical placement of rigid plastic sheeting. Thirty cm of the sheet height was buried below the sediment surface with 10 cm of sheet height above the sediment surface. Porewater collection wells, screened from 5 cm to 30 cm below the sediment surface, were placed at the bottom of each plot (experimental porewater wells) and 3 (natural porewater wells) on each lateral side of every block (Fig. 1). A diffuser plate was buried in the sand 10 cm from the upland planting edge to help disperse the introduced solution.

2. Experimental methodology and sampling

On June 21, 2010 and June 30, 2011 fluorescein tracing tests were conducted by releasing 15 L of a 20 mg L^{-1} fluorescein solution into each plot. Fluorescein was used because it is not actively removed through biological processes at high rates; therefore, the only factor that can change the fluorescein concentration is dilution [26]. These attributes of fluorescein allow it to be used to assess travel time and dilution rates of groundwater plumes through the plots [26]. The fluorescein solution was released during receding tides when the tide line was at the upland edge of the plots. It took approximately 15 min for the 15 L of solution to disperse through the diffuser plate. This quick pulse was intended to mimic groundwater inputs from an intense rain event over porous sediments and was equivalent to a 1.5 cm rainfall event drained off of a 1 hectare area through 100 linear meters of fringing marsh over a 12 hour period. Porewater samples were taken from each well in 15 min intervals after tracer release for 90 minutes and stored on ice in a cooler for transport back to Dauphin Island Sea Lab for analysis. All fluorescein samples were analyzed on a Turner Designs-700 fluorometer.

To determine how effective the marsh designs were at removing nutrient pollution from a fast flowing groundwater solution, four rounds of NO_3^- solution release and subsequent sampling were performed after the initial planting (June 27, July 8, 9, and 13, 2010) and 1 year post-planting (July 14, 21, 29 and August 2, 2011). The releases were done during receding tides when the tide line was at the upland edge of the plots (i.e., the same timing as the fluorescein tracing tests). For each round we released 15 L of a 200 μM NO_3^- solution into each plot. Water samples were taken from each input container and porewater well at the bottom of the plots. The sampling timeline for porewater samples was determined by the fluorescein experiments (i.e., sample at time fluorescein peak for each plot). To determine background nutrient levels, we also took samples from the natural porewater wells located outside of each block on every sampling day. Porewater samples were filtered in the field and transported to the lab on ice for analysis. NO_3^- concentrations were analyzed using cadmium reduction azo dye assays [27].

3. Calculations and statistical analyses

3.1. Flow rates. Flow rates were analyzed by recording the time when peak fluorescein concentration was observed in each plot. These peak times were analyzed using an ANOVA (treatment × year × block). If block and the interaction between treatment and year were found to be insignificant factors, data were pooled across block and reanalyzed with an ANOVA (treatment × year). For all statistical analyses, tests were conducted using Sigma Stat 3.5 and significance was considered at p<0.05 [28].

3.2. NO_3^- removal. As the introduced solution travels through the plots, it will be subjected to dilution through mixing with natural porewater (i.e., not subjected to the introduced solution) with lower background concentrations of NO_3^-. To determine the $[NO_3^-]$ in the portion of the porewater derived from the introduced solution, a dilution correction must be applied that accounts for the $[NO_3^-]$ found in natural porewater. Applying a dilution correction allows for the removal of NO_3^- from the introduced solution to be calculated as the introduced solution (*Input*) travels to the downland edge of the marsh. To calculate removal of NO_3^- from the introduced solution, we subtracted the dilution corrected $[NO_3^-]$ at the downland porewater well from the NO_3^- concentration in the input using the following equation: *Removal = Input - (Downland well – Natural × Dilution) ÷ (1-Dilution)*. The previous equation will be referred to as equation 1 throughout the manuscript. The term *Downland well* in equation 1 is the $[NO_3^-]$ measured at restored marsh porewater wells subjected to the simulated pollution plume. The *Natural* term is the $[NO_3^-]$ in the porewater wells outside of the restored marsh that was not subjected to the simulated pollution plume (Fig. 1). The *Dilution* term in equation 1 is the proportion of sample derived from natural porewater and was calculated as the proportional decrease in [fluorescein] from the input to the porewater collection well (i.e., peak % tracer contribution in Fig. 2). NO_3^- removal was converted to percent removal by dividing it by $[NO_3^-]$ of the input. Similar to the flow rate study, percent NO_3^- removal was first analyzed using an ANOVA (treatment × year × block). If block and the interaction between treatment and year were found to be insignificant factors,

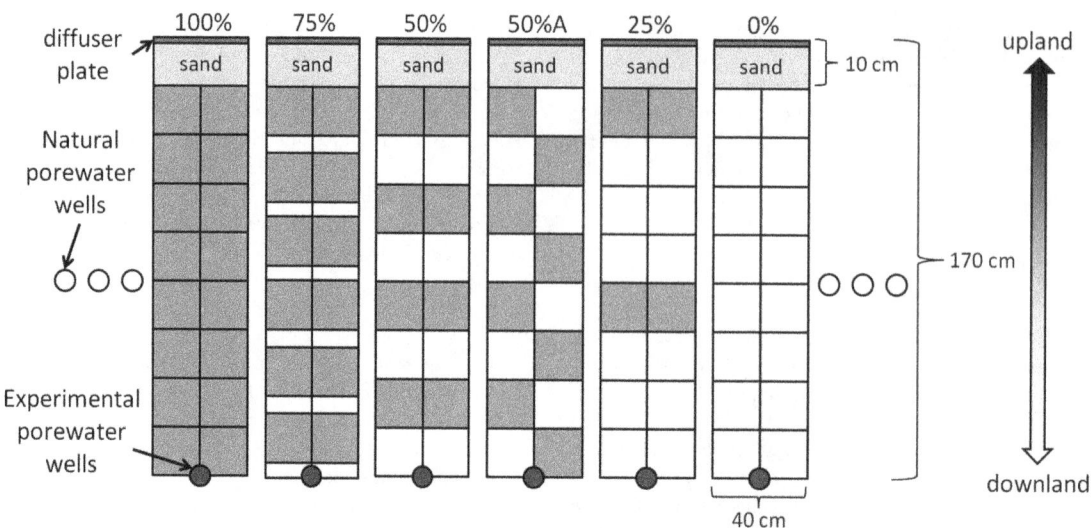

Figure 1. Schematic of 1 block of 6 marsh restoration designs (0%, 25%, 50%, 50%A, 75% and 100%). Shaded squares represent planted sods. There were a total of 3 blocks with each block consisted of a randomized arrangement of all 6 restoration designs. The groundwater solution was introduced at the diffuser plate and flowed down the plots toward the porewater collection well.

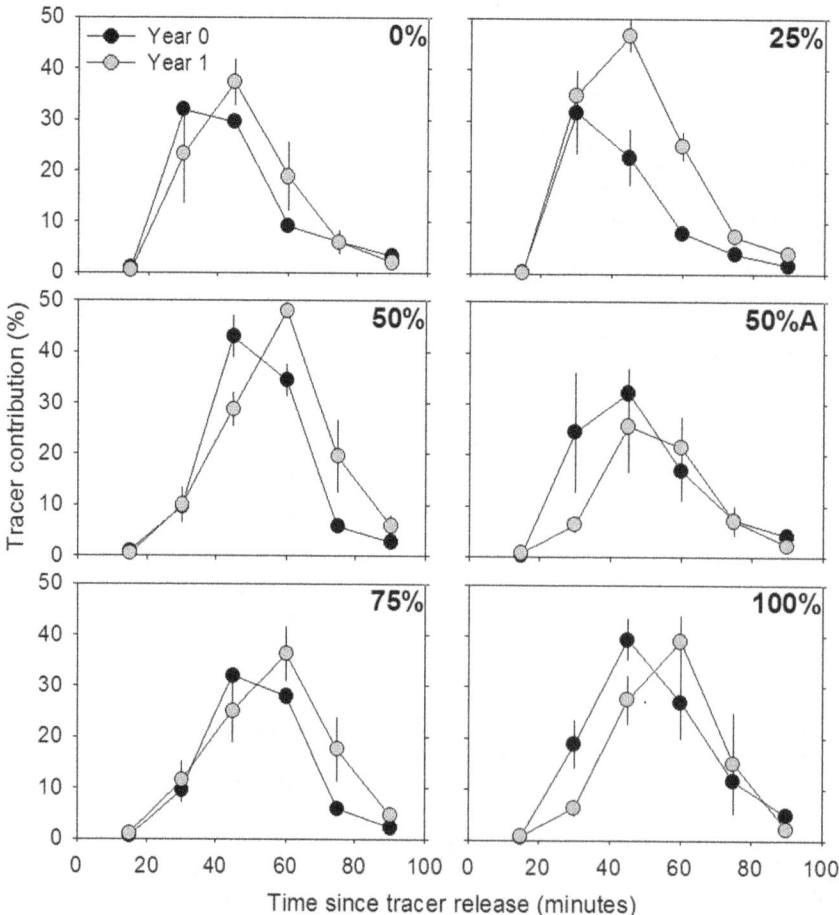

Figure 2. Fluorescein tracer porewater contribution (%) at the downland well through time. Black circles represent year 0 and grey circles represent year 1 samples. Percentages in the top right portion of each plot represents the planting density (0%, 25%, 50%, 50%A, 75% and 100%). Error bars indicate ±1 SE.

the data was pooled across block and reanalyzed with an ANOVA (treatment × year).

Results

For flow rate and NO_3^- removal, block was never a significant factor (flow rate - p = 0.40; NO_3^- removal - p = 0.14) and there were no significant interactions between treatment and year (flow rate - p = 0.34; NO_3^- removal - p = 0.99). Therefore, data were pooled across blocks and analyzed with an ANOVA for the effects of treatment and time (Table 1). Only the results of the ANOVA on the data pooled across blocks will be further discussed.

Most of the fluorescein solution traversed the plots in less than one hour across all treatments (Fig. 2). In general, the flow of the introduced solution decreased with increasing planting density (Table 1), as indicated by the later peaks in the dilution curves (Fig. 2). Furthermore, it took longer for the solution to cross the plots one year after planting than two weeks after planting (Table 1; Fig. 2). Longer retention time of the introduced solution within the plots at one year after planting than two weeks after planting implies that flow rates of the introduced solution, through the plots, decreased over time.

Concentrations of NO_3^- in porewater collections wells outside the plots ranged from 0.5 μM to 1.5 μM (i.e., natural or ambient porewater) and were low when compared to concentrations within the plots (i.e., subjected to the introduced solution). As expected, the input had $[NO_3^-]$ of 200 μM ± 10 μM, whereas wells at the downland edge of the plots had lower $[NO_3^-]$ ranging from 44 μM to 126 μM. When combining the observed changes in $[NO_3^-]$ with a dilution factor (equation 1), our calculations showed only a small percentage of NO_3^- was removed from the introduced solution (3–7%; Table 2). These small percentages of NO_3^- removal were similar across all treatments (Table 1) and sampling years (Table 1; Fig. 3). While NO_3^- processing increased slightly over time and with plant cover (Table 2) these differences were not statistically significant (Table 1).

Discussion

In this study, we found that the introduced groundwater traveled slowest through the most vegetated planting designs and slower one year after planting than immediately after planting. These results offer evidence that the presence of marsh plants increases the time required for groundwater to flow through the restoration area, likely through the presence of the marsh rhizosphere and accumulation of finer grained sediments [29]. As the planted plots mature and density increases, they will likely continue to decrease groundwater flow rates through binding sediments, expansion of the marsh rhizosphere [30] and reduction

of wave energy that aids in the accumulation of finer grained sediments [31].

Increases in plant density and the accumulation of finer grained sediment are conducive to increased nutrient removal in marshes [15]. We did find some suggestive evidence that increases in vegetated area increased NO_3^- removal (Table 2), albeit not statistically significant (Table 1). Our range in planting density was large (0%–100%) and we measured NO_3^- removal in all of these planting densities twice over one year. If planting density was a primary driver of NO_3^- removal from this fast flowing solution, we would have captured it with this sampling design. Given no effect of planting density and overall findings that only a small portion of the input NO_3^- was removed across all of the planting designs through time (3–7%), it appears that the groundwater was traveling too quickly through the marsh rhizosphere for plants to have an impact on NO_3^- removal. In studies where groundwater traveled slower through the marsh rhizosphere, marsh plants had time to uptake and facilitate bacterial processes that fueled large removals of nitrogen [17]. Comparing these studies to our study suggests that flow rate can influence how effective marshes are at removing NO_3^- from groundwater. With these results, it is likely that nutrient processing in our restored marsh will remain similar to the unvegetated plots for several years to come [5]. An additional factor contributing to the negligible increases in NO_3^- removal over time is the typical slow growth of black needlerush [21]. This slow growth pattern was evident by visual observations of marginal increases in vegetated area (<5%) for the vegetated designs and no colonization of the 0% planting design at one year after planting. A timeframe when these marshes will decrease groundwater flow enough to allow for large proportions of input nutrients to be removed is unknown; however, previous studies have indicated restoring other ecosystem functions in restored marshes to natural levels takes many years [2,32,33].

A probable explanation to the high groundwater flow rates and low nutrient filtration is that the sediment at the restoration site was sandy, which is also the case for many restored marshes (e.g., Jamaica Bay marsh islands in New York, USA, Grand Bay National Estuarine Research Reserve boat launch marsh in Mississippi, USA, and the Labranche wetlands in Louisiana, USA). Groundwater flows more quickly through sand than sediments with higher mud and silt content typical of mature marshes [34]. Most other studies of nutrient filtration in salt marshes have been conducted mature natural marshes with fine sediments (mainly mud and silt). Finer sediments have smaller interstitial spaces and slow down groundwater flow in relation to the flow in sandy sediments [12–14,35]. Slower flow rates increase contact time between nutrients and reactive areas of the sediment and rhizosphere, thereby increasing opportunities for nutrient processing [19]. In addition, sandy sediments are primarily composed of inorganic material (e.g. quartz) and typically contain

Table 1. Results of ANOVA for flow rates and NO_3^- removal.

Test	Effect	Degrees of Freedom	F value	P value
Flow rates	Treatment	5	21.545	<0.001
	Year	1	76.818	<0.001
	Treatment × Year	5	1.200	0.339
NO_3^- removal	Treatment	5	1.195	0.317
	Year	1	0.197	0.658
	Treatment × Year	5	0.020	0.999

Table 2. Mean percent NO_3^- removal across 6 restoration designs directly after planting and 1 year after planting (± 1 SE).

	Treatment					
Year	**0%**	**25%**	**50%**	**50% A**	**75%**	**100%**
0	3.22 (± 0.87)	4.29 (± 0.94)	5.44 (± 1.81)	6.23 (± 1.54)	5.95 (± 1.22)	6.47 (± 2.18)
1	3.73 (± 1.19)	4.28 (± 1.25)	5.99 (± 1.27)	6.27 (± 2.68)	6.89 (± 1.12)	6.97 (± 2.49)

little organic matter, which limits the biological processing of incoming nutrients [15].

While we did not find evidence for a strong role of these marsh planting designs as filters of runoff nutrient pollution, they provide other important services such as habitat and shoreline stabilization. We did not quantify these services, but we would expect these services to be small in magnitude for these young marshes and increase as the marshes age [6]. Similarly, we expect that these restored marshes will increase vegetated area and become effective nutrient filters through time.

Conclusions

In conclusion, our results suggest that groundwater flow rates and sediment type should be considered when planning marsh restoration efforts with the specific goal of runoff pollution

removal. Despite finding slower groundwater flow rates with increasing vegetation, our introduced groundwater did still flow quickly through the restored plots (30–60 minutes). This fast flow is most likely attributable to the coarse texture of the sediment. We calculated that on average only 3–7% of the NO_3^- entering the restored marshes was processed by the marshes within one year since planting, and it appears that NO_3^- processing by these restored marshes will remain similar to the unvegetated plots for many years to come. However, the slower groundwater flow in the more vegetated plots suggests they will likely become effective nutrient filters prior to the less vegetated plots. Understanding the nature, control, and extent of nutrient processing by restored marshes requires additional research, such as more restoration designs across many different marsh environments, in order to help managers decide on best restoration practices given budget constraints.

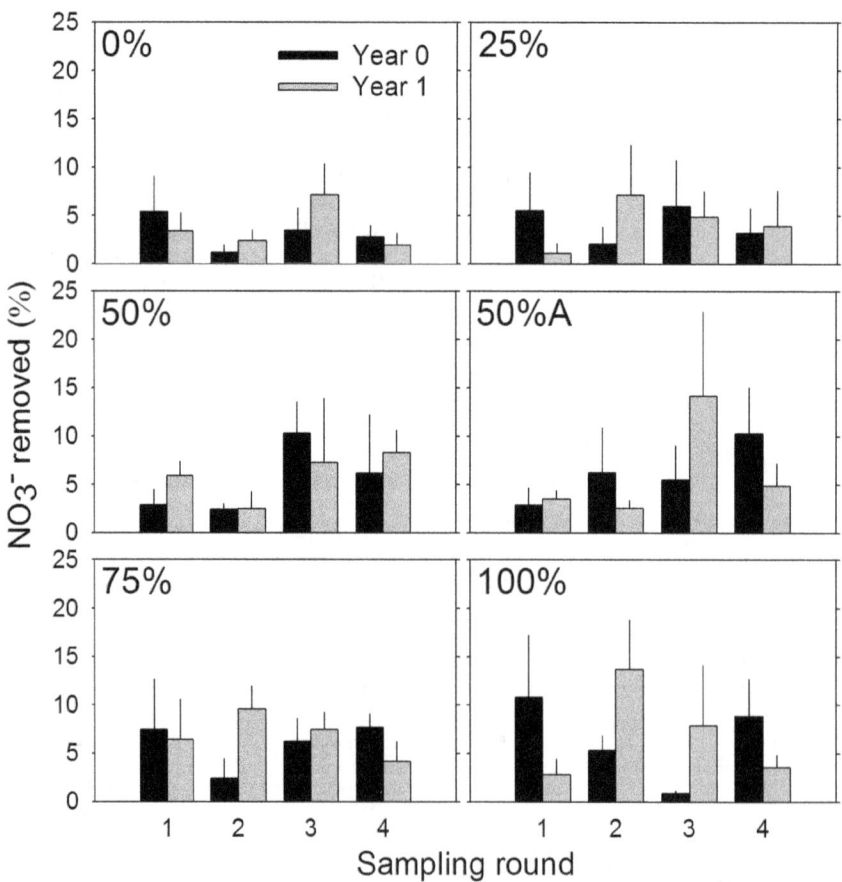

Figure 3. Percentage of NO_3^- removed from introduced solution directly after planting and 1 year post-planting for each sampling round. Black bars represent year 0 and grey bars represent year 1 samples. Percentages in the top left portion of each plot represent initial planting densities. Error bars indicate ± 1 SE.

Acknowledgments

We would like to thank Jason Howard, Jelani Reynolds, Jennifer Hemphill and Amanda Pratt of the Ecosystems Lab at the Dauphin Island Sea Lab for their field assistance.

Author Contributions

Conceived and designed the experiments: ELS JC SMS. Performed the experiments: ELS SMS. Analyzed the data: ELS. Contributed reagents/materials/analysis tools: JC. Contributed to the writing of the manuscript: ELS JC.

References

1. Bromberg Gedan K, Silliman BR, Bertness MD (2009) Centuries of Human-Driven Change in Salt Marsh Ecosystems. Annu Rev of Mar Sci 1: 117–141.
2. Sparks EL, Cebrian J, Biber PD, Sheehan KL, Tobias CR (2013) Cost-effectiveness of two small-scale salt marsh restoration designs. Ecol Eng 53(2013): 250–256.
3. Chapman MG, Underwood AJ (2000) The need for a practical scientific protocol to measure successful restoration. Wetlands (Australia) 19(1): 28–49.
4. Costanza R, d'Arge R, de Groot R, Farber S, Grasso M, et al. (1997) The value of the world's ecosystem services and natural capital. Nature 387: 253–260.
5. Ehrenfeld JG (2000) Defining the Limits of Restorations: The Need for Realistic Goals. Restor Ecol 8: 2–9.
6. Hilderbrand RH, Watts AC, Randle AM (2005) The myths of restoration ecology. Eco Soc 10: (online) URL: http://www.ecologyandsociety.org/vol10/iss1/art19/.
7. Beck M, Heck K Jr, Able K, Childers D, Eggleston D, et al. (2001) The identification, conservation, and management of estuaries and marine nurseries for fish and invertebrates. Biosci 51(8): 633–641.
8. Chmura GL, Anisfeld SC, Cahoon DR, Lynch JC (2003) Global carbon sequestration in tidal, saline wetland soils. Glob Biogeochem Cycles 17(44): 1–22.
9. Moeller I, Spencer T, French JR (1996) Wind wave attenuation over saltmarsh surfaces: Preliminary results from Norfolk, England. J Coast Res 12: 1009–1016.
10. Valiela I, Cole ML (2002) Comparative Evidence that Salt Marshes and Mangroves May Protect Seagrass meadows from Land-derived Nitrogen Loads. Ecosystems 5(1): 92–102.
11. Hammersley MR, Howes BL (2005) Coupled nitrification-denitrification measured in situ in a Spartina alterniflora marsh with a 15NH4 tracer. Mar Ecol Prog Ser 299: 123–135.
12. Tobias C, Macko S, Anderson I, Canuel E, Harvey J (2001) Tracking the fate of a high concentration nitrate plume through a fringing marsh: A combined groundwater tracer and in-situ isotope enrichment study. Limnol Oceanogr 46(8): 1977–1989.
13. Drake DC, Peterson BJ, Galvan KA, Deegan LA, Hopkinson C, et al. (2009) Salt marsh ecosystem biogeochemical responses to nutrient enrichment: a paired ^{15}N tracer study. Ecology 90(9): 2535–2546.
14. Tobias CR, Anderson IC, Canuel EA, Macko SA (2001) Nitrogen cycling through a fringing marsh-aquifer ecotone. Mar Ecol Prog Ser 210: 25–39.
15. Tobias CR, Neubauer SC (2009) Salt marsh biogeochemistry: An overview in Perillo G, Wolanski E, Cahoon D, Brinson M (eds). Coastal wetlands: An integrated ecosystem approach. Elsevier. p. 445–492.
16. Lehrter JC, Cebrian J (2009) Uncertainty propagation in an ecosystem nutrient budget. Ecol Appli 20: 508–524.
17. Tobias C, Harvey JW, Anderson IC (2001) Quantifying groundwater discharge through fringing wetlands to estuaries: Seasonal variability, methods comparison, and implication for wetland-estuary exchange. Limnol Oceanogr 46(3): 604–615.
18. Valiela I, Costa J, Foreman K, Teal JM, Howes B, et al. (1999) Transport of groundwater-borne nutrients from watersheds and their effects on coastal waters. Biodegradation 10(3): 177–197.
19. Barling RD, Moore ID (1994) Role of Buffer Strips in Management of Waterway Pollution: A Review Environ manag 18(4): 543–558.
20. Stout JP, Heck KL Jr, Valentine JF, Dunn SJ, Spitzer PM (1998) Preliminary Characterization of Habitat Loss: Mobile Bay National Estuary Program, MESC Contribution Number 301.
21. Eleuterius LN (1976) The distribution of Juncus roemerianus in the salt marshes of North America. Earth Environ Sci 17(4): 289–292.
22. Turner RE (1990) Landscape development and coastal wetland losses in the northern Gulf of Mexico Am Zool 30(1): 89–105.
23. LaSalle MW (1996) Assessing the functional level of a constructed intertidal marsh in Mississippi. Tech Rep WRP-RE-15, US Army Corps of Engineers, Waterways Experiment Station, Vicksburg, MS.
24. Lewis RR (1982) Creation & Restoration of Coastal Plant Communities. CRC Press, Boca Raton: 153–171.
25. Turner RE, Streever B (2002) Approaches to Coastal Wetland Restoration: Northern Gulf of Mexico. SPB Academic Publishing, Hague, The Netherlands.
26. Corbett DR, Dillon K, Burnett W (2000) Tracing groundwater flow on a barrier island in the north-east Gulf of Mexico. Estuar Coast Shelf Sci 51(2): 227–242.
27. Maynard DG, Kalra YP (1993) Nitrate and extractable ammonium nitrogen. In M.R. Carter, Ed. Soil Sampling and Methods of Analysis. Lewis Publishers, Boca Raton, FL, 25–38.
28. Quinn GP, Keough MJ (2002) Experimental design and data analysis for biologists. Experimental design and data analysis for biologists. Cambridge University Press, Cambridge, UK.
29. Vukovic M, Soro A (1992) Determination of Hydraulic Conductivity of Porous Media from Grain-Size Composition. Littleton, Colorado: Water Resources Publications.
30. Stumpf RP (1983) The process of sedimentation on the surface of a salt marsh. Estuar Coast Shelf Sci 17: 495–508.
31. Harrell J, Blatt H (1978) Polycrystallinity; effect on the durability of detrital quartz. J Sediment Res 48(1): 25–30.
32. Wilkins S, Keith DA, Adam P (2003) Measuring success: evaluating the restoration of a grassy eucalypt woodland on the Cumberland Plain, Sydney, Australia. Restor Ecol 11: 489–503.
33. Zedler JB, Callaway JC (1999) Tracking wetland restoration: do mitigation sites follow desired trajectories? Restor Ecol 7: 69–73.
34. Dingman SL (2008) Physical Hydrology. Waveland Press Inc, Long Grove, IL, USA.
35. Valiela I, Teal JM, Volkmann S, Shafer D, Carpenter EJ (1978) Nutrient and particulate fluxes in a salt marsh ecosystem: Tidal. Limnol Oceanogr 23(4): 798–812.

3

Rapid Losses of Surface Elevation following Tree Girdling and Cutting in Tropical Mangroves

Joseph Kipkorir Sigi Lang'at[1,2], James G. Kairo[1], Maurizio Mencuccini[3,7], Steven Bouillon[4], Martin W. Skov[5], Susan Waldron[6], Mark Huxham[2]*

1 Kenya Marine and Fisheries Research Institute, Mombasa, Kenya, 2 School of Life, Sport and Social Sciences, Edinburgh Napier University, Edinburgh, United Kingdom, 3 School of Geosciences, University of Edinburgh, Crew Building, Edinburgh, United Kingdom, 4 Department of Earth and Environmental Sciences, Katholieke Universiteit Leuven, Leuven, Belgium, 5 School of Ocean Sciences, Bangor University, Menai Bridge, Anglesey, United Kingdom, 6 School of Geographical and Earth Sciences, University of Glasgow, Glasgow, South Lanarkshire, United Kingdom, 7 Centre for Ecological Research and Forestry Applications, Universitat Autònoma de Barcelona, Barcelona, Spain

Abstract

The importance of mangrove forests in carbon sequestration and coastal protection has been widely acknowledged. Large-scale damage of these forests, caused by hurricanes or clear felling, can enhance vulnerability to erosion, subsidence and rapid carbon losses. However, it is unclear how small-scale logging might impact on mangrove functions and services. We experimentally investigated the impact of small-scale tree removal on surface elevation and carbon dynamics in a mangrove forest at Gazi bay, Kenya. The trees in five plots of a *Rhizophora mucronata* (Lam.) forest were first girdled and then cut. Another set of five plots at the same site served as controls. Treatment induced significant, rapid subsidence (-32.1 ± 8.4 mm yr^{-1} compared with surface elevation changes of $+4.2 \pm 1.4$ mm yr^{-1} in controls). Subsidence in treated plots was likely due to collapse and decomposition of dying roots and sediment compaction as evidenced from increased sediment bulk density. Sediment effluxes of CO_2 and CH_4 increased significantly, especially their heterotrophic component, suggesting enhanced organic matter decomposition. Estimates of total excess fluxes from treated compared with control plots were 25.3 ± 7.4 tCO$_2$ ha^{-1} yr^{-1} (using surface carbon efflux) and 35.6 ± 76.9 tCO$_2$ ha^{-1} yr^{-1} (using surface elevation losses and sediment properties). Whilst such losses might not be permanent (provided cut areas recover), observed rapid subsidence and enhanced decomposition of soil sediment organic matter caused by small-scale harvesting offers important lessons for mangrove management. In particular mangrove managers need to carefully consider the trade-offs between extracting mangrove wood and losing other mangrove services, particularly shoreline stabilization, coastal protection and carbon storage.

Editor: Vanesa Magar, Centro de Investigacion Cientifica y Educacion Superior de Ensenada, Mexico

Funding: This work was funded by grants from the UK Natural Environment Research Council (Grant No. NE/G009589/1), Earthwatch Institute and AVIVA Ltd, London, UK. The funders had no role in study design, data collection and analysis, decision to publish, or preparation of the manuscript, with the exception of volunteers from Earthwatch Institute who helped in field work.

Competing Interests: The authors can confirm that they received funding from the charitable wing of Aviva Ltd.

* Email: m.huxham@napier.ac.uk

Introduction

Mangrove forests are highly productive systems and often allocate a large proportion of their energy budget to root production [1–3]. Because of the presence of aerial roots, mangroves trap allochthonous organic matter in sediment, with carbon (C) sequestration rates exceeding those of terrestrial tropical forests by a factor of ~6 [4]. Unlike terrestrial forest soils, mangrove sediments do not attain C saturation because of continued sediment accumulation and vertical accretion [5] and hence the size of the C store continues to increase over time [6]. Anoxia, low levels of nutrients and the high lignin content of the roots result in slow decomposition of below-ground organic matter [7–9] and the accumulation of large reserves of peat and C-rich sediments [10–12]. Mangroves are thus amongst the most carbon dense of all forests, with C stocks sometimes exceeding 1000 tonnes C ha^{-1} [12–14], and hence play an important role in global carbon storage [15–17].

As a result of continued vertical accretion and below-ground root growth, surface elevation of mangroves increases over time at rates of up to 4.8 mm yr^{-1} e.g. [5]. These increases in surface elevation are important in allowing mangrove recovery after natural disturbances and are considered essential for many mangroves to survive projected sea level rise of 1.7 to 3.3 mm yr^{-1} this century [18,19].

Human disturbances such as wood harvesting and clearing threaten to impair these important ecological processes. Trends in mangrove loss are alarming, with an estimated 30–50% of forests lost over the past half century [20,21]. Although rates of loss may be declining [22] they remain high, typically 0.7–3% y^{-1}, partly because of high levels of poverty and dense human populations along many tropical coasts [13]. While a recent estimate of the impact of such losses on the total mangrove carbon sink suggested that mangrove destruction could contribute up to 10% of the annual GHG emissions from land use change [12], little understanding exists of the impact of mangrove harvesting on surface elevation dynamics. Reductions in surface elevation after harvesting may be caused by a combination of processes. Following harvesting, root growth and expansion stops, whilst decomposition of dead roots and old organic matter may be accelerated by higher temperatures in the exposed substrate and increased sediment oxidation. Sediment erosion may also increase and the lack of aerial roots may prevent continued accumulation of allochthonous sediment. Additionally, the aerenchymatous tissues of the dying roots may shrink leading to increased bulk density but lower elevation. Finally, leaching of dissolved inorganic and organic C and lateral transport to the sea may occur [2,3,17,23,24].

Few studies have directly measured the impacts of tree removal on below-ground carbon storage and surface elevation. Two studies report on the effects of hurricanes [25,26] and one on total deforestation [27]. Under such extreme conditions carbon losses can be large with resulting 'peat collapse' and coastal erosion. The impacts of smaller scale tree loss are even less well known, although work from Micronesia after non-experimental tree felling [28] and Florida following lightning strikes [29] shows that surface elevation losses might be rapid. Whilst healthy forests are likely to be resilient and show recovery, those under anthropogenic pressure may experience longer term change. Mangroves of the Western Indian Ocean (WIO) region experience small-scale but widespread degradation from indiscriminate harvesting [30,31]. A recent assessment of mangrove decline in Kenya has indicated an annual loss of 0.7%, which underestimates the anthropogenic stressors since it records only total canopy removal [32]. Here we document the first controlled experiment testing the impacts of small-scale cutting, the most common form of mangrove exploitation in the WIO region, on sediment carbon losses and surface elevation. Increasing attention is being paid to avoiding deforestation by Reducing Emissions from Deforestation and Degradation and conservation of forest ecosystems (REDD+) schemes and to managing forests for a range of services, not only wood production. It is essential therefore that the impacts of forest management scenarios on key ecosystem services such as coastal protection and carbon sequestration are analysed to understand possible trade-offs between these and extractive uses.

Mangrove forests of the WIO region are utilized by the local communities for construction and fuelwood [33]. In Kenya mangroves are the only natural forests currently licensed by the Kenya Forest Service (KFS) for wood harvesting [30,34]. In combination with widespread illegal harvesting this has left many mangrove areas either degraded or completely denuded of vegetation [30,35]. Since mangroves meet ~70% of the wood requirements of the coastal population [36], there is always a ready market for mangrove poles, especially in major coastal towns.

We used a controlled experiment to explore the impacts of tree harvesting on: a) surface elevation dynamics; b) the main processes affecting changes in surface elevation i.e., sediment accretion, sediment properties such as bulk density, %C and sediment moisture, and organic matter decomposition; and c) sediment surface C efflux in a natural mangrove forest, here at Gazi bay, Kenya. The C flux from forest floors comes from root (autotrophic) respiration and sediment/soil organic matter decomposition (heterotrophic respiration) [37]. The relative contributions of these sources have not been distinguished in mangroves; such partitioning is useful in understanding carbon stocks and flows in forests. Girdling trees stops the flow of photosynthates to the root system and thereby stops metabolic activities in the roots whilst maintaining the tree canopy. Hence it can be used to separate the components of soil respiration, since CO_2 emitted

from the sediment shortly after girdling is assumed to be primarily from organic matter decomposition e.g., [38–40]. Here we also report on the first time this approach has been used for mangroves.

Results

Surface elevation and sediment accretion

At the conclusion of monitoring after 760 days the control plots had gained mean surface elevation of $+11.1\pm10.5$ mm, at a mean rate of $+4.2\pm1.4$ mm yr^{-1}, while the treatment plots showed a subsidence of -51.3 ± 12.0 mm, at a mean rate of -32.1 ± 8.4 mm yr^{-1} (Figure 1). For a period of 110 days after setting up the horizon marker, both control and treated plots experienced similar trends in sediment accretion, ranging from 5.5 to 8.0 mm (controls, mean 6.4 ± 1.4 mm) and 2.5 to 11.7 mm (treated, mean 5.2 ± 4.6 mm). Disturbance of the horizon marker by crab activities in the control plots did not allow further monitoring beyond four months after set up.

Sediment Surface Carbon Fluxes

Approximately 30 days after girdling, CO_2-C emissions in treated plots increased and remained higher than in controls throughout the girdled period (Table 1 and Figure 2). For the first two months after cutting, CO_2-C emissions in the treated plots were similar to controls, but then increased again for three months before dropping to levels similar to the controls by the end of the sampling period (3.2 ± 0.9 vs. 3.9 ± 1.8 gCO_2–C m^{-2} d^{-1}, respectively; Figure 2). At ~30 days after girdling, the $\delta^{13}C$ signature of the sediment respired CO_2 from the treated plots was significantly more ^{13}C-depleted than in the controls (Figure 3). Methane emissions were highly variable and increased in the treated plots during the girdled period only (Figure 2). Mean emissions of both CO_2–C and CH_4–C were significantly higher in treated than control plots during the girdled period (Table 1). Throughout the treatment period, the mean sediment temperature in the treated plots was higher than the control plots by 0.9 to 5.8°C.

Separation of sediment autotrophic and heterotrophic respiration components

Girdling was not effective in separating the components of sediment respiration since the girdled trees began to lose leaves

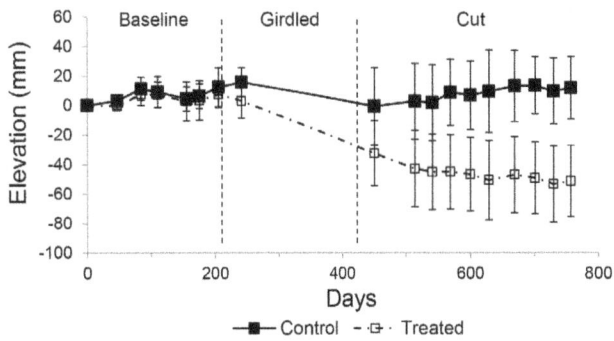

Figure 1. Trends in surface elevation change in control and treated sites in *R. mucronata* forest at Gazi bay, Kenya. Error bars are 95% CI. Vertical broken lines indicate periods when trees were girdled and cut in the treatment plots. Baseline, girdled and cut periods ran from March 2009 to October 2009 (205 days), December 2009 to May 2010 (189 days) and May 2010 to April 2011 (343 days), respectively. The controls and the treatment consisted of five replicates each.

Table 1. Nested design ANOVA for carbon fluxes in *R. mucronata* secondary forest, with mean data for each of six chambers per plot nested within treatment; the data for CO_2 were log-transformed.

Period	Variable	Source of Variation	DF	MS	F	P
Girdled	CO_2	Temperature	1	0.017	0.420	0.518
		Burrows	1	0.017	0.45	0.506
		Treatment	1	0.152	4.73	0.036
		Plot (Treatment)	8	0.025	0.64	0.736
		Error	48	0.039		
	CH_4	Temperature	1	46.140	2.4	0.128
		Burrows	1	2.660	0.14	0.712
		Treatment	1	745.800	20.04	0.000
		Plot (Treatment)	8	55.700	2.9	0.01
		Error	48	19.230		
Clear-cut	CO_2	Temperature	1	0.005	0.16	0.695
		Burrows	1	0.044	1.26	0.267
		Treatment	1	0.001	0.02	0.897
		Plot (Treatment)	8	0.028	0.81	0.6
		Error	48	0.035		
	CH_4	Temperature	1	24.758	3.43	0.07
		Burrows	1	5.73	0.79	0.378
		Treatment	1	1.458	0.19	0.667
		Plot (Treatment)	8	10.834	1.5	0.182
		Error	48	7.225		

and die within one month of girdling; this was much faster than recorded for terrestrial trees e.g., [40]. Therefore, an indirect regression method of estimating autotrophic respiration was used [38]. In each of the control plots, live root biomass was determined after the final sampling in April 2011 (530 days after treatment) by excavations made beneath the pre-marked positions of each chamber. The regression of CO_2 fluxes, measured at the final sampling for each control plot chamber, against the live root biomass and sediment surface temperature adjacent to each chamber, was significant:

Multiple regression: $lnR_S = -3.093+0.0002*$live root biomass $+ 0.127*$temperature; $R^2 = 0.37$; $P = 0.044$, 0.011 and 0.012 for the constant, live root biomass and sediment temperature, respectively.

Heterotrophic respiration (R_H) for final sample values was calculated by applying the equation above, while setting the value of root biomass to zero [38]. Autotrophic respiration (R_A), obtained by subtracting the values for R_H from the measured total sediment respiration (R_S), contributed a mean ($\pm95\%$ CI) of $41.5\pm11.8\%$ to R_S at the final sampling. Across the entire sampling period, R_A contributed an average of $40.5\pm7.0\%$ to R_S, which was not significantly different from that obtained at the final sampling (t-test, $t = -0.16$, $P = 0.874$). The partitioning of the components of total R_S in the control plots allowed a comparison of the treated plot CO_2 fluxes (where autotrophic R was zero by definition) and control plot R_H. Sediment respiration of the treated plots was higher than control plots R_H throughout the treatment period (during both girdling and cutting) by 0.6 to 3.7 g CO_2–C m^{-2} d^{-1} (Figure 2).

Estimated C losses from the sediment

Combining data on elevation changes with sediment characteristics provided one estimate of total carbon losses of 35.7 ± 76.9 t C ha^{-1} from the treated compared to control plots.

The net belowground carbon losses due to treatment estimated from sediment surface flux data amounted to 14.2 ± 10.3 tCO_2 ha^{-1} (mean rate of 9.8 ± 7.1 t CO_2 ha^{-1} yr^{-1}) over a period of 530 days after treatment. When only R_H in control plots was considered, the net additional C loss amounted to 36.7 ± 10.7 t CO_2 ha^{-1}, with a mean rate of 25.3 ± 7.4 t CO_2 ha^{-1} yr^{-1}.

Root Decomposition

After 267 days root-bags in the treated plots had lost significantly more mass than those in the controls, with rates of 0.19 ± 0.02 and $0.16\pm0.03\%$ dry weight loss day^{-1}, respectively (t-test, $t = -2.06$, $P = 0.049$). The trend in decomposition rate in the treated plots indicated that it was increasing with time, while the rate in the control plots remained constant between 156 and 267 days.

Belowground Biochemical Characteristics

The plant roots had a similar $\delta^{13}C$ signature to that of the sediment carbon (Table 2). Treated plots showed significant reductions in % C and sediment moisture and significant increases in bulk density. The sediment carbon stocks to a depth of 1 m ranged from 414.1 to 610.7 t C ha^{-1} for controls and 457.3 to 586.3 t C ha^{-1} for treated plots. Overall the control plots tended to have higher mean C stocks than treated plots, with mean ($\pm95\%$ CI) of 524.1 ± 88.8 vs. 488.4 ± 68.6 tC ha^{-1}, respectively, but the difference was not significant due to the very large small-scale variability in sediment properties.

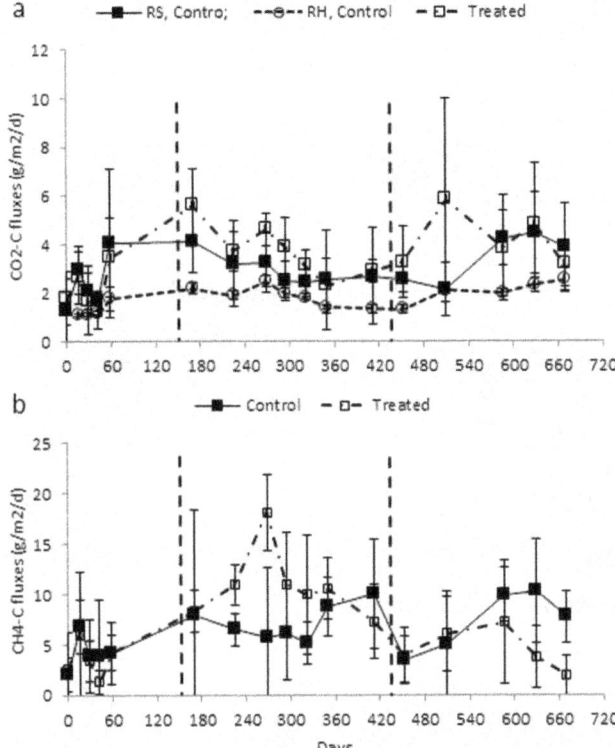

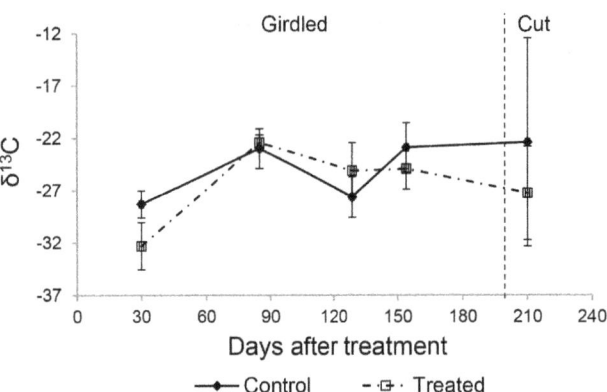

Figure 3. Trends in $\delta^{13}C$ of sediment respired CO_2 in control and treated sites in *R. mucronata* forest at Gazi bay, Kenya. Values are means ±95% CI. Vertical broken line indicates when the trees were clear-cut in treatment plots. The controls and the treatment consisted of five replicates each.

Figure 2. Mean (±95% CI) Carbon emissions. a) CO_2 fluxes of total sediment respiration (R_S) (solid line with filled squares), heterotrophic respiration (R_H) (dashed line with open circles) in control plots and CO_2 fluxes from treated plots (broken line with open squares) and b) CH_4 emissions in *R. mucronata* forest at Gazi bay Kenya. Vertical broken lines indicate periods when trees were girdled and cut in the treatment plots. Sampling for baseline, girdled and cut periods were done from June 2009 to August 2009 (84 days), December 2009 to May 2010 (189 days) and May 2010 to April 2011 (343 days), respectively. The controls and the treatment consisted of five replicates each.

Discussion

Tree death induced rapid and significant subsidence. This was likely due to rapid decomposition of fine roots exacerbated by the absence of new root growth, as shown by the significant difference in root bag decomposition. In addition, sediment compaction and the collapse of arenchymatic tissues due to consolidation of air spaces (as shown by the increase in bulk density), and the significant loss of sediment moisture caused by the treatment, may also have contributed. Rates of sediment accretion did not differ significantly between control and treated plots hence the subsidence was not due to enhanced erosion. The rate of subsidence was surprisingly high given the relatively small scale of the treatments; massive hurricane damage, which led to 'peat-collapse' in Honduran mangroves, caused elevation losses only around ~0.3 as fast as those recorded here (−11 compared with −32 mm yr⁻¹ [25]).

Interestingly, the C loss of around 25.3±7.4 t CO_2 ha⁻¹ yr⁻¹ reported here due to small-scale cutting was similar to that reported for mangrove forests impacted by large-scale clearing (29 tCO_2 ha⁻¹ yr⁻¹) [27], and that inferred from peat collapse due to hurricane damage (15 t CO_2 ha⁻¹ yr⁻¹) [25]. Much of the C loss occurred within the first year after treatment (mean ±SD? rates of 13.22±9.71 and 7.86±6.77 t CO_2 ha⁻¹ yr⁻¹ during the girdled and cut periodsrespectively) and by ~1.5 years the C losses

induced by treatment began to drop. A similar pattern was recorded in clear-cut mangroves in Belize in which the C emissions in disturbed areas declined with time [27]. However, there was evidence in our study that decomposition of sediment organic carbon (SOC) - not only newly-killed root material - was enhanced by treatment, and that rates of SOC decomposition might be increasing with time. Buried root bags recorded significantly higher rates of decomposition in treated plots (0.19±0.02 vs. 0.16±0.02% dry weight loss day⁻¹, respectively) with most of the difference occurring after the first set of root bags were retrieved (150 days after burial). This was probably due to enhanced sediment surface temperatures in the cut plots due to canopy removal, since the treated plots experienced increases in sediment surface temperatures of 0.8 to 5.9°C compared to the control plots. Therefore, these results highlight the potential impact of physico-chemical changes on C losses in cut forests, which are separate from and additional to the losses from root death per se.

Although our estimates suggest high rates of below-ground C loss caused by tree death (similar to those seen following much larger impacts such as hurricanes) they are likely to represent an underestimate, since surface fluxes of CO_2 cannot account for below-ground, lateral flows of carbon in dissolved inorganic carbon (DIC) [2,3,24].

CH_4 emissions were significantly enhanced during the girdled period, possibly because of the addition of easily fermentable substrates [41–43] from dying roots. However, emissions in both treatment periods (girdled and cut respectively 0.7–1.5 and 0.2–0.9 mmol m⁻² d⁻¹), did not exceed those reported for a number of pristine mangrove forests worldwide, 0.01–5.0 mmol m⁻² d⁻¹ [40]. Sediment-respired CO_2 collected shortly after girdling showed significant ^{13}C-depletion (−32.3 ‰; Fig. 3). Such depleted signatures are unlikely to arise from respiration of existing organic matter alone, as here $\delta^{13}C$ would be similar to the control. Rather, as coincident with increased CH_4 emissions, the most parsimonious explanation is that oxidation of methane comprised a component of the CO_2 efflux. Assuming a typical $\delta^{13}C$ value for methane in marine environments of −60 ‰ [44], the measured ^{13}C-depletion would represent an additional 12% CO_2 contribution from methane oxidation.

Overall, our experiment showed that small-scale harvesting for wood production as is typical in the countries around the Western

Table 2. Belowground roots and sediment biochemical characteristics in control and cut plots in *R.mucronata* forest at Gazi bay, Kenya.

Variable	Control	Cut
% OC$_R$	37.2±1.7	-
δ^{13}C$_R$	−27.0±0.3	-
Sediment moisture content (%)	46.5±4.0	42.0±5.7
Bulk Density (g cm^{-3})	0.84±0.08	0.88±0.10
Sediment C concentration (g C cm^{-3})	0.052±0.008	0.052±0.008
% N	0.38±0.04	0.36±0.11
TOC/TN	18.59±0.8	17.6±0.7
δ^{13}C$_S$	−27.2±0.2	−27.4±0.1
Sediment C stocks* (t C ha^{-1})	524.1±62.7	488.4±48.4

Values are means ±95% CI, OC$_R$ = organic carbon content of mangrove roots, δ^{13}C$_R$ and δ^{13}C$_S$ = carbon isotopic value of mangrove roots and sediment organic matter, respectively, and TOC and TN = total carbon and nitrogen content of the sediment organic matter, respectively.
*Sediment C stocks to 1 m depth.

Indian Ocean caused significant sediment subsidence, which was caused primarily by sediment compaction, loss of sediment moisture and increased organic matter decomposition (both aerobic and anaerobic). Study of lightning damage in Florida mangroves, creating forest gaps larger than our experimental plots, showed even more subsidence (of up to 61 mm), but also suggests that undisturbed forests can recover [29]. Such evidence should not induce complacency since pressures on mangroves in many areas are increasing. Sea level is projected to rise this century [18], with current projections varying between 1.7 and 3.3 mm yr^{-1} [19]. Employing the lower estimate of sea level rise of 1.7 mm yr^{-1}, and the conservative assumption that our cumulative surface elevation loss of 62.4 mm will not increase further in the future, the impact of the loss we observed is equivalent to increasing sea level rise for ~37 years. Hence our study raises serious concerns that the combination of effects caused by global change and small-scale forest harvesting on mangrove surface elevation may be highly damaging to these sensitive ecosystems.

Conclusions

Kenyan mangroves are being lost at a rate of 0.7% cover per year [45], mainly because of the demands of the large and growing population for wood fuel and timber [34,36].

A new approach towards the sustainable management of mangrove forests uses payments for ecosystem services (PES) schemes such as the proposed REDD+ programme [46]. Such schemes should allow combining income from provisioning and regulating ecosystem services (such as timber and coastal protection or carbon sequestration), but only if the trade-offs between them are understood and managed. The present work demonstrates the susceptibility of mangroves to rapid subsidence (with consequent enhanced vulnerability to sea level rise and erosion) and at least short-term carbon loss following relatively small-scale and controlled canopy removal. With this understanding, management regimes aiming to conserve carbon stocks and promote climate resilience should be wary of clear-cutting, particularly in areas that may be exposed to erosion, and should emphasise instead selective cutting, rapid replacement of the lost canopy and the maintenance of un-cut buffer strips on seaward fringes to avoid the risks of erosion.

Materials and Methods

Study site

The study was carried out at Gazi Bay (4° 25′ S and 4° 27′ S; 39° 50′ E and 39° 50′ E), ~55 km south of Mombasa, Kenya. Gazi Bay is a creek system with a total area of 615 ha mangrove forest [47], dominated by *Rhizophora mucronata* (Lam), *Ceriops tagal* (Perr.) C. B. Robinson and *Avicennia marina* (Forsk.) Vierh. The mean annual precipitation of Gazi bay ranges from 1000–1600 mm [47]. The bay receives freshwater from two semi-permanent rivers: Kidogoweni to the north, which discharges in to the Kidogoweni creek, and Mkurumuji river to the south, discharging to the mouth of the bay. The forest is government-owned and permission to use the site was granted by the Kenya Marine and Fisheries Research Institute; our experiments did not affect endangered or protected species, and the individual featured in the accompanying image gave written informed consent (as outlined in the PLOS consent form) to publish this image.

Experimental Design

Five pairs of 12 m×12 m plots were established in March 2009 at high water mark (height above mean sea-level: 2.91±0.01 m) within a *Rhizophora mucronata* dominated forest (specific location: 4°24′32″S 39°31′23″E). In October 2009, five plots (one from each pair) were randomly allocated to 'treatment' and all the trees within them were girdled at ~20 cm above the highest prop root. Girdling is a method that has been used in a number of terrestrial forests to estimate the contribution of root respiration to total sediment gas flux. The rationale is to prevent the flow of carbohydrates from the tree canopy to their roots (thus stopping root respiration) whilst leaving the above-ground components relatively undisturbed; trees may retain foliage for many months after girdling [39,40]. The other plots in the five pairs served as the controls. In May 2010, all the trees in the girdled plots were cut at ~20 cm above the highest stilt roots and all the debris, excepting small fragments, was removed. The treated plots were allowed to stabilize after disturbance for approximately three weeks; thereafter sampling was resumed. During each treatment operation saplings and seedlings were cut down. Hence the experiment consisted of three sampling periods: a) baseline (pre-treatment) (June 2009 to August 2009, 84 days), b) girdled period (December

2009 to May 2010, 159 days) and c) cut period (May 2010 to April 2011, 343 days).

Surface elevation and sediment accretion

Surface elevation dynamics were monitored using surface elevation stations, consisting of two stainless steel rods (6 mm by 1 m) and a horizon marker (kaolin) set up in a 20 cm×20 cm quadrat in each plot. The rods were installed leaving a height of 20 cm above the ground and at opposite corners of the quadrat such that measurements would be made diagonally across it. Height measurements vertically from the ground surface to the heights of the rods were made at seven clearly marked points along a wooden board placed across the rods. Subsequent measurements were made at the same points along the wooden board. All data were averaged to give a single measurement per plot per time. Sediment accretion was determined from measurements of height above the horizon marker. In each quadrat, at least four sediment blocks of 2 cm×2 cm were carefully removed with a sharp knife, the height of sediment above the horizon marker was noted and the block was then carefully replaced in its original position. This approach allows the separation of total surface elevation/ subsidence (which depends on both the accretion or erosion of new sediments and on below-ground processes such as root growth and expansion) from accretion/erosion [25].

Samples for sediment physico-chemical analysis were taken from each plot during February 2010 and August 2012. A sediment core was taken in the centre of the plot using a plastic corer (diameter 6 cm, length 3 m) in February 2010 and again in August 2012. Subsamples of this large core were taken with a small stainless steel corer (diameter 3 cm, length 5 cm) at depths of 0, 2, 4, 8, 10, 20, 30 and 40 cm (February 2010) and depths of 0, 10, 20, 30, 50 and 100 cm (August 2012). To minimize compression of the sediment the coring was done in a series of stages according to the depth profiles for sub-sampling. Sediment samples were oven-dried at 80°C to constant dry weight and bulk density was determined at six depths down to 1 m. The oven-dried samples were transferred to the Department of Earth and Environmental Sciences, KU Leuven, Belgium for analysis of % OC (organic carbon), % N and $\delta^{13}C$ of OC. The concentrations of OC, total N, and $\delta^{13}C$ values of sediment OC were measured on subsamples weighed into Ag cups, acidified with dilute HCl to remove inorganic C, and analysed with a Thermo Flash HT elemental analyser coupled to a Thermo Delta V Advantage IRMS (Conflo IV interface). Data were calibrated with IAEA-C6, and internally calibrated acetanilide and leucine. From the sediment bulk density and OC content, the carbon density and hence the sediment carbon stocks down to 100 cm were derived for the control and treated plots.

Gas fluxes and stable carbon isotope signatures

Gas flux (CO_2, CH_4) samples were collected at approximately monthly intervals at low tide during spring tides using six chambers per plot; some sampling times that were missed due to loss or damage to bags and other equipment. Each chamber was inserted ~5 cm in to the sediment, occupying an area of 0.064 m² with an internal volume of 0.011 m³. The samples from each chamber were taken 20 minutes after closure. Using a 60 ml syringe, at least 240 ml of gas were transferred from each chamber to labelled airtight gas-bags (Cali-5-bond gas bags, Calibrated Instruments Inc. USA). A gas sample of ambient concentration was taken from each chamber before closure; ambient air concentration samples for each plot were collected in one gas bag. Linearity checks were performed by repeatedly sampling the chamber gas for periods of about 60 minutes. They showed that a linear approximation over a 20 minutes period resulted in ~15% underestimation of the slope of gas concentration increase over time. This systematic downward bias was not corrected for. Sediment surface temperature measurements were made beside each chamber with a temperature probe inserted to ~1 cm in to the sediment. The number of crab burrows within the area enclosed by the chamber was noted. The positions of the chambers were marked for subsequent sampling; chambers were always returned to the same sampling positions within plots. Samples for $\delta^{13}C$ analysis of CO_2 were transferred from the chambers to 12 ml pre-evacuated exetainers (Labcoexetainer, Labco Ltd., High Wycombe, UK). Gas flux samples were analysed at the Institute of Atmospheric and Environmental Sciences, University of Edinburgh, UK. For CO_2, the samples were analysed by gas chromatography (GC) using a Perkin Elmer Model 310 with a thermal conductivity detector (TCD). Concentrations of CH_4 were measured using GC (Hewlett Packard 5890 GC, Hewlett Packard Ltd, Stockport, Cheshire, UK) equipped with a flame ionisation detector (FID) and a digital integrator. The $\delta^{13}C$ analysis samples were transferred to the Department of Earth and Environmental Sciences, KU Leuven, Belgium and analysed using a Sercon 20– 20 isotope ratio mass spectrometer (IRMS) interfaced with a cryofocussing unit.

In May 2011, at the end of the experiment, all roots beneath each chamber in the control plots were excavated to a depth of 60 cm, washed of sediments and separated in to live root and dead plant materials. Dead roots were differentiated based on the loss of structural integrity, colour and signs of decomposition [5]. The samples were oven-dried at 80°C before weighing.

Decomposition in Root-bags

In September 2010, live roots were excavated within the *R. mucronata* forest contiguous to the experimental plots. Nylon mesh (1 mm) bags each containing ~30 g fresh roots were buried to ~20 cm depth at six random points within each plot. Three bags were retrieved from each plot 156 days after burial, whilst the other three were retrieved 267 days after burial. The contents of each root-bag were rinsed and oven-dried at 80°C for 24 h before weighing. The rate of root decay (% weight loss day^{-1}) was calculated as the % weight loss divided by the number of days buried, using wet-dry weight conversion factors derived from representative samples of live roots, oven-dried at 80°C until constant dry weight.

Statistical analysis

The data for CO_2 were log-transformed and the analysis for each gas was executed using MINITAB 14 software package. Initial analyses included time in repeated measures models, however there were multiple significant interactions preventing legitimate conclusions and so data were separated into the three experimental periods. The gas flux values for each chamber were pooled across each period for the controls and treated plots, giving six single values per period for each plot. For each period, nested ANOVA was carried out for each gas, with variation among plots (i.e. 6 chambers) nested within treatment and sediment surface temperature and crab burrows as covariates. Estimates for respired $\delta^{13}C$ were derived from the Miller-Tans mixing model combined with geometric regression [48,49]. Kayler et al., [49] found that the combination of geometric regression and Miller-Tans mixing model gave the most accurate and precise estimate of $\delta^{13}C_S$ (S = sediment respired CO_2). The gas mixing models are based on the conservation of mass given as [48]:

$$\delta_{obs}[CO_2]_{\mathrm{obs}} = \delta_{bg}[CO_2]_{bg} + \delta_s[CO_2]_s$$

This equation describes the gas observed (obs) as coming from two sources: background atmosphere (bg) and source of respiration (s), where δ refers to the isotopic value of each component. Details of the Miller-Tans mixing model combined with geometric regression are discussed by Kayler et al., [49]. The mean $\delta^{13}C$ of the respired CO_2 for each plot was analysed using two-sample t-tests.

To examine the autotrophic contribution to sediment fluxes, stepwise multiple regressions (forward and backward elimination) were performed, with the final CO_2 fluxes measured in each control chamber in April 2011 as the dependent variable and the live root biomass, sediment surface temperature and crab burrows for each chamber as the independent variables. The equation takes the form of $y = k + a*roots + b*temperature + c*burrows$; where $k =$ constant, a, b and c are the coefficients of the estimators. The number of crab burrows was not significant, and hence this term was omitted from the equation. The significant factors were used in estimating the autotrophic respiration from the final CO_2 flux data. First, the heterotrophic respiration (R_H) was calculated as the value of the 'y' when live root biomass $= 0$, i.e. $R_H = k + b*temperature$. Then the autotrophic respiration (R_A) was obtained as the difference between total sediment respiration (R_S) and the heterotrophic component (R_H) and expressed as a percentage of R_S (i.e. $\%R_A = (R_S - R_H)/R_S*100$). To estimate the contribution of R_A across the entire sampling period, the equation was applied to the CO_2 flux data, together with the sediment surface temperature for each control chamber at each sampling time. The mean R_A contribution across the sampling period was then compared with that obtained from the final sampling time.

The total additional C emissions due to treatment (treatment-induced C emissions) were estimated as the area under the response curve based on the trapezoidal rule [50]:

$$Ce = \sum_{i=l}^{n} \frac{(m_i + m_i + 1)}{2} t_i$$

where $C_e =$ treatment induced C emissions, $n =$ number of measurements, $m =$ individual measurements and $t =$ time difference between any two consecutive measurements.

The difference in the rates of root decomposition in the control and treated plots was tested using a two-sample t-test. Sediment carbon concentration (g C cm^{-3}) was calculated as the product of bulk density (BD) and % organic C of the sediment. Thereafter, the sediment carbon stocks down to a depth of 100 cm for each treatment were calculated as the product of carbon concentration and the depth and expressed as t C ha^{-1}: $C_S = C_C * (100 + E_c)$, where, $C_S =$ sediment C stocks, $C_C =$ C concentration and $E_c =$ elevation change. Since the control plots gained 1.1 cm and the treated plots lost 5.1 cm in surface elevation (see Figure 2), the depth for each treatment was adjusted to reflect these changes, i.e. $100 + E_c = 101.1$ and 94.9 cm for control and treated plots, respectively.

Acknowledgments

Our gratitude goes to our field assistants Laitani Suleiman and Tom Kisiengo Peter, KMFRI staff at Gazi village, the Gazi community and to the many Earthwatch International volunteers who assisted in the field. We would also like to thank Robert Howard (School of GeoSciences, University of Edinburgh, UK) for assisting in laboratory analysis of CO_2 and CH_4 gases. Ken Krauss gave helpful advice on an early draft.

Author Contributions

Conceived and designed the experiments: JL JK MM SB MS SW MH. Performed the experiments: JL JK MS MM MH. Analyzed the data: JL MM SB SW MH. Contributed reagents/materials/analysis tools: SW MM SB. Wrote the paper: JL JK MM MS SB SW MH.

References

1. Twilley RR, Chen RH, Hargis T (1992) Carbon sinks in mangroves and their implications to carbon budget of tropical coastal ecosystems. Water Air Soil Poll 64: 265–288.
2. Bouillon S, Borges A, Castañeda-Moya E, Diele K, Dittmar T, et al. (2008) Mangrove production and carbon sinks: A revision of global budget estimates. Global Biogeochem Cycles 22: GB2013, doi:2010.1029/2007GB003052.
3. Kristensen E, Bouillon S, Dittmar T, Marchand C (2008) Organic carbon dynamics in mangrove ecosystems: A review. Aquat Bot 89: 201–219.
4. Breithaupt JL, Smoak JM, Smith TJ III, Sanders CJ, Hoare A (2012) Organic carbon burial rates in mangrove sediments: Strengthening the global budget. Global Biogeochem Cycles: doi:10.1029/2012GB004375, in press.
5. McKee KL, Cahoon DR, Feller I (2007) Caribbean mangroves adjust to rising sea level through biotic controls on change in soil elevation. Glob Ecol Biogeogr 16: 545–556.
6. Chmura GL, Anisfeld SC, Cahoon DR, Lynch JC (2003) Global carbon sequestration in tidal, saline wetland soils. Global Biogeochem Cycles, 11: 1111–11120.
7. Middleton BA, McKee KL (2001) Degradation of mangrove tissues and implications for peat formation in Belizean island forests. J Ecol 89: 818–828.
8. Gleason SM, Ewel KC (2002) Organic matter dynamics on the forest floor of a Micronesian mangrove forest: An investigation of species composition shifts. Biotropica, 34: 190–198.
9. Huxham M, Lang'at J, Tamooh F, Kennedy H, Mencuccini M, et al. (2010) Decomposition of mangrove roots: Effects of location, nutrients, species identity and mix in a Kenyan forest. Estuar Coast Shelf Sci 88: 135–142.
10. Golley F, Odum HT, Wilson RF (1962) The structure and metabolism of a Puerto Rican red mangrove forest in May. Ecology, 43: 9–19.
11. Fujimoto K, Imaya1 A, Tabuchi R, Kuramoto S, Utsugi H, et al. (1999) Belowground carbon storage of Micronesian mangrove forests. Ecol Res 14: 409–413.
12. Donato DC, Kauffman J, Murdiyarso D, Kurnianto S, Stidham M, et al. (2011) Mangroves among the most carbon-rich forests in the tropics. Nat Geosci 4: 293–297.
13. Alongi DM (2012) Carbon sequestration in mangrove forests. Carbon Manage 3: 313–322.
14. Trumper K, Bertzky M, Dickson B, van der Heijden G, Jenkins M, et al. (2009) The Natural Fix? The role of ecosystems in climate mitigation. United Nations Environment Programme, UNEPWCMC (Cambridge, UK). 68 pp.
15. Laffoley DA, Grimsditch G (2009) The management of natural coastal carbon sinks. IUCN. 53 pp.
16. Nellemann C, Corcoran E, Duarte CM, Valdés L, De Young C, et al. (2009) Blue Carbon. United Nations Environment Programme. 80 p.
17. McLeod E, Chmuira GL, Bouillon S, Salm R, Björk M, et al. (2011) A blueprint for blue carbon: toward an improved understanding of the role of vegetated coastal habitats in sequestering CO_2. Front Ecol Environ 9: 552–560.
18. IPCC (2007) Climate Change 2007: Synthesis Report. Contribution of Working Groups I, II and III to the Fourth Assessment Report of the Intergovernmental Panel on Climate Change. IPCC (Geneva, Switzerland). 104 p.
19. Nicholls RJ, Cazenave A (2010) Sea-level rise and its impact on coastal zones. Science, 328: 1517–1520.
20. Valiela I, Bowen JL, York JK (2001) Mangrove forests: one of the world's threatened major tropical environments. BioScience, 51: 807–815.
21. Valiela I, Kinney E, Culbertson J, Peacock E, Smith S (2009) Global losses of mangroves and salt marshes, In Duarte, C. M. [ed.], Global Loss of Coastal Habitats: Rates, Causes and Consequences. Fundacion BBVA. pp. 107–133.
22. FAO (Food and Agriculture Organization) (2007) The World's Mangroves, 1980–2005: A Thematic Study in the Framework of the Global Forest Resources Assessment 2005. Forestry Paper 153, FAO, Rome. ix+77 p.
23. Couwenberg J, Dommain R, Joosten H (2010) Greenhouse gas fluxes from tropical peatlands in south-east Asia. Global Change Biol 16: 1715–1732.
24. Alongi DM, de Carvalho N, Amaral A, Costa A, Trott L, et al. (2012) Uncoupled surface and below-ground soil respiration in mangroves: implications for estimates of dissolved inorganic carbon export. Biogeochemistry 109: 151–162.

25. Cahoon DR, Hensel P, Rybczyk J, McKee K, Proffitt E, et al. (2003) Mass tree mortality leads to mangrove peat collapse at Bay Islands, Honduras after Hurricane Mitch. J Ecol 91: 1093–1105.
26. Barr JG, Engel V, Smith TJ, Fuentes JD (2012) Hurricane disturbance and recovery of energy balance, CO$_2$ fluxes and canopy structure in a mangrove forest of the Florida Everglades. Agric Fore Meteor 153: 54–66.
27. Lovelock CE, Ruess RW, Feller IC (2011) CO$_2$ efflux from cleared mangrove peat. PLoS ONE, 6: e21279.
28. Krauss K, Cahoon D, Allen J, Ewel K, Lynch J, et al. (2010) Surface elevation change and susceptibility of different mangrove zones to sea-level rise on Pacific High Islands of Micronesia. Ecosystems 13: 129–143.
29. Whelan K (2005) The successional dynamics of lightning-initiated canopy gaps in the mangrove forests of shark river, Everglades National Park, USA. PhD thesis, Florida International University.
30. Kirui KB, Kairo JG, Bosire J, Viergever KM, Rudra S, et al. (2012) Mapping of mangrove forest land cover change along the Kenya coastline using Landsat imagery. Ocean Coastal Manage: doi:10.1016/j.ocecoaman.2011.1012.1004.
31. Abuodha P, Kairo JG (2001) Human-induced stresses on mangrove swamps along Kenya coast. Hydrobiologia 458: 255–265.
32. Dahdouh-Guebas F, Van Pottelbergh I, Kairo JG, Cannicci S, Koedam N (2004) Human-impacted mangroves in Gazi (Kenya): predicting future vegetation based on retrospective remote sensing, social surveys, and distribution of trees. Mar Ecol Prog Ser 272: 77–92.
33. Taylor M, Ravilious C, Green EP (2003) Mangroves of East Africa. UNEP World Conservation Monitoring Centre, (Cambridge, UK). 25 p.
34. Dahdouh-Guebas F, Mathenge C, Kairo JG, Koedam N (2000) Utilization of mangrove wood products around Mida Creek (Kenya) amongst subsistence and commercial users. Econ Bot, 54: 513–527.
35. Bosire JO, Dahdouh-Guebas F, Kairo JG, Koedam N (2003) Colonization of non-planted mangrove species into restored mangrove stands in Gazi Bay, Kenya. Aquat Bot 76: 267–279.
36. Wass P (1995) Kenya's Indigenous Forests: Status, Management and Conservation. IUCN. Xii+250 pp. 252-8317-0292-8315.
37. Alongi DM (2009) The Energetics of Mangrove Forests. (Springer, Netherlands). 216 p.
38. Hanson PJ, Edwards NT, Garten CT, Andrews JA (2000) Separating root and soil microbial contributions to soil respiration: A review of methods and observations. Biogeochemistry, 48: 115–146.
39. Hogberg P, Nordgren A, Buchmann N, Taylor AFS, Ekblad A, et al. (2001) Large-scale forest girdling shows that current photosynthesis drives soil respiration. Nature, 411: 789–792.
40. Andersen C, Nikolov I, Nikolova P, Matyssek R, Häberle K-H (2005) Estimating "autotrophic" belowground respiration in spruce and beech forests: decreases following girdling. Europ J Fore Res 124: 155–163.
41. Blodau C (2002) Carbon cycling in peatlands - A review of processes and controls. Environ Rev 10: 111–134.
42. Goreau TJ, de Mello WZ (2007) Mininmizing net greenhouse gas sources from mangrove and wetland soils, In Tateda, Y. [ed.], Greenhouse Gas and Carbon Balances in Mangrove Coastal Ecosystems. pp. 239–248.
43. Kristensen E (2007) Carbon balance in mangrove sediments: The driving processes and their controls, In Tateda, Y. [ed.], Greenhouse gas and carbon balances in mangrove coastal ecosystems. Gendai Tosho. pp. 61–78.
44. Reeburgh WS (2007) Oceanic methane biogeochemistry. Chem Rev 107: 486–513.
45. Kirui KB, Kairo J, Bosire J, Viergever K, Rudra S, et al. (2012) Mapping of mangrove forest land cover change along the Kenya coastline using Landsat imagery. Ocean Coast Manage: doi:10.1016/j.ocecoaman.2011.1012.1004.
46. Locatelli B, Brockhaus M, Buck A, Thompson I (2010) Forests and adaptation to climate change: challenges and opportunities, In Mery G, et al. [eds.], Forests and Society-Responding to Global Drivers of Change. International Union of Forest Research Organizations (IUFRO). pp. 21–42.
47. UNEP (United Nations Environment Program) (1998) Eastern Africa Atlas of Coastal Resources 1: Kenya. (EAF-14) UNEP, 119 p.
48. Miller JB, Tans PP (2003) Calculating isotopic fractionation from atmospheric measurements at various scales. Tellus B, 55: 207–214.
49. Kayler Z, Ganio L, Hauck M, Pypker T, Sulzman EW, et al. (2010) Bias and uncertainty of δ^{13}CO$_2$ isotopic mixing models. Oecologia, 163: 227–234.
50. Cerone P, Dragomir SS (2000) Trapezoidal-type rules from an inequalities point of view, In Anastassiou, G. [ed.], Handbook of Analytic-Computational Methods in Applied Mathematics. CRC Press. pp. 65–134.

4

Spatio-Temporal Patterns of Major Bacterial Groups in Alpine Waters

Remo Freimann[1,2]*, Helmut Bürgmann[3], Stuart E. G. Findlay[4], Christopher T. Robinson[2]

1 Institute of Molecular Health Sciences, Professorship of Genetics, ETH Zurich, Zurich, Switzerland, 2 Department of Aquatic Ecology, Swiss Federal Institute of Aquatic Science and Technology, Eawag, Dübendorf, Switzerland and Institute of Integrative Biology, ETH-Zurich, Zurich, Switzerland, 3 Department of Surface Waters – Research and Management, Swiss Federal Institute of Aquatic Science and Technology, Eawag, Kastanienbaum, Switzerland, 4 Cary Institute of Ecosystem Studies, Millbrook, New York, United States of America

Abstract

Glacial alpine landscapes are undergoing rapid transformation due to changes in climate. The loss of glacial ice mass has directly influenced hydrologic characteristics of alpine floodplains. Consequently, hyporheic sediment conditions are likely to change in the future as surface waters fed by glacial water (kryal) become groundwater dominated (krenal). Such environmental shifts may subsequently change bacterial community structure and thus potential ecosystem functioning. We quantitatively investigated the structure of major bacterial groups in glacial and groundwater-fed streams in three alpine floodplains during different hydrologic periods. Our results show the importance of several physico-chemical variables that reflect local geological characteristics as well as water source in structuring bacterial groups. For instance, *Alpha-*, *Betaproteobacteria* and *Cytophaga-Flavobacteria* were influenced by pH, conductivity and temperature as well as by inorganic and organic carbon compounds, whereas phosphorous compounds and nitrate showed specific influence on single bacterial groups. These results can be used to predict future bacterial group shifts, and potential ecosystem functioning, in alpine landscapes under environmental transformation.

Editor: Jack Anthony Gilbert, Argonne National Laboratory, United States of America

Funding: Funding provided by Swiss National Science Foundation (No. 31003A-119735) to CR HB SF RF. The funders had no role in study design, data collection and analysis, decision to publish, or preparation of the manuscript.

Competing Interests: The authors have declared that no competing interests exist.

* Email: remofreimann@gmail.com

Introduction

Heterotrophic bacteria are key players in the functional ecology of aquatic ecosystems, alpine waters in particular [1]. Their high metabolic capacity, phylogenetic variation and abundance enable bacterial assemblages to process and retain nutrients and chemical compounds under varying environmental conditions. Bacteria also represent an integral component within trophic food webs and global carbon cycling [2]. Glaciated alpine floodplains are an important source of fresh water as they are regions of high precipitation that is stored as snow/ice in winter and released during warm periods. Alpine environments not only modulate flow patterns and affect water chemistry but also provide microbe-mediated ecosystem services to waters used intensely by humans at lower elevations [3,4]. Furthermore, alpine headwaters can have a high variability in species diversity within relative small geographical distance and thus contribute to the maintenance of microbial and functional diversity in the fluvial network [5,6].

Members of bacterial groups are associated with metabolic traits and occupy habitat niches depending on the physico-chemical environment and apparent bacterial (single cell) functional plasticity [7]. Glaciated alpine floodplains comprise a mosaic of groundwater-fed (krenal) and glacial (kryal) streams differing in physico-chemical characteristics and thus different habitat niches [8]. The hyporheic zone of streams, in particular, is known as a biological hotspot of microbial abundance and functioning within these landscapes [9]. Environmentally-induced recession of glaciers and current shifts in precipitation patterns will affect alpine waters and associated habitats by reducing kryal systems at the landscape scale [10,11]. This reduction includes shifts in quality, quantity and timing of glacially-released organic matter and nutrients as well as changes in flow regimes that can alter hyporheic sediment characteristics [12]. Thus, ecological shifts in bacterial structure in conjunction with potential functioning are likely to occur. The underlying mechanics are dependent on the magnitude and rate of environmental change and are at least partially determined by the properties of the contemporary bacterial assemblage [13]. The magnitude of ecological change will ultimately depend on the degree of functional redundancy/plasticity and is likely manifest in a combination of changes involving cell abundances, single cell metabolic activities and shifts in bacterial composition [7].

In previous studies, we documented a strong linkage between bacterial structure and function in hyporheic sediments [6]. We showed how ecological strategies of communities (generalists vs. specialists) influence the trajectory of community changes due to altered physico-chemical properties [13]. However, these earlier results only focused on the structure of bacterial assemblages without taking into account phylogenetic identities. In the present study, we quantitatively examined the phylogeny of major freshwater bacterial groups present in hyporheic sediments via catalyzed reporter deposition fluorescence in-situ hybridization (CARD-FISH) within three alpine glacial catchments. The catchments differed in the degree of deglaciation and thus covered a wide range of habitat types within krenal and kryal systems. These different physic-chemical conditions were associated with differences in bacterial groups and the broad range in habitat and bacterial attributes probably encompass future scenarios for alpine landscapes under transformation.

Material and Methods

We collected hyporheic sediments (~10 to 20 cm depth) from 45 stream sites in three alpine floodplains, Val Roseg (VR, 9°53′53″E, 46°29′24″N), Loetschental (L, 07°49′03″E, 46°25′08″N) and Macun (M, 10°07′31″E, 46°43′51″N), that differed in their degree of deglaciation (% of the catchment area covered by ice, data provided by the Federal Office for Environment and the Swiss National Park) and general landscape features such as the presence of interconnected lakes (Table 1, Figure 1). Sampling sites along streams were categorized into krenal and kryal systems, depending on the connectivity to glaciers/glacial meltwater and based on water chemistry patterns (Figure 1) [6]. Macun had a perennially-reduced glacial water input, and more specific characterizations of these catchments can be found elsewhere [6]. Sites were sampled during three distinct hydrological periods: glacial ablation in summer (August: A), winter stagnation (October: O) and snowmelt input in spring (June: J) in Val Roseg and Loetschental and for A and O in Macun. No specific permission was required for the sampling in VR and L. Permission for the sampling campaign in M was issued by the Swiss National Park. The study did not involve endangered or protected species.

A 0.5 ml aliquot of collected sediment was suspended in 1.11 ml paraformaldehyde (2%, final concentration) in an Eppendorf tube and fixed for 24 h at 4°C followed by three washing steps with 1× PBS and 5 min centrifugation at 10,000 g between washing steps. Samples were then stored at −20°C in a 1:1 mix of PBS/ethanol until further processing [14]. Cell detachment was done by sonication (Branson Digital Sonifier 250, Danbury, USA, 5-mm tapered microtip, actual output of 20 W, 30 s). Catalyzed reporter deposition fluorescence in-situ hybridization (CARD-FISH) was performed following the protocol of Pernthaler et al. [15] paired with a high throughput imaging system [16]. Horseradish labeled FISH probes (Biomers Inc, Ulm) EUB I-III targeting the domain

Bacteria (EUBI-III) [17], Alf968 and Bet42a affiliated with classes of Alphaproteobacteria (Alph) and Betaproteobacteria (Bet) of the phylum Proteobacteria, respectively [18,19], and CF319a assigned to the class of Cytophaga-Flavobacteria (CF) within the phylum Bacteroidetes [20] were used to quantify microbial groups within the stream sediments. The domains and classes are expressed as percentage of total bacterial abundance as assessed by counter-staining bacterial cells with 4′,6-diamidino-2-phenylindole (DAPI) (Sigma-Aldrich Co) [21].

Specific conductance (μS cm^{-1} at 20°C) and temperature were measured in the field with a conductivity meter (LF323, WTW, Weilheim, Germany). Surface water samples were analyzed for dissolved organic carbon (DOC), particulate organic carbon (POC), total inorganic carbon (TIC), ammonium (NH_4-N), nitrite (NO_2-N), nitrate (NO_3-N), dissolved organic nitrogen (DON), particulate organic nitrogen (PN), phosphate (PO_4-P), dissolved phosphorus (DP) and particulate phosphorus (PP) according to standard protocols detailed in Tockner et al. [22]. Sediments were analyzed for pH [23] and organic matter content (OM) as ash free dry mass. Grain size distribution of sediment was assessed by sieving with mesh sizes of 6.3, 2.0, 1.0, 0.5, 0.25, 0.125 and 0.063 mm. The D90/D10 gradation index was then calculated using GRADISTAT software [24]. The raw physico-chemical data have been published elsewhere [6].

To assess the explained variance of physico-chemical variables on CARD-FISH based bacterial community composition, we performed a redundancy analysis (RDA) based on forward selected environmental variables [25]. Significance of physico-chemical constraining variables and constrained RDA axes were tested by permutation tests (999 permutations) [26]. The relative contribution of physico-chemical variables to the constraint variation of single RDA axis was assessed via canonical correlation coefficients. Factor (catchment, water source and season) and vector (i.e. physico-chemical variables) fitting was performed on the first two RDA axes to assess their significance and relationship (r^2) to the phylogenetic patterns shown on these two axes. We tested all levels of interactions for the factor fitting, i.e. the influence of a single factor and all double and triple interactions of the respective factors. Pearson's product moment correlation was used to correlate physico-chemical variables with single bacterial groups and correlations were tested using Algorithm AS 89 [27]. Lastly, comparison of hybridization rates between water source, seasons and catchments were done using ANOVAs followed by Tukey's honest significance text (Tukey's HSD). Analyses were based on arcsin($\sqrt{x}$) transformed percentage values of CARD-FISH results

Table 1. Basic characteristics of the three catchments.

Catchment	Val Roseg	Loetschental	Macun
Coordinates	9°53′53″E, 46°29′24″N	07°49′03″E, 46°25′08″N	10°07′31″E, 46°43′51″N
Altitude [m a.s.l.]	1766–4049	1375–3200	2616–3046
Catchment area [km², (% glaciated)	66.5 (30.1)	77.8 (36.5)	3.6 (18.8)*
Annual precipitation [m]	1.6	1.1	0.9
Mean discharge [m³ s⁻¹]	28.5	37.2	ND
Mean water temperature of main channel [°C] (range)	3.6 (1–12)	4 (0.1–10.9)	2.9 (0.1–19.2)
Interconnected lakes	No	No	Yes
Geology, dominating minerals	Crystalline bedrock, diorite, granite	Crystalline bedrock, amphibolite, gneiss	Crystalline bedrock, ortho-gneiss

Abbreviation: ND, no data; *rock glaciers.

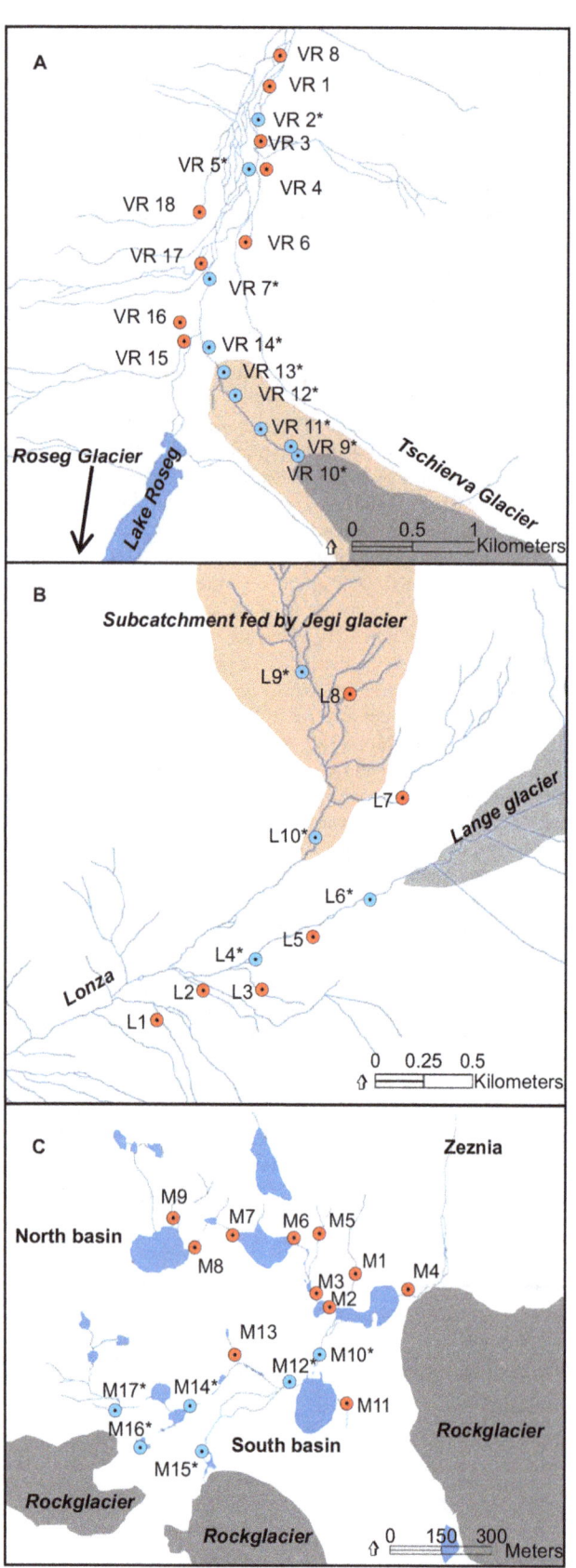

moraine area in VR and the sub-catchment in L are depicted in light grey and orange, respectively.

and log +1 transformed physico-chemical variables. All analysis were done using the R statistical environment [28].

Results

Sediments from L and VR had a higher mean *EUBI-III* hybridization rate than M (Figure 2, Table 2, Tukey's HSD: P< 0.001). *EUBI-III hybridization rate* in Macun showed proportionally the smallest overlap with the further defined lower taxonomic level (classes, Figure 2). We refer to the not further defined *Bacteria* group (i.e., %EUBI-III − (%Alph + %Bet + %CF)) as *EUBI-III(undef)* (Figure 3). There was a higher *EUBI-III*

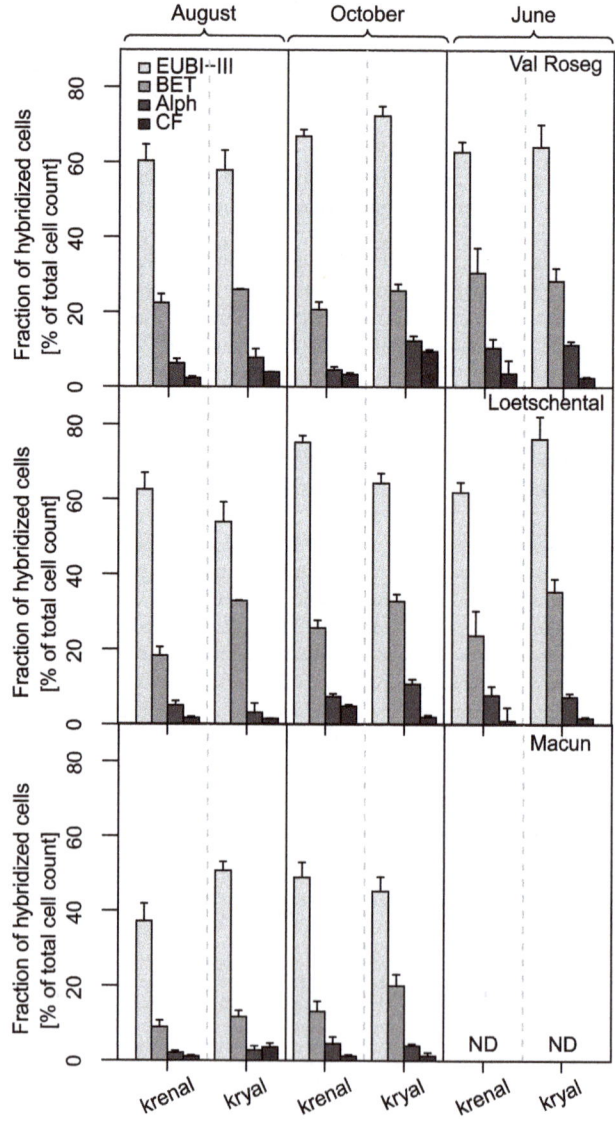

Figure 1. Map of the study sites in the three catchments, A: Val Roseg (VR), B: Loetschental (L) and C: Macun (M). Kryal sites are depicted in blue and annotated with an asterisk. Glaciers and the

Figure 2. Bar plots (+1SE) of the relative abundance of major bacterial groups in the Val Roseg (A), Loetschental (B) and Macun (C) catchment. Bar groups are split within a panel by season and water source. ND: No data.

Table 2. ANOVA results (F values) for bacterial group abundance.

F-statistic parameter	Catchment (C)	Sampling date (S)	Water source (W)	C×S	C×W	S×W	C×S×W
EUB-III	7.49***	6.68**	4.55*	1.19	3.26*	3.89*	3.1*
VR		4.68*	0.4			0.85	
L		4.14*	0.11			5.08*	
M		0.16	0.49			1.64	
BET	20.22***	1.54	5.78*	1.39	4.18*	0.79	3.1
VR		3.49*	1.02			1.06	
L		0.67	12.84**			0.48	
M		5.44*	3.14			0.59	
Alph	5.57**	5.98**	4.25*	1.06	1.23	1.51	1.05
VR		2.73	5.79*			2.3	
L		6.48**	0.11			1.65	
M		3.35	0			0.37	
CF	8.26***	3.18*	5.37*	4.18**	4.68*	0.02	7.38***
VR		10.31***	10.89**			9.61***	
L		1.87	0.76			1.4	
M		3.13	4.04			3.45	

Catchment (C), Sampling date (S) and Water source (W) were used as independent variables.

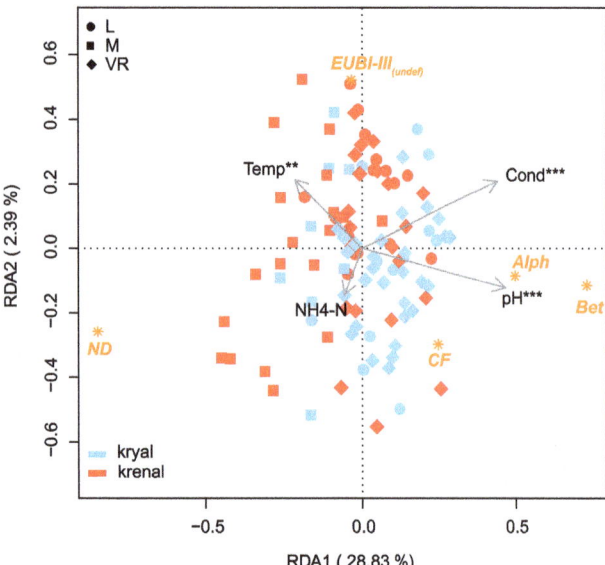

Figure 3. Biplot of the redundancy analysis based on the CARD-FISH data. Red symbols correspond to krenal sampling sites, whereas blue symbols depict kryal sites. Centroids of the respective probes are given: beta-*proteobacteria* (*Bet*), alpha-*proteobacteria* (*Alph*), Cytophaga-Flavobacteria (*CF*), Eubacteria excluding *Alph*, *Bet* and *CF* (*EUBI-III(undef)*) and not-defined DAPI positive cells (*ND*). Arrows depict the forward selected physico-chemical variables. Asterisks depict significantly tested variables (* p<0.05, ** p<0.01). Explained variation for the first two constraint axes is given.

hybridization rate in winter compared to summer (Tukey's HSD: P<0.01).

Bet was the most abundant group according to CARD-FISH and was lowest in M compared to the other two catchments (Tukey's HSD: p<0.001). *Bet* had on average highest abundance in kryal sediments in L (Tukey's HSD: P<0.001). *Alph* showed highest abundance in kryal sediments in VR (Tukey's HSD: p< 0.05) and higher abundance in winter compared to spring in L (Tukey's HSD: p<0.01). *CF* was least abundant and VR had higher *CF* abundance compared to the other two catchments (Tukey's HSD: p<0.001) and showed a peak in kryal sediments in winter (Tukey's HSD: p<0.001). See also table 2 for detailed results of ANOVAs.

RDA based on forward selected environmental variables explained 31.9% of the total variation and revealed a differentiation of the three catchments concerning their phylogenetic group structuring (Figure 3). Catchment and water source were significantly (p<0.05) fitted on the RDA biplot ($r^2 = 0.09$ and 0.05, respectively). This was also true for the interaction term of catchment, sampling date and water source (P<0.001, $r^2 = 0.30$). The ordination ellipses in figure 4 depict the standard error of weighted average scores (confidence limit = 0.95) of the interaction sampling date × water source split by catchments. This figure is based on the RDA biplot of figure 3. Significant differences between water sources within a specific season can be expected when the respective ellipses do not overlap. The pH, conductivity, temperature and NH$_4$-N contributed 23.2%, 2.1%, 1.7% and 1.8% to the total 28.8% explained variation on the first RDA axis (Figure 3). The pH, conductivity and temperature could also be fitted a-posteriori as independent variables on the biplot ($r^2 = 0.53$, 0.46 and 0.13, respectively, p<0.01). NH$_4$-N could not be significantly fitted on the RDA (p = 0.73) (Figure 3). Additionally,

we fitted all non-forward selected environmental variables on the first two RDA axes of the forward selected model to assess their relative importance in structuring the bacterial communities (Table S1). OM, TIC and PP showed significant a-posteriori fitting on the first two RDA axes (p<0.01). The abundances of the different bacterial groups could be linked to several physico-chemical parameters (Figure 5). All bacterial groups were negatively correlated with OM and temperature, and *Bet* also was negatively correlated with POC and PN. TIC, Cond and pH were positively correlated with *Alph*, *Bet* and *CF*. PP was positively correlated with *Bet* and *CF*, whereas *Alph* was correlated with DN. PO$_4$-P and NO$_2$-N showed a positive Pearson correlation with *CF*, whereas *Bet* was correlated with NO$_3$-N.

Discussion

Our results showed that bacterial group composition in hyporheic sediments of streams in glaciated alpine floodplains have a strong spatio-temporal dynamic. Physico-chemical differences between catchments, such as pH, conductivity and temperature, were related to geological and geographical characteristics, and dictate the coarse-scale boundary conditions on bacterial group composition. These factors have been described previously to drive assemblage composition and diversity in soils as well as in stream sediments, although not at the group level assessed here [6,29,30]. Importantly, these previously documented variables affecting bacterial structural patterns at a higher phylogenetic level (i.e. automated ribosomal intergenic spacer analysis) within these catchments were largely congruent to the variables structuring the lower phylogenetic level structures (i.e. CARD-FISH) in this study [13]. This congruence was also true for drivers (i.e. OM, POC, PP, TIC and NO$_2$-N) linked more specifically to single bacterial groups (see Figure 5). Furthermore, enzymatic activities measured in the above mentioned study were used as constraints variables in an RDA analysis based on the CARD-FISH data and revealed significant correlations to the phylogenetic group structuring (29.6% total explained variation, see Figure S1). Results from the automated ribosomal intergenic spacer analysis and the enzyme activities were also fitted a-posteriori on the RDA biplot produced here (Figure S2). Taken together, these results underpin the separation of the different catchments based on the CARD-FISH data and indicate a substantial coupling of bacterial groups and their potential metabolic capabilities.

Most of the examined variables differed between the two water sources (kryal vs krenal) [6] and partly induced the observed seasonal turnover in assemblage structure (Figure 4). Kryal systems are rich in PP during summer ablation and favored *Bet* and *CF* within kryal habitats in VR and L (Figure 5) [22]. Catchment M showed reduced abundance of *Bet*, which was correlated to reduced PP input into kryal streams here [6]. Also, NH$_4$-N and NO$_2$-N levels are high in summer in kryal waters with the latter being positively correlated with the abundance of *Alph*, *Bet* and *CF* [13]. During winter, kryal systems become physico-chemically more equal to krenal systems as glacial water input diminishes. Nevertheless, there was no distinct shift in kryal group composition towards a krenal one at the floodplain scale within the three catchments, suggesting relatively resistant local bacterial assemblages despite the aforementioned coupling of structure and function (see dispersion ellipses in figure 4). Experiments where kryal and krenal bacterial communities were cross-transplanted between their natural habitats revealed a high structural resistance along with a pronounced functional plasticity in response to physico-chemical disturbance [13]. Nevertheless, at this higher

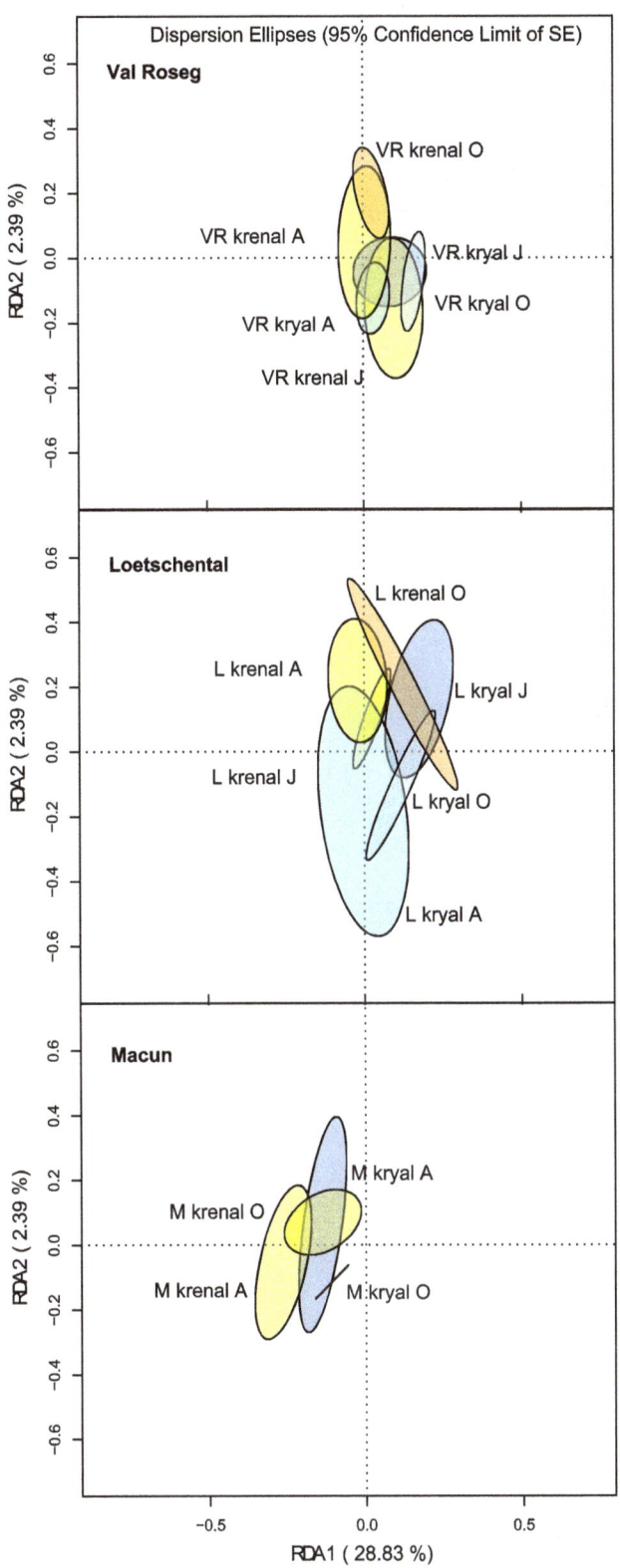

Figure 4. Dispersion ellipses fitted on the biplot of the redundancy analysis based on the CARD-FISH data constraint by physico-chemical variables (Figure 3). Dispersion ellipses split by catchments for different water sources and seasons are shown and depict the standard error of weighted average scores (confidence limit = 0.95).

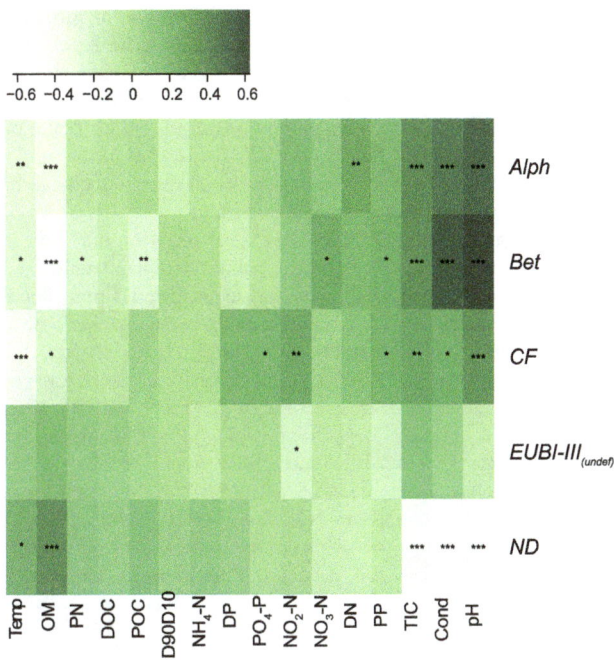

Figure 5. Heat-map of the Pearson correlations of beta-*proteobacteria* (*Bet*), alpha-*proteobacteria* (*Alph*), Cytophaga-Flavobacteria (*CF*), Eubacteria excluding *Alph*, *Bet* and *CF* (*EUBI-III(undef)*) and not-defined DAPI positive cells (ND) to physico-chemical variables. Asterisk indicate the p-value (*: p< 0.05; **: p<0.01; ***: p<0.001) of correlations between paired samples.

phylogenetic level and at the landscape scale, there was a difference in coupling of bacterial structure and functions within krenal and kryal systems with the latter having higher congruency [6]. At the phylogenetic level studied here, it seems that functional shifts, which depend on apparent redundancy/plasticity, are linked to structural changes of similar extent in both water systems. This discrepancy is likely due to non-detected seasonal changes of bacterial species within bacterial groups. That is, kryal systems can be seen as dominated by specialists whereas krenal systems tend to harbor generalists (on the species level) [6]. Thus, kryal community composition changes in response to new functional requirements, since the level of plasticity/redundancy in the community is low. Krenal communities, on the other hand, don't shift as strongly, since the communities plasticity/redundancy is high. Such differences in community shifts between the systems cannot be detected at the level of bacterial groups, if the community shift occurs mainly within and not across groups. Regardless, the coupling of bacterial groups to functions becomes evident when comparing different catchments.

A dominance of *Proteobacteria* has been shown before in snowpacks as well as in different glacial habitats [31,32]. Thus, the low total abundance of *Proteobacteria* in catchment M may be due to an interactive effect of decreased glacial water input and a generally harsh environment due to the high altitude (Table 1). *Bet* and partly *Alph* are often predominant in sediments in lower elevation streams [33,34]. *Bet* has been described as a diverse group dominating freshwater systems of different oligotrophic states [35,36] and are highly competitive at the initial state of biofilm development [37]. This trait may be the reason that they can compete well in kryal sediments in VR and L, as these habitats experience mechanic abrasion induced by flow as well as low OM. *Bet* also have been shown to be involved in the degradation of pollutants, and thus may provide beneficial ecosystem functions within alpine floodplains. Indeed, precipitation driven pollutant inputs may favor *Bet* as proposed by Brümmer et al. [35]. Biofilm bacterial assemblages in urban rivers have been shown to be dominated by *Bet* and *CF*, which may be linked to their pollution load [38]. Similar to our findings in VR, Araya *et al.* also found a peak of *CF* during the winter season in urban rivers which can be linked to lower temperatures (Figure 5) [8,38].

In summary, ongoing (and rapid) glacial recession and shift in water source (i.e. physico-chemical habitat template) will likely influence bacterial group composition in glaciated alpine floodplains. Differences between kryal and krenal systems were not as distinct at this taxonomic resolution (Figure 3, Figure S2) such that a dramatic change in bacterial group diversity at the landscape scale is expected. At a longer time scale, there will be other environmental changes concomitant to the glacial mass loss induced physico-chemical habitat shifts, such as changes in terrestrial vegetation, increasing OM inputs and increased water temperatures. These changes will likely induce subtle shifts in major bacterial groups towards a krenal assemblage composition and thus a homogenization of bacterial assemblages. Ultimately, alpine bacterial assemblage structure may become more similar to present lower elevation stream bacterial assemblages, thereby affecting ecosystem functioning and services.

Supporting Information

Figure S1 Biplot of the redundancy analysis based on the CARD-FISH data constrained by enzymatic activities measured in a previous study. Dark grey dots correspond to krenal sampling sites, whereas light grey dots depict kryal sites. Centroids of the respective probes are given: Beta*proteobacteria* (*Bet*), Alpha*proteobacteria* (*Alph*), Cytophaga-Flavobacteria (*CF*), Eubacteria excluding *Alph*, *Bet* and *CF* (*EUBI-III$_{(undef)}$*) and not-defined DAPI positive cells (ND). Arrows depict the forward selected Hellinger transformed enzyme activities (expressed as nmol m^2 h^{-1}). Asterisks depict significantly tested enzymes (* p< 0.05, ** p<0.01). Dispersion ellipses for the catchments and

different water sources are shown and depict the standard error of weighted average scores (confidence limit = 0.95). Explained variation for the first two constraint axes is given.

Figure S2 Biplot of the redundancy analysis based on the CARD-FISH data constrained by physico-chemical variables (not shown but equal to Figure 3). Dark grey dots correspond to krenal sampling sites, whereas light grey dots depict kryal sites. Centroids of the respective probes are given: Beta*proteobacteria* (*Bet*), Alpha*proteobacteria* (*Alph*), Cytophaga-Flavobacteria (*CF*), Eubacteria excluding *Alph*, *Bet* and *CF* (*EUBI-III$_{(undef)}$*) and not-defined DAPI positive cells (ND). Arrows depict the a posteriori fitted enzymatic activities (p< 0.05) and operational taxonomic units (p<0.01) from a previous study. Dispersion ellipses for the catchments and different water sources are shown and depict the standard error of weighted average scores (confidence limit = 0.95). Explained variation for the first two constraint axes is given.

Table S1 Vector fitting of physico-chemical variables on the first two axis of the RDA-Biplot (Figure 3).

Acknowledgments

We thank the Swiss National Park for access to the Macun catchment, Prof. Dr. Jakob Pernthaler and Dr. Michael Zeder for providing the infrastructure and in-house software for conducting CARD-FISH analysis in the most rapid way, and Simone Blaser and Christa Jolidon for field and laboratory assistance.

Author Contributions

Conceived and designed the experiments: RF HB SF CR. Performed the experiments: RF HB SF CR. Analyzed the data: RF. Contributed reagents/materials/analysis tools: RF SF SF CR. Wrote the paper: RF HB SF CR.

References

1. Battin TJ, Kaplan LA, Findlay S, Hopkinson CS, Marti E, et al. (2008) Biophysical controls on organic carbon fluxes in fluvial networks. Nature Geoscience 1: 95–100.
2. Pollarda PC, Ducklow H (2011) Ultrahigh bacterial production in a eutrophic subtropical Australian river: Does viral lysis short-circuit the microbial loop? Limnol Oceanogr 56: 1115–1129.
3. Battin TJ (1999) Hydrologic flow paths control dissolved organic carbon fluxes and metabolism in an alpine stream hyporheic zone. Water Resour Res 35: 3159–3169.
4. Hood E, McKnight DM, Williams MW (2003) Sources and chemical character of dissolved organic carbon across an alpine/subalpine ecotone, Green Lakes Valley, Colorado Front Range, United States. Water Resour Res 39: 1188–1199.
5. Besemer K, Singer G, Quince C, Bertuzzo E, Sloan W, et al. (2013) Headwaters are critical reservoirs of microbial diversity for fluvial networks. Proc R Soc B Biol Sci 280.
6. Freimann R, Bürgmann H, Findlay SEG, Robinson CT (2013) Bacterial structures and ecosystem functions in glaciated floodplains: Contemporary states and potential future shifts. ISME J 7: 2361–2373.
7. Comte J, del Giorgio PA (2011) Composition influences the pathway but not the outcome of the metabolic response of bacterioplankton to resource shifts. PLoS ONE 6: e25266.
8. Tockner K, Malard F, Uehlinger U, Ward JV (2002) Nutrients and organic matter in a glacial river-floodplain system (Val Roseg, Switzerland). Limnol Oceanogr 47: 266–277.
9. Hendricks SP (1993) Microbial ecology of the hyporheic zone: A perspective integrating hydrology and biology. J N Am Benthol Soc 12: 70–78.
10. IPCC (2007) Climate change 2007: The physical science basis. New York.
11. Horton P, Schaefli B, Mezghani A, Hingray B, Musy A (2006) Assessment of climate-change impacts on alpine discharge regimes with climate model uncertainty. Hydrol Process 20: 2091–2109.

12. Ward JV (1994) Ecology of alpine streams. Freshwat Biol 32: 277–294.
13. Freimann R, Bürgmann H, Findlay SEG, Robinson CT (2013) Response of lotic microbial communities to altered water source and nutritional state in a glaciated alpine floodplain. Limnol Oceanogr 58: 951–965.
14. Pernthaler J, Glöckner FO, Schönhuber W, Amann R (2001) Fluorescence in situ hybridization (FISH) with rRNA-targeted oligonucleotide probes. In: Paul JH, editor. Methods in Microbiology: Academic Press. pp. 207–226.
15. Pernthaler A, Pernthaler J, amann R (2004) Sensitive multicolour fluorescence in situ hybridization for the identification of environmental organisms. Molecular Microbial Ecology Manual 2: 711–726.
16. Zeder M, Pernthaler J (2009) Multispot live-image autofocusing for high-throughput microscopy of fluorescently stained bacteria. Cytometry Part A 75: 781–788.
17. Daims H, Brühl A, Amann R, Schleifer KH, Wagner M (1999) The domain-specific probe EUB338 is insufficient for the detection of all bacteria: Development and evaluation of a more comprehensive probe set. Syst Appl Microbiol 22: 434–444.
18. Neef A, Amann R, Schleifer KH (1997) Spezifischer und schneller in situ - Nachweis von Mikroorganismen aus Aerosolen mit Gensonden. Zentralbl Hyg Umweltmed 199: 410.
19. Manz W, Amann R, Ludwig W, Wagner M, Schleifer KH (1992) Phylogenetic oligodeoxynucleotide probes for the major subclasses of proteobacteria: Problems and solutions. Syst Appl Microbiol 15: 593–600.
20. Manz W, Amann R, Ludwig W, Vancanneyt M, Schleifer KH (1996) Application of a suite of 16S rRNA-specific oligonucleotide probes designed to investigate bacteria of the phylum cytophaga-flavobacter-bacteroides in the natural environment. Microbiology 142: 1097–1106.
21. Porter KG, Feig YS (1980) The use of DAPI for identifying and counting aquatic microflora. Limnol Oceanogr 25: 943–948.

22. Tockner K, Malard F, Burgherr P, Robinson CT, Uehlinger U, et al. (1997) Physico-chemical characterization of channel types in a glacial floodplain ecosystem (Val Roseg, Switzerland). Archiv für Hydrobiologie 140: 433–463.

23. Schofield RK, Taylor AW (1955) The measurement of soil pH. Soil Sci Soc Am J 19: 164–167.

24. Simon J. Blott KP (2001) GRADISTAT: A grain size distribution and statistics package for the analysis of unconsolidated sediments. ESPL 26: 1237–1248.

25. Blanchet FG, Legendre P, Borcard D (2008) Forward selection of explanatory variables. Ecology 89: 2623–2632.

26. Legendre P, Oksanen J, ter Braak CJF (2011) Testing the significance of canonical axes in redundancy analysis. Methods in Ecology and Evolution 2: 269–277.

27. Best DJ, Roberts DE (1975) Algorithm AS 89: The Upper Tail Probabilities of Spearman's Rho. Journal of the Royal Statistical Society Series C (Applied Statistics) 24: 377–379.

28. R (2014) R Development Core Team (2014). R: A language and environment for statistical computing. 3.1.0 ed. Vienna, Austria: R core team.

29. Fierer N, Jackson RB (2006) The diversity and biogeography of soil bacterial communities. Proc Natl Acad Sci U S A 103: 626–631.

30. Fierer N, Morse JL, Berthrong ST, Bernhardt ES, Jackson RB (2007) Environmental controls on the landscape-scale biogeography of stream bacterial communities. Ecology 88: 2162–2173.

31. Amato P, Hennebelle R, Magand O, Sancelme M, Delort AM, et al. (2007) Bacterial characterization of the snow cover at Spitzberg, Svalbard. FEMS Microbiol Ecol 59: 255–264.

32. Xiang SR, Shang TC, Chen Y, Jing ZF, Yao T (2009) Dominant bacteria and biomass in the Kuytun 51 Glacier. Appl Environ Microbiol 75: 7287–7290.

33. Kloep F, Manz W, Röske I (2006) Multivariate analysis of microbial communities in the River Elbe (Germany) on different phylogenetic and spatial levels of resolution. FEMS Microbiol Ecol 56: 79–94.

34. Brablcová L, Buriánková I, Badurová P, Rulík M (2013) The phylogenetic structure of microbial biofilms and free-living bacteria in a small stream. Folia Microbiol (Praha) 58: 235–243.

35. Brümmer I, Fehr W, Wagner-Döbler I (2000) Biofilm community structure in polluted rivers: Abundance of dominant phylogenetic groups over a complete annual cycle. Appl Environ Microbiol 66: 3078–3082.

36. Manz W, Szewzyk U, Ericsson P, Amann R, Schleifer K, et al. (1993) In situ identification of bacteria in drinking water and adjoining biofilms by hybridization with 16S and 23S rRNA-directed fluorescent oligonucleotide probes. Appl Environ Microbiol 59: 2293–2298.

37. Manz W, Wendt-Potthoff K, Neu TR, Szewzyk U, Lawrence JR (1999) Phylogenetic composition, spatial structure, and dynamics of lotic bacterial biofilms investigated by fluorescent in situ hybridization and confocal laser scanning microscopy. Microb Ecol 37: 225–237.

38. Araya R, Tani K, Takagi T, Yamaguchi N, Nasu M (2003) Bacterial activity and community composition in stream water and biofilm from an urban river determined by fluorescent in situ hybridization and DGGE analysis. FEMS Microbiol Ecol 43: 111–119.

5

Phytogeographical Implication of *Bridelia* Will. (Phyllanthaceae) Fossil Leaf from the Late Oligocene of India

Gaurav Srivastava*, R.C. Mehrotra

Birbal Sahni Institute of Palaeobotany, Lucknow, India

Abstract

Background: The family Phyllanthaceae has a predominantly pantropical distribution. Of its several genera, *Bridelia* Willd. is of a special interest because it has disjunct equally distributed species in Africa and tropical Asia i.e. 18–20 species in Africa-Madagascar (all endemic) and 18 species in tropical Asia (some shared with Australia). On the basis of molecular phylogenetic study on *Bridelia*, it has been suggested that the genus evolved in Southeast Asia around 33±5 Ma, while speciation and migration to other parts of the world occurred at 10±2 Ma. Fossil records of *Bridelia* are equally important to support the molecular phylogenetic studies and plate tectonic models.

Results: We describe a new fossil leaf of *Bridelia* from the late Oligocene (Chattian, 28.4–23 Ma) sediments of Assam, India. The detailed venation pattern of the fossil suggests its affinities with the extant *B. ovata, B. retusa* and *B. stipularis*. Based on the present fossil evidence and the known fossil records of *Bridelia* from the Tertiary sediments of Nepal and India, we infer that the genus evolved in India during the late Oligocene (Chattian, 28.4–23 Ma) and speciation occurred during the Miocene. The stem lineage of the genus migrated to Africa via "Iranian route" and again speciosed in Africa-Madagascar during the late Neogene resulting in the emergence of African endemic clades. Similarly, the genus also migrated to Southeast Asia via Myanmar after the complete suturing of Indian and Eurasian plates. The emergence and speciation of the genus in Asia and Africa is the result of climate change during the Cenozoic.

Conclusions: On the basis of present and known fossil records of *Bridelia*, we have concluded that the genus evolved during the late Oligocene in northeast India. During the Neogene, the genus diversified and migrated to Southeast Asia via Myanmar and Africa via "Iranian Route".

Editor: Navnith K.P. Kumaran, Agharkar Research Institute, India

Funding: These authors have no support or funding to report.

Competing Interests: The authors have declared that no competing interests exist.

* Email: gaurav_jan10@yahoo.co.in

Introduction

The family Phyllanthaceae has a predominantly pantropical distribution (with a few temperate elements) [1] (Figure 1) consisting of morphologically diverse, ~60 genera and 2000 species. The family was separated from the Euphorbiaceae *s.l.* (*sensu lato*) on the basis of molecular data [2]. The pollen evidence indicates that the family became well diversified by the Eocene [3,4]. Of its several genera, *Bridelia* Willd. is of a special interest because it has disjunct equally distributed species in Africa and tropical Asia i.e. 18–20 species in Africa and Madagascar (all endemic) and 18 species in tropical Asia (some shared with Australia) [5–7]. On the basis of molecular phylogenetic study on *Bridelia*, it has been suggested that the genus evolved around 33±5 Ma (i.e. the stem age) and radiated by the 10±2 Ma (i.e. the crown group) [8]. Fossil records of such taxon are equally important to support the molecular phylogenetic studies.

In the present paper we describe a new leaf impression/compression of *Bridelia* from the late Oligocene (Chattian, 28.4–23 Ma; [9]) sediments of Makum Coalfield (27°15'–27° 25' N), Assam, India (Figure 2) which was located at low palaeolatitude (i.e. 10°–15° N) during the period [10]. The suturing of the Indian plate with the Eurasian plate, during the aforesaid period, was not complete to facilitate the plant migration (Figure 3) [11–13]. An attempt has also been made to discuss its origin and dispersal.

Regional geology

The Makum Coalfield is a well known basin having exposure of the late Oligocene sediments. The coalfield is important because (i) it is one of the largest coal producing basins in northeast India and (ii) it contains a well diversified low latitude palaeoflora [14–22]. Infact, there is no other Oligocene sedimentary basin in the Indian sub-continent which contains such a rich and diversified assemblage of fossil plants. The basin was situated at a low palaeolatitude i.e. ~10°–15° N (Figure 3) [10] and the sediments were deposited in a deltaic, mangrove or lagoonal environment [15,23–25]. The coalfield is made up of Baragolai, Ledo, Namdang, Tikak, Tipong and Tirap collieries, lies in between

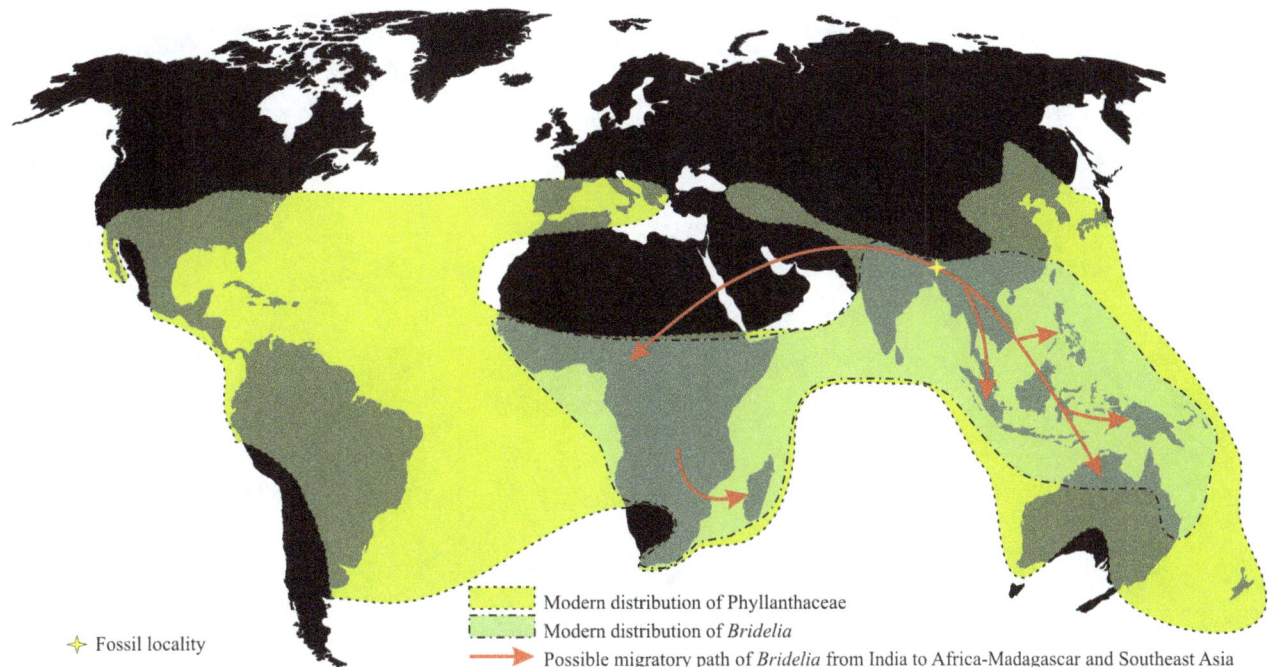

Fossil locality
Modern distribution of Phyllanthaceae
Modern distribution of *Bridelia*
Possible migratory path of *Bridelia* from India to Africa-Madagascar and Southeast Asia

Figure 1. Map showing the modern distribution of the family Phyllanthaceae and *Bridelia* [1, 64].

the latitudes 27° 15′–27° 25′ N and longitudes 95° 40′–95° 55′ E (Figure 2) and is located along the outermost flank of the Patkai range. On the southern and southeastern sides are the hills which rise abruptly to heights of 300–500 m from the alluvial plains of the Buri Dihing and Tirap rivers, respectively.

The fossils collected for the present study belong to the Tikak Parbat Formation being considered to be late Oligocene (Chattian, 28.4–23 Ma; [9]) in age on the basis of regional lithostratigraphy [26], remote sensing [27] and biostratigraphic controls [24].

The Tikak Parbat Formation constitutes alternations of sandstone, siltstone, mudstone, shale, carbonaceous shale, clay and coal seams [28]. However, the plant remains are mainly confined to the grey carbonaceous and sandy shales. The formation is underlain by 300 m of predominantly massive, micaceous or ferruginous sandstones that incorporate the Baragolai Formation, which is successively underlain by 1100–1700 m of thin-bedded fine-grained quartzitic sandstones with thin shale and sandy shale partings that constitute the Naogaon Formation [29]. Together the three formations represent the

Barail Group (Figure 2). In this group, there is an upward trend of marine to non-marine palaeoenvironment which symbolizes the infilling of a linear basin on the eastern edge of the Indian plate. The detailed sedimentary information of the Tirap mine section was given by Kumar et al. [24].

Materials and Methods

Material for the present study was collected from the Tirap colliery of the Makum Coalfield, Tinsukia District, Assam. The prior permission was taken from the General Manager, Northeastern Coalfield, Margherita, Assam, India for the collection of fossil plants. The specimen was first cleared with the help of a fine chisel and hammer and then photographed in natural low angled light using 10 megapixel digital camera (Canon SX110). The terminology used in describing the fossil leaf is based on Hickey [30], Dilcher [31] and Ellis et al. [32]. Attempts were made to extract cuticle from the leaf but it did not yield. The permission was taken from the Directors, Forest Research Institute, Dehradun and the Botanical Survey of India, Kolkata for the herbarium consultation. The fossil plant was identified with the help of herbarium sheets of the extant plant available there. The type specimen (no. BSIP 40115) is housed in the museum of the Birbal Sahni Institute of Palaeobotany, Lucknow, India.

Nomenclature

The electronic version of this article in Portable Document Format (PDF) in a work with an ISSN or ISBN will represent a published work according to the International Code of Nomenclature for algae, fungi, and plants, and hence the new names contained in the electronic publication of a PLOS ONE article are effectively published under that Code from the electronic edition alone, so there is no longer any need to provide printed copies. The online version of this work is archived and available from the following digital repositories: PubMed Central, LOCKSS.

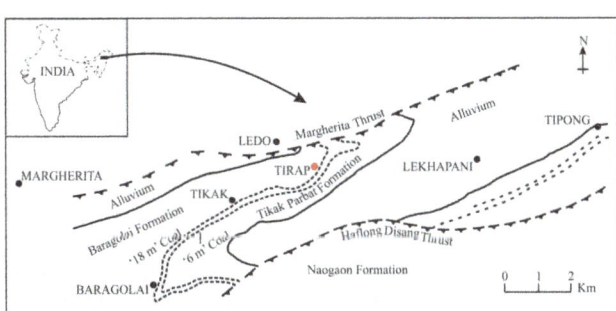

Figure 2. Simplified geological map of the Makum Coalfield, Assam, India showing the fossil locality (red circle) [65].

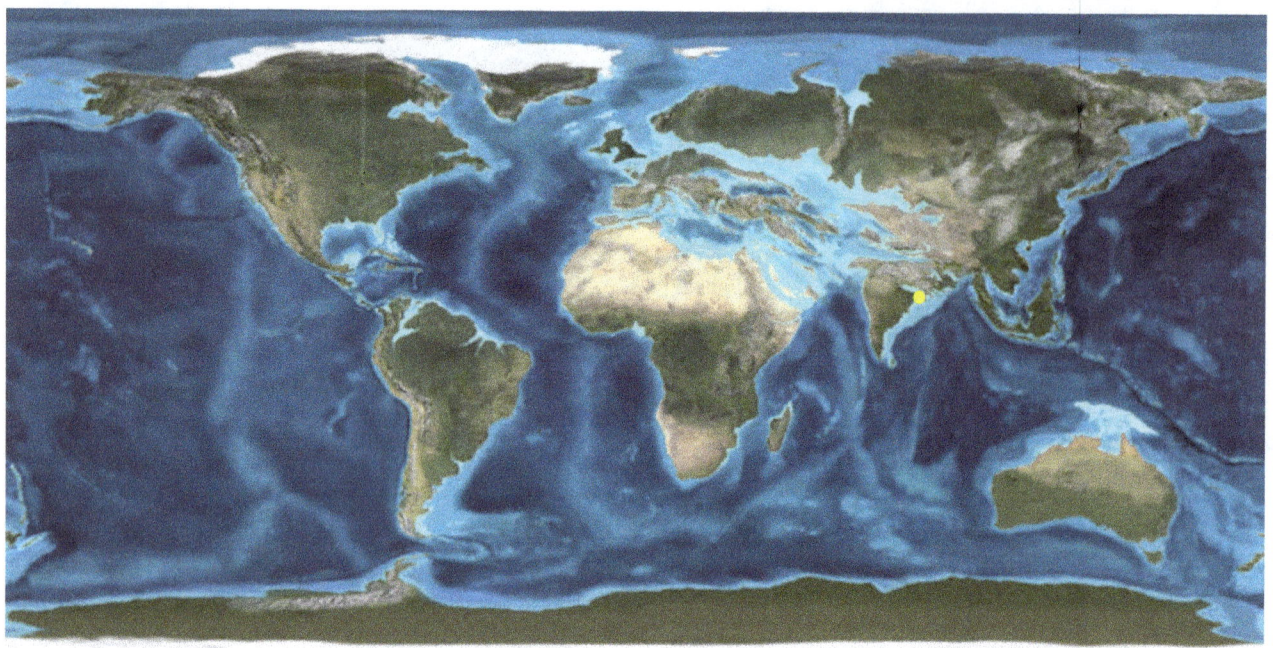

Figure 3. Map showing the fossil locality (yellow dot) in palaeogeographic map during the Oligocene [66].

Results

Systematic description

Order. Malpighiales Juss. (1820) [33]

Family. Phyllanthaceae Martinov (1820) [34]

Subfamily. Phyllanthoideae Asch. (1864) [35]

Tribe. Bridelieae Müll. Arg. (1864) [36]

Genus. *Bridelia* Willd. (1806) [37]

Species. *B. makumensis* Srivastava and Mehrotra, sp. nov.

Figures. 4A, B, D; 5A, D

Holotype. Specimen No. BSIP 40115

Horizon. Tikak Parbat Formation

Locality. Tirap Colliery, Tinsukia District, Assam (27° 17′ 20″ N; 95° 46′ 15″ E)

Age. Late Oligocene (Chattian, 28.4–23 Ma)

Number of specimens studied. One.

Description. Leaf nearly complete, simple, symmetrical, microphyll, elliptic, preserved lamina length 5.2 cm (estimated lamina length 7.3 cm), maximum width near the middle portion 2.4 cm; apex not preserved but seems to be acute-obtuse; base slightly broken, asymmetrical, acute, normal; margin entire but slightly crenate seen on the distal portion, seemingly wavy; texture appearing chartaceous; attachment with petiole not preserved; venation pinnate, simple craspedodromous to eucamptodromous; primary vein stout in thickness, curved; secondary veins 11 pairs visible, 0.2–0.4 cm apart, predominantly alternate, angle of divergence moderate acute (48°–62°), smoothly and sometimes abruptly curved up near the margin, attachment with the primary vein normal, rarely decurrent; intersecondary veins absent; tertiary veins simple percurrent, recurved, forked, oblique to mid-vein, angle of origin AO, AA, AR, predominantly alternate, close; marginal ultimate venation fimbriate; quaternary veins orthogonal.

Affinities. The characteristic features of the fossil leaf such as elliptic shape, crenate margin, craspedodromous to eucampto-dromous venation, moderate acute angle of divergence of secondary veins, percurrent to recurved tertiary veins and fimbriate marginal venation suggest its affinity with *Bridelia* of the family Phyllanthaceae. A large number of species of *Bridelia* such as *B. assamica* Hook.f., *B. cinnamomea* Hook.f., *B. glauca* Blume, *B. insulana* Hance, *B. ovata* Decne. (syn. *B. burmanica* Hook.f.), *B. retusa* (L.) A. Juss. (syn. *B. squamosa*), *B. stipularis* (L.) Blume (syn. *B. scandens*), *B. tomentosa* Blume and the species of its sister genus *Cleistanthus* Hook.f. ex Planch such as *C. collinus* (Roxb.) Benth. ex Hook.f., *C. malabaricus* Müll.Arg. and *C. monoicus* (Lour.) Müll.Arg. were studied and compared in the herbarium of the Forest Research Institute, Dehradun and the Central National Herbarium, Howrah.

In *B. assamica*, *B. cinnamomea*, *B. glauca*, *B. scandens* and *Cleistanthus monoicus* the angle of divergence of secondary veins is narrow-moderate acute which is in contrast to the present fossil. In *B. insulana*, *B. tomentosa*, *Cleistanthus collinus* and *C. malabaricus* the distance between the two secondary veins is greater than the present fossil. In the venation pattern the fossil shows maximum similarity with *B. retusa* (Figure 4C, E) and *B. stipularis* (Figure 5B, C, E) but differs from them in having asymmetrical base. In having asymmetrical base our fossil shows resemblance with *B. ovata* where base varies from asymmetrical ([38], Plate 14, Figure 2) to symmetrical (CNH Herbarium sheet no. 400497). However, angle of divergence of secondary veins is more acute in *B. ovata* than that of our fossil. It appears that our fossil shows a combination of characters found in *B. ovata*, *B. retusa* and *B. stipularis*. The comparable species are distributed throughout the hotter parts of India, along the foot of the Himalaya, south India, Malacca, Malayan Peninsula, Myanmar, Sri Lanka, Philippines and tropical Africa [39].

As far as fossil leaf records of *Bridelia* are concerned they are known mainly from Nepal and India. Two fossil species of the genus, namely *B. mioretusa* and *B. siwalica* are known from the

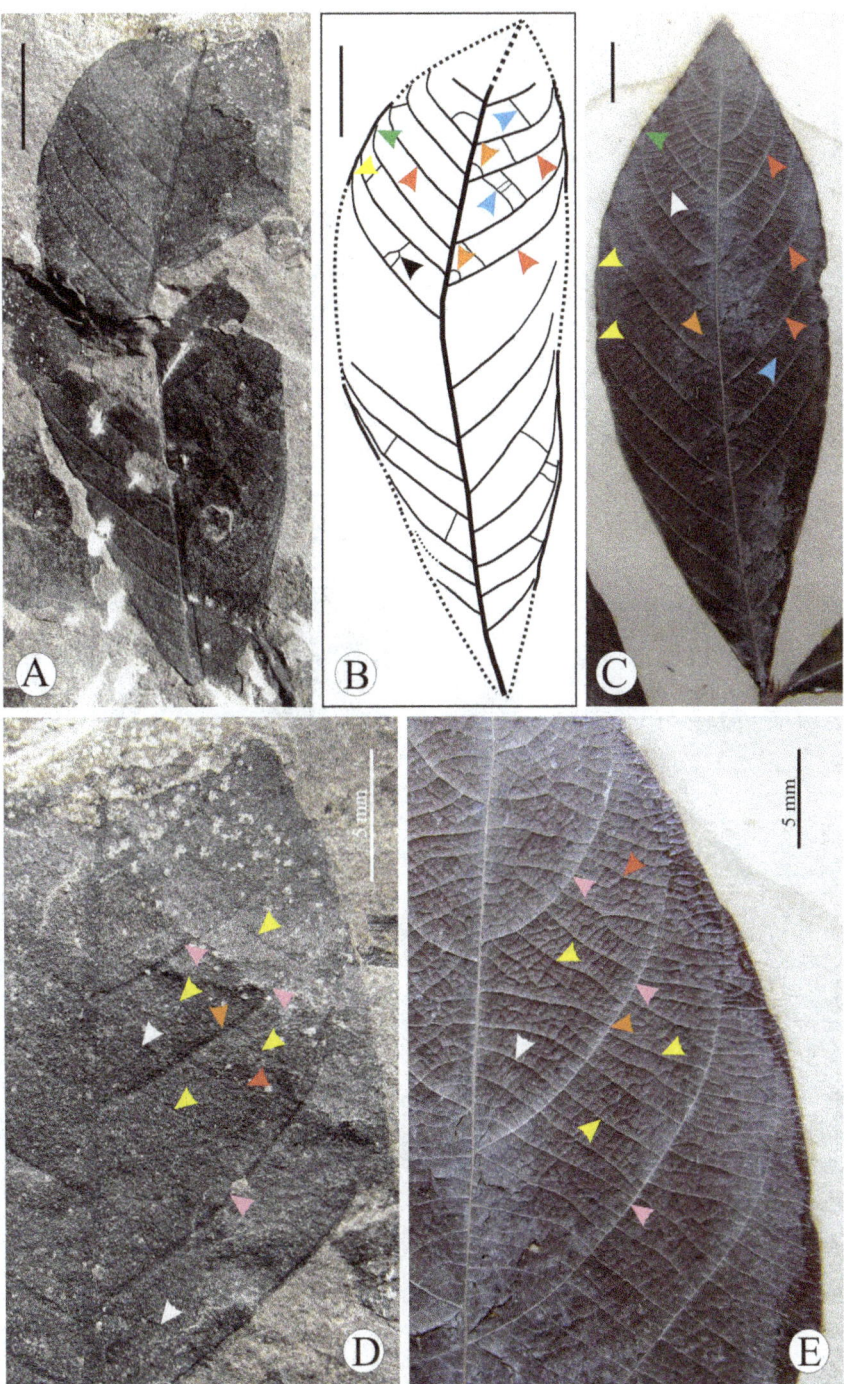

Figure 4. *Bridelia* leaves. A. Fossil leaf of *Bridelia makumensis* sp. nov. showing shape, size and venation pattern. B. Text diagram of the fossil leaf showing craspedodromous and eucamptodromous venation (yellow and green arrows), secondary veins (red arrows) and percurrent, recurved and forked tertiary veins (blue, orange and black arrows). C. Modern leaf of *Bridelia retusa* showing craspedodromous and eucamptodromous venation (yellow and green arrows), secondary veins (red arrows) and percurrent and recurved tertiary veins (blue and orange arrows). D. Enlarged portion of the fossil leaf showing secondary veins (pink arrows), percurrent, recurved and forked tertiary veins (yellow, white and red arrows); predominantly alternate tertiary veins (orange arrow). E. Modern leaf of *Bridelia retusa* showing secondary veins (pink arrows); percurrent, recurved and forked tertiary veins (yellow, white and red arrows) Scale bar = 1 cm, unless mentioned.

Siwalik sediments (late Miocene) of Surai Khola, western Nepal [38]. Three more fossil species of *Bridelia*, namely *B. stipularis* and *B. verucosa* have been described from the Middle Siwalik sediments of Darjeeling, West Bengal [40], while another species viz., *B. oligocenica* is known from the late Oligocene sediments of

Assam [15]. All the aforesaid fossils are different from the present fossil in a combination of characters (Table 1). Under such circumstances a new species, *B. makumensis* Srivastava and Mehrotra, sp. nov. is created and the specific epithet is after the fossil locality.

Figure 5. *Bridelia* leaves. A. Enlarged apical portion of the fossil leaf showing craspedodromous and eucamptodromous venation (orange and blue arrows), secondary veins (yellow arrows) and percurrent tertiary veins (red arrows). B. Apical portion of the modern leaf of *B. stipularis* showing similar craspedodromous and eucamptodromous venation (orange and blue arrows) as found in the fossil and secondary veins (yellow arrows). C. Modern leaf of *B. stipularis* showing shape, size and venation pattern. D. Basal portion of the fossil leaf showing course of secondary veins (yellow arrows). E. Basal portion of the modern leaf of *B. stipularis* showing similar course of secondary veins as found in the fossil (yellow arrows).

Table 1. Comparative chart of the known fossil leaves of *Bridelia* from the Cenozoic sediments.

Fossil taxa	Modern Comparable Forms	Lamina Apex	Base	Margin	Shape	Balance	Venation pattern	2° veins	3° veins
B.oligocenica Awasthi and Mehrotra [15]	*B. retusa*	NP	Acute	Entire	Elliptic	Symmetrical	Eucamptodromous	Narrow acute	?Percurrent
Bridelia mioretusa Prasad and Pandey 30]	*B. retusa*	Acute	NP	Entire	Elliptic	Symmetrical	Craspedodromous-eucamptodromous	Narrow acute	Percurrent
B. siwalika Prasad and Pandey [30]	*B.ovata*	NP	Obtuse, asymmetrical	Entire	Obovate	Asymmetrical	Eucamptodromous	Wide acute	Percurrent
B. stipularis Pathak 31]	*B. stipularis*	Acute-Obtuse	Round	Entire	Elliptic	Symmetrical	Eucamptodromous	Wide acute	NG
B. verucosa Pathak 31]	*B. verucosa*	Obtuse	Obtuse	Entire	Elliptic	Symmetrical	Eucamptodromous	Narrow acute	NG
B. makumensis Srivastava and Mehrotra	*B. ovata, B. retusa* and *B. stipularis*	Seemingly acute-obtuse	Acute, asymmetrical	Entire-slightly crenate	Elliptic	Symmetrical	Craspedodromous-eucamptodromous	Moderate acute	Percurrent

NP = Not preserved;
NG = Not given.

Discussion

Fossil wood record of *Bridelia*

As far as the fossil records of *Bridelia* are concerned, they are known in the form of leaves and woods. The leaf fossil records have been discussed in the affinities (see section affinities). Fossil woods of *Bridelia* are known from various Tertiary sediments of central South Asia, temperate Asia, tropical Africa, Northern Africa, Europe, America and Australia [41] but their affinities with modern *Bridelia* are uncertain because of the homogeneity in wood characters of various genera of Euphorbiaceae *s.l.*

Bailey [42] instituted the genus *Paraphyllanthoxylon* for the fossil woods resembling Phyllanthoid Euphorbiaceae that includes the genus *Bridelia* also. However, he was not sure of its affinities. Wheeler et al. [43] after studying various species of *Paraphyllanthoxylon* concluded that the genus can not be assigned with certainty to the Euphorbiaceae because of its similarities to other families.

Ramanujam [44] instituted the genus *Bischofioxylon* for the fossil woods resembling *Bischofia* and described *Bischofioxylon miocenicum* from south India. Mädel [45] had suggested its affinities with *Bridelia* and merged it into another organ genus *Bridelioxylon* Mädel. Therefore, Bande [46] established *Bischofinium* for the fossil woods resembling *Bischofia*. Awasthi [47] suggested that neither *B. miocenicum* Ramanujam nor *Biscofinium* Bande belongs to *Bischofia* or *Bridelia*. The generic diagnosis of *Bridelioxylon* is within the range of generic diagnosis of *Paraphyllanthoxylon* Bailey. In our opinion due to the homogeneity in wood characters, it is difficult to separate *Bridelia* from other genera of the Euphorbiaceae.

Disjunct distribution pattern and possible migratory path of *Bridelia* from India to Africa-Madagascar and Southeast Asia

The disjunct phytogeography of *Bridelia* with equal distribution of species in Africa and tropical Asia and endemism in the African-Madagascar species are interesting. Based on the molecular data, Li et al. [8] inferred that *Bridelia* separated from its sister genus *Cleistanthus* at 33±5 Ma and suggested that the genus evolved in Southeast Asia and later migrated westward to India and Africa and eastward to Australia. However, this hypothesis didn't get support from the fossil records of the genus. The present fossil record from the late Oligocene sediments of northeast India is important because during the late Oligocene, suturing of the Indian plate with the Eurasian plate was not complete to facilitate the plant migration between India and Southeast Asia (Figure 3) [11–13]. In the light of the oldest fossil evidence of *Bridelia* from northeast India we suggest that the genus evolved most likely during the late Oligocene in northeast India, while the speciation must have occurred during the Miocene followed by the dispersal of the genus from India to Southeast Asia via Myanmar as the suturing of both the aforesaid plates completed during the early Miocene [48,49]. Our fossil shows affinity with three species of *Bridelia*, namely *B. ovata*, *B. retusa* and *B. stipularis*; this again suggests that the speciation must have occurred after the late Oligocene and most likely during the Miocene as suggested by the molecular data [8] and supported by the diversity in fossil records from the Siwalik of Nepal and India [38,40]. The climatic conditions also favoured in the evolution of *Bridelia* because the late Oligocene was the time of last significant globally warm climate during the Cenozoic [50] under which the genus evolved and the speciation of the genus in Asia most likely to have occurred during the Middle Miocene Climatic Optimum (MMCO) [50]. All the above facts indicate that *Bridelia* evolved in India during the late Oligocene and after the complete suturing of Indian and Eurasian plates the genus migrated to Southeast Asia via Myanmar and then to Australia, along with several other plant taxa such as *Alphonsea* of the family Annonaceae [20], *Mangifera* [51] and *Semecarpus* of the family Anacardiaceae [18] (Figure 1). Similarly, the stem lineage of *Bridelia* also migrated to Africa via "Iranian Route" [52] during the late Miocene to early Pleistocene and this can be explained on the basis of plate tectonic model. Africa was isolated from Eurasia during the mid-Cretaceous to early Miocene [53]. By the early Miocene, Africa made land connections with east Eurasia via "Iranian Route" (Iranian and Arabian block) [52], and *Bridelia* most likely to have migrated through this route during the Miocene (Figure 1). After reaching to Africa, the stem lineage of the genus speciosed locally due to the availability of free niche and less competition which resulted in the local endemism of the genus in Africa. The speciation of stem lineage of *Bridelia* in Africa again coincides with the climatic condition in Africa i.e. occurrence of aridity in Africa during the late Neogene [54–56]. The aforesaid view also gets supports from the molecular phylogenetic study which suggests that the African clade speciosed at *ca* 3±1 Ma [8].

The corridor via "Iranian Route" was not only for *Bridelia* but also common for faunal exchange [52,57] and several other African plant taxa reported from the Pliocene of western India [58]; this suggests that the migration was in between Africa and east Eurasia.

The migration of *Bridelia* from India to Africa and Southeast Asia again supports the "Out of India" hypothesis [59].

Palaeofloristics and Palaeoclimate of the Makum Coalfield, Assam

The Makum Coalfield is important in view of its high diversity of plant fossils. The late Oligocene was the time of last significant globally warm period during the Cenozoic and the fossil locality was situated at 10°–15° N palaeolatitude [10]. The known floristic diversity indicates that the family Fabaceae was the most dominant followed by Anacardiaceae, Clusiaceae, Combretaceae, Arecaceae, Annonaceae, Lauraceae and Sapindaceae etc. Most of the aforesaid families have pantropical distribution, while the abundance of palms indicates high water availability.

The families like Annonaceae, Burseraceae, Clusiaceae, Combretaceae, Lecythidaceae, Myristicaceae and Rhizophoraceae are typical pantropical megatherm families [60] whose presence in the palaeoflora provides an evidence that the CMT (mean temperature of the coldest month) was at least not less than 18°C [25]. Similarly, the presence of most dominant family Fabaceae [17] whose abundance and richness covary with the temperature [61], indicates warm climate. The occurrence of families Avicenniaceae and Rhizophoraceae is also very significant in terms of the depositional environment. These families are highly indicative of deltaic, mangrove or lacustrine deposition of coal seams and associated sediments in the Makum Coalfield [15]. The abun-

dance of palms like *Nypa* [23] provides clear evidence of a coastal plain environment where both temperature and humidity remain high throughout the year [62]. Quantitative palaeoclimate reconstruction based on CLAMP analysis on the Makum Coalfield palaeoflora was made by Srivastava et al. [25]. They have used 80 different leaf morphotypes, from the Tirap colliery of the Makum Coalfield, which were analysed by following standard protocol of CLAMP analysis [63]. The CLAMP analysis indicates mean annual temperature (MAT) 28.3±3.7°C, warm month mean temperature (WMMT) 34.2±5.2°C and cold month mean temperature (CMMT) 23.6±5.5°C. The precipitation estimates suggest a marked seasonality in the rainfall pattern showing a wet season with 20 times the rainfall of the dry season. The similar pattern can be seen in Sunderbans lying in the modern Ganges/Brahmaputra/Meghna delta. Therefore, it is suggested that the South Asian Monsoon was already established by the late Oligocene time at an intensity similar to that of today [25].

Conclusions

Fossil evidences, along with the molecular data are important in studying the evolution and speciation of an organism. In the present paper we have reported fossil leaf of *Bridelia* from the late Oligocene sediments of northeast India and suggested its affinity with *B. ovata*, *B. retusa* and *B. stipularis*. Our fossil data, along with the known fossil records of *Bridelia* from the Neogene sediments of Nepal and India suggests that the genus evolved during the late Oligocene and migration and speciation occurred from India to Southeast Asia via Myanmar and from India to Africa via "Iranian Route" during the Miocene. The present finding fits well with the molecular phylogenetic analysis and plate tectonic models.

Acknowledgments

The authors are thankful to the authorities of the Coal India Limited (Northeastern Region), Margherita for permission to collect plant fossils from the Makum Coalfield. Thanks are also due to the Directors, Botanical Survey of India, Kolkata and the Forest Research Institute, Dehradun for permitting us to consult the herbarium. The authors are also thankful to Prof. Sunil Bajpai, Director, Birbal Sahni Institute of Palaeobotany, Lucknow for providing necessary facilities and permission to carry out this work. They are thankful to Prof. E.A. Wheeler of North Carolina State University, USA for sending her valuable reprints. The authors are also grateful to Prof. Khum Paudiyal, one anonymous reviewer and Prof. Navnith K.P. Kumaran for their valuable suggestions in improving the manuscript.

Author Contributions

Conceived and designed the experiments: GS RCM. Performed the experiments: GS RCM. Analyzed the data: GS RCM. Contributed reagents/materials/analysis tools: GS RCM. Contributed to the writing of the manuscript: GS RCM.

References

1. Stevens PF (2001 onwards). Angiosperm Phylogeny Website. Version 12, July 2012 [and more or less continuously updated since]. Available: http://www.mobot.org/MOBOT/research/APweb/. Accessed 2013 May 28.

2. APG II (2003) An update of the Angiosperm Phylogeny Group classification for the orders and families of flowering plants. APG II. Bot J Linn Soc 141: 399–436.

3. Muller J (1992) Fossil pollen records of extant angiosperms. Bot Rev 47: 1–142.

4. Gruas-Cavagnetto C, Köhler E (1992) Pollen fossils d'Euphorbiacées de l'Eocéne français. Grana 31: 291–304.

5. Dressler S (1996) The genus *Bridelia* Willd. (Euphorbiaceae) in Malesia and Indochina: a regional revision. Blumea 41: 263–331.

6. Forster PI (1999) A taxonomic revision of *Bridelia* Willd. (Euphorbiaceae) in Australia. Austrobaileya 5: 405–419.

7. Li PT, Dressler S (2008) Euphorbiaceae. In: Wu ZY, Raven PH, Hong DY, eds. Flora of China. Beijing, China: Science Press/St Louis: Miss Bot Garden Press. pp. 172–177.

8. Li Yongquan, Dressler S, Zhang D, Renner SS (2009) More Miocene dispersal between Africa and Asia- the case of *Bridelia* (Phyllanthaceae). Syst Bot 34(3): 521–529.

9. Gradstein FM, Ogg JG, Smith A (2004) A Geologic Time Scale. Cambridge, UK: Cambridge University Press.

10. Molnar P, Stock JM (2009) Slowing of India's convergence with Eurasia since 20 Ma and its implications for Tibetan mantle dynamics. Tectonics 28: TC3002.

11. Lakhanpal RN (1970) Tertiary flora of India and their bearing on the historical geology of the region. Taxon 19: 675–694.

12. Bande MB, Prakash U (1986) The Tertiary flora of Southeast Asia with remarks on its palaeoenvironment and phytogeography of the Indo-Malayan region. Rev Palaeobot Palynol 49: 203–233.

13. Srivastava G, Mehrotra RC (2010) Tertiary flora of northeast India vis-á-vis movement of the Indian plate. Mem Geol Soc India 75: 123–130.

14. Awasthi N, Mehrotra RC, Lakhanpal RN (1992) Occurrence of *Podocarpus* and *Mesua* in the Oligocene sediments of Makum Coalfield, Assam, India. Geophytology 22: 193–198.

15. Awasthi N, Mehrotra RC (1995) Oligocene flora from Makum Coalfield, Assam, India. Palaeobotanist 44: 157–188.

16. Mehrotra RC, Dilcher DL, Lott TA (2009) Notes on elements of the Oligocene flora from the Makum Coalfield, Assam, India. Palaeobotanist 58: 1–9.

17. Srivastava G, Mehrotra RC (2010) New legume fruits from the Oligocene sediments of Assam. J Geol Soc India 75: 820–828.

18. Srivastava G, Mehrotra RC (2012) Oldest fossil of *Semecarpus* L.f. from the Makum Coalfield, Assam, India and comments on its origin. Curr Sci 102(3): 398–400.

19. Srivastava G, Mehrotra RC, Bauer H (2012) Palm leaves from the Late Oligocene sediments of Makum Coalfield, Assam. J Earth Syst Sci 121(3): 747–754.

20. Srivastava G, Mehrotra RC (2013) First fossil record of *Alphonsea* Hk. f. & T. (Annonaceae) from the late Oligocene sediments of Assam, India and comments on its phytogeography. PLoS ONE 8(1): e53177.

21. Srivastava G, Mehrotra RC (2013) Endemism due to climate change: evidence from *Poeciloneuron* Bedd. (Clusiaceae) leaf fossil from Assam, India. J Earth Syst Sci 122(2): 283–288.

22. Srivastava G, Mehrotra RC (2013) Further contribution to the low latitude leaf assemblage from the late Oligocene sediments of Assam and its phytogeographical significance. J Earth Syst Sci 122(5): 1341–1357.

23. Mehrotra RC, Tiwari RP, Mazumber BI (2003) *Nypa* megafossils from the Tertiary sediments of northeast India. Geobios 36: 83–92.

24. Kumar M, Srivastava G, Spicer RA, Spicer TEV, Mehrotra RC, et al. (2012) Sedimentology, palynostratigraphy and palynofacies of the late Oligocene Makum Coalfield, Assam, India: a window on lowland tropical vegetation during the most recent episode of significant global warmth. Palaeogeogr Palaeoclimatol Palaeoecol 342–343: 143–162.

25. Srivastava G, Spicer RA, Spicer TEV, Yang J, Kumar M, et al. (2012) Megaflora and palaeoclimate of a late Oligocene tropical delta, Makum Coalfield, Assam: evidence for the early development of the South Asia monsoon. Palaeogeogr Palaeoclimatol Palaeoecol 342–343: 130–142.

26. Pascoe EH (1964) A Manual of the Geology of India and Burma. Calcutta: Geol Surv India.

27. Ganju JI, Khare BM, Chaturvedi JS (1986) Geology and hydrocarbon prospects of Naga Hills south of 27° latitude. Bull Oil Nat Gas Comm 23: 129–145.

28. Misra BK (1992) Tertiary coals of Makum Coalfield, Assam, India; petrography, genesis and sedimentation. Palaeobotanist 39(3): 309–326.

29. Mishra HK, Ghosh RK (1996) Geology, petrology and utilization of some Tertiary coals of the northeastern region of India. Int J Coal Geol 30: 65–100.

30. Hickey LJ (1973) Classification of the architecture of dicotyledonous leaves. Am J Bot 60: 17–33.

31. Dilcher DL (1974) Approaches to the identification of angiosperm leaf remains. Bot Rev 40: 1–157.

32. Ellis B, Daly DC, Hickey LJ, Johnson KR, Mitchell JD, et al. (2009) Manual of leaf architecture. USA: Cornell University Press.

33. Jussieu ALD (1820) O Prirozenosti Rostin 225.

34. Martinov II (1820) Tekhno-Botanicheskiĭ Slovar': na latinskom i rossĭiskom ăzykakh. Sanktpeterburgie: V tip. Imperatorskoĭ rossĭiskoĭ akademiĭ.

35. Ascherson PFA (1864) Flora der Provinz Brandenburg 1(2): 59.

36. Müller Argoviensis J (1864) Botanische Zeitung 22: 324.

37. Willdenow CLv (1806) Species Plantarum. Berlin: Editio Quarta 4.

38. Prasad M, Pandey SM (2008) Plant diversity and climate during Siwalik (Miocene-Pliocene) in the Himalayan foot-hills of western Nepal. Palaeontographica B 278: 13–70.

39. Hooker JD (1885) The flora of British India 5. Kent: Reeve & Company.

40. Pathak NR (1969) Megafossils from the foot-hills of Darjeeling District, India. In: Santapau H, et al., eds. J Sen Memorial volume. Calcutta, India: Bot Soc of Bengal. pp. 379–384.

41. Gregory M, Poole I, Wheeler EA (2009) Fossil dicot wood names: an annotated list with full bibliography. IAWA (suppl 6): 220.

42. Bailey IW (1924) The problem of identifying the wood of Cretaceous and later dicotyledons: *Paraphyllanthoxylon arizonense*. Ann Bot 38: 439–451.

43. Wheeler EA, Lee M, Matten LC (1987) Dicotyledonous woods from the Upper Cretaceous of southern Illinois. Bot J Linn Soc 94: 111–126.

44. Ramanujam CGK (1960) Silicified woods from the Tertiary rocks of South India. Palaeontographica B 106: 99–140.

45. Mädel E (1962) Die fossilen Euphorbiaceen-Hölzer mit besonderer Berücksichtigung neuer Funde aus der Oberkreide Süd-Afrikas. Senckenberg. Lethaea 43: 283–321.

46. Bande MB (1974) Two fossil woods from the Deccan Intertrappean beds of Mandla District, Madhya Pradesh. Geophytology, 4(2): 189–195.

47. Awasthi N (1989) Occurrence of *Bischofia* and *Antiaris* in Namsang beds (Miocene-Pliocene) near Deomali, Arunachal Pradesh, with remarks on the identification of fossil woods referred to *Bischofia*. Palaeobotanist 37: 147–151.

48. Chatterjee S, Scotese CR (1999) The breakup of Gondwana and the evolution and biogeography of the Indian plate. Proc Indian Natl Sci Acad A 65: 397–425.

49. Chatterjee S, Goswami A, Scotese CR (2013) The longest voyage: tectonic, magmatic, and paleoclimatic evolution of the Indian plate during its northward flight from Gondwana to Asia. Gond Res 23: 238–267.

50. Zachos J, Pagani M, Sloan L, Thomas E, Billups K (2001) Trends, rhythms, and aberrations in global climate 65 Ma to present. Science 292: 686–693.

51. Mehrotra RC, Dilcher DL, Awasthi N (1998) A Palaeocene *Mangifera*- like leaf fossil from India. Phytomorphology 48: 91–100.

52. Gheerbrant E, Rage J-C (2006) Paleobiogeography of Africa: how distinct from Gondwana and Laurasia? Palaeogeogr Palaeoclimatol Palaeoecol 241: 224–246.

53. McLoughlin S (2001) The breakup history of Gondwana and its impact on pre-Cenozoic floristic provincialism. Aust J Bot 49: 271–300.

54. Griffin David L (2002) Aridity and humidity: two aspects of the late Miocene climate of North Africa and the Mediterranean. Palaeogeogr Palaeoclimatol Palaeoecol 182: 65–91.

55. Bobe R (2006) The evolution of arid ecosystem in eastern Africa. J Arid Environ 66: 564–584.

56. Bonnefille R (2010) Cenozoic vegetation, climate changes and hominid evolution in tropical Africa. Global Planet Change 72: 390–411.

57. Bibi F (2011) Mio-Pliocene faunal exchanges and African biogeography: the record of fossil bovids. PLoS ONE 6(2): e16688.

58. Shukla A, Mehrotra RC, Guleria JS (2012) African elements (fossil woods) from the upper Cenozoic sediments of western India and their palaeoecological and phytogeographical significance. Alcheringa 37(1): 1–18.

59. Bossuyt F, Milinkovitch MC (2001) Amphibians as indicators of early Tertiary "Out-of-India" dispersal of vertebrates. Science 292: 93–95.

60. van Steenis CGGJ (1962) The land bridge theory in Botany. Blumea 11: 235–372.

61. Punyasena SW, Eshel G, McElwain JC (2008) The influence of climate on the spatial patterning of neotropical plant families. J Biogeogr 35: 117–130.

62. Tomlinson PB (1990) The Structural Biology of Palms. Oxford, UK: Clarendon Press.

63. Wolfe JA (1993) A method of obtaining climatic parameters from leaf assemblages. U S Geol Surv Bull 2040: 1–73.

64. Global Biodiversity Information Facility Website. Available: http://data.gbif.org/species/3076094/. Accessed 2013 May 28.

65. Ahmed M (1996) Petrology of Oligocene coal, Makum coalfield, Assam, northeast India. Int J Coal Geol 30: 319–325.

66. Colorado Plateau Geosystems, Inc. Available: http://cpgeosystems.com/paleomaps.html. Accessed 2013 May 28.

6

Sediment Delivery Ratio of Single Flood Events and the Influencing Factors in a Headwater Basin of the Chinese Loess Plateau

Mingguo Zheng[1], Yishan Liao[2], Jijun He[3]*

1 Key Laboratory of Water Cycle and Related Land Surface Processes, Institute of Geographic Sciences & Natural Resources Research, Chinese Academy of Sciences, Beijing, China, **2** Guangdong Institute of Eco-environment and Soil Sciences, Guangzhou, China, **3** Base of the State Laboratory of Urban Environmental Processes and Digital Modelling, Capital Normal University, Beijing, China

Abstract

Little is known about the sediment delivery of single flood events although it has been well known that the sediment delivery ratio at the inter-annual time scale is close to 1 in the Chinese Loess Plateau. This study examined the sediment delivery of single flood events and the influencing factors in a headwater basin of the Loess Plateau, where hyperconcentrated flows are dominant. Data observed from plot to subwatershed over the period from 1959 to 1969 were presented. Sediment delivery ratio of a single event (SDR_e) was calculated as the ratio of sediment output from the subwatershed to sediment input into the channel. It was found that SDR_e varies greatly for small events (runoff depth < 5 mm or rainfall depth <30 mm) and remains fairly constant (approximately between 1.1 and 1.3) for large events (runoff depth >5 mm or rainfall depth >30 mm). We examined 11 factors of rainfall (rainfall amount, rainfall intensity, rainfall kinetic energy, rainfall erosivity and rainfall duration), flood (area-specific sediment yield, runoff depth, peak flow discharge, peak sediment concentration and flood duration) and antecedent land surface (antecedent precipitation) in relation to SDR_e. Only the peak sediment concentration significantly correlates with SDR_e. Contrary to popular belief, channel scour tends to occur in cases of higher peak sediment concentrations. Because small events also have chances to attain a high sediment concentration, many small events (rainfall depth <20 mm) are characterized by channel scour with an SDR_e larger than 1. Such observations can be related to hyperconcentrated flows, which behave quite differently from normal stream flows. Our finding that large events have a nearly constant SDR_e is useful for sediment yield predictions in the Loess Plateau and other regions where hyperconcentrated flows are well developed.

Editor: Vanesa Magar, Centro de Investigacion Cientifica y Educacion Superior de Ensenada, Mexico

Funding: Financial support for this research was provided by Non-profit Industry Financial Program of MWR (201201083; www.cws.net.cn) and National Natural Science Foundation of China (41401302 and 41271306; www.nsfc.gov.cn). The funders had no role in study design, data collection and analysis, decision to publish, or preparation of the manuscript.

Competing Interests: The authors have declared that no competing interests exist.

* Email: hejiun_200018@163.com

Introduction

Sediment yield represents the total quantity of sediment observed at a certain point in a landscape or a river system, such as the watershed outlet, in a specified time interval. The sediment yield prediction is of key interest in stream and watershed management due to increasing concerns on water quality, aquatic habitat, biodiversity and life of man-made structures (dams, bridges, harbors, and water supply systems) [1,2]. The concept of sediment delivery ratio (SDR), commonly defined as the ratio of sediment yield to gross erosion [3,4], provides a convenient way to estimate sediment yield to a point of interest. In equation form SDR is expressed as

$$SDR = SY/SE, \qquad (1)$$

where SY (mass per unit time) represents sediment yield at a point of interest, and SE (mass per unit time) represents gross erosion rate of the area upstream of that point. It is believed that if the relationship between SDR and its influential factors is well established, equation (1) would be greatly helpful for estimating sediment yields in ungauged locations. Because of the simplicity in concept and the ability to link on-site erosion with downstream sediment yield, studies of SDR have received much attention [4–6]. SDR is affected by fluvial processes operating at a variety of spatial scales from slopes to channels. Factors affecting SDR almost includes all variables representing hydrological regime (e.g. flood and rainfall) and watershed prosperities (e.g. topography, vegetation and land use). Due to the multitude of the influencing factors and their interactions, it is difficult to identify the dominant controls on SDR [7–9]. As a result, the established relationships between SDR and the influencing factors are largely empirical and can hardly extrapolate beyond the data range with confidence [7–10].

The Chinese Loess Plateau is famous for its high-intensity soil erosion, which frequently exceeds 10, 000 t km^{-2} a^{-1}. Gong and Xiong [11] proposed that SDR is as high as 1 in the Loess Plateau. Mou and Meng [12] subsequently found that almost all sediments (>95%) are moved as wash load as a result of the fine texture of the loess in combination with the strong sediment transport capacity [13] of hyperconcentrated flows, which are well developed in the Loess Plateau [14] and behave quite differently from normal sediment-laden streamflow [15–17]; this mechanism physically enables a SDR as high as 1. Nowadays, it has been widely accepted that SDR in the Loess Plateau is close to 1 over a wide range of basin sizes at inter-annual time scale [5,7,14,18–20]. Nevertheless, knowledge of the sediment delivery and the influencing factors is currently lacking at the time scale of the flood event in the loess Plateau. Among more than 100 rainstorm events over a single year, only 2–7 rainstorm events are erosive in the Loess Plateau [21]. Research efforts are, therefore, needed to investigate sediment delivery processes at the event time scale.

One of great concerns in determining SDR is the enormous uncertainty introduced by estimating gross erosion [1,3], i.e. the denominator in Equation (1). To guarantee a reasonable estimation of the gross erosion and in turn, the SDR, we limited our study to the Tuanshangou subwatershed (Fig. 1), a headwater basin of the first-order channel in the Loess Plateau. The subwatersheds, where eroded sediments are primarily sourced, are the endmember unit for soil conservation practices in the Loess Plateau. The object of this study is to examine sediment delivery processes of single events in the Tuanshangou subwatershed, hoping to further the knowledge of fluvial processes under the control of hyperconcentrated flows. We firstly calculated the SDR for single events and then, examined a number of factors in relation to sediment delivery, including factors of rainfall, flood and antecedent land surface.

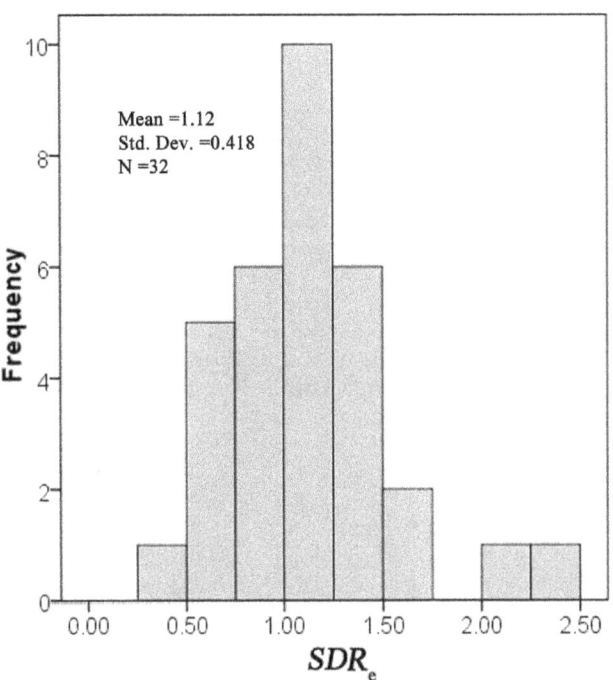

Figure 1. Histograms of *SDR*$_\text{e}$ for 32 flood events observed at the Tuanshangou station.

Study Area and Data Source

The Tuanshangou Creek (latitude 37°41′N, longitude 109°58′E; See Fig. 1 in [22] for the location) drains an area of 0.18 km^2. Typical of the Loess Plateau, the local loess mantle is thicker than 100 m. As wind-borne dust in Quaternary times, loess is loosely compact and highly erodible. The climate is typically semi-arid. During the monitoring period (1959–1969), the mean average annual precipitation is approximately 450 mm and the maximum 30-min rainfall intensity is as high as 2.17 mm min^{-1}. The mean slope of the Tuanshangou subwatershed is as high as 26.8°. The valley side slope is particularly steep (>35°), allowing active mass wasting such as slumping, sliding and collapsing. Approximately 80% of the area was under arable without soil conservation practices. Other lands were abandoned due to the precipitous topography.

Observations at the subwatershed outlet (i.e. the Tuanshangou station) show that the annual sediment yield varied from 200 to 72 000 t km^{-2} a^{-1} with a mean of 19 700 t km^{-2} a^{-1} during the monitoring period. The instantaneous sediment concentrations of storm runoff frequently exceeded 1 000 kg m^{-3} and the mean sediment concentration of flood events was 742 kg m^{-3} (See Table 4 in [22]), much higher than the concentration threshold between the normal sediment-laden flow and the hyperconcentrated flow in the Loess Plateau (200 kg m^{-3} [23] or 300–400 kg m^{-3} [14]).

Data and Methods

Data

Unless stated otherwise, all data used were obtained from the Yellow River Water Conservancy Commission (YRWCC). The YRWCC stream-gauging crews conducted all measurements following national standard procedures of China [24], which have been described in [22] and [25].

This study primarily used data observed at three experimental sites: the Tuanshangou station and two runoff plots within the Tuanshangou subwatershed (i.e. Plots 7 and 9 in [22] and [25]). Both plots were under arable. Crops varied between years, including millet, potato, mung bean, clover, sorghum and wheat. The plots are composed of two parts: hill slope and valley side slope. Such slopes are conventionally termed as the entire slope in Chinese literatures. The plot lengths were 126 m and 164 m, respectively. Detailed information of the plots is available in Table 2 in [22]. Hyetograph data were obtained using a rainfall gauge near Plot 7.

Calculations of SDR

This study defined the gross erosion, i.e. the denominator in Equation (1), as sediment input into stream channels, as did in [18,26]. Such a definition in effect reflects the sediment transport efficiency of stream channels [27]. Plots 7 and 9 were large enough for gullies to develop, through which overland flows drain into the Tuanshangou Creek (See Fig. 2 in [25]). Plot 7 was in the lower part and Plot 9 was in the upper part of the Tuanshangou Creek. Hence, we can use the average of the collective discharges of sediment at Plots 7 and 9 to estimate the sediment input into the creek. We calculated SDR of a single flood event (SDR$_e$, dimensionless) as follows:

$$SDR_e = SSY/(0.5E_7 + 0.5E_9) \qquad (2)$$

where SSY (t km^{-2}) represents observed area-specific sediment yield at the Tuanshangou station, and E_7 and E_9 (t km^{-2})

represents erosion intensity at Plots 7 and 9 for a rainfall event, respectively. Comparisons between E_7 and E_9 on the same rainfall day produced a regression coefficient very close to 1 (0.94) and a R^2 of 0.93, suggesting that the erosion intensity may not vary greatly among entire slopes within the Tuanshangou subwatershed. We thus believe that the average of E_7 and E_9, i.e. the denominator of Equation (2), reasonably represents the sediment discharge into the Tuanshangou Creek per unit area.

The SDR_e values obtained from Equation (2) can be larger or smaller than 1. A SDR_e larger than 1 indicates channel degradation, while a SDR_e smaller than 1 indicates channel aggradation. When SDR_e is equal to 1, stream channels are in equilibrium.

Factors influencing sediment delivery

Factors influencing sediment delivery can be grouped into three categories: rainfall factors, flood factors and antecedent land surface factors. SDR is related to not only flow discharge but also the rheologic and fluid proprieties of flows, which largely depends on suspended load in flows. Flood factors we examined thus include five factors: SSY (t km^{-2}), h (runoff depth of a flood event, mm), q_{max} (peak flow discharge of a flood event, m^3 s^{-1}), C_{max} (maximum sediment concentration of a flood event, kg m^{-3}) and T_f (flood duration, min). All flood factors were measured at the Tuanshangou station.

Rainfall factors we examined also included 5 factors: P (rainfall depth, mm), I_{30} (the maximum 30-min rainfall intensity, mm min^{-1}), E (rainfall kinetic energy, J m^{-2}), EI_{30} (the product of E and I_{30}) and T_p (rainfall duration, min). EI_{30}, the rainfall erosivity index of the Universal Soil Loss Equation (USLE) [28], is the most common rainfall erosivity index. E was calculated using the following relationship:

$$E = \sum_{r=1}^{m} e_r p_r, \qquad (3)$$

where e_r is the rainfall kinetic energy per unit depth of rainfall per unit area (J m^{-2} mm^{-1}), and p_r is the depth of rainfall (mm) for the rth interval among m intervals of the storm hyetograph. e_r is calculated using an empirical equation for the Loess Plateau [29]:

$$e_r = 28.95 + 12.3 \lg i_r, \qquad (4)$$

where i_r (mm min^{-1}) represents the mean rainfall intensity for the rth interval. Equation (4), built on measurements of the drop size distribution of 195 storms, is almost identical to the rainfall intensity-energy equation of the USLE [28] after unit conversion.

Pre-event factors we examined only includes the antecedent precipitation index (P', mm), which is a surrogate for pre-event soil moisture and an important factor affecting runoff yield and soil erodibilities [30]. The vegetation cover rarely exceeded 25% in the Tuanshangou subwatershed. Hence, we do not take the vegetation cover into consideration. P' is defined as [31,32]:

$$P' = \sum_{i=1}^{n} k^i P_i, \qquad (5)$$

where n is the number of antecedent days, P_i (mm) is the daily precipitation for the ith day prior to the event, and k (dimensionless) is the decay constant representing the outflow of the regolith. In practices, k generally lie between 0.80 and 0.98 and n is typically 5, 7 or 14 days [31,33]. Here, k is set at 0.9

following [34]. The gauging crews made measurements of soil moisture near Plot 7. The correlation between P' and the observed soil moisture (top 30 cm) increases asymptotically with increasing n. Because the correlation varies little when n exceeds 11 days [35], n is taken as 11 days in this study. The resultant P' is well correlated with the observed soil moisture ($r = 0.85$, $p < 0.001$).

Results and Discussion

The calculation results of SDR

A total of 36 storm events were well monitored simultaneously at the three sites: the Tuanshangou station and Plots 7 and 9. We calculated SDR_e for all of the events. The bedrock is exposed at the channel bed of the Tuanshangou Creek. Because the bedrock is more prone to runoff production than loess slopes, runoff and sediment are primarily sourced from the channel bed in cases of small rainfall intensities. Among 11 small events with h smaller than 1 mm, four had a SDR_e (2.8, 5.7, 6.3 and 23.7 respectively) distinctively higher than others (the maximum is 1.2), implying that these events essentially conveyed pre-event sediment storage on the channel bed rather than soils eroded from uplands. The four events all had a runoff depth not greater than 0.1 mm at Plots 7 and 9, implying that the sediment discharge into the channel was small enough to be neglected. We removed the four events from the subsequent analyses and the subsequent calculation of SDR_e involves 32 events.

Histograms of SDR_e for the 32 events are presented in Fig. 1. SDR_e ranges from 0.46 to 2.39 with a mean of 1.12 and a median of 1.1. Twenty events have a SDR_e larger than 1. As shown in Fig. 2, SDR_e varies greatly for small events (approximately $h < 5$ mm or $P < 30$ mm) and remains fairly constant for large events ($h > 5$ mm or $P > 30$ mm). Only in cases of small events can significant channel degradation or aggradation occurs. For major sediment-producing events ($SSY > 5000$ t km^{-2}, $h > 5$ mm or $P > 30$ mm), SDR_e essentially lies between 1.1–1.3 (Fig. 2(a), (c) and (f)). Interesting to note is that many small events ($P < 20$ mm; Fig. 2(f)) have a SDR_e much larger than 1.

Flood factors in relation to sediment delivery

Among 11 factors we examined, only C_{max} are significantly correlated with SDR_e ($p = 0.004$; Fig. 2). Nevertheless, the considerable scatters, as shown in Fig. 2(b), prevent C_{max} from being a good predictor of SDR_e.

Contrary to popular belief, SDR_e increase with increasing C_{max} (Fig. 2(b)), a phenomenon also reported in [18]. When C_{max} is higher than 700 kg m^{-3}, most of the events (16 out of 19) have a SDR_e larger than 1, indicating channel scour. In contrast, most of the events (9 out of 13) correspond to a SDR_e smaller than 1 when C_{max} is smaller than 700 kg m^{-3}, indicating channel fill. This observation can be related with hyperconcentrated flows. Different from normal sediment-laden flows, the energy expenditure on suspended-load motion of hyperconcentrated flows decreases with increasing sediment load, as evidenced by laboratory experiments and field observations [14,17,36]. As a result, the sediment transport capacity of hyperconcentrated flows would increase with sediment concentrations and thus, the channel scour is more likely to occur at high rather than at small C_{max}, as has also been observed in the main stream of the Yellow River (See Fig. 3 in [14]).

Sediment delivery along a stream channel depends on not only sediment transport capacity of flows, but also sediment availability within channels. The time interval between large flood events, which occurs relatively infrequently, is generally longer than small ones. Numerous preceding runoff events can prepare a large

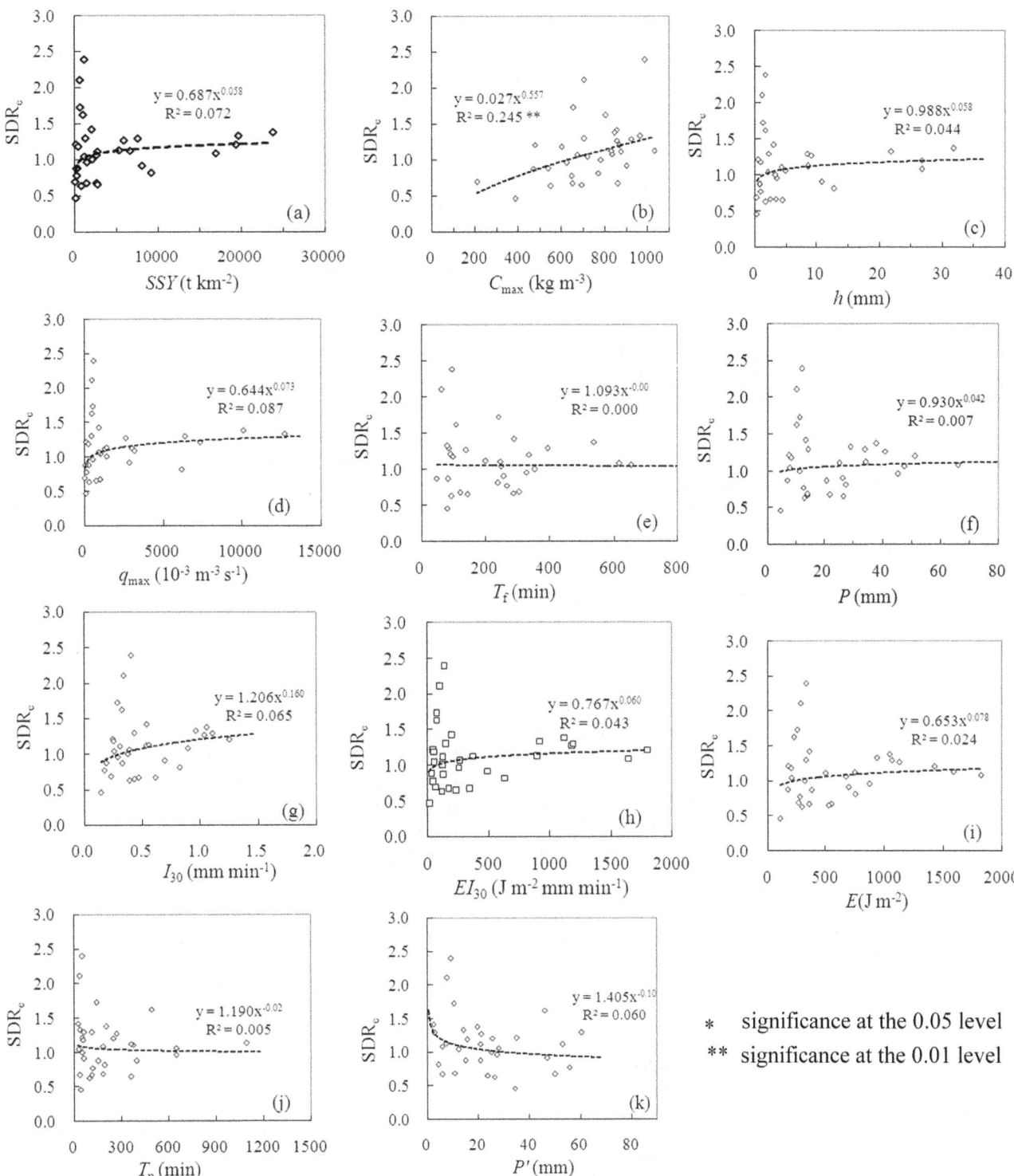

Figure 2. SDR_e in relation to flood factors (a–e), rainfall factors (f–j) and the pre-event factor (k).

amount of sediment storage within channels prior to a large flood event. Mass wasting also occurs more readily during large events. Hence, large flood events generally have a SDR_e larger than 1.

As indicated in Fig. 2(a), (c) and (d), SDR_e is larger than 1 not only for large events but also for many small events, as opposed to that observed in the Murray Darling Basin of Australia [7]. No direct relationship exists between sediment concentration and

water discharge for hyperconcentrated flows [13]. Small events also have chances to attain a high level of sediment concentration (Fig. 3) and thus, a high sediment transport capacity. In addition, antecedent sediment storage within channels may contribute a large part of the event sediment yield considering the minuscule sediment yield of a small event. In contrast, antecedent sediment storage within channels can hardly form the major sediment

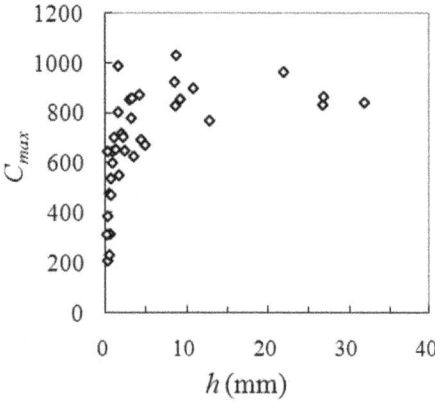

Figure 3. The relationship between C_{max} and h at the Tuanshangou station, showing that small runoff events also have chances to achieve a high level of sediment concentration.

source for large events due to their tremendous sediment yields. Only for small events, hence, can SDR_e be distinctively greater than 1. In contrast, SDR_e for large events falls within a narrow range between approximately 1.1 and 1.3.

Rainfall factors and antecedent precipitation in relation to sediment delivery

None of Rainfall properties show correlation with SDR_e (Fig. 2(f)–(j)). Raindrops splash soil particles and provide sediment for overland flows. Rainfall impact also increases the turbulence of sheet flow and thus, enhances its ability to detach soil and to transport sediment. In both ways, rainfall exerts direct effects on sediment delivery. However, rill erosion and gully erosion are strongly dominant over splash erosion in the Loess Plateau [37,38]. Meanwhile, raindrop impact has no effect on rill flows or other concentrated flows because the turbulent effect of raindrop impact is attenuated with increasing water depth [39–41]. Consequently, the both direct mechanisms that rainfall affects sediment delivery become ineffective. Though the large rainfall event corresponds to a high q_{max} resulting a high sediment transport capacity, the small event can achieve a high transport capacity by achieving a high sediment concentration. Consequently, the indirect mechanism also poses no impact on SDR_e.

For the same reason as above, it cannot be expected that a high antecedent soil moisture results in a high SDR_e by increasing q_{max}. Due to high vulnerability to water erosion at high antecedent soil moistures, sediment concentration and thus sediment transport capacity of hyperconcentrated flows may be enhanced. However, this enhancement should be obscured in a site where mass wasting events are active. Small-sized mass wasting events, such as bank failure and knickpoint retreat, even act as an important agent for rill development [42–44]. Indeed, there is no correlation between P' and C_{max} ($p = 0.37$). As a result, SDR_e are totally independent of P' (Fig. 2(k)). Similarly, due to intensive mass wastings in the Loess Plateau, vegetation and slope land measures for soil conservation, such as terraces and ridges, have no effect on sediment concentrations at the watershed outlet [45].

Conclusions

This study calculated the sediment delivery ratio of single flood events (SDR_e) and examined the factors of rainfall, flood and antecedent land surface in relation to SDR_e in a headwater basin of the Loess Plateau, where hyperconcentrated flows dominate the fluvial processes. SDR_e were calculated as the ratio of sediment output from the subwatershed to sediment input into the channel. Due to distinct behaviours of hyperconcentrated flows, the sediment delivery process of the examined subwatershed is quite different from that under the control of the normal stream flow:

1) SDR_e varies greatly for small events ($h<5$ mm or $P<30$ mm) and remains fairly constant for large events ($h>5$ mm or $P>30$ mm). Most of the examined events (20 out of 32) have a SDR_e higher than 1, implying channel degradation. Such high sediment transfer efficiency can be related to hyperconcentrated flows, which have very strong capacity to transport sediment.

2) Due to decreasing energy expenditure on suspended-load motion with increasing sediment load for hyperconcentrated flows, SDR_e show increasing trends with the increasing level of sediment concentration, as indexed by the maximum sediment concentration of a flood event (C_{max}). Channel degradation primarily occurs when C_{max} exceed 700 kg m^{-3}. Otherwise, channel aggradation occurs. Among 11 factors we examined, only C_{max} is correlated with SDR_e ($p<0.01$).

3) Small events also have chances to attain a high C_{max}, thereby leading to a SDR_e higher than 1. Moreover, the extremely large SDR_e always corresponds to small events because pre-event sediment storage within channels can hardly form the major sediment source for large events. Because both large and small events are capable of achieve a high SDR_e, the peak flow discharge is poorly correlated with SDR_e.

4) Both rainfall factors (including rainfall amount, rainfall intensity, rainfall kinetic energy, rainfall erosivity and rainfall duration) and antecedent precipitation show no correlation with SDR_e. Due to poor correlations and considerable scatters, any factors we examined cannot be expected to be a good predictor of SDR_e. Nevertheless, our finding that large flood events ($h>5$ mm or $P>30$ mm) has similar values of SDR_e in a narrow range between approximately 1.1 and 1.3 should be a valuable aid to the sediment yield prediction in the Loess Plateau given the fact that large events contribute almost all sediments.

Acknowledgments

We appreciate the suggestions of the anonymous reviewer and the editor. Thanks the Data Sharing Infrastructure of Earth System Science-Data Sharing Infrastructure of Loess Plateau for providing our data.

Author Contributions

Conceived and designed the experiments: MGZ JJH. Performed the experiments: MGZ JJH YSL. Analyzed the data: MGZ JJH YSL. Contributed reagents/materials/analysis tools: MGZ YSL. Wrote the paper: MGZ JJH.

References

1. Rovira A, Batalla RJ (2006) Temporal distribution of suspended sediment transport in a Mediterranean basin: The Lower Tordera (NE SPAIN). Geomorphology 79(1): 58–71.

2. Shi ZH, Ai L, Fang NF, Zhu HD (2012) Modeling the impacts of integrated small watershed management on soil erosion and sediment delivery: a case study in the Three Gorges Area, China. J Hydrol 438: 156–167.

3. Lane LJ, Hernandez M, Nichols M (1997) Processes controlling sediment yield from watersheds as functions of spatial scale. Environ. Modell Softw 12(4): 355–369.

4. Alatorre LC, Beguería S, García-Ruiz JM (2010) Regional scale modeling of hillslope sediment delivery: A case study in the Barasona Reservoir watershed (Spain) using WATEM/SEDEM. J Hydrol 391: 109–123.

5. Walling DE (1983) The sediment delivery problem. J Hydrol 65(1): 209–237.

6. De Vente J, Poesen J, Arabkhedri M, Verstraeten G (2007) The sediment delivery problem revisited. Prog Phys Geog, 31(2): 155–178.

7. Lu H, Moran CJ, Prosser IP (2006) Modelling sediment delivery ratio over the Murray Darling Basin. Environ. Modell Softw 21(9): 1297–1308.

8. Shi ZH, Ai L, Li X, Huang XD, Wu GL, et al. (2013) Partial least-squares regression for linking land-cover patterns to soil erosion and sediment yield in watersheds. J Hydrol 498: 165–176.

9. Yan B, Fang NF, Zhang PC, Shi ZH (2013) Impacts of land use change on watershed streamflow and sediment yield: an assessment using hydrologic modelling and partial least squares regression. J Hydrol 484: 26–37.

10. Ferro V, Minacapilli M (1995) Sediment delivery processes at basin scale. Hydrol Sci J, 40:6, 703–717.

11. Gong SY, Xiong GS (1979) The origin and the regional distribution of sediment of the Yellow River. Yellow River (1): 7–11. (in Chinese).

12. Mou JZ, Meng QM (1982) Sediment delivery ratio as used in the computation of the watershed sediment yield. J Sediment Res (2): 223–230. (in Chinese).

13. Pierson TC (2005) Hyperconcentrated flow-transitional process between water flow and debris flow. In: Jakob M, Hungr O, editors. Debris-flow Hazards and Related Phenomena. Berlin Heidelberg: Springer. pp. 159–201.

14. Xu JX (1999) Erosion caused by hyperconcentrated flow on the Loess Plateau. Catena 36: 1–19.

15. Engelund F, Wan ZH (1984) Instability of hyperconcentrated flow. J Hydraul Eng 110(3): 219–233.

16. Pierson TC, Scott KM (1985) Downstream dilution of a lahar: Transition from debris flow to hyperconcentrated stream flow. Water Resour Res 21: 1511–1524.

17. Hessel R (2006) Consequences of hyperconcentrated flow for process-based soil erosion modelling on the Chinese Loess Plateau. Earth Surf. Processes Landforms 31: 1100–1114.

18. Cai QG, Wang GP, Chen YZ (1998) Processes of soil erosion and sediment yield and the related simulation for small catchments on the Loess Plateau. Beijing: Science Press. (in Chinese).

19. Jing K, Cheng YZ, Li FX (1993) Sediment and environment in the Huanghe River. Beijing: Science Press. (in Chinese).

20. Walling DE (1999) Linking land use, erosion and sediment yields in river basins. Hydrobiologia 410: 223–240.

21. Zhou PH, Wang ZL (1992) Study on Rainstorm causing erosion in the Loess Plateau. Journal of soil and water conservation 6(3): 1–5. (In Chinese).

22. Zheng MG, Yang JS, Qi DL, Sun LY, Cai QG (2012) Flow-sediment relationship as functions of spatial and temporal scales in hilly areas of the Chinese Loess Plateau. Catena 98: 29–40.

23. Wan ZH, Wang ZY (1994) Hyperconcentrated Flow, IAHR monograph series. Balkema: Rotterdam.

24. Ministry of Water Conservancy and Electric Power, PRC (1962) National Standards for Hydrological Survey of China. Beijing: Industry Press. (in Chinese).

25. Zheng MG, Qin F, Yang JS, Cai QG (2013) The spatio-temporal invariability of sediment concentration and the flow-sediment relationship for hilly areas of the Chinese Loess Plateau. Catena 109: 164–176.

26. Walling DE (1988) Erosion and sediment yield research-some recent perspectives. J Hydrol 100(1): 113–141.

27. Goudie A (2004) Encyclopedia of geomorphology (Vol. 2). Psychology Press. PP. 932–933.

28. Wischmeier WH, Smith DD (1978) Predicting rainfall erosion losses: a guide to conservation planning. USDA Handbook 537, Washington, DC.

29. Jiang ZS, Song WJ, Li XY (1983) Studies of the raindrop characteristics for Chinese loess area. Soil and Water Conservation in China (3): 32–36. (in Chinese).

30. Kinnell PIA (2010) Event soil loss, runoff and the Universal Soil Loss Equation family of models: A review. J Hydrol 385: 384–397.

31. Anctil F, Michel C, Perrin C, Andréassian V (2004) A soil moisture index as an auxiliary ANN input for stream flow forecasting. J Hydrol 286: 155–167.

32. Ma T, Li C, Lu Z,Wang B (2014) An effective antecedent precipitation model derived from the power-law relationship between landslide occurrence and rainfall level. Geomorphology 216: 187–192.

33. Heggen RJ (2001) Normalized antecedent precipitation index. J Hydrol Eng 6 (5): 377–381.

34. Li Q (1989) Variation of the decay constant of soil moisture and calculation of runoff yield in loess areas of China. Yellow River (3): 18–23. (in Chinese).

35. Cheng XA (2010) Study on soil erosion and erosion empirical model in hilly loess region on the Loess Plateau-as an example to Cheabagou. M.D. Dissertation. Wuhan: Huazhong Agricultural University, pp. 54–56. (in Chinese).

36. Chien N, Wan ZH (1999) Mechanics of sediment transport. Reston: ASCE Press.

37. Wang L, Shi ZH, Wang J, Fang NF, Wu GL, et al. (2014) Rainfall kinetic energy controlling erosion processes and sediment sorting on steep hillslopes. J Hydrol 512: 168–176.

38. Liu QJ, Shi ZH, Fang NF, Zhu HD, Ai L (2013) Modeling the daily suspended sediment concentration in a hyperconcentrated river on the Loess Plateau, China, using the Wavelet–ANN approach. Geomorphology 186: 181–190.

39. Foster GR, Lambaradi F, Moldenhauer WC (1982) Evaluation of rainfall-runoff erosivity factors for individual storms. Trans AM Soc Agric Eng 25:124–129.

40. Zhang KL (1999) Hydrodynamic characteristics of rill flow on loess slopes. J Sediment Res (1): 55–60. (in Chinese).

41. Schiettecatte W, Verbist K, Gabriels D (2008) Assessment of detachment and sediment transport capacity of runoff by field experiments on a silt loam soil. Earth Surf Processes Landforms 33, 1302–1314.

42. Chen YZ, Jing K, Cai QG (1988) Modern Erosion and Management in Loess Plateau. Beijing: Science Press. (in Chinese).

43. Han P, Ni JR, Wang XK (2003) Experimental study on gravitational erosion process. J Basic SCI Eng (1): 51–56. (in Chinese).

44. Wirtza S, Seegerb M, Riesa JB (2012) Field experiments for understanding and quantification of rill erosion processes. Catena 91: 21–34.

45. Zheng MG, Cai QG, Chen H (2007) Effect of vegetation on runoff-sediment yield relationship at different spatial scales in hilly areas of the Loess Plateau, North China. Acta Ecologica Sinica 27: 3572–3581.

7

Stone Anvil Damage by Wild Bearded Capuchins (*Sapajus libidinosus*) during Pounding Tool Use: A Field Experiment

Michael Haslam[1]*, Raphael Moura Cardoso[2], Elisabetta Visalberghi[3], Dorothy Fragaszy[4]

1 Research Laboratory for Archaeology and the History of Art, University of Oxford, Oxford, United Kingdom, **2** Institute of Psychology, University of São Paulo, São Paulo, Brazil, **3** Istituto di Scienze e Tecnologie della Cognizione, Consiglio Nazionale delle Ricerche, Roma, Italy, **4** Department of Psychology, University of Georgia, Athens, Georgia, United States of America

Abstract

We recorded the damage that wild bearded capuchin monkeys (*Sapajus libidinosus*) caused to a sandstone anvil during pounding stone tool use, in an experimental setting. The anvil was undamaged when set up at the Fazenda Boa Vista (FBV) field laboratory in Piauí, Brazil, and subsequently the monkeys indirectly created a series of pits and destroyed the anvil surface by cracking palm nuts on it. We measured the size and rate of pit formation, and recorded when adult and immature monkeys removed loose material from the anvil surface. We found that new pits were formed with approximately every 10 nuts cracked, (corresponding to an average of 38 strikes with a stone tool), and that adult males were the primary initiators of new pit positions on the anvil. Whole nuts were preferentially placed within pits for cracking, and partially-broken nuts outside the established pits. Visible anvil damage was rapid, occurring within a day of the anvil's introduction to the field laboratory. Destruction of the anvil through use has continued for three years since the experiment, resulting in both a pitted surface and a surrounding archaeological debris field that replicate features seen at natural FBV anvils.

Editor: Roscoe Stanyon, University of Florence, Italy

Funding: MH was supported by the OUP John Fell Fund, an UK Arts and Humanities Research Council Early Career Fellowship and European Research Council Starting Grant #283959 (PRIMARCH). RMC was supported by a CNPq scholarship (process: CNPq 143014/2009-9). EV was supported by the Short-Term Mobility program funded by the CNR. DF was funded by the University of Georgia, the National Geographic Society, and the L.S.B. Leakey Foundation. The funders had no role in study design, data collection and analysis, decision to publish, or preparation of the manuscript.

Competing Interests: The authors have declared that no competing interests exist.

* Email: michael.haslam@rlaha.ox.ac.uk

Introduction

Stone tool use is currently known to be habitual or customary among members of three wild non-human primate species: western chimpanzees (*Pan troglodytes verus*) in West Africa, Burmese long-tailed macaques (*Macaca fascicularis aurea*) in Thailand, and bearded capuchin monkeys (*Sapajus libidinosus*) in Brazil [1–3]. All three species use hand-held stones as pounding tools to access embedded food, and stone surfaces (including cobbles, boulders and outcrops) are among the natural substrates used as anvils by each species to support the pounded item. The use of stone for percussive tasks means that both hammers and anvils survive for a considerable period of time and can be used repeatedly, and the forceful impact associated with percussive strikes can damage the stones through fracture and abrasion. One result is the formation of pits in the surface of both hammers and anvils, which have been noted as an indicator of pounding tool use for both non-human primates and hominins [4–10]. Stone anvil fracture has also been posited as a potential path to the creation of sharp-edged tools through intentional stone flaking, a trait that appears confined to the hominin lineage [11,12].

Anvil damage or use-wear is one of the primary means by which an anvil stone may be distinguished from other naturally occurring stones and outcrops [e.g., 13,14–16]. In order to interpret anvil damage correctly, however, we must first understand the process by which it occurs. In instances where anvil use has not been directly observed and recorded, basic questions such as the duration and intensity of past use can only be addressed through analysis of damage patterns. To help answer such questions, we present here an experimental study of use-wear formation on a stone anvil used by wild bearded capuchins, at the Fazenda Boa Vista (FBV) site in Brazil.

Fazenda Boa Vista

FBV is located in the southern Parnaíba Basin (S 09° 39′ 49.6″, W 45° 25′ 22.5″) in Piauí, Brazil. Details of the local environment are provided in [7]. The climate is seasonally dry, with 1,290 mm of annual rainfall, and 25 mm rainfall during the dry season, May to September [17].

Capuchins at FBV use a variety of stone materials as hammers to crack open resistant palm nuts (89% of tool use episodes), as well as other encased foods [17]. Stone hammers range in weight from hundreds of grams to more than two kilograms [7], with adult individuals producing a maximum kinetic energy of 7–12 J per strike in one experiment [18]. Stone tool use occurs at a median rate of about one episode per 10 h for each tool user, accounting for around 1% of the total time budget for the group [17]. Both

Table 1. Age, sex and body mass of monkeys in the studied group, May 2011.

Individual	Age	Sex	Mass (kg)*
Piaçava	Adult	F	1.98
Teninha	Adult	F	2.18
Chuchu	Adult	F	1.96
Amaralinha	Adult	F	1.63
Dita	Adult	F	2.09
Mansinho	Adult	M	3.30
Teimoso	Adult	M	3.34
Jatoba	Adult	M	3.84
Tomate	Immature	M	1.80
Catu	Immature	M	1.81
Congaceiro	Immature	M	1.83
Pati	Immature	M	1.68
Coco	Immature	M	1.14
Doree	Immature	F	1.37
Pamonha	Immature	F	1.23
Paçoca	Immature	F	1.18
Chani	Infant	F	0.46
Thais	Infant	F	0.42
Presente	Infant	F	0.24

* Body mass was obtained using a voluntary weighing system described in [26].

stone and wood anvils are used at FBV, and most of them are located in the transition zone between large sandstone mesas and flat open woodland [7,19]. FBV stone anvils are formed from relatively soft sandstones and siltstones, which shear from the mesas and can present large, almost horizontal surfaces that average around 1.9 m^2 [7]. Stone anvils used by capuchins around the mesa that includes the experimental area have an average Rx rebound value of 29.6, with anvil hardness across the study area reflecting the hardness of the prevailing sedimentary rock [7].

Almost all stone anvils at FBV are pitted, as a result of repeated abrasion and compression forces when foods are placed on their surface and forcibly pounded. Most anvils have fewer than 10 pits, although anvils with up to several dozen pits are present, and the maximum number of pits recorded for one anvil to date is 83 [7; MH unreported data]. Anvils on the north and east sides of the mesas are used more frequently at FBV than those with southerly or westerly aspects [16].

Methods

Our experiment was conducted in May 2011 with the one group of wild bearded capuchins [20], which at the time consisted of 19 individuals: 3 adult males, 5 adult females, 5 immature males and 6 immature females (Table 1). The capuchins are habituated to human presence, and the experiment took place in part of the monkeys' natural range that had previously been used for other experiments [e.g., 21,22–24]. This 'field laboratory' is a flat area approximately 15 m in diameter with good visibility and several stone and wood anvils, as well as a variety of quartzite and siltstone hammers (Figure 1a). The site is located on the north-east edge of the nearest sandstone mesa.

To investigate the entire use-wear process, we set up an unused sandstone block at the field laboratory for use as an anvil (Figure 2). For ease of reference, the block was designated

'Bigorna Nova' (BN, literally new anvil). This block had a minimally-weathered, flat, undamaged fracture plane measuring 49×31 cm, which was found close to vertical and therefore had not been previously used by the capuchins. The fracture plane formed the horizontal upper surface of the anvil once we transferred the block to the field laboratory, which was 29 m to the north-west of the block's original location (Figure 2a). Once stabilized and leveled at its new location the BN upper surface was 24–25 cm above the slightly uneven ground, and the base of the anvil was 58×40 cm. The BN sandstone was the same colour and composition of other anvils at the field laboratory and the surrounding area.

We recorded all interactions between the monkeys and BN for four days, covering the anvil with a tarpaulin to prevent capuchin access while the researchers were absent. All monkeys' activities with BN were recorded using a digital video camera set up 4.45 m from the anvil, as well as *ad libitum* digital photography (Figure 3). Three quartzite potential hammer stones were initially provided next to the anvil, weighing 0.46 kg, 1.05 kg and 2.08 kg, along with two species of nuts collected from wild plants at FBV: piaçava (*Orbignya* sp.) and tucum (*Astrocaryum campestre*). The former nut is much harder and bigger than the latter [25]. Soon after the experiment began, we restricted hammer use at BN to the 1.05 kg stone, to better control this variable. The capuchins were free to use the stones and nuts provided, or to bring additional nuts to the anvil, and they could approach and use the anvil from all sides. Additional palm nuts were provided to facilitate use of BN, and other anvils and hammers were always available in the field laboratory away from the BN study.

Each time that the monkeys used a stone hammer to strike a nut they had placed on the BN anvil, we used the video record to note the position of that strike on the anvil upper surface. At a minimum of every 20 strikes, which usually involved multiple nuts

Figure 1. The FBV field laboratory. (a) The experimental area beside a steep mesa, with sandstone (SS) and wood (W) anvils, and the Bigorna Nova (BN). (b) A sandstone anvil at the field laboratory in 2003, and (c) the same anvil in 2014, showing the erosive effect of capuchin pounding.

Figure 2. Setting up the Bigorna Nova (BN). (a) The original location and position of the stone SE of the field laboratory, with a 10 cm scale on the tilted face; (b) BN in position at the start of the experiment; the upper surface has chalked crosses every 5 cm to aid in determining strike positions.

being cracked, we recorded the presence of macroscopically-observed pits in the anvil surface. We recorded pit location, and maximum dimensions of length, width (perpendicular to length) and depth. We recorded the latter by placing a plasticine ball within the pit, then laying a ruler across the top of the pit to compress the plasticine, and measuring the resulting thickness with calipers. For each strike, we also recorded the individual monkey, the nut type, and whether the nut was whole or partially broken (where this was visible on the video, or mentioned in the video by the experimenter).

We also recorded whether or not the monkey removed loose material from an existing pit (making the accessible part of the pit deeper), and/or swept loose anvil and nut debris from the surface of the anvil onto the surrounding ground. Both behaviors were labeled cleaning actions. Cleaning removed fragments of the anvil, including pieces from mm to cm in size, that had detached from the main anvil body but had remained *in situ* either within or around surface pits (Fig. 3). Previous observations indicate that this material would otherwise form a barrier between the nut and anvil, and its removal accelerates pit formation and overall damage by exposing the underlying anvil surface.

Following one week of complete monitoring, BN was left uncovered and monkey use was no longer continually recorded. For the subsequent three years on an annual basis we recorded the anvil cross section in two perpendicular planes to assess overall changes in height and surface shape, and photographed the anvil.

To provide controlled comparative data, MH created use-wear pits in a separate sandstone anvil by repeatedly dropping a 1.05 kg

quartzite hammer onto positioned piaçava and tucum nuts. The hammer weight was chosen based on the reported average weight of hammers at Boa Vista of 1.096 kg [7]. The hammer stone was

Figure 3. The Bigorna Nova experiment. (a) Bigorna Nova (BN) experiment in progress, with RMC operating the video camera; (b) BN surface damage during the experiment, with tucum shells and quartzite hammer stone, scale is 10 cm; (c) BN at the end of the period of continuous monitoring, note the pitted surface and surrounding debris, scale is 10 cm; (d) BN 16 months after the experiment, surrounding stone and nut debris is extensive, scale is 5 cm.

Table 2. Summary data for the BN experiment.

Individual	Strikes	Nuts	% Cleaning*	# pits started
Jatoba	169	49	39.0	9
Mansinho	119	32	36.6	1
Teimoso	159	36	4.9	3
Piaçava	130	39	4.9	1
Dita	50	14	7.3	0
Chuchu	73	4	4.9	0
Immature	99	54	2.4	0

*% of all cleaning events that were performed by this individual.

dropped from a height of 33 cm each time; this value was based on averaged data for maximum vertical hammer height during nut-cracking by four Boa Vista capuchins [18]. Capuchins may add force to each downward strike, which was absent in this experiment to err on the conservative side. Each drop fell squarely on the nut, and was counted as a strike, with the dimensions (width, length, depth) of the resulting pits measured every 10 strikes up to a total of 100 strikes. The anvil used in the stone drop experiment was collected from close to the BN original location. A further aim of this study was to assess variation in pit formation between tucum and piaçava that may allow for discrimination of these nuts via pit data from anvils at FBV. Specifically, we hypothesized that the rounded tucum would produce pits with a greater depth, relative to pit length and width, than the broader piaçava nuts. Pit measurements were taken as for the BN.

We calculated the odds ratio to assess whether whole or partial nuts were preferentially placed within pits.

Ethics statement

Permission to work in Brazil was granted to EV and DF by Instituto Brasileiro do Meio Ambiente e dos Recursos Naturais Renováveis (IBAMA) and Conselho Nacional de Desenvolvimento Científico e Tecnológico (CNPq). The study was conducted on private land, owned by the family of Marino Gomes Oliveira. This research was approved by the IACUC of the University of Georgia (A2010 04-067 and A2013 03-001) and complied with all institutional guidelines for the ethical participation of non-human animals in research.

Results

During the period of continual observation, a total of six adults (3 males, 3 females) used tools on the BN anvil, along with a number of immature individuals (Table 2). The latter were not identified to individual, and were analysed collectively. We recorded strikes on 67 tucum (n = 320 strikes) and 161 piaçava nuts (n = 479 strikes), for a total of 799 strikes. Note that these values do not necessarily reflect successful nut-cracking, and so should not be considered measures of efficiency. The capuchins created 14 identifiable pits during this period, with 579 strikes located within pits and the remainder on other parts of the anvil surface.

The capuchins performed 341 strikes in which the nut was placed into a pit and for which the whole or partial nature of the nut could be ascertained; 261 of these strikes were on whole nuts, and 80 on partial nuts. Of the 134 strikes in which the nut was placed elsewhere than in a pit, and for which we could ascertain whether the nut was whole or partial, 21 were on whole nuts, and

113 on partial nuts. We calculated the odds ratio of a whole nut being placed in a pit as 17.5 (χ^2 p<0.001), indicating that monkeys preferentially place whole nuts in pits, and partial nuts outside these depressions.

Pit size data (Figure 4) indicate that pit width increases linearly as length increases ($r^2 = 0.857$), while depth also increases but at a slower rate (Figure 4a). Pit depths reach a plateau typically less than 25 mm, which reflects the fact that beyond that depth the monkeys often strike the surrounding stone surface rather than the nut, resulting in the abrasion and local fracture of the anvil surface. On occasion, this can actually result in pit depth decreasing slightly as the other dimensions increase, even though the overall trend is towards an increase in all three dimensions. Because of this process, the ratio of length to depth decreases over time (Figure 4b), producing larger but shallower pits. For comparison, and to ensure that data from BN reflected natural occurrences, we also measured a sample of pits on FBV anvils that were surveyed over several years and reported in [16] (Figure 4a). The surveyed anvil pits showed similar sizes to the BN data, although they tend to be slightly wider (by a few mm) for pit lengths below 50 mm.

Most of the cleaning behavior was performed by adult male monkeys (33 cleaning events from 117 nuts) (Table 2). Adult females cleaned the anvil seven times (from 57 nuts) and immature individuals once (from 51 nuts). The current and former alpha males (Jatoba and Mansinho, respectively) were most active in this process, collectively cleaning the anvil on 38% of their visits. From these data, the main agents for accelerating anvil damage through cleaning are dominant males. Adult females play a minor role, and immature individuals very rarely engage in cleaning behavior. Monkeys cleaned almost twice as often when cracking tucum rather than piaçava nuts (27% to 14%).

Initiation of a new pit occurred when a monkey placed a nut outside of the already established pits, and cracked nuts repeatedly in that location until a macroscopically visible pit formed. Of the 14 pits created during the experiment, nine were initiated by the alpha male (Jatoba), one by the former alpha male (Mansinho), three by another subordinate male (Teimoso), and one by the alpha female (Piaçava) (Table 2). Immature monkeys never started a new pit. On average, a new pit was initiated after 10 nuts were cracked, corresponding to an average of 38.5 strikes per new pit. We also looked at whether individuals would preferentially re-use the pit that had been used by the previous monkey at the anvil, and found that the capuchins re-used the same pit 40% of the time, with similar frequency seen in this behavior between immature monkeys, adult females and adult males (44%, 46% and 37% respectively; total n = 144 events).

From May 2011 to May 2014, the BN anvil continued to be used opportunistically by the FBV capuchins (Figures 3d, 5).

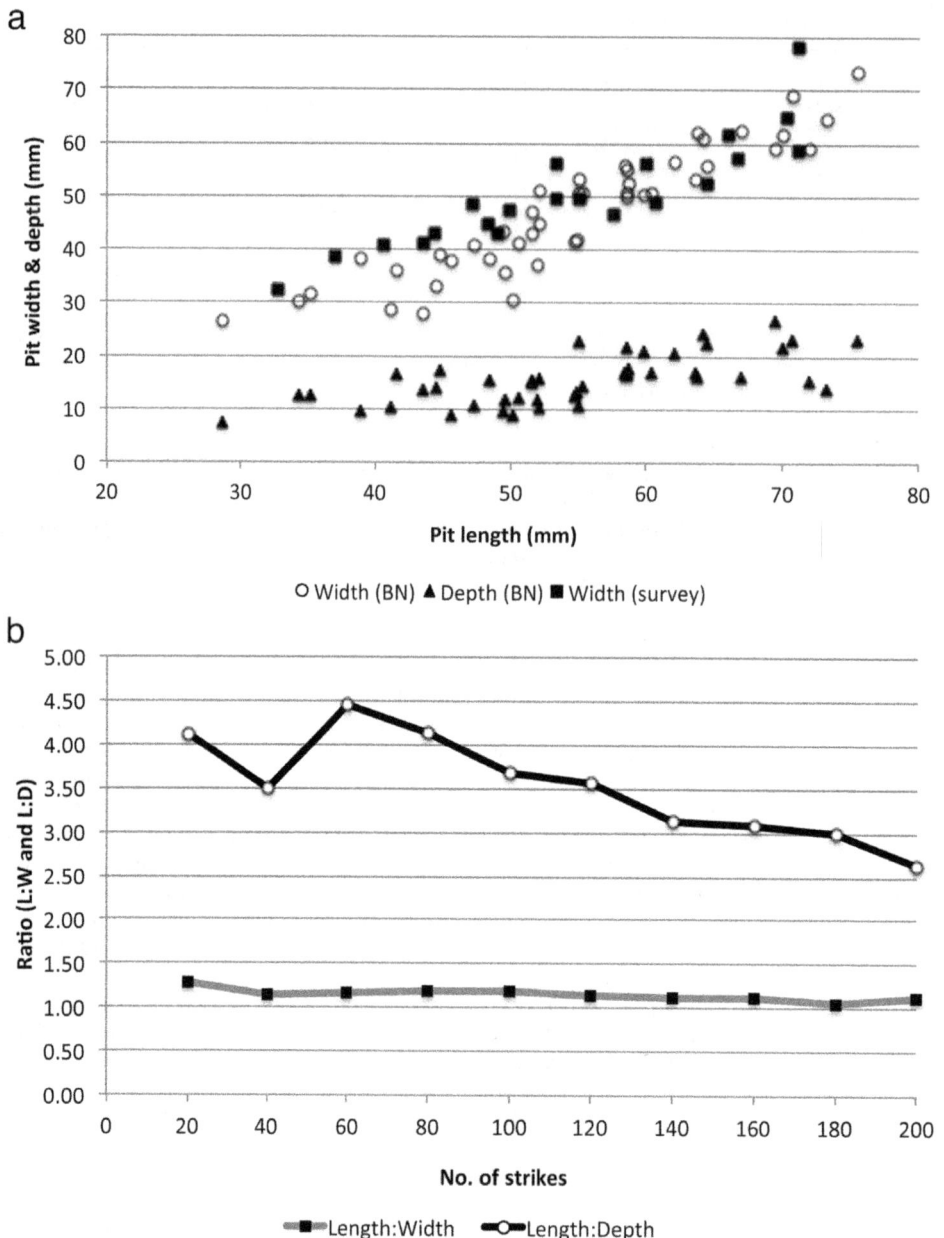

Figure 4. BN pit measurements, with anvil survey data for comparison: (a) BN pit width and depth relative to length, and surveyed anvil pit widths relative to length; (b) ratios of pit length:width and length:depth formed on the BN surface during the experiment.

Through pounding activities, its height decreased relative to the original anvil surface by at minimum of 2.6 cm, and a maximum of 13 cm (i.e., more than half the original anvil height). The erosion of the anvil upper portion left a considerable debris field in the immediate vicinity, chiefly within 50 cm of the anvil base. During the initial one-week period of observation, the anvil decreased in height by 0.4–4.8 cm. This was 41.7% of the total mass lost, as estimated by averaging loss at points taken 5 cm apart in two cross-sections, indicating that capuchin use of the anvil during that period was considerably more intensive than in the subsequent three years. We suggest that the latter period is more representative of normal use patterns, although the change in intensity affects only the overall rate of wear, not the mechanisms involved. The anvil upper surface remained pitted throughout the three-year period.

Results of the stone drop experiment performed by a human are very similar to those from the BN study involving capuchins. Three pits were created while processing piaçava, and these show consistent formation rates (Figure 6a). One piaçava nut, measuring 61 ×40 mm, was used for this experiment (it did not crack); the final pit sizes exceed these values, likely resulting from slight nut movement at the moment of impact. The tucum nuts fractured after an average of 33 strikes for each nut, which is a much lower fracture rate than that typically seen among the monkeys [23], supporting the conservative assumption of lower energy input from the stone drop experiment. One aspect that differed between the BN and stone drop studies is that the stone drop protocol did not also involve striking nuts placed beside the pits, which explains the perhaps unrealistic greater maximum pit depth (close to 40 mm) and the high number of strikes to crack tucum attained in

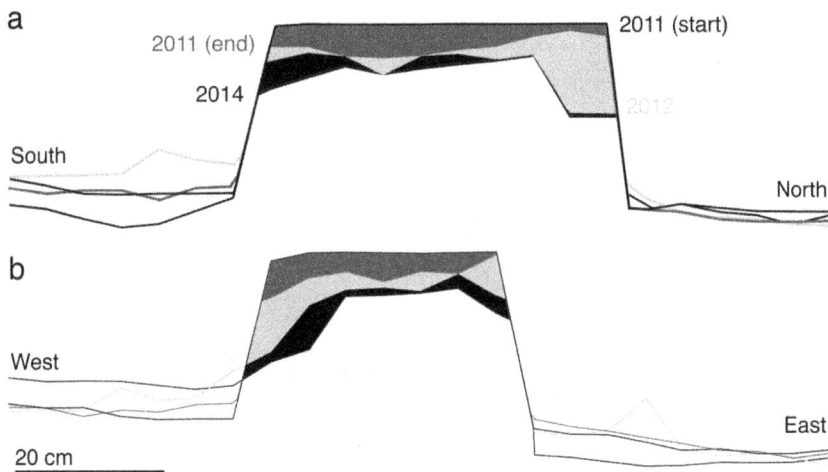

Figure 5. Schematic BN cross-sections, 2011–2014: (a) South to North; and (b) West to East. The uppermost level in each cross-section is the anvil shape at the start of the experiment (2011 start) and subsequent levels were taken at the end of the initial experimental period (2011 end), in September 2012, and in May 2014.

the latter study. Placing nuts beside the pits would likely have abraded that surface and therefore slightly increased the length and width of the pits already present, while decreasing their relative depth. Although they did not attain the same maximum size in this study, pits resulting from tucum processing were not distinguishable via measurement ratios from those created by piaçava nuts (Figure 6b).

Discussion

Damage clearly identifiable as resulting from capuchin tool use occurred within a day of establishing the BN at FBV. This damage included rapid formation of the distinct pits seen at most anvils used by wild capuchins in the local area, along with break-up and removal of the bedded sandstone surface. Initiation of new pits was dominated by the activities of the alpha male, potentially as a result of his greater strength and body mass [26].

On the small BN anvil, the maintenance of more than five or six pits appeared difficult, as pits would start to join together or be lost through abrasion and destruction of the surrounding surface. Larger and harder anvils than the BN stone could no doubt sustain a greater number of pits, as evidenced by the anvil diversity seen across the FBV landscape [7,16]. The published average density of pits for stone anvils measured at FBV is 6.6 pits/m^2 (anvils average 7.8 pits and 1.89 m^2; range 0–43 pits and 0–43.4 pits/m^2), and anvils around the same mesa as the field laboratory average 9.5 pits/m^2 (averages of 9.8 pits and 1.29 m^2) [7]. BN had a maximum pit density of 39.5 pits/m^2 (6 pits and 0.15 m^2), at the higher end of previously recorded values, but within the natural range. The two highest pit densities published at FBV occur at the same mesa as the field laboratory (MM23: 31.1 pits/m^2; MM30: 43.4 pits/m^2), as does a previously unpublished and heavily-pitted anvil with a density of 16.9 pits/m^2 (83 pits and 4.89 m^2). These elevated densities may relate either to intensity of use or the relative softness of the sandstone in this part of the FBV site.

New pits were formed on the BN anvil with approximately every 10 nuts cracked, or 38 strikes. While these data provide an initial guide for interpreting the abundance of pits at anvils formed from similar sandstone at FBV, we caution against uncritically applying this metric to other types of stone surface (e.g. harder sandstones), to other capuchin sites, or to other primate sites.

Additional factors that may mediate the formation of pits must be considered, such as weathering rates, the processing of different food types to those seen here, and the size of the anvil (the closeness of pits to the edge of small anvils most probably affects the likelihood of edge fractures affecting or removing those pits). These data also do not apply to the formation of use-damage on wooden anvils, which require separate study. The main criterion used here to measure anvil damage, percussion pitting, will be less useful in areas with more resistant anvils [27], or in circumstances where particular environments alter anvil characteristics more readily (e.g., the inter-tidal zones exploited by tool-using long-tailed macaques in Thailand [8,28]).

The stone hammer material may also influence the rate of anvil pit formation. Specifically, use of softer rocks as hammers likely results in absorption of more of the force of a blow into the hammer itself, deforming or even breaking the hammer rather than the nut and underlying substrate. The dominant use of quartzite in this experiment precludes our testing this hypothesis, but we would expect the use of softer hammers to lengthen the time required for pit formation, rather than eliminating pit formation altogether.

Our data from the initial period of observation allow us assess the approximate rate of BN use over the subsequent three years. The initial processing of 228 nuts produced 42% of the estimated mass lost between 2011 and 2014, so we estimate that the monkeys have processed around 550 nuts in total on the BN anvil. Total strikes are estimated at over 1900 in that time. Under natural conditions (outside the initial study period), we therefore estimate 106 nuts and 373 strikes per year at BN. It is possible that BN became less appealing following its initial use as an anvil, due to its reduced size and damage, but we have no data to test this hypothesis.

The debris field of sandstone anvil fragments and broken nut shells surrounding BN is distinctive, and composed of durable materials that constitute an archaeological signature [16]. Similar debris fields surround other anvils at FBV, and reduction in anvil volume and height has been qualitatively observed among the anvils at the field laboratory since commencement of research in 2003 (Figure 1b,c). At BN, sediment levels adjacent to the anvil have seen both decreases and increases, with the most notable change being the accretion of around 4 cm sediment to the anvil's

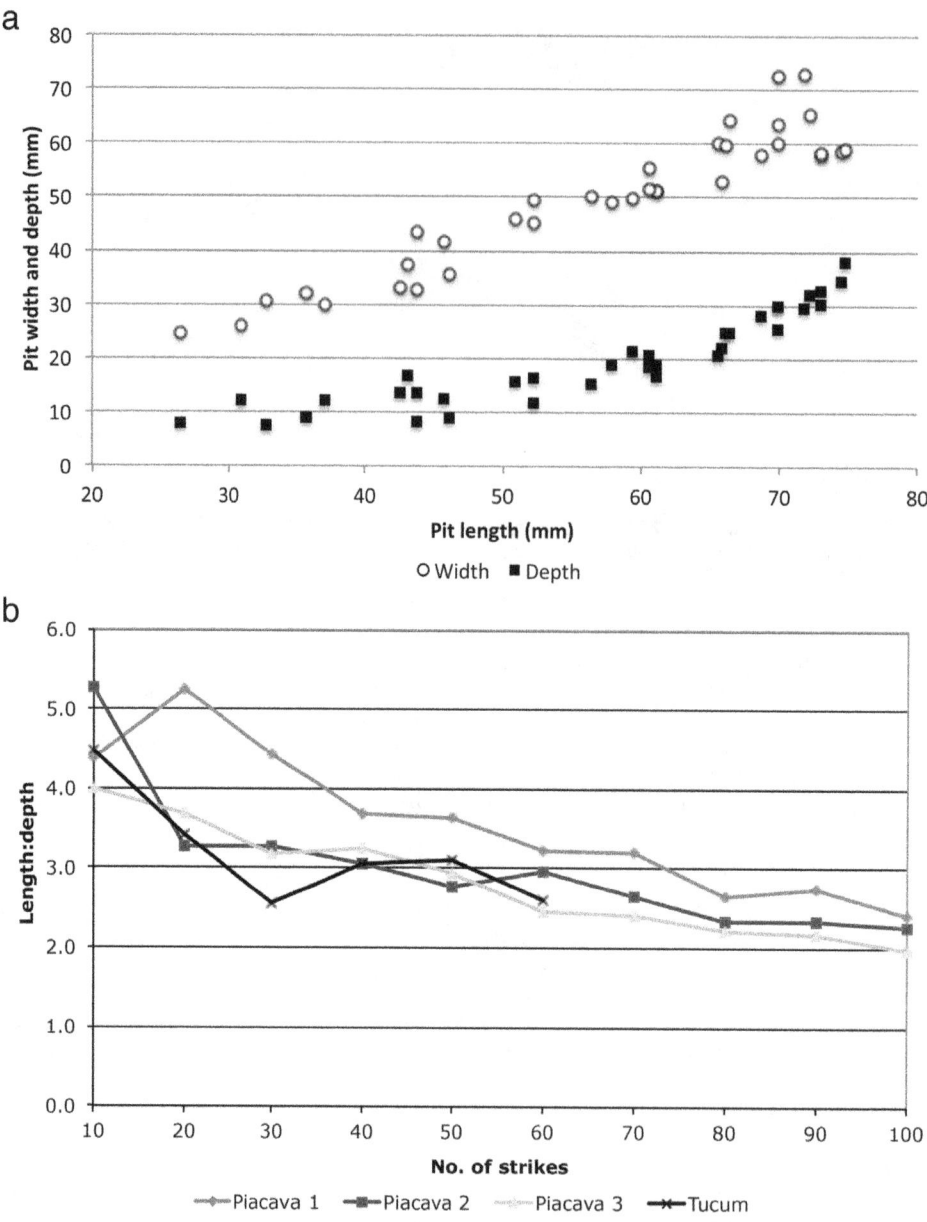

Figure 6. Results of the stone-drop use-wear experiment: (a) pit width and depth relative to length (piaçava and tucum results combined); (b) ratio of pit length:depth for three piaçava nuts, and one tucum nut (the tucum nut has sixty strikes, the others 100 strikes).

south and west. South-west is the direction of the nearby mesa, which is the main sediment source at the field laboratory.

Whole nuts were preferentially placed within existing pits, while partially cracked nuts were more often placed outside of pits. We propose that this behavior may result from (a) the fragmented partial nuts being smaller and therefore less accessible if placed within pits, (b) the rounded whole nuts being more prone to move during striking, and be lost, if not placed within a controlling pit [see 24,26], and/or (c) partially cracked nuts being more stable when placed on a flat surface, because they are typically placed and struck with a flat facet downwards once initially broken. Neither adults nor immature monkeys consistently used existing pits as opposed to other exposed portions of the anvil surface or the most recently used pit. If pits form part of a constructed material niche [29], then we could say that the BN experiment

influenced the behavior of immature and adult individuals to a similar extent in this experiment.

Adult males primarily engaged in cleaning, and in particular the current and former alpha males, although monkeys of all ages and sexes conducted this behavior. Cleaning activities continually exposed the surface of the anvil directly to impact with the processed nuts, and assisted in the concurrent build-up of sandstone and nut debris surrounding the anvil. These activities appear important therefore for the archaeological recognition of anvils as activity centers, both as surface evidence of tool use and in the formation of a subsurface material record.

Previous studies of pit formation and interpretation from primatological and paleoanthropological contexts have typically not considered the impact of anvil maintenance on wear rates and types [4–7,9], which the capuchin data indicate may be a useful complement to existing research. Use-wear research also generally does not consider how demographic factors may influence damage patterns or rates, but the influence of large male capuchins as opposed to females or immature monkeys at FBV marks this as another area worthy of further investigation. For example, cleaning by alpha males may be an example of a socially-partitioned role [e.g., 30], or it may be that extra force employed by some males generates additional debris, requiring more frequent cleaning to maintain a stable nut position. The fact that one lower-ranking adult male engaged in cleaning only at low levels, similar to the adult females, does not allow us to distinguish between these options at present.

Finally, we note that some archaeological sites in northeast Brazil (e.g., in Serra da Capivara National Park) contain pitted sandstone surfaces likely resulting from human activity (MH, personal observation). These may be the outcome of unintentional anthropogenic use-damage, or they may be cupules resulting from deliberate human manufacture. Data on both the formation and shape of anvil use-damage from capuchin monkey nut-cracking activities are required to properly assess the origins of cupules or use-damaged pits found at archaeological sites in this region.

Acknowledgments

Our thanks to the Oliviera family for logistical assistance and permission to work at Fazenda Boa Vista.

Author Contributions

Conceived and designed the experiments: MH RMC EV DF. Performed the experiments: MH RMC. Analyzed the data: MH RMC DF. Wrote the paper: MH RMC EV DF.

References

1. Malaivijitnond S, Lekprayoon C, Tandavanittj N, Panha S, Cheewatham C, et al. (2007) Stone-tool usage by Thai long-tailed macaques (Macaca fascicularis). Am J Primatol 69: 227–233.
2. McGrew WC (1992) Chimpanzee Material Culture: Implications for Human Evolution. Cambridge: Cambridge University Press.
3. Ottoni E, Izar P (2008) Capuchin monkey tool use: overview and implications. Evol Anthropol 17: 171–178.
4. Boesch C, Boesch H (1982) Optimization of nut-cracking with natural hammers by wild chimpanzees. Behaviour 83: 265–286.
5. Mercader J, Panger M, Boesch C (2002) Excavation of a chimpanzee stone tool site in the African rainforest. Science 296: 1452–1455.
6. Goren-Inbar N, Sharon G, Melamed Y, Kislev M (2002) Nuts, nut cracking, and pitted stones at Gesher Benot Ya'aqov, Israel. Proc Natl Acad Sci USA 99: 2455–2460.
7. Visalberghi E, Fragaszy D, Ottoni E, Izar P, de Oliveira M, et al. (2007) Characteristics of hammer stones and anvils used by wild bearded capuchin monkeys (Cebus libidinosus) to crack open palm nuts. Am J Phys Anthropol 132: 426–444.
8. Haslam M, Gumert M, Biro D, Carvalho S, Malaivijitnond S (2013) Use-wear patterns on wild macaque stone tools reveal their behavioural history. PLoS One 8: e72872.
9. Carvalho S, Cunha E, Sousa C, Matsuzawa T (2008) Chaînes opératoires and resource-exploitation strategies in chimpanzee (Pan troglodytes) nut cracking. J Hum Evol 55: 148–163
10. Haslam M (2012) Towards a prehistory of primates. Antiquity 86: 299–315.
11. Marchant LF, McGrew WC (2005) Percussive technology: chimpanzee baobab smashing and the evolutionary modeling of hominid knapping. In: Roux V, Bril B, editors. Stone Knapping: The Necessary Conditions for a Uniquely Hominin Behaviour Cambridge: McDonald Institute for Archaeological Research. pp.341–350.
12. Haslam M, Hernandez-Aguilar A, Ling V, Carvalho S, de la Torre I, et al. (2009) Primate archaeology. Nature 460: 339–344.
13. Canale GR, Guidorizzi CE, Kierulff MCM, Gatto CAFR (2009) First record of tool use by wild populations of the yellow-breasted capuchin monkey (Cebus xanthosternos) and new records for the bearded capuchin (Cebus libidinosus). Am J Primatol 71: 366–372.
14. Ferreira R, Emidio R, Jerusalinsky L (2010) Three stones for three seeds: natural occurrence of selective tool use by capuchins (Cebus libidinosus) based on an analysis of the weight of stones found at nutting sites. Am J Primatol 72: 270–275.
15. de la Torre I, Benito-Calvo A, Arroyo A, Zupancich A, Proffitt T (2013) Experimental protocols for the study of battered stone anvils from Olduvai Gorge (Tanzania). J Archaeol Sci 40: 313–332.
16. Visalberghi E, Haslam M, Spagnoletti N, Fragaszy D (2013) Use of stone hammer tools and anvils by bearded capuchin monkeys over time and space: construction of an archeological record of tool use. J Archaeol Sci 40: 3222–3232.
17. Spagnoletti N, Visalberghi E, Verderane M, Ottoni E, Izar P, et al. (2012) Stone tool use in wild bearded capuchin monkeys, Cebus libidinosus. Is it a strategy to overcome food scarcity? Anim Behav 83: 1285–1294.
18. Liu Q, Simpson K, Izar P, Ottoni E, Visalberghi E, et al. (2009) Kinematics and energetics of nut-cracking in wild capuchin monkeys (Cebus libidinosus) in Piauí, Brazil. Am J Phys Anthropol 138: 210–220.
19. Visalberghi E, Spagnoletti N, Ramos da Silva E, de Andrade F, Ottoni E, et al. (2009) Distribution of potential suitable hammers and transport of hammer tools and nuts by wild capuchin monkeys. Primates 50: 95–104.
20. Spagnoletti N, Visalberghi E, Ottoni E, Izar P, Fragaszy D (2011) Stone tool use by adult wild bearded capuchin monkeys (Cebus libidinosus). Frequency, efficiency and tool selectivity. J Hum Evol 61: 97–107.
21. Visalberghi E, Addessi E, Truppa V, Spagnoletti N, Ottoni E, et al. (2009) Selection of effective stone tools by wild bearded capuchin monkeys. Curr Biol 19: 213–217.
22. Massaro L, Liu Q, Visalberghi E, Fragaszy D (2012) Wild bearded capuchin (Sapajus libidinosus) select hammer tools on the basis of both stone mass and distance from the anvil. Anim Cogn 15: 1065–1074.
23. Fragaszy D, Greenberg R, Visalberghi E, Ottoni E, Izar P, et al. (2010) How wild bearded capuchin monkeys select stones and nuts to minimize the number of strikes per nut cracked. Anim Behav 80: 205–214.
24. Liu Q, Fragaszy D, Wright B, Wright K, Izar P, et al. (2011) Wild bearded capuchin monkeys (Cebus libidinosus) place nuts in anvils selectively. Anim Behav 81: 297–305.
25. Visalberghi E, Sabbatini G, Spagnoletti N, Andrade F, Ottoni E, et al. (2008) Physical properties of palm fruits processed with tools by wild bearded capuchins (Cebus libidinosus). Am J Primatol 70: 884–891.
26. Fragaszy D, Pickering T, Liu Q, Izar P, Ottoni E, et al. (2010) Bearded capuchin monkeys' and a human's efficiency at cracking palm nuts with stone tools: field experiments. Anim Behav 79: 321–332.
27. Emidio R, Ferreira R (2012) Energetic payoff of tool use for capuchin monkeys in the caatinga: variation by season and habitat type. Am J Primatol 74: 332–343.
28. Gumert M, Kluck M, Malaivijitnond S (2009) The physical characteristics and usage patterns of stone axe and pounding hammers used by long-tailed macaques in the Andaman Sea region of Thailand. Am J Primatol 71: 594–608.
29. Fragaszy D, Biro D, Eshchar Y, Humle T, Izar P, et al. (2013) The fourth dimension of tool use: temporally enduring artefacts aid primates learning to use tools. Phil Trans R Soc B 368: doi:10.1098/rstb.2012.0410
30. Hockings K, Anderson J, Matsuzawa T (2006) Road crossing in chimpanzees: a risky business. Curr Biol 16: R668–R670.

8

High Phylogenetic Diversity of Glycosyl Hydrolase Family 10 and 11 Xylanases in the Sediment of Lake Dabusu in China

Guozeng Wang[1,2◑], Xiaoyun Huang[1◑], Tzi Bun Ng[3], Juan Lin[1,2]*, Xiu Yun Ye[1,2]*

1 College of Biological Science and Technology, Fuzhou University, Fuzhou 350108, P.R. China, **2** National Engineering Laboratory for High-efficiency Enzyme Expression, Fuzhou 350002, P. R. China, **3** School of Biomedical Sciences, Faculty of Medicine, The Chinese University of Hong Kong, Shatin, New Territories, Hong Kong, China

Abstract

Soda lakes are one of the most stable naturally occurring alkaline and saline environments, which harbor abundant microorganisms with diverse functions. In this study, culture-independent molecular methods were used to explore the genetic diversity of glycoside hydrolase (GH) family 10 and GH11 xylanases in Lake Dabusu, a soda lake with a pH value of 10.2 and salinity of 10.1%. A total of 671 xylanase gene fragments were obtained, representing 78 distinct GH10 and 28 GH11 gene fragments respectively, with most of them having low homology with known sequences. Phylogenetic analysis revealed that the GH10 xylanase sequences mainly belonged to Bacteroidetes, Proteobacteria, Actinobacteria, Firmicutes and Verrucomicrobia, while the GH11 sequences mainly consisted of Actinobacteria, Firmicutes and Fungi. A full-length GH10 xylanase gene (*xynAS10-66*) was directly cloned and expressed in *Escherichia coli*, and the recombinant enzymes showed high activity at alkaline pH. These results suggest that xylanase gene diversity within Lake Dabusu is high and that most of the identified genes might be novel, indicating great potential for applications in industry and agriculture.

Editor: Chunxian Chen, USDA/ARS, United States of America

Funding: This research was supported by the National Natural Science Foundation of China (31301406), the Oceanic Public Welfare Industry Special Research Project of China (201305015) and the Priming Scientific Research Foundation of Fuzhou University (XRC-1326). The funders had no role in study design, data collection and analysis, decision to publish, or preparation of the manuscript.

Competing Interests: The authors have declared that no competing interests exist.

* Email: ljuan@fzu.edu.cn (JL); xiuyunye@fzu.edu.cn (XYY)

◑ These authors contributed equally to this work.

Introduction

Xylan is the major hemicellulose component of the plant cell wall, which is the second most abundant polysaccharide on earth after cellulose [1]. Xylan is composed of a homopolymeric backbone chain of β-1, 4-linked xylopyranose units with substituted side chains at different positions, and its complete hydrolysis requires a group of enzymes including endo-1,4-β-D-xylanase, β-D-xylosidase, α-D-glucuronidase, α-L-arabinofuranosidase, acetyl xylanesterase, and arylesterase [2,3]. Endo-1,4-β-D-xylanase (EC 3.2.1.8) is a crucial component because it catalyzes the hydrolysis of xylan to short xylooligosaccharides of varying lengths [2,3]. Xylanases have been classified into glycosyl hydrolase (GH) families (http://www.cazy.org/fam/acc_GH.html; [4]) 5, 7, 8, 10, 11 and 43 [3] based on sequence similarities of the catalytic domain. Among these, GH10 and 11 xylanases are the most abundant, which have distinct three-dimensional structures [5], mechanisms of action [6] and substrate specificity to xylan.

Although xylanases are produced by diverse organisms, including bacteria, algae, fungi, protozoans, gastropods and anthropods [3], microbial xylanases are the focus of intense research owing to their significant application in various industrial processes. Specifically, they are used to improve the digestibility of animal feedstock, enhance filterability in brewing, increase dough volume and improve the textural and staling properties of bread, and for fruit juice and wine clarification, the bioconversion of lignocellulosic materials into fermentative products, and facilitation of the release of lignin from the pulp [1,2,7,8]. Many microbial xylanases have been purified and characterized, and the genes encoding xylanases have been cloned and expressed in heterologous systems, and the structures of a number of enzymes have been determined [3,8].

Most microbial xylanases reported to date have acidic or neutral pH optima. Alkaline xylanases are of great interest because xylan is more readily soluble under alkaline pH than neutral pH. Additionally, the application of alkaline xylanases in the paper and pulp industry can reduce the use of chlorine, which is very attractive from an economical and technical point of view [9,10]. Consequently, studies are continually conducted in attempts to identify novel xylanases with potential applications in the pulp and paper industries. Although alkaline xylanases have been reported from microorganisms isolated from nonalkaline environments [11,12], xylanases produced by microorganisms isolated from extreme alkaline environments have attracted increasing attention. These organisms are exposed to hostile environments, resulting in their evolution and accumulation of a variety of adaptive features for activity and stability under these conditions [13–15].

Natural alkaline environments are not common, and soda lakes and deserts are the most stable naturally occurring alkaline ecosystems on earth [16,17]. Soda lakes have high carbonate alkalinity, a pH of 9 to 11, and moderate to extremely high salinity [17]. Despite their extreme environmental conditions, soda lakes harbor extremely productive microbial communities. A number of alkaliphilic microorganisms that play key metabolic roles in soda lakes have previously been studied, and many novel alkaliphilic microorganisms have been isolated from these unique ecosystems [18]. Molecular biological techniques that do not depend on culture have revealed a high diversity and novelty of microbial communities in soda lake environments [19–21].

Although the microbial diversity of soda lakes has been studied extensively, the functional diversity of genes in such systems has not. Lin et al. analyzed the methane monooxygenase genes in Mono Lake and suggested that increased methane oxidation activity was correlated with changes in methanotroph community structure [22]. Kovaleva et al. explored the diversity of RuBisCO and ATP citrate lyase genes in the sediments from six soda lakes of the Kulunda Steppe [23]. However, few studies have focused on the functional gene diversity of plant material dehygrolysis in soda lake environments. We explored the genetic diversity of GH10 xylanase in diverse soil environments using culture-independent molecular methods and found that pH was one of the most important factors influencing the xylan degrading microbial community [24].

In this study, we focused on the genetic diversity of GH10 and GH11 xylanases in the natural Lake Dabusu, an alkaline (pH 10.2) and saline (101 g/liter) lake situated in northeastern China. Partial xylanase genes were amplified directly from the metagenomic DNA of the soda lake sediment. Sequence analysis showed that most xylanase gene fragments had low homologies to known xylanases, suggesting a large number of uncharacterized xylanase genes in the soda lake. Additionally, phylogenetic diversity analysis suggested a surprising diversity of xylanase genes in this harsh environment. A novel GH10 full-length xylanase gene was directly cloned from the metagenomic DNA and expressed in *Escherichia coli*. The recombinant xylanases showed high activity at alkaline pH. Our study provides new insight into the genetic diversity and distribution of microbial xylanases in the soda lake ecosystem, which will help us understand their roles in this microenvironment.

Materials and Methods

Ethics statement

We declare that no living animals were used in this research. No specific permissions were required for the described field studies. The sediment sample was collected from the location that is not privately-owned or protected in any way. The field studies did not involve endangered or protected species because this study only concentrated on the sediment sample.

Sample site

Lake Dabusu is located in the southwestern part of the Qian'an County (Jilin Province, China), in the center of the depressed belt of Songliao Basin. Lake Dabusu is a closed inland alkaline lake located 122 meters above sea level, with an average water depth of about 0.90 meters and an area of approximately 38 km^2 in the rainy season. Because of strong evaporation and a lack of outflow within the closed basin, alkaline materials accumulate in the lake water. Lake Dabusu is a typical soda lake with salinity of 62.34 g L^{-1} to 347.34 g L^{-1} and a pH of 10 to 11 depending on season [25].

Table 1. Physicochemical characters of the alkaline saline lake sediment and xylanase fragment sequences obtained.

Location	T (°C)	pH	Total organic carbon (mg/g)	Total nitrogen (mg/g)	C/N ratio	GH family	Clones sequenced	Sequences recovered	OTUs[a]
44 48 20 N 123 40 32 E	20	10.2	1.10	0.35	3.14	GH10	550	467	78
						GH11	250	204	28

[a]5% dissimilarity as cutoff.

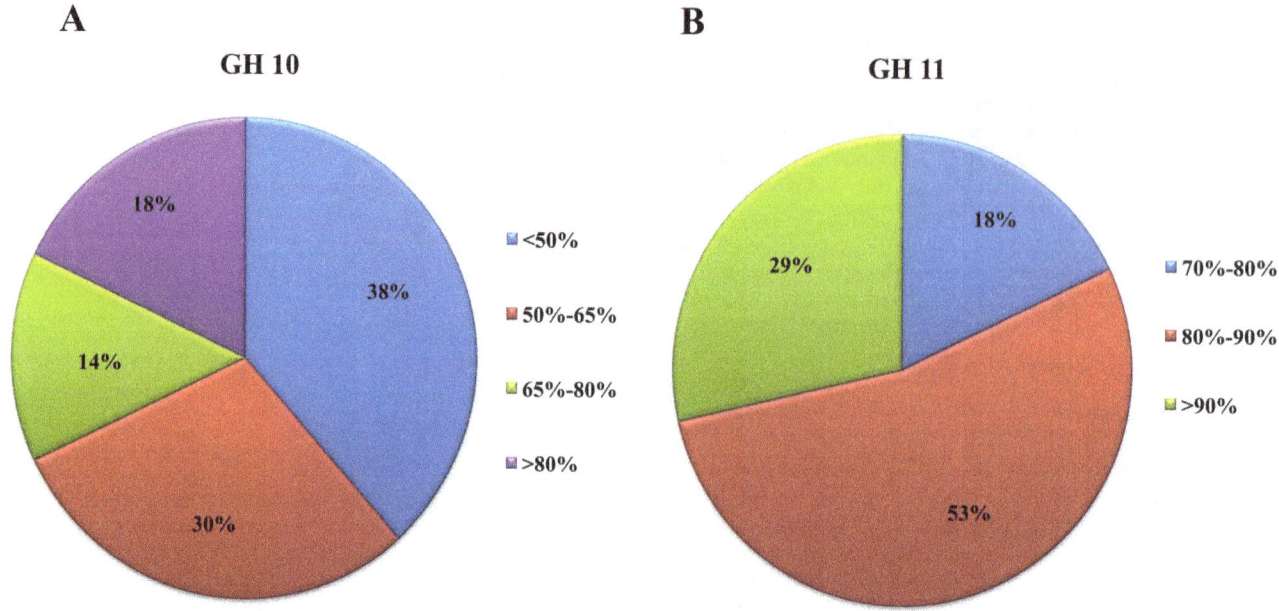

Figure 1. Amino acid sequence homologies of GH10 and GH11 xylanase gene fragments from sediment metagenomic DNA to known xylanases. Each sequence was analyzed by NCBI BLASTp (version 2.2.29) against the GenBank nr database. An E-score (expect value) cutoff of 10^{-10} (default) was applied and the top BLASTp hit to the known xylanases was collected.

Sample collection and DNA extraction

On June 12, 2013, superficial (0 to 10 cm depth) sediment samples were collected and transported into a sterile sampling bag and immediately shipped to the laboratory on ice packs. Upon arrival, portions of the samples were stored at 4°C for physicochemical characterization and −80°C for metagenomic DNA extraction. Details regarding the location and physicochemical properties of the lake sediment are provided in Table 1. By using a modified protocol specific for high molecular weight DNA from environmental samples [26], the metagenomic DNA of the sediment was extracted as described in detail previously [27]. The sediment metagenomic DNA was purified using an Omega Gel Extraction Kit (Norcross, GA) and stored at −20°C until use.

Xylanase gene fragments amplification, library construction and sequencing

GH10 and GH11 xylanase gene fragments were amplified with the purified metagenomic DNA as a template by touchdown PCR using the CODEHOP primers X10-F and X10-R specific for GH10 xylanase and X11-F and X11-R specific for GH11 xylanase [27], respectively. PCR products were then visualized on an agarose gel and purified using a Qiaquick gel extraction kit (Qiagen, Valencia, CA). The purified PCR products were then ligated into the PMD 19-T vector (TaKaRa, Tokyo, Japan) and electroporated into *Escherichia coli* DH5a (TaKaRa, Tokyo, Japan) following the procedure recommended by the manufacturer to construct the clone library for each xylanase family. Positive transformants (white colonies) from each library were randomly picked for further confirmation by PCR with primers M13F (GTAAAACGACGGCCAGT) and M13R (GGATAA-CAATTTCACACAGGA). They were then sequenced by Life Technologies using the Sanger method with an ABI-3730 automatic sequencer (Life Technologies, Carlsbad, CA).

Phylogenetic analysis

By using the Figaro software [28] (http://sourceforge.net/apps/mediawiki/amos/index.php?title=Figaro), vector sequences introduced by automated Sanger sequencing machines were removed. The sequences were analyzed by NCBI BLASTx (version 2.2.29) searches against the GenBank nr database, with an E-score (expect value) cutoff of 10^{-10}. Nucleotide sequences identified as xylanase gene fragments were translated into amino acids by EMBOSS Transeq (http://www.ebi.ac.uk/Tools/st/emboss_transeq/). Then, they were aligned with known sequences in the GenBank database at the protein level using ClustalW. Redundant amino acid sequences were removed using Cd-hit [29] with a 95% sequence identity cutoff.

The protein sequence similarities were assessed using the BLASTp program (http://www.ncbi.nlm.nih.gov/BLAST/; until January 15, 2014). Phylogenetic trees were constructed with MEGA 4.1 [30] using the neighbor-joining method [31]. Confidence for tree topologies was estimated by bootstrap values based on 1,000 replicates. A total of 39 and 17 representative sequences were selected and used as references for GH10 and GH11 phylogenetic tree constructions, respectively.

Abundance analysis

The abundance of each GH family was estimated using the distance-based operational taxonomic unit and richness determination (DOTUR) software [32]. By using PHYLIP software (http://evolution.genetics.washington.edu/phylip.html), distance matrices of the fragment sequences were calculated at the protein level with the default parameters of protdist. Based on UPGMA (average linkage clustering) implemented in DOTUR with default parameters of precision (0.01) and 1000 bootstrap replicates, sequences were then assigned to OTUs.

Cloning of full-length xylanase gene

Fragment AS10-66, which showed phylogenetic novelty (Figure 1, Table S1) and was most abundant in the GH10 library (see

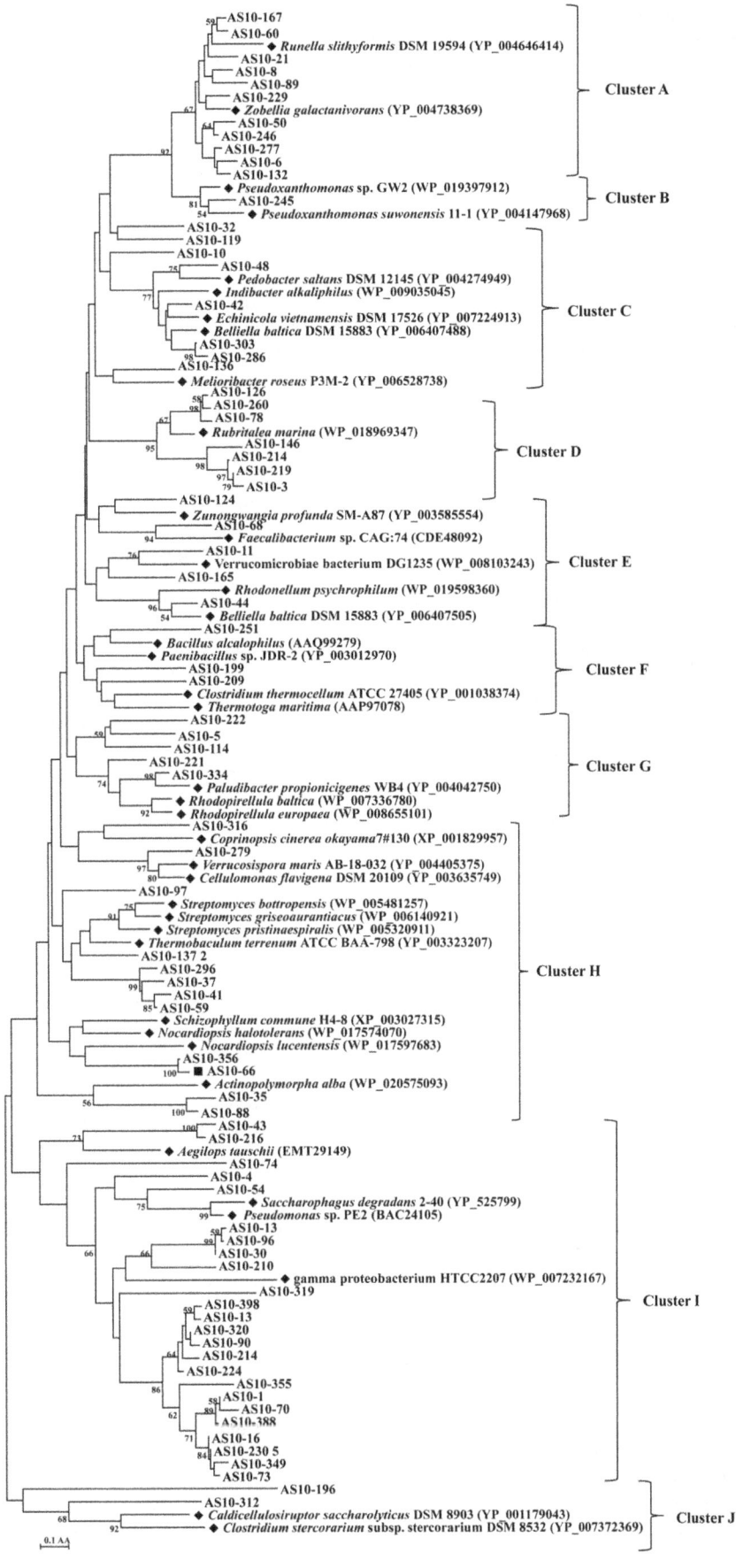

Figure 2. Phylogenetic analysis based on the partial amino acid sequences of GH10 xylanase genes detected in the Lake Dabusu sediment metagenomic DNA. The tree was constructed using the neighbor joining method (MEGA 4.1). The lengths of the branches indicate the relative divergence among amino acid sequences. The reference sequences are marked with a closed diamond (♦) with source strains and GenBank accession numbers in parentheses. The gene fragments (AS10-66) used for full length cloning were marked with a solid square (■). The numbers at the nodes indicate bootstrap values based on 1,000 bootstrap replications and bootstrap values (>50) are displayed. The scale bar represents 0.1 amino acid substitutions per position.

Table S1), was subjected to full-length gene cloning. The flanking regions of the xylanase gene fragments were cloned by using a modified TAIL-PCR with eight specific primers (Table S3) following the protocol [33]. PCR products of the expected size appeared between the third and fourth rounds of amplification were purified, cloned into the PMD 19-T vectors, sequenced, and then assembled with the known fragment sequence. The full-length xylanase gene was designated as *xynAS10-66*. The signal peptide sequence of XynAS10-66 was predicted with SignalP [34] (http://www.cbs.dtu.dk/services/SignalP/). The DNA and protein sequence identities/similarities were assessed with the BLASTn and BLASTp programs (http://www.ncbi.nlm.nih.gov/BLAST/), respectively.

Xylanase expression and activity assay

Using two primer sets (Table S3), the coding sequence of the xylanase gene *xynAS10-66* without the signal peptide was amplified and cloned into vector pET-22b(+), and then transformed into *E. coli* BL21 (DE3) competent cells for recombinant expression. The positive transformant harboring pET-*xynAS10-66* was grown in LB medium containing 100 μg mL^{-1} ampicillin at 37°C to an A_{600} of 0.6. Protein expression was induced by addition of isopropyl-β-D-1-thiogalactopyranoside (IPTG) at a final concentration of 1 mM, and the culture was incubated at 30°C for additional 12 h.

Xylanase activity was determined by measuring the release of reducing sugar from substrate by the 3, 5-dinitrosalicylic acid method [35]. To accomplish this, reactions containing 0.1 ml of appropriately diluted enzyme and 0.9 ml of 1% (w/v) beechwood xylan as substrate. After incubation at 55°C for 10 min, the reaction was stopped with 1.5 ml DNS reagent and boiled for 5 min, and the absorbance at 540 nm (A_{540}) was measured. Finally, using a standard curve generated with D-xylose, the absorbance was converted into moles of reducing sugars produced. One unit (U) of xylanase activity was defined as the amount of enzyme that released 1 μmol of reducing sugar equivalent to xylose per minute.

Purification and partial characterization of recombinant XynAS10-66

To purify the His-tagged recombinant proteins (rXynAS10-66), culture supernatant was collected after centrifugation (12,000×*g*, 4°C for 15 min). Then the culture supernatant was concentrated with an ultrafiltration membrane (PES5000; Sartorius Stedim Biotech, Germany) and then loaded onto a Ni^{2+}-NTA agarose gel column (Qiagen, Germany) with an imidazole gradient of 20–200 mM in Tris-HCl buffer (20 mM Tris-HCl, 500 mM NaCl, 10% glycerol, pH 7.6). Sodium dodecyl sulfate-polyacrylamide gel electrophoresis (SDS-PAGE) was used to determine the purity and apparent molecular mass of rXynAS10-66.

The optimal pH for xylanase activity of the purified rXynAS10-66 was determined at 37°C with pH ranging from 4.0 to 11.0. The buffers used were McIlvaine buffer (0.2 M Na$_2$HPO$_4$/0.1 M citric acid) for pH 4.0–7.0, 0.1 M Tris-HCl for pH 7.0–9.0, and 0.1 M glycine-NaOH for pH 9.0–11.0. The optimal temperature for purified rXynAS10-66 activity was determined over the range of 40–90°C in Tris-HCl buffer (pH 9.0). The K_m and V_{max} values for rXynAS10-66 were determined in Tris-HCl buffer (pH 7.0) containing 1–10 mg mL^{-1} beechwood xylan at 55°C, respectively. K_m and V_{max} were determined from a Lineweaver-Burk plot using the non-linear regression computer program GraFit (Erithacus, Horley, Surrey, UK).

Nucleotide sequence accession numbers

The GH10 and GH11 xylanase gene fragments were deposited into the GenBank database under accession numbers KJ463250–KJ463327 and KJ463328–KJ463355, respectively. Accession number KJ463356 was assigned to the full-length xylanase gene *xynAS10-66*.

Results and Discussion

Soda lakes are one of the most stable naturally occurring alkaline and saline environments that harbor numerous novel microorganisms [16,17,21]. Although the microbial diversity of this unique ecosystem has been thoroughly investigated, only a few studies have considered the functional diversity [22,23]. Functional gene diversity based on the metagenomic sequences can provide insight into functional diversity and metabolic potential at the community level. Xylanases play key roles in the initial steps of plant cell wall breakdown and have great potential for industrial and agricultural applications. Thus xylanase genes were targeted for diversity analysis in this study. Because many microorganisms cannot be cultured in the laboratory owing to the extremely harsh conditions of soda lakes, culture-independent molecular approaches were used to explore the xylanase gene diversity of Lake Dabusu.

Abundance of GH10 xylanase in Lake Dabusu sediment

Using the CODEHOP primers X10-F and X10-R specific for GH10 xylanases [27], PCR product of about 260 bp was amplified from the metagenomic DNA of the sediment. The product was purified and used to construct a clone library. The positive transformants were picked and then confirmed by PCR with primers M13F and M13R. Overall, 550 clones were sequenced, 467 sequences of which were identified as GH10 xylanase gene fragments based on BLASTx analysis and presence of the Asn residue in the protein sequence, which is conserved among GH10 xylanases [36].

After removing redundant sequences using the CD-hit program [29], 78 sequences showed divergence (<95% homology, Table S1). Based on BLASTp analysis, about 68% (53/78) of the sequences had low similarities (<65%) with known xylanases in GenBank (Figure 1), implying that they may be novel xylanases. Abundance analysis using the distance based operational taxonomic unit and richness determination (DOTUR) software [32] showed that AS10-66 was the predominant operational taxonomic unit (OTU), representing 117 sequences. Forty OTUs contained only one sequence (Table S1).

Culture dependent and culture-independent methods revealed the presence of numerous novel microorganisms in soda lake environments [16,21]. In the present study, GH10 xylanase gene

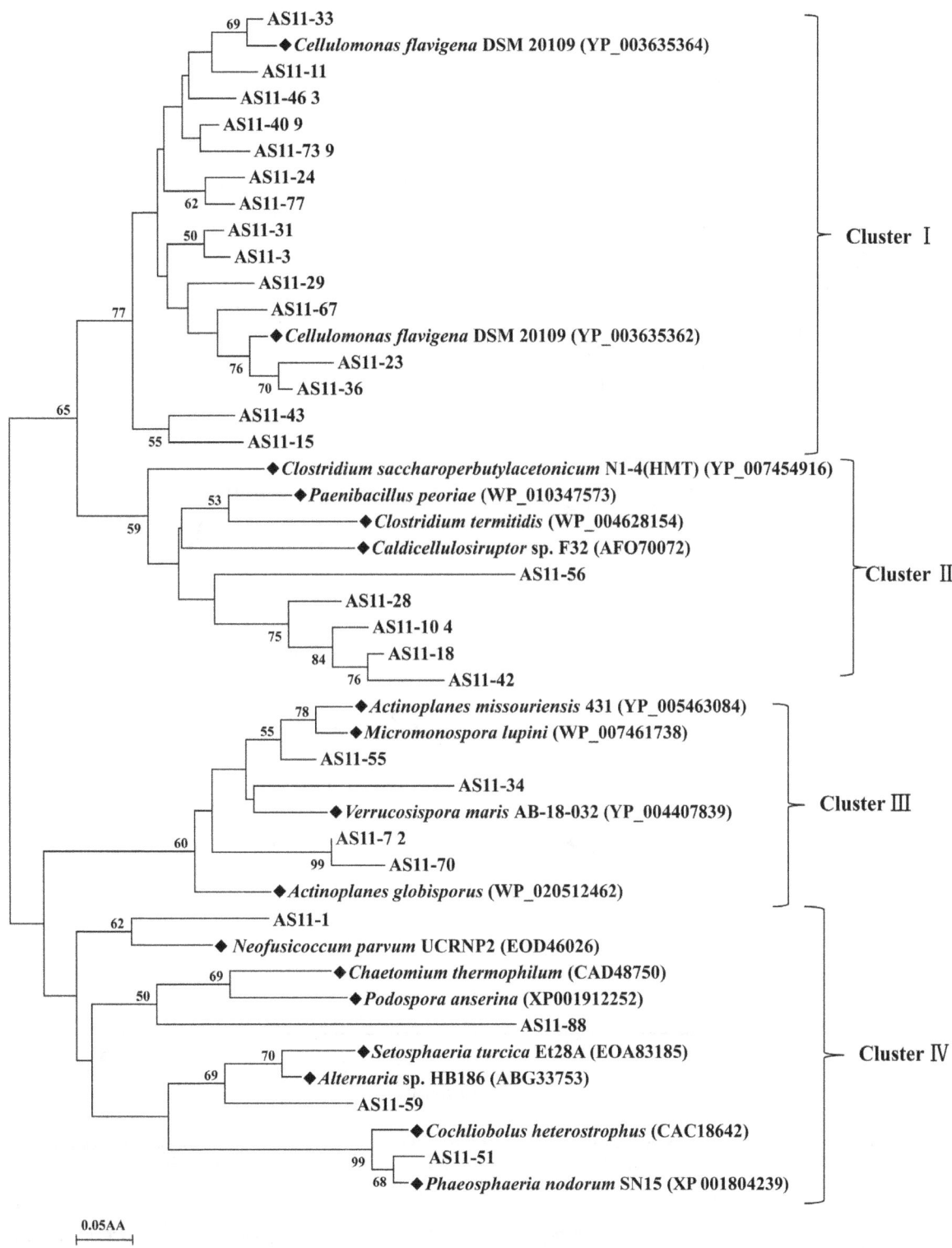

Figure 3. Phylogenetic analysis based on the partial amino acid sequences of GH11 xylanase genes detected in the Lake Dabusu sediment metagenomic DNA. The tree was constructed using the neighbor joining method (MEGA 4.1). The lengths of the branches indicate the relative divergence among amino acid sequences. The reference sequences are marked with a closed diamond (♦) with source strains and GenBank accession numbers in parentheses. The numbers at the nodes indicate bootstrap values based on 1,000 bootstrap replications and bootstrap values (>50) are displayed. The scale bar represents 0.05 amino acid substitutions per position.

fragment sequences obtained directly from the sediment metage-nomic DNA had low homology with known xylanases. Specifical-ly, none of the xylanase fragments amplified in this study had greater than 90% homology with known xylanases in the

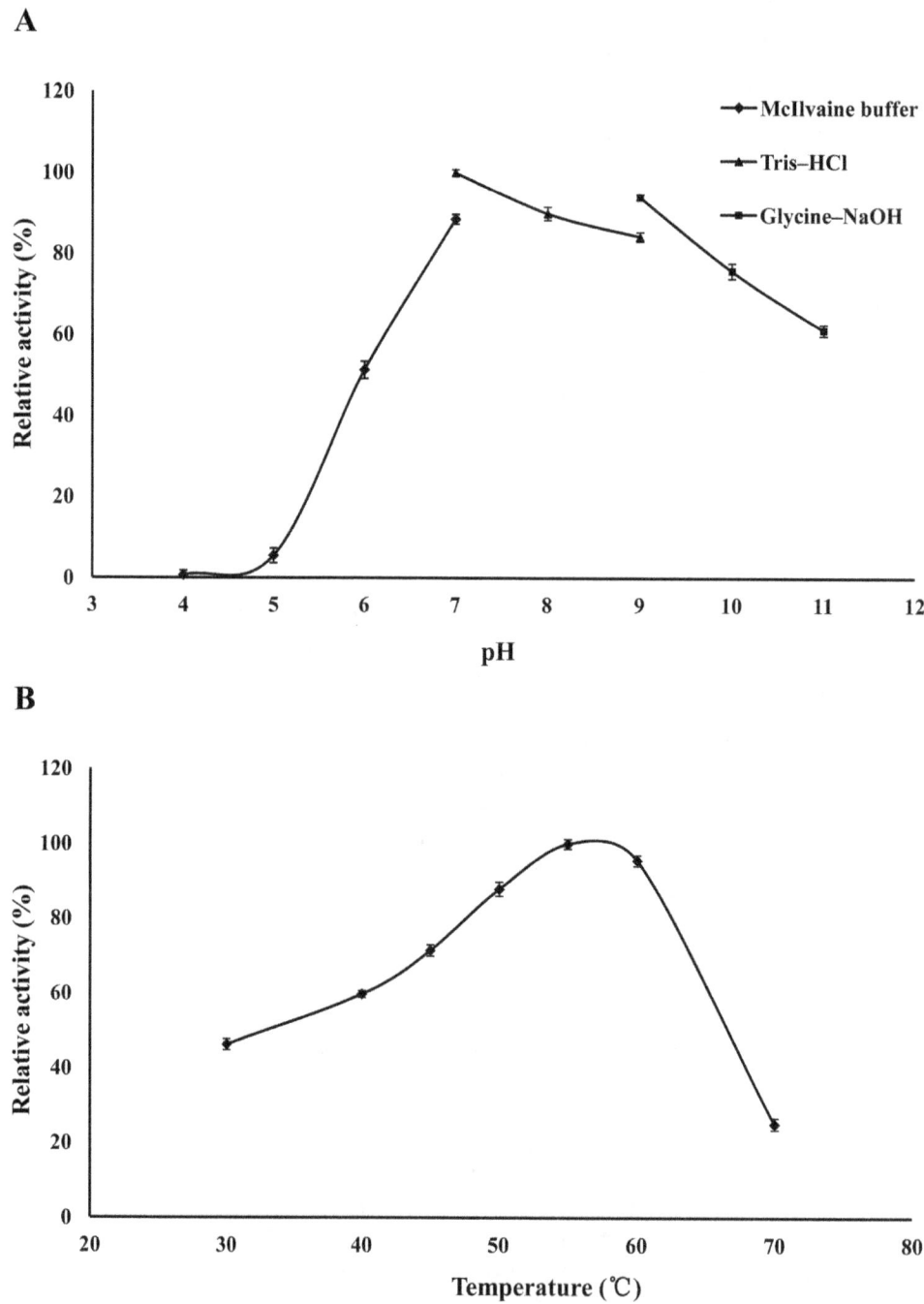

Figure 4. pH and temperature activity profiles of purified recombinant XynAS10-66. A Effect of pH on XynAS10-66 activity. Activities at various pHs were assayed at 30°C in different buffer. **B** Effect of temperature on XynAS10-66 activity in Tris-HCl buffer (pH 9.0).

GenBank database (Table S1). The lowest similarities were observed for AS10-35 and AS10-88, which both had 35% homology with xylanase from *Nocardiopsis* sp. CNS639 (WP_019610727), while AS10-8 showed the greatest homology of 87% with xylanase from *Cellulophaga algicola* DSM 14237 (YP_004163134). Moreover, almost 70% of the obtained sequences had low similarity (<65%) with known xylanases. This is much higher than that from the other soil environments we previously investigated [24]. Overall, these findings suggested that there may be abundant novel xylan-degrading microorganisms in this environment.

High genetic diversity of GH10 xylanase in the alkaline soda lake sediment

Using 78 divergent sequences from the GH10 clone library and 39 reference sequences from the GenBank Database, an unrooted protein-level phylogenetic tree of GH10 xylanases was constructed. All sequences were confined to ten clusters, denoted as Cluster A to Cluster J, indicating substantial diversity among GH10 xylanases in the sediment (Figure 2). The presence of many clades without close relatives suggests their novelty, which might be because of the large portion of unidentified microorganisms in this environment.

Cluster A, C, E and G contained a total of 28 sequences from the soda lake sediment and 14 reference sequences of genera belonging to Bacteroidetes including *Runella slithyformis* DSM 19594, *Pedobacter saltans*, *Indibacter alkaliphilus* and *Rhodopirellula baltica*. Cluster B and cluster I contained the sequences of 23 clones and reference sequences from the phylum Proteobacteria, including gamma proteobacterium HTCC2207, *Pseudomonas* sp. PE2, *Saccharophagus degradans* 2–40 and *Pseudoxanthomonas suwonensis* 11-1. Twelve sequences and 11 reference sequences from different genera in the phylum Actinobacteria, including *Actinopolymorpha*, *Streptomyces* and *Nocardiopsis,* formed cluster H. Three sequences in cluster F and two in cluster J were closely related to xylanase produced by members of the phylum Firmicutes, including *Bacillus alcalophilus*, *Paenibacillus* sp. JDR-2, *Clostridium thermocellum* ATCC 27405, *Caldicellulosiruptor saccharolyticus* DSM 8903 and *Clostridium stercorarium* subsp. stercorarium DSM 8532. Seven sequences in cluster D shared the highest identity with xylanase from *Rubritalea marina*.

As shown in Fig. 2, phylogenetic analysis revealed that Bacteroidetes, Proteobacteria, Actinobacteria, Firmicutes and Verrucomicrobia are the main xylanolytic bacteria to produce xylanases in the sediment of Lake Dabusu. These findings are consistent with the conclusions of previous studies [16,17,20,37]. However, the composition of the xylanolytic community differs from that of the microbial community. Microbial community analysis suggested that Firmicutes are the most abundant microorganisms in soda lake environments, followed by Proteobacteria and Actinobacteria. In the present study, xylanases were found to primarily belong to Bacteroidetes, which harbored more than 35% (28/78) of all GH10 fragment sequences. One reason for this may be that the xylantic microbial community differs from that of the total microbial community. Soda lakes harbor abundant and diverse microorganisms which are critical in decomposition of organic matter and cycling of carbon, nitrogen, phosphorus, and sulphur [38]. Second, the physicochemical properties of Lake Dabusu differ from those of other soda lakes [16], which play important role in the composition of the microbial community [20,39].

Xylanase fragment sequences related to those from Proteobacteria were the second most abundant, accounting for about 30% (23/78) of all GH10 xylanase sequences. All reference sequences of gamma proteobacterium HTCC2207, *Pseudomonas* sp. PE2, *Saccharophagus degradans* 2–40 and *Pseudoxanthomonas suwonensis* 11-1 belonged to the Gammaproteobacteria. This is concurrent with the finding that Gammaproteobacteria is one of the most abundant and diverse groups in soda lakes [37]. Moreover, we found that although these sequences were grouped in a cluster, they were not closely related to each other (Figure 2), suggesting that these sequences might be novel.

Amplification and sequence analysis of GH11 xylanase gene fragments

PCR product with a size of about 210 bp was obtained from the metagenomic DNA of sediment using CODEHOP primers X11-F and X11-R specific for GH11 xylanases [24]. Overall, 250 clones were randomly selected from the clone library constructed using the PCR products and sequenced. As a result, 204 sequences showed 70–98% amino acid identity with known GH11 xylanases (Table S2). Additionally, Glu (catalytic residue), which is highly conserved in GH11 xylanases, was found in all protein sequences (Table S2). Thus, these sequences were considered to be partial GH11 xylanases. After removing the redundant sequences using the CD-hit program, 28 sequences showed divergence (sharing < 95% identity) (Table S2). Abundance analysis using DOTUR

software showed that AS11-42 was the predominant OTU, occurring in 39 sequences. Nine OTUs contained only one sequence (Table S2).

Phylogenetic diversity of GH11 xylanase in lake sediment

The 28 distinct partial sequences of GH11 xylanases were used to construct an unrooted phylogenetic tree with 17 reference sequences (Figure 3). Four clusters (I, II, III, and IV) were formed based on high bootstrap values. Some clades formed without closely related references, suggesting that these sequences differed from known xylanases. Nineteen sequences shared the highest identity with xylanases from *Cellulomonas flavigena* DSM 20109, *Actinoplanes missouriensis* 431, *A. globisporus Micromonospora lupine* and *Verrucosispora maris* AB-18-032, and fell into clusters I and III. Five sequences and those of *Clostridium saccharoperbutylacetonicum* N1-4(HMT), *C. Paenibacillus peoriae*, and *Caldicellulosiruptor* sp. F32 were grouped into cluster II. Four sequences closely related to xylanases from *Neofusicoccum parvum* UCRNP2, *Podospora anserine*, *Setosphaeria turcica* Et28A, *Alternaria* sp. HB186 and *Phaeosphaeria nodorum* SN15 fell into cluster IV.

Unlike GH10 family xylanase genes, GH11 xylanase genes from Lake Dabusu were less diverse and showed a different microbial distribution. Sequences of the GH11 xylanase fragments were found to be related to Actinomycetes, Firmicutes and Fungi. Actinomycetes are important degraders of organic matter in most habitats. An alkaline xylanase was purified and characterized from alkaliphilic *Micrococcus* sp. AR-135, which was isolated from an alkaline soda lake in Ethiopia [14]. Upon phylogenetic analysis, more than 67% (19/28) of the sequences were grouped into a cluster containing reference sequences from different genera in Actinobacteria. Along with the 12 sequences of GH10 xylanase, a total 31 fragment sequences were found to be associated with Actinobacteria, indicating the high diversity of xylanolytic actinobacteria in Lake Dabusu.

Using culture and culture-independent methods, the bacterial community in the soda lake was found to be dominated by clones affiliated with Firmicutes. Moreover, several xylanases were characterized from the isolated Firmicutes with high activity at alkaline pH [13,15]. In this study, xylanase fragments related to xylanases of different genera Firmicutes were also found. Specifically, five GH10 xylanase fragment sequences were found to be closely related to *Bacillus alcalophilus*, *Clostridium thermocellum* ATCC 27405 and *Caldicellulosiruptor saccharolyticus* DSM 8903, while five GH11 fragments were closely related to *Caldicellulosiruptor* sp. F32 (AFO70072), *Clostridium termitidis* (WP_004628154) and *Clostridium saccharoperbutylacetonicum* N1-4(HMT) (YP_007454916), suggested the diversity of xylanolytic Firmicutes in this environment.

When compared with Bacteria and Archaea, little is known about the diversity, abundance and activity of micro-eukaryotes in soda lakes. Recently, a high diversity of micro-eukaryotes in soda lakes located in the Ethiopian Rift Valley was revealed by high-throughput sequencing [21]. However, no functional genetic diversity has been reported regarding fungi in soda lake environments. In the present study, four sequences were found to be related to xylanases from Ascomycota, the dominant fungi in soda lakes, suggesting that xylantic fungi are present in these systems.

Direct cloning and novelty of xylanase gene XynAS10-66

Until recently, the majority of known xylanases were obtained from pure cultures. Only a few were cloned using culture-independent approaches, such as construction and screening of metagenomic libraries and PCR-based methods. In our lab, we developed a fast and efficient modified TAIL-PCR method to obtain full-length genes directly from metagenomic DNA based on fragment sequences [33]. In this study, the fragment sequence AS10-66 was selected to obtain the full-length gene because it was the most abundant of all the sequences, representing about one quarter of all sequences (117/467). Moreover, AS10-66 has been reported to have low homology with known xylanases (Table S1).

Based on the fragment sequences of AS10-66 and using modified TAIL-PCR [33], a full-length xylanase gene (*xynAS10-66*) was direct cloned from metagenomic DNA of the soda lake sediment. The complete sequence of *xynAS10-66* contained an open reading frame of 1092 bp encoding a 363-residue polypeptide with a typical signal peptide (residues 1–24). The calculated mass of XynAS10-66 was 38.7 kDa, and the theoretical isoelectric point was 4.74. Sequence similarity searches showed that deduced XynAS10-66 shared the highest homology (44%) with the thermostable GH10 xylanase from *Thermomonospora alba* ULJB1, followed by *Thermobispora bispora* DSM 43833 (44%), an isolate recovered from decaying manure and *Nocardiopsis halotolerans* (43%), an isolate recovered from salt marsh soil in Kuwait [40].

Partial biochemical characterization of purified recombinant XynAS10-66

The gene encoding the mature proteins was expressed in *E. coli* BL21 (*DE3*). Following induction with IPTG at 30°C for 12 h, substantial xylanase activity was detected in the culture supernatant of recombinant cells. Using beechwood xylan as the substrate, rXynAS10-66 showed the highest activity at pH 7.0, while it retained more than 95% activity at pH 9.0, and >60% of the maximum activity at pH 7.0–11.0 (Figure 4A). Using beechwood xylan as the substrate, the K_m and V_{max} values were 1.5 ± 0.04 mg $\cdot$ mL^{-1}, 1102 ± 9.54 µmol $\cdot$ mg^{-1} $\cdot$ min^{-1}, respectively. Although the optimum temperature of the recombinant XynAS10-66 is neutral, it has substantial activity at alkaline pH, with more than 80% and 60% activity being retained at pH 10 and pH 11, respectively. This characterization differs from those of alkaline xylanases obtained from *Bacillus* and *Micrococcus* sp. isolated from other soda lakes, which showed optimum activity at pH 8 to 9 [14,15]. The optimal temperature for the enzyme activity of XynAS10-66 was 55°C (Figure 4B), while more than 60% activity was retained at 45°C to 65°C. The recombinant XynAS10-66 had the same optimum temperature as the alkaline xylanase from alkaliphilic *Micrococcus* sp. AR-135 [14]. However, it had a lower

activity than that of xylanases from *Bacillus halodurans* S7 [13] and two thermostable alkaline xylanases from an alkaliphilic *Bacillus* sp. [15]. BLAST analysis revealed that XynAS10-66 shared high similarity with a thermostable endo-beta-1,4-xylanase of *Thermomonospora alba* ULJB1 [41], which showed good activity at up to 95°C. Sequence alignment revealed that XynAS10-66 lacks a carbohydrate binding module (CBM) at the C terminal, which may have resulted in the lower temperature optimum of XynAS10-66. Future studies will be conducted to investigate this phenomenon. Moreover, other fragment sequences will be used to generate more full-length genes directly from the metagenomic DNA, and any genes generated will be characterized in subsequent studies.

In conclusion, culture-independent molecular methods led to recovery of abundant GH10 and GH11 xylanase genes from the sediment of the Lake Dabusu. Similarities among these sequences with known xylanases were low, and they were distantly related based on phylogenetic analysis. These results suggest that xylanase gene diversity within Lake Dabusu is high and most of them might be novel. Our study provides new insight into the genetic diversity and distribution of xylanases in soda lake environments and a rapid culture-independent molecular method for the retrieval of the source of xylanase genes with potential industrial applications.

Supporting Information

Table S1 GH10 xylanase gene fragments detected in the sediment of Lake Dabusu and their closest relative based on amino acid sequence identity and similarity.

Table S2 GH11 xylanase gene fragments detected in the sediment of Lake Dabusu and their closest relative based on amino acid sequence identity and similarity.

Table S3 Primers used for gene cloning and expression.

Author Contributions

Conceived and designed the experiments: JL XY. Performed the experiments: GW XH. Analyzed the data: GW TBN JL XY. Contributed reagents/materials/analysis tools: JL XY. Contributed to the writing of the manuscript: GW XH TBN JL XY.

References

1. Kulkarni N, Shendye A, Rao M (1999) Molecular and biotechnological aspects of xylanases. FEMS Microbiol Rev 23: 411–456.
2. Sunna A, Antranikian G (1997) Xylanolytic enzymes from fungi and bacteria. Crit Rev Biotechnol 17: 39–67.
3. Collins T, Gerday C, Feller G (2005) Xylanases, xylanase families and extremophilic xylanases. FEMS Microbiol Rev 29: 3–23.
4. Cantarel BL, Coutinho PM, Rancurel C, Bernard T, Lombard V, et al. (2009) The Carbohydrate-Active EnZymes database (CAZy): an expert resource for Glycogenomics. Nucl Acids Res 37: D233–238.
5. Biely P, Vrsanska M, Tenkanen M, Kluepfel D (1997) Endo-beta-1,4-xylanase families: differences in catalytic properties. J Biotechnol 57: 151–166.
6. Jeffries TW (1996) Biochemistry and genetics of microbial xylanases. Curr Opin Biotechnol 7: 337–342.
7. Beg QK, Kapoor M, Mahajan L, Hoondal GS (2001) Microbial xylanases and their industrial applications: a review. Appl Microbiol Biotechnol 56: 326–338.
8. Polizeli ML, Rizzatti AC, Monti R, Terenzi HF, Jorge JA, et al. (2005) Xylanases from fungi: properties and industrial applications. Appl Microbiol Biotechnol 67: 577–591.
9. Martin-Sampedro R, Rodriguez A, Ferrer A, Garcia-Fuentevilla LL, Eugenio ME (2012) Biobleaching of pulp from oil palm empty fruit bunches with laccase and xylanase. Bioresour Technol 110: 371–378.

10. Khandeparkar R, Bhosle NB (2007) Application of thermoalkalophilic xylanase from *Arthrobacter* sp. MTCC 5214 in biobleaching of kraft pulp. Bioresour Technol 98: 897–903.
11. Verma D, Kawarabayasi Y, Miyazaki K, Satyanarayana T (2013) Cloning, expression and characteristics of a novel alkalistable and thermostable xylanase encoding gene (Mxyl) retrieved from compost-soil metagenome. PLoS ONE 8: e52459.
12. Simkhada JR, Yoo HY, Choi YH, Kim SW, Yoo JC (2012) An extremely alkaline novel xylanase from a newly isolated Streptomyces strain cultivated in corncob medium. Appl Biochem Biotechnol 168: 2017–2027.
13. Mamo G, Delgado O, Martinez A, Mattiasson B, Hatti-Kaul R (2006) Cloning, sequence analysis, and expression of a gene encoding an endoxylanase from *Bacillus halodurans* S7. Mol Biotechnol 33: 149–159.
14. Gessesse A, Mamo G (1998) Purification and characterization of an alkaline xylanase from alkaliphilic *Micrococcus* sp AR-135. Journal of Industrial Microbiology and Biotechnology 20: 210–214.
15. Gessesse A (1998) Purification and properties of two thermostable alkaline xylanases from an alkaliphilic bacillus sp. Appl Environ Microbiol 64: 3533–3535.
16. Antony CP, Kumaresan D, Hunger S, Drake HL, Murrell JC, et al. (2013) Microbiology of Lonar Lake and other soda lakes. ISME J 7: 468–476.

17. Jones BE, Grant WD, Duckworth AW, Owenson GG (1998) Microbial diversity of soda lakes. Extremophiles 2: 191–200.

18. Zhao B, Chen S (2012) Alkalitalea saponilacus gen. nov., sp. nov., an obligately anaerobic, alkaliphilic, xylanolytic bacterium from a meromictic soda lake. Int J Syst Evol Microbiol 62: 2618–2623.

19. Rees HC, Grant WD, Jones BE, Heaphy S (2004) Diversity of Kenyan soda lake alkaliphiles assessed by molecular methods. Extremophiles 8: 63–71.

20. Foti MJ, Sorokin DY, Zacharova EE, Pimenov NV, Kuenen JG, et al. (2008) Bacterial diversity and activity along a salinity gradient in soda lakes of the Kulunda Steppe (Altai, Russia). Extremophiles 12: 133–145.

21. Lanzen A, Simachew A, Gessesse A, Chmolowska D, Jonassen I, et al. (2013) Surprising prokaryotic and eukaryotic diversity, community structure and biogeography of Ethiopian soda lakes. PLoS ONE 8: e72577.

22. Lin JL, Joye SB, Scholten JC, Schafer H, McDonald IR, et al. (2005) Analysis of methane monooxygenase genes in mono lake suggests that increased methane oxidation activity may correlate with a change in methanotroph community structure. Appl Environ Microbiol 71: 6458–6462.

23. Kovaleva OL, Tourova TP, Muyzer G, Kolganova TV, Sorokin DY (2011) Diversity of RuBisCO and ATP citrate lyase genes in soda lake sediments. FEMS Microbiol Ecol 75: 37–47.

24. Wang G, Meng K, Luo H, Wang Y, Huang H, et al. (2012) Phylogenetic diversity and environment-specific distributions of glycosyl hydrolase family 10 xylanases in geographically distant soils. PLoS ONE 7: e43480.

25. Shen J, Cao JT, Wu YH (2001) Paieoclimatic changes in Dabusu Lake. Chin J Oceanol Limnol 19: 91–96.

26. Brady SF (2007) Construction of soil environmental DNA cosmid libraries and screening for clones that produce biologically active small molecules. Nat Protoc 2: 1297–1305.

27. Wang G, Wang Y, Yang P, Luo H, Huang H, et al. (2010) Molecular detection and diversity of xylanase genes in alpine tundra soil. Appl Microbiol Biotechnol 87: 1383–1393.

28. White JR, Roberts M, Yorke JA, Pop M (2008) Figaro: a novel statistical method for vector sequence removal. Bioinformatics 24: 462–467.

29. Li W, Godzik A (2006) Cd-hit: a fast program for clustering and comparing large sets of protein or nucleotide sequences. Bioinformatics 22: 1658–1659.

30. Tamura K, Dudley J, Nei M, Kumar S (2007) MEGA4: Molecular Evolutionary Genetics Analysis (MEGA) Software Version 4.0. Mol Biol Evol 24: 1596–1599.

31. Saitou N, Nei M (1987) The neighbor-joining method: a new method for reconstructing phylogenetic trees. Mol Biol Evol 4: 406–425.

32. Schloss PD, Handelsman J (2005) Introducing DOTUR, a computer program for defining operational taxonomic units and estimating species richness. Appl Environ Microbiol 71: 1501–1506.

33. Huang H, Wang G, Zhao Y, Shi P, Luo H, et al. Direct and efficient cloning of full-length genes from environmental DNA by RT-qPCR and modified TAIL-PCR. Appl Microbiol Biotechnol 87: 1141–1149.

34. Petersen TN, Brunak S, von Heijne G, Nielsen H (2011) SignalP 4.0: discriminating signal peptides from transmembrane regions. Nat Methods 8: 785–786.

35. Miller GL, Blum R, Glennon WE, Burton AL (1960) Measurement of carboxymethylcellulase activity. Analytical Biochemistry 1: 127–132.

36. Solomon V, Teplitsky A, Shulami S, Zolotnitsky G, Shoham Y, et al. (2007) Structure-specificity relationships of an intracellular xylanase from *Geobacillus stearothermophilus*. Acta Crystallographica Section D 63: 845–859.

37. Ma Y, Zhang W, Xue Y, Zhou P, Ventosa A, et al. (2004) Bacterial diversity of the Inner Mongolian Baer Soda Lake as revealed by 16S rRNA gene sequence analyses. Extremophiles 8: 45–51.

38. Joshi AA, Kanekar PP, Kelkar AS, Shouche YS, Vani AA, et al. (2008) Cultivable bacterial diversity of alkaline Lonar lake, India. Microb Ecol 55: 163–172.

39. Pagaling E, Wang H, Venables M, Wallace A, Grant WD, et al. (2009) Microbial biogeography of six salt lakes in Inner Mongolia, China, and a salt lake in Argentina. Appl Environ Microbiol 75: 5750–5760.

40. Al-Zarban SS, Abbas I, Al-Musallam AA, Steiner U, Stackebrandt E, et al. (2002) *Nocardiopsis halotolerans* sp. nov., isolated from salt marsh soil in Kuwait. Int J Syst Evol Microbiol 52: 525–529.

41. Blanco J, Coque JJ, Velasco J, Martin JF (1997) Cloning, expression in Streptomyces lividans and biochemical characterization of a thermostable endo-beta-1,4-xylanase of *Thermomonospora alba* ULJB1 with cellulose-binding ability. Appl Microbiol Biotechnol 48: 208–217.

9

Quantifying Fish Assemblages in Large, Offshore Marine Protected Areas

Nicole A. Hill[1]*, Neville Barrett[1], Emma Lawrence[2], Justin Hulls[1], Jeffrey M. Dambacher[3], Scott Nichol[4], Alan Williams[5], Keith R. Hayes[3]

1 Institute for Marine and Antarctic Studies, University of Tasmania, Hobart, Tasmania, Australia, 2 Digital Productivity Flagship, Commonwealth Scientific and industrial Research Organisation (CSIRO), Brisbane, QLD, Australia, 3 Digital Productivity Flagship, Commonwealth Scientific and industrial Research Organisation (CSIRO), Hobart, Tasmania, Australia, 4 Geoscience Australia, Canberra, ACT, Australia, 5 Oceans and Atmosphere Flagship, Commonwealth Scientific and industrial Research Organisation (CSIRO), Hobart, Tasmania, Australia

Abstract

As the number of marine protected areas (MPAs) increases globally, so does the need to assess if MPAs are meeting their management goals. Integral to this assessment is usually a long-term biological monitoring program, which can be difficult to develop for large and remote areas that have little available fine-scale habitat and biological data. This is the situation for many MPAs within the newly declared Australian Commonwealth Marine Reserve (CMR) network which covers approximately 3.1 million km^2 of continental shelf, slope, and abyssal habitat, much of which is remote and difficult to access. A detailed inventory of the species, types of assemblages present and their spatial distribution within individual MPAs is required prior to developing monitoring programs to measure the impact of management strategies. Here we use a spatially-balanced survey design and non-extractive baited video observations to quantitatively document the fish assemblages within the continental shelf area (a multiple use zone, IUCN VI) of the Flinders Marine Reserve, within the Southeast marine region. We identified distinct demersal fish assemblages, quantified assemblage relationships with environmental gradients (primarily depth and habitat type), and described their spatial distribution across a variety of reef and sediment habitats. Baited videos recorded a range of species from multiple trophic levels, including species of commercial and recreational interest. The majority of species, whilst found commonly along the southern or south-eastern coasts of Australia, are endemic to Australia, highlighting the global significance of this region. Species richness was greater on habitats containing some reef and declined with increasing depth. The trophic breath of species in assemblages was also greater in shallow waters. We discuss the utility of our approach for establishing inventories when little prior knowledge is available and how such an approach may inform future monitoring efforts within the CMR network.

Editor: David Mark Bailey, University of Glasgow, United Kingdom

Funding: This work was undertaken for the Marine Biodiversity Hub, a collaborative partnership supported through funding from the Australian Government's National Environmental Research Program (NERP). NERP Marine Biodiversity Hub partners include the Institute for Marine and Antarctic Studies, University of Tasmania; CSIRO, Geoscience Australia, Australian Institute of Marine Science, Museum Victoria, Charles Darwin University and the University of Western Australia. The funders had no role in study design, data collection and analysis, decision to publish, or preparation of the manuscript.

Competing Interests: The authors have declared that no competing interests exist.

* Email: Nicole.Hill@utas.edu.au

Introduction

Marine ecosystems face pressures from a range of sources, including pollution, fishing, habitat destruction, invasive species and a changing climate [1]. These pressures threaten the essential processes and resources that marine ecosystems provide, such as climate regulation, nutrient cycling and the provision of food, as well as threatening their intrinsic natural and cultural value [2]. As a consequence, management strategies have been adopted to safeguard marine ecosystems and their associated biodiversity. These strategies include the establishment of Marine Protected Areas (MPAs) which offer varying levels of protection according to their zoning or IUCN designation [3]. Countries such as the United States, Australia, South Africa, New Zealand, Kenya and the Philippines have a long history of establishing MPAs [4,5]. However, the absence of a truly global system of MPAs prompted

the Convention of Biological Diversity [6] and the Jakarta Mandate [7] to provide a framework and renewed international impetus for protecting marine biodiversity. This framework has resulted in a significant increase in recent years in the declaration of MPAs worldwide, with many forming part of large networks or systems of MPAs [8].

The increasing global commitment to the sustainability of healthy marine ecosystems is encouraging, but the declaration alone of MPAs does not ensure the conservation of biodiversity or ecosystems [4]. Management authorities must: clearly articulate the objectives of individual MPAs and MPA networks, both of which may contain multiple IUCN zones; formulate and enforce management plans [9] and periodically revisit and evaluate their management plans, which in turn requires that they establish a long-term biological monitoring program [4]. Developing monitoring programs for individual MPAs within networks of MPAs is often a difficult task. There is generally a mismatch between the spatial scales of information used to design MPA networks and those required to manage individual MPAs. Inventories of the types and distribution of habitats, communities and species within individual MPAs are often required before monitoring programs can be designed. In addition, for offshore or difficult to access MPAs, there are significant logistical and statistical challenges to working in remote environments.

Australia, a signatory to the CBD, provides a good example of the challenges involved in ultimately developing long-term monitoring programs in large MPA networks. In 2012, the Australian government announced an expansion of its MPA network to include 33 new MPAs. This makes a total of 60 MPAs covering approximately 3.1 million square kilometres, divided between six planning bioregions [10]. Although called the "Commonwealth Marine Reserve" or "CMR" network, areas with IUCN zones ranging from 1a (Sanctuary Zone) to VI (General Use Zone) are distributed throughout the network. From this point forward we adopt the above terminology and refer to MPAs within this network as reserves regardless of their IUCN zoning. All of the reserves in the CMR are however, located in offshore waters, most are large and encompass a broad depth range, and many are remote.

The Australian CMR network aims to meet the principles of being Comprehensive, Adequate and Representative (the CAR principles) [11] and utilised the national Integrated Marine and Coastal Regionalisation of Australia (IMCRA; [12]). The IMCRA bio-regionalisation is based on breaks in the distribution of demersal fish communities (together with some physical datasets) and is the best available continental-scale regionalisation of Australian marine fauna. Hence it was the most appropriate mechanism for delineating management regions and informing the placement of reserves. When designing biological monitoring programs for individual reserves however, much finer scale information is needed on the spatial distribution of important habitats (such as reefs) and the biological communities they contain. For many reserves in the new network this level of information is simply unavailable. It is therefore necessary to provide a robust inventory of the biological communities and key species within the reserves, delineate their spatial distribution and quantify the relationship between these distributions and key environmental drivers to enable the development of monitoring programs that will ultimately evaluate the effectiveness of each reserve.

Inventory, and ultimately monitoring, of large, deep reserves is challenging and relatively expensive. It requires vessels large enough to conduct surveys safely in offshore waters, together with a suite of non-extractive survey methods that can be deployed at depth, and sampling designs that are efficient and flexible enough to accommodate multiple (and potentially changing) objectives. A suite of non-extractive survey methods are currently available including multibeam sonar (MBS) for characterising the seafloor and associated habitat types [13], and image-based methods, such as towed camera systems, cameras attached to autonomous or remotely operated vehicles and baited cameras, for observing benthic fauna. Here we trialled MBS and Baited Remote Underwater Video (BRUVs), deployed within a spatially balanced design, for inventorying demersal fish in the multiple use zone (IUCN VI) of one reserve in the Australian network, the Flinders CMR. We use BRUVs because of their demonstrated ability to survey fish in deep waters [14] and because they have been extensively used in Australian inshore environments for monitoring and other ecological applications [15–17]. We trialled a spatially-balanced (i.e. evenly spread out in space) design known as

the Generalised Random- Tessellation Stratified (GRTS) design [18] because it is flexible and provides a good way to obtain a representative sample that respects the spatial distribution of habitats and communities in the target population [19].

In this study we: identify patterns in demersal fish assemblages across the continental shelf (a multiple-use zone, IUCN VI) of the Flinders CMR; quantify these patterns in relation to environmental gradients; examine the spatial distribution of assemblages across the reserve; and describe how our approach and the information gained may be useful to the development of a long-term biological monitoring program. We focus on fish communities because: broad-scale distributional records of demersal fish were instrumental in the marine bioregionalisation that underpins the design of the reserve network [20]; many species are of recreational and commercial value and are therefore likely to respond to current and future management actions within the reserve; and demersal fish are key components of benthic ecosystems.

Methods

Survey Area

The Flinders Commonwealth Marine Reserve (CMR) was established in 2007 and lies about 25 km offshore of northern Tasmania (Fig. 1). The reserve is 26,975 km^2 in size and extends as a west-east corridor from 35 m to > 3,000 m water depths. It consists of two zones: a multiple use zone (IUCN Category VI) that covers the majority of the continental shelf and slope; and a marine national park zone (IUCN Category II) that extends from the continental slope to the edge of the reserve at the 200 nautical mile limit of Australia's Exclusive Economic Zone (Fig. 1 inset). Our study area is the continental shelf, part of the multiple use zone (IUCN Category VI), where activities that impact on benthic habitats are prohibited, including demersal trawling, Danish seining and scallop dredging, or subject to permit requirements, for example mining activities [21]. Benthic habitats on the Flinders CMR shelf consist of sediment plains with patches of low profile and sand-inundated reefs, and steep rocky outcrops where canyon heads incise the shelf break (Lawrence, unpublished data). Within this environment, reefs and rocky outcrops have been identified as features likely to contain enhanced benthic diversity [21].

The Flinders CMR shelf straddles two provincial biogeographical regions: the Southeast Shelf Transition in the north and the Tasmanian Shelf Province in the South [12]. This region is characterised by variable but generally high exposure and strong tidal currents, especially in shallow areas between Flinders Island and the Tasmanian mainland [22,23]. The region is also influenced by southwards incursions of the East Australian Current (EAC) which brings warmer waters on to the shelf in summer [24]. The flora of the region is moderately rich and contains species common in cold temperate waters as well as low abundances of species common to warmer temperate waters [20].

Sampling design and methods

In contrast to the Flinders CMR continental slope, which has been comprehensively mapped using multibeam sonar (from which seafloor habitats have been inferred), very little spatially explicit habitat or biological information is available for the shelf. This represents a challenge for designing targeted sampling programs. Interpolated bathymetry for the Flinders shelf exists, but is at a gridded resolution of 250 m [25] (Fig. 1) which does not allow the identification of fine scale features of interest on the shelf, such as reefs. As a consequence of our limited *a priori* knowledge of the area, the inventory and description of fish assemblages

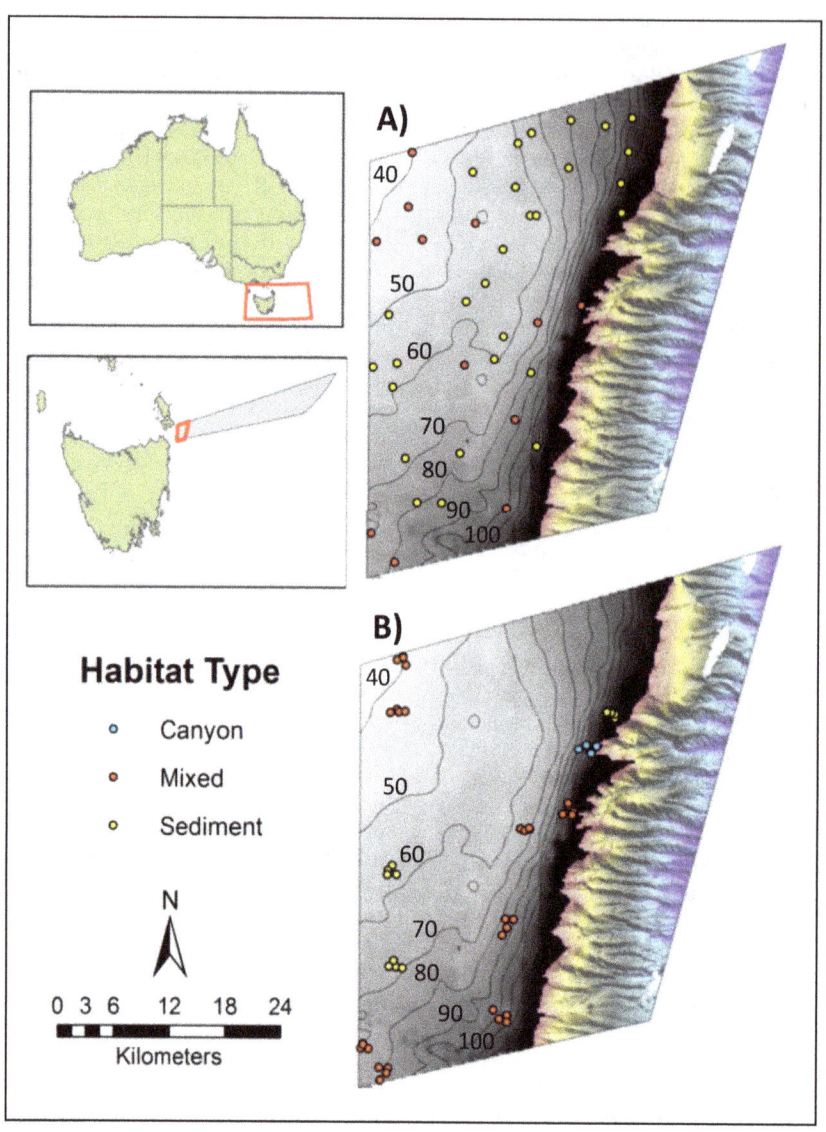

Figure 1. Location of the Flinders Commonwealth Marine Reserve (CMR) in Tasmania, Australia. Panels A and B show the CMR's Multiple Use Zone (IUCN VI) where survey work was conducted. The grey zone is the continental shelf (less than 200 m depth) where there was little pre-existing mapping data. Coarse bathymetry data (gridded at 250 m horizontal resolution) sourced from Geoscience Australia is shown with 10 m contour intervals overlain. The coloured area to the right shows the relatively steep and highly incised upper continental slope that had been mapped previously with multibeam sonar extending from 200 to 1500 m. A) Location of the 40 sites surveyed for habitat type in phase one of the sampling program. B) Location of the clustered sites where Baited Remote Underwater Videos (BRUVs) were deployed in phase two. Sites are coloured according to the broad habitat type: sediment (yellow); mixed, low-profile reef and sediments (red); canyon head (blue) recorded during phase one of sampling.

presented here forms one component of a multi-objective and multi-phase study within the Flinders CMR. Additional objectives of the broader study include examining the distribution and estimating the area of the different habitat types within the reserve, as well as quantifying benthic invertebrate communities using imagery captured by towed video and AUV, and these results will be reported elsewhere. To accommodate the objectives of the different components of the study, we used the probabilistic design, Generalised Random –Tessellation Stratified sampling technique (GRTS; [18]). GRTS ensures sampling sites are well spread out across the survey area (spatially-balanced), a desirable property for spatial sampling that enhances estimation efficiency and representativeness [18]. Other common sampling designs include simple or stratified random sampling or systematic

sampling. However, simple random sampling often clumps sites [19] and stratified random sampling (based on habitat in our case) was not possible without more detailed knowledge of the study area. Systematic sampling prevents clumping but an accurate design-based variance estimator does not exist; a factor that was important for other aspects of the broader study [18,26]. provide a good discussion of the relative merits of various spatial sampling strategies. Another property of GRTS is that it produces an ordered list of sample sites and any set of consecutive sampling sites also maintain spatial balance, offering flexibility to adaptively change sample sizes in the field, if for example, sampling takes less time than expected. To date GRTS has primarily been used in natural resource assessment and monitoring in the United States e.g. [27].

Because of our lack of knowledge on the distribution of habitat types, we implemented a two-phase sampling program. In phase one, the distribution of habitat types across the reserve was quantified by characterising the first forty, 200 m square, sites from a master list of GRTS sites (essentially an ordered list of all possible GRTS sites in the study area. See [28] for a detailed discussion on the use of master lists) as either 'soft' sediments or 'mixed', low profile patchy reef using MBS and footage obtained from a drop camera (Fig. 1; Lawrence, unpublished data). In phase two of the sampling program, the subject of this paper, we sampled the fish assemblages using BRUVs. Ideally we would have revisited all forty of the phase one GRTS sites. The large size of the study area and the soak time required for each BRUV deployment however, meant that it was not possible to do so in the ship time available. Instead, we sampled clusters of sites surrounding a subset of the phase one GRTS- sites. In phase two, we sampled mixed reef habitats more intensively than sediment habitats because shelf reef systems within the reserve are recognised as an important biodiversity feature and a *priori* we expected that reefs would harbour a greater diversity of assemblage types than sediments. We utilised the fact that consecutive GRTS sites maintain spatial balance and selected (from the ordered list of phase one sites) the first eight mixed habitat sites and the first three sediment sites of the forty phase one sites as the basis of the phase two clusters (Fig. S1. shows the site numbers of the forty phase ones sites and lists those used as the basis of phase two clusters.) This amounts to selecting phase one sites for the BRUV sampling that are spatially balanced *within* each of the phase one habitat types, mixed reef and sediments, and presumably therefore also spread across the environmental gradients in the region. In addition, a site was added near the head of a canyon, to ensure representation of another identified feature of the reserve which occupies a relatively small proportion of space and is therefore less likely to be sampled in a probabilistic sampling design (Fig. 1B). Clusters of sites within 1 km of the subset of phase one sites were selected again using the GRTS master list, resulting in five sites per cluster (Fig. 1B).

At each site, one BRUV was deployed using systems and deployment conditions widely used in Australia [29]. Each stereo-BRUV consisted of a frame that houses a stereo camera pair, a bait bag attached to an arm within field of view of the cameras, a diode for synchronising imagery between camera pairs. In water deeper than approximately 70 m, a light with a blue filter was also attached [15]. Stereo- BRUVs were baited with 1 kg of crushed pilchards and deployed for 60 min (soak time). Appropriate ethics (University of Tasmania Animal Ethics Permit: A12514) and fieldwork (Australian Government Director of National Parks Approval of Research Activities in the Southeast Commonwealth Marine Reserve Network. Ref No 07/10622) approvals were obtained for this work.

Analysis

Imagery collected using stereo-BRUVs was scored using standard metrics including scoring the maximum number of fish occurring in any one frame for each species (MaxN) [15,30]. Scoring was completed with Event Measure software [31].

To determine the types of fish assemblages present at sites sampled on the Flinders CMR shelf, fuzzy clustering was performed on multivariate fish composition data using the 'cluster' package [32] in R [33]. Multivariate data were square root transformed to reduce the influence of the most abundant species and a Bray Curtis dissimilarity matrix generated. The number of fuzzy clusters was set to six, following preliminary analysis using several different clustering methods, and the membership expo-

nent (which controls the 'fuzziness' of clusters) set to 1.3 after examining silhouette profiles of group membership.

A constrained ordination (Canonical Analysis of Principle Coordinates; CAP) [34] was used to test if the assemblage groups determined by fuzzy clustering were significantly different from each other (using a permutational test). In addition, jack-knife sampling was used to examine the overall and individual classification accuracy of the six groups. To enable the ecological interpretation of assemblages, the trophic level and habitat preference of species was defined. Species were assigned one of seven habitat preference categories: 1) pelagic; 2) wide-ranging demersal; 3) sediment associated; 4) reef and sediment associated; 5) reef and seagrass associated; 6) reef associated; and 7) demersal generalists (found in more than two habitats) based on species attributes described in [35,36]. The trophic level of species was extracted from FishBase [37]. Differences in the trophic level of species comprising the six groups was assessed with a one factor Analysis of Variance (ANOVA) after transforming data to satisfy the assumptions of ANOVA. To understand which species were the key contributors to each of the six assemblage groups, a SIMPER analysis was performed. SIMPER decomposes the Bray-Curtis dissimilarity matrix between all pairs within a group into the percentage contributions from each species [34].

Correlations between fish assemblages and their environment were examined against three easily quantifiable environmental variables that, from the literature, may influence benthic assemblages in our system; latitude, depth and the substratum type. At each cluster of five sites, only the central site had been surveyed with MBS and drop camera in phase one of the sampling program. Therefore, substratum type was derived from the BRUV field of view and characterised as either: 'sediment' if no hard substratum or no hard substratum-associated organisms (i.e. sponges etc.) were present; 'mixed' if some hard substratum was visible or if some hard substratum-associated organisms were present; and 'reef' if the majority of the field of view contained hard substratum or hard substratum-associated organisms. Latitude was taken from the recorded location of BRUV drops and depth was derived using the 250 m resolution bathymetry grid available from Geoscience Australia [25]. Relationships between assemblage groups and environmental variables were inferred by examining correlations with CAP axes and overlaying vectors of the environmental variables on CAP plots. Ordination and SIMPER analysis were performed in PRIMER + with PERMANOVA [34].

The influence of environmental variables in determining patterns in species richness (the number of species present) was examined using Generalised Linear Mixed Models (GLMMs). Observations were modelled using a Poisson distribution with a log link function and sampling cluster was included as a normally distributed random factor. Mixed models were fit in R [33] with penalised quasi-likelihood, using the glmmPQL function in the package MASS [38].

Results

Distribution of habitats across the CMR

The results of phase one sampling indicated that the majority (70%) of the 40 sites on the shelf were sedimentary habitats, and the remainder 'mixed' reef and sediment habitat. Mixed reef habitat occurred in the north-west, south-west, and near the edge of the shelf of the study area (Fig. 1A; Lawrence, unpublished data).

Table 1. List of species recorded in Baited Remote Underwater Video deployments across the continental shelf in the Flinders Commonwealth Reserve.

Family	Species name	Common name	Across CMR			Mean (SE) MaxN by group					
			Total MaxN	Max MaxN in any drop	Prevalence (% drops)	Group 1	Group 2	Group 3	Group 4	Group 5	Group 6
Blenniidae, Gobiidae, Tripterygiidae	Blenniidae, Gobiidae, Tripterygiidae	Blennies, Gobies, Triplefins	7	3	10	0.38 (0.18)	0.4 (0.31)	0	0	0	0
Callorhinchidae	Callorhinchus milii	Elephant shark	1	1	2	0	0.1 (0.1)	0	0	0	0
	Nemadactylus douglasii	Grey morwong	7	2	10	0	0.2 (0.13)	0.33 (0.24)	0.33 (0.33)	0	0
	Nemadactylus macropterus	Jackass morwong	379	43	59	0	10.7 (2.04)	23.22 (4.78)	0	5.25 (1.23)	0
Cyttidae	Cyttus australis	Silver dory	6	1	12	0	0.1 (0.1)	0	0	0.42 (0.15)	0
Diodontidae	Diodon nicthemerus	Globe fish	1	1	2	0	0	0	0	0.08 (0.08)	0
Gempylidae	Thyrsites atun	Barracouta	14	3	20	0.38 (0.37)	0.1 (0.1)	0.11 (0.11)	0	0.75 (0.22)	0
Gerreidae	Parequula melbournensis	Silverbelly	79	10	31	0	0.5 (0.27)	0	0	5.83 (0.6)	0.67 (0.67)
Heterodontidae	Heterodontus portusjacksoni	Port Jackson shark	4	1	8	0	0.3 (0.15)	0	0	0.08 (0.08)	0
Labridae	Ophthalmolepis lineolatus	Southern maori wrasse	1	1	2	0	0	0.11 (0.11)	0	0	0
	Pseudolabrus rubicundus	Rosy wrasse	72	13	25	0	1.4 (0.88)	0.67 (0.67)	0	4.33 (1.38)	0
Lamnidae	Isurus oxyrinchus	Shortfin Mako shark	1	1	2	0.12 (0.12)	0	0	0	0	0
Latridae	Latris lineata	Striped trumpeter	95	27	22	0	0.2 (0.13)	10.11 (3.47)	0.33 (0.33)	0	0
Loliginidae	Sepioteuthis australis	Southern calamari	21	3	29	0	0.2 (0.13)	0	0	1.5 (0.23)	0.17 (0.17)
Monacanthidae	Acanthaluteres vittiger	Toothbrush leatherjacket	15	3	18	0	0	0	0	1.25 (0.28)	0
	Meuschenia freycineti	Six-spine leatherjacket	3	1	6	0	0.2 (0.13)	0	0	0.08 (0.08)	0
	Meuschenia scaber	Velvet leatherjacket	97	11	53	0	3.3 (0.87)	0.78 (0.43)	0	4.25 (0.97)	1 (0.63)
	Meuschenia venusta	Stars and stripes leatherjacket	1	1	2	0	0.1 (0.1)	0	0	0	0
	Thamnaconus degeni	Degen's leatherjacket	471	89	41	0	1 (0.42)	0.11 (0.11)	0	36.92 (7.37)	2.83 (1.9)
Moridae	Pseudophycis barbata	Bearded cod	8	2	12	0	0.1 (0.1)	0.78 (0.28)	0	0	0
Mullidae	Upeneichthys vlamingii	Southern goatfish	16	8	16	0	0.3 (0.21)	0.11 (0.11)	0	1 (0.65)	0
Neosebastidae	Neosebastes scorpaenoides	Common gurnard perch	44	4	41	0	1.1 (0.31)	0	0	1.5 (0.42)	2.5 (0.5)
Ommastrephidae	Ommastrephidae sp.	Squid	9	2	12	0.5 (0.27)	0	0	0.83 (0.4)	0	0
Ostraciidae	Aracana aurita	Shaw's cowfish	7	2	12	0	0.1 (0.1)	0	0	0.5 (0.19)	0
Ostraciidae	Aracana ornata	Ornate cowfish	1	1	2	0	0.1 (0.1)	0	0	0	0
Palinuridae	Jasus edwardsii	Southern rocklobster	2	1	4	0	0	0.22 (0.15)	0	0	0
Paraulopidae	Paraulopus nigripinnis	Blacktip cucumberfish	47	12	22	2 (0.65)	0.1 (0.1)	0	5 (2.16)	0	0
Pinguipedidae	Parapercis allporti	Barred Grubfish	36	5	35	2.12 (0.52)	0.8 (0.25)	0.44 (0.44)	0.5 (0.34)	0	0.67 (0.42)
Platycephalidae	Platycephalus bassensis	Sand flathead	33	6	24	0.12 (0.12)	0	0	0.17 (0.17)	0.83 (0.42)	3.5 (0.96)
	Platycephalus richardsoni	Tiger flathead	21	4	25	0	0.2 (0.13)	0	1.67 (0.56)	0.17 (0.17)	1.17 (0.31)
Pristiophoridae	Pristiophorus cirratus	Longnose sawshark	2	1	4	0	0	0	0	0.08 (0.08)	0.17 (0.17)
Rajidae	Dentiraja lemprieri	Thornback skate	2	1	4	0	0	0	0	0.17 (0.11)	0

Table 1. Cont.

Family	Species name	Common name	Across CMR			Mean (SE) MaxN by group					
			Total MaxN	Max MaxN in any drop	Prevalence (% drops)	Group 1	Group 2	Group 3	Group 4	Group 5	Group 6
	Spiniraja whitleyi	Melbourne skate	9	1	18	0	0.3 (0.15)	0	0.33 (0.21)	0.33 (0.14)	0
Scyliorhinidae	Cephaloscyllium laticeps	Draughtboard shark	63	4	57	0.12 (0.12)	1.7 (0.5)	0.44 (0.34)	1.33 (0.49)	2.33 (0.31)	0.83 (0.48)
Sebastidae	Helicolenus percoides	Red gurnard perch	65	12	31	0.25 (0.25)	2.5 (0.81)	4.22 (1.28)	0	0	0
Serranidae	Caesioperca lepidoptera/rasor	Butterfly and Barber perch	140	60	14	0	6.1 (5.99)	2.11 (1.42)	0	5 (2.79)	0
Squalidae	Squalus acanthias	Spiny dogfish	1	1	2	0	0	0	0.17 (0.17)	0	0
	Squalus megalops	Shortnose spurdog	15	3	14	0.75 (0.49)	0	0	1.5 (0.43)	0	0
Triakidae	Mustelus antarcticus	Gummy shark	18	2	33	0.25 (0.16)	0	0	1.17 (0.17)	0.25 (0.13)	1
Triglidae	Lepidoperca pulchella	Tasmanian perch	11	5	10	0	0.1 (0.1)	1.11 (0.56)	0	0	0
	Lepidotrigla sp	Gurnards	3	1	6	0	0.1 (0.1)	0.11 (0.11)	0.17 (0.17)	0	0
Urolophidae	Urolophus cruciatus	Banded stingaree	2	1	4	0	0.1 (0.1)	0	0	0	0.17 (0.17)
	Urolophus paucimaculatus	Sparsely spotted stingaree	2	1	4	0	0.2 (0.13)	0	0	0	0
	Urolophidae sp.	Stingarees	4	1	8	0.12 (0.12)	0.2 (0.13)	0	0	0	0.17 (0.17)

The Total number of individuals recorded at Max N across all deployments, the greatest MaxN recorded in any single deployment and the prevalence (the percentage of deployments where a species was recorded) are presented as well as the mean (and standard error) MaxN for each assemblage group.

Table 2. Comparison between the number of sites assigned to each of the six groups using fuzzy clustering and Canonical Analysis of Principle coordinates (CAP).

	CAP Classified Group						Total	Correct
	1	**2**	**3**	**4**	**5**	**6**		**(%)**
Fuzzy Group								
1	8	0	0	0	0	0	8	100
2	0	9	1	0	0	0	10	90
3	0	1	8	0	0	0	9	89
4	0	0	0	5	0	1	6	83
5	0	0	0	0	12	0	12	100
6	0	0	0	0	0	6	6	100

Jack knife validation in the CAP procedure was used to assess overall and group-wise accuracy of the six groups determined by fuzzy clustering. Classification accuracy of all groups is good, indicating that the groups determined by the fuzzy clustering are robust.

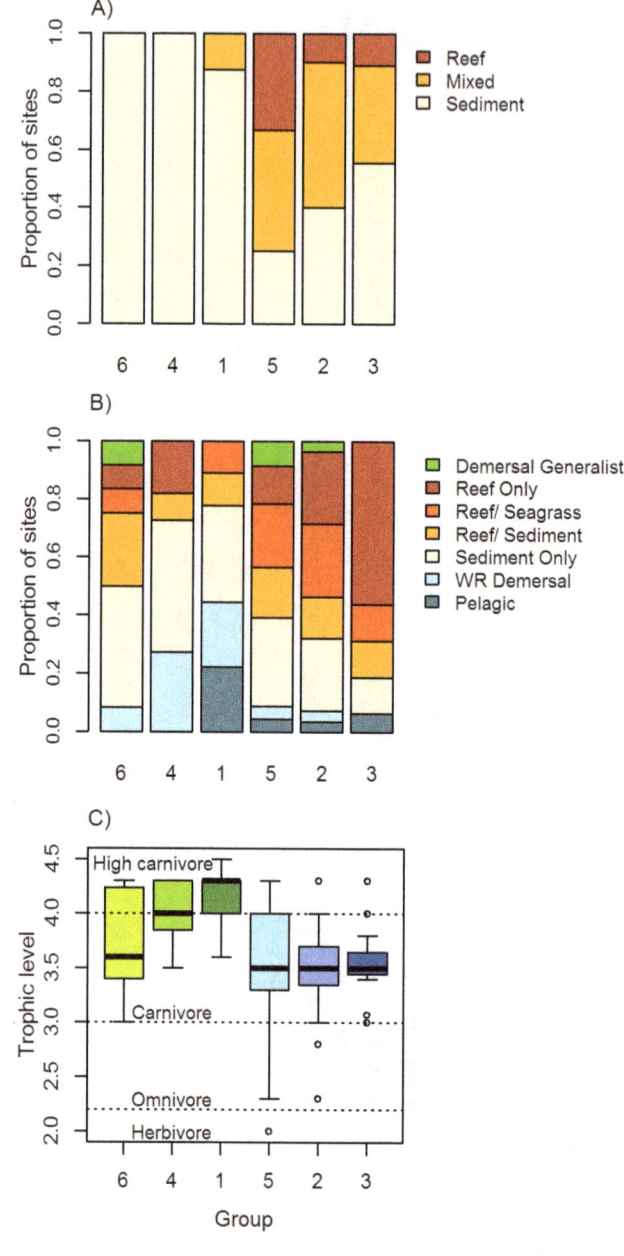

Figure 2. Characteristics of the six assemblage groups identified using fuzzy-clustering. A) Proportion of sites within each group classified as sediment, mixed or reef habitat based on the BRUVs footage; B) the habitat preference of species within each assemblage group; C) boxplot of the trophic level of species within each assemblage group; the box represents the 1st and 3rd quartiles, and circles denote potential outliers. In all panels, groups are ordered based on dominant habitat contained within groups (sediment versus mixed reef/reef) and from shallow to deep within each broad habitat type (based on spatial distribution of groups presented in Fig. 5) In plot C) sediment- associated groups are coloured green and reef-associated groups, blue.

Fish assemblages present in the CMR

Sixty BRUV deployments were completed in phase two, of which 51 yielded video footage of sufficient quality for scoring fish assemblage composition. Overall, 45 species were observed, with a total of 1,837 individuals recorded (at Max N) (Table 1). Of these species, 66% are endemic to Australia. Leatherjackets (Mona-canthidae) were the most numerically abundant group, and

Sediment Assemblage

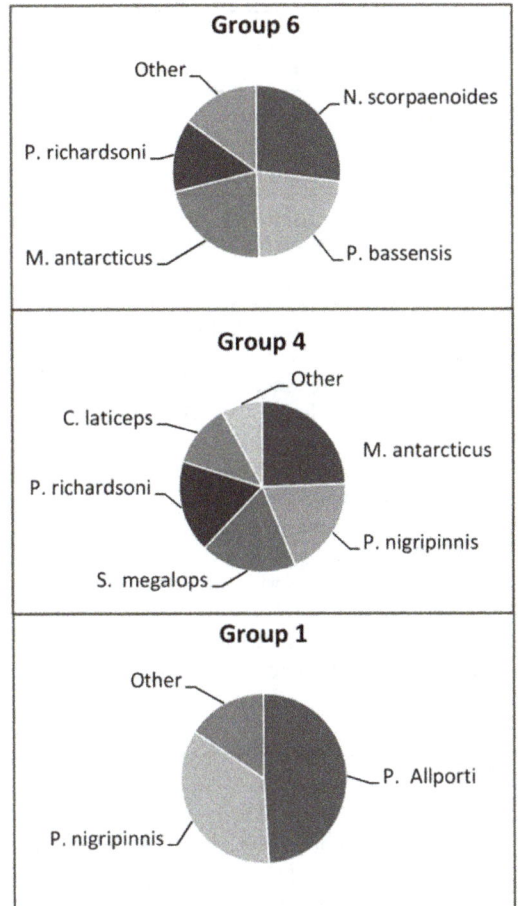

Reef Assemblage

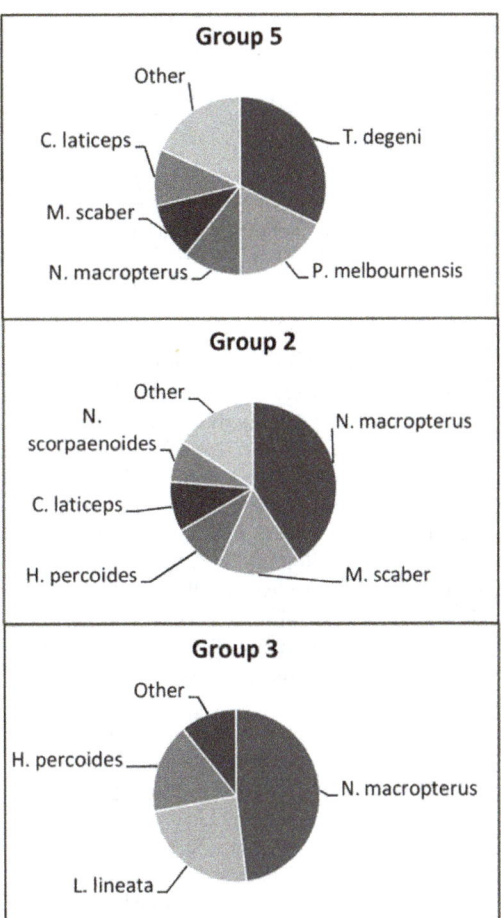

Figure 3. Species contributing up to 80% towards the similarity of the predominantly sediment- and predominantly reef-associated groups identified using fuzzy clustering.

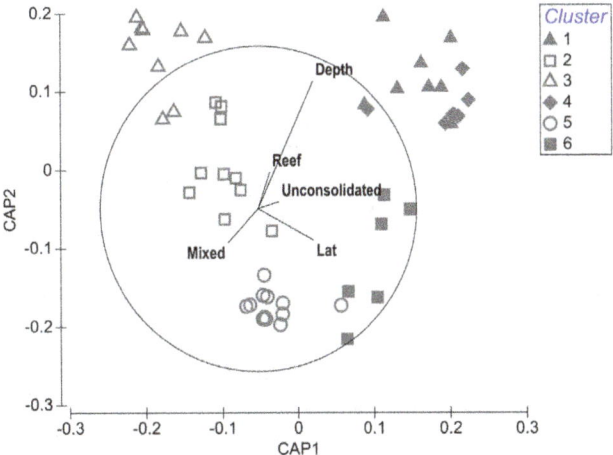

Figure 4. Canonical Analysis of Principle coordinates (CAP) plot discriminating sites based on fuzzy clustering. Each site was assigned to the cluster for which it had the highest probability of membership. Sites are shown as symbols: filled grey symbols represent sand-associated assemblages and open symbols represent reef-associated assemblages. Vectors of all environmental variables are overlaid and are proportional to their correlation with either CAP axis one or two.

comprised mostly *Thamnaconus degeni* (Degan's leatherjacket) and *Meuschenia scaber* (velvet leatherjacket) with, respectively, a total of 471 and 97 individuals recorded and MaxN of 89 and 11. *Nemadactylus macropterus* (jackass morwong), and *Caesioperca Lepidoptera and C. razor* (butterfly and barber perch) were the next most abundant species (379 and 140 individuals recorded and MaxN of 89 and 60). Other abundant species included *Latris lineata* (striped trumpeter), *Parequula melbournensis* (silverbelly), *Pseudolabrus rubicundus* (rosy wrasse), *Helicolenus percoides* (reef ocean perch), and *Cephaloscyllium laticeps* (draughtboard shark). The five most prevalent species were *N. macropertus* (59% of sites), *C. laticeps* (57% of sites), *M. scaber* (53% of sites), *Neosebastes scorpaenoides* (common gurnard perch; 41% of sites) and *T. degeni* (41% of sites).

The structure of fish assemblages surveyed was adequately represented by six groups. The overall classification accuracy of the six assemblage groups was high (94%) and a majority of sites (82%) were also assigned to groups with a high degree of confidence (greater than 0.85 probability of membership). Groups four and three were the least certain, but still had a high classification accuracy of 83% and 89%, respectively (Table 2). In addition, these six groups were statistically meaningful with significantly different multivariate community composition (P = 0.001).

Table 3. Results of Poisson GLMM for species richness recorded in BRUV deployments.

Factor	Co-efficient	Std Error	df	t-value	p-value
Depth	−0.277	0.067	36	−4.1489	0.0002
Mixed Habitat	0.330	0.092	36	3.575	0.0010
Reef Habitat	0.382	0.121	36	3.169	0.0031

Predictor variables have been scaled and centered and co-efficients are on the scale of the link function (log). Spatial clusters introduced as part of the sampling design were included as a random effect and the effects of habitat type (sediment, mixed or reef) are presented relative to sediment habitat.

The six assemblage groups were broadly aligned with habitats observed in the BRUV footage. Sites belonging to three groups (groups 1, 4, 6) contained exclusively or predominately sediment habitat (Fig. 2A) and contained greater numbers of species with a preference for sediment habitats. However, species with a strong affinity for reef were also found in the predominantly sediment-dominated habitat group, with groups 4 and 6 containing disproportionately more reef species than suggested by the available habitat (Fig. 2B). Sediment- associated assemblages (particularly groups 1 and 4) also contained a relatively high proportion of wide-ranging demersal species. Groups 2, 3, and 5 contained moderate to high proportions of mixed reef and reef habitat with the highest proportion of reef-associated species observed in group 3 (Fig. 2A, B).

Species representing a range of trophic levels were recorded in BRUV footage. Carnivores were the most numerous species, as expected given the use of baits. Even so, the average trophic level of species differed between the six assemblage groups (df = 5, F = 5.46, P<0.001; Fig. 2C). Groups 1 and 4 (sediment groups) contained species with significantly higher trophic level (higher carnivores) than groups 2 and 3 (mixed-reef) which contained primarily carnivores. The remaining two groups (5 and 6) contained species from a relatively wider range of trophic levels (Fig. 2C) with group 5 represented by the largest range in values and the only herbivore observed.

The characteristic species of each assemblage group (making the highest contributions to within-group similarity) re-enforced the broad habitat-related patterns described above (Fig. 3). Sediment-associated species such as flathead (*Platycephalus bassensis*, *Neoplatycephalus richardsoni*) and sharks (*Mustelus antarcticus*, *Squalus megalops*) as well as the common gurnard perch (*Neosebastes scorpaenoides*) and cucumber fish (*Paraulopus nigripinnis*) were characteristic of groups 4 and 6. Group 1 also consisted of sediment-associated fish (eastern- barred grubfish - *Parapercis allporti*- as well as *P. nigripinnis*), but was differentiated from groups 4 and 6 by relatively low abundances and very few species. Of the predominantly mixed-reef associated groups, groups 2 and 3 were both dominated by jackass morwong (*Nemadactylus macropterus*). Group 2 was also composed of a variety of others species including the velvet leatherjacket (*Meuschenia scaber*), reef ocean perch (*Helicolenus percoides*) and draughtboard sharks (*Cephaloscyllium laticeps*). Group 3 was further characterised by relatively high abundances of striped trumpeter (*Latris lineata*) as well as reef ocean perch. Group 5 was differentiated from other groups by large numbers of leatherjackets (primarily *Thamnaconus degeni*), but was also characterised by silverbelly (*Parequula melbournensis*) and jackass morwong.

Relationships with environmental variables

Environmental factors correlated well with the fish assemblage groupings (Fig. 4). Depth delineated groups most strongly (a

correlation with CAP axis 2 of 0.78) and groups 1, 4 and 3 were generally found at greater depth. Substratum type also influenced assemblages, with sediment-associated groups (1, 4 and 6) falling on the right hand-side of the CAP plot, confirming the patterns seen in the composition of species above. Latitude was also moderately correlated with axis one (0.35) and appears to separate the mixed/reef associated groups 2, 3 and 5.

Species richness was also correlated with environmental factors. Richness decreased with depth and mixed and reef substratum also supported more species than sediments (Table 3). Latitude however, was not correlated with species richness (Table 3). This resulted in spatial patterns as depicted in Fig. 5, where assemblages are generally richer towards the shallow, western side of the reserve.

Spatial distribution of fish assemblages across CMR

Fish assemblages showed distinct spatial patterns across the Flinders CMR (Fig. 5). When combined with depth, the assemblages can be broadly categorised as: shallow, reef-associated (group 5); intermediate reef-associated (group 2); deeper, reef-associated (group 3); shallow sediment-associated (group 6); and deeper, reef-associated (groups 1 and 4). While sites within the same spatial cluster often belonged to the same fish assemblage, this was not always the case. Clusters of sites close to the shelf edge and in the south of the reserve were often more heterogeneous, composed of two to three different assemblages within a 1 km radius.

Discussion

This study trialled the use of BRUVs and a spatially- balanced survey design to provide the first quantitative description of demersal fish species and assemblages occurring the multiple use zone of the Flinders CMR shelf. Non-extractive sampling with BRUVs identified characteristic species (including jackass morwong, leatherjackets, draughtboard sharks, reef ocean perch and common gurnard perch), several species of commercial interest (including gummy sharks, reef ocean perch, striped trumpeter and jackass morwong) and species from trophic groups ranging from omnivores through to higher predators. While many of the species present in the study area are widely distributed across southern and/or south-eastern Australia, the majority are endemic to Australian waters. This highlights the unique biodiversity represented in temperate Australia, primarily as a result of its relatively isolated evolutionary history [39,40], and reinforces the global importance of conserving representative fauna.

Spatial and environmental patterns in fish assemblages

Six assemblages with a distinct spatial pattern were observed at sites across the reserve. Many species contributed to more than one assemblage and differences in the relative abundance of

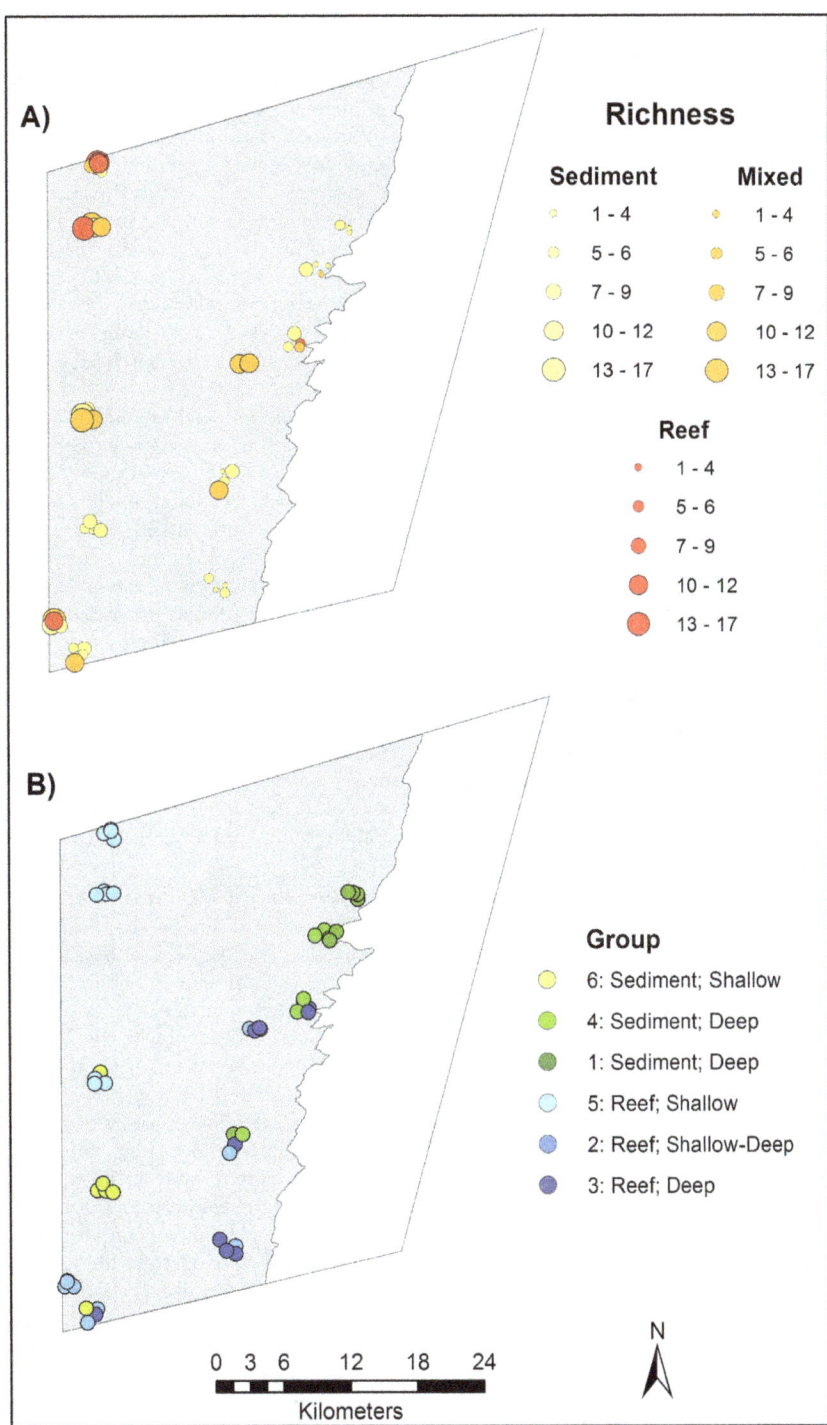

Figure 5. Spatial distribution of A) species richness B) fish assemblage groups on the Flinders CMR shelf (shaded grey). Increasing size of symbols in A) indicate increasing species richness. Symbols are colour coded according to the observed substratum type in BRUV footage: yellow = sediment; orange = mixed; red = reef. Assemblages in B) are coded by colour, with predominantly sediment-associated assemblages coloured green and predominantly reef-associated assemblages coloured blue.

species defined the differences between several assemblages. Assemblages were structured by depth and habitat type, and could be heterogeneous at relatively small scales (i.e. within a sampling cluster).

Depth-related assemblage transitions were most distinct and occurred in waters greater than approximately 80 m for sediment-associated assemblages and approximately 60 m for reef-associated assemblages, which is largely consistent with other studies on the southeast Australian shelf [41]. While some species were found across almost the full depth range of the study area (e.g. jackass morwong and the flathead, *Platycephalus richardsoni*) several species characterised particular substratum and depth zones. Leatherjackets (*T. degeni* in particular) for example, were characteristic of shallow reef sites; striped trumpeter were indicative of deeper, reef-associated assemblages; and cucumber fish were indicative of deeper, sediment-associated assemblages. These results accord with other studies where depth has a major influence on the distribution of fish assemblages [17,42].

As well as influencing assemblage composition, depth also influenced the diversity and ecology of assemblages. Species richness generally declined with depth consistent with observations elsewhere [39,43,44] and the trophic range of species occurring in shallow assemblages was generally greater than deeper assemblages. In reef-associated assemblages for example, larger trophic ranges of the shallow water assemblage were due to the co-occurrence of species such as leatherjackets (many of which are omnivorous) and draughtboard sharks (which are carnivores). Similarly, other studies have noted trophic-related changes with depth, such as increases in body size [42] and a transition towards higher −order predators [34].The combination of assemblage level patterns and changes in ecology and diversity with depth as seen here, have been used to suggest that the edge of the continental shelf forms an important faunal break. Consequently the shelf and slope environments should be considered separate management units [8,31] and not subsumed within the same management zone as currently occurs within the Flinders CMR.

Although a ubiquitous pattern, the processes involved in the depth structuring of benthic fish assemblages can be varied. In the shelf environments of the Flinders CMR, we suggest these are partly attributable to the strong tidal currents which affect the shallow, western section on the reserve [22]. In this region, low-profile reefs are dominated by tall biogenic structures with flexible forms such as sea whips and erect branching sponges (personal observation) that are indicative of high energy environments. This 3D biogenic structure is much reduced towards the deeper, eastern extent of the study area. These differences in reef structure [45], as well as physical limitations posed by increased current velocities [46], may have a large influence on the assemblages found in shallow regions of the reserve. Other physical and oceanographic conditions also co-vary with depth. For example, light availability decreases with depth, which affects primary productivity and therefore the range of food sources available to fish and appears to correspond with the observed changes in the trophic composition of assemblages. In addition, the oceanography in this region is complex, driven largely by inter-annual and inter-seasonal variability in the southward penetration of the warm-water, East Australian Current (EAC). The EAC can incur on the shelf or form fronts at the shelf break [24] and affects processes such as recruitment [47] and primary productivity [48] that are in turn likely influence the composition of fish assemblages, particularly on the outer shelf. Regardless of the proximal processes involved, the depth structuring of assemblages observed in the study area suggests that future sampling would benefit by breaking the region into shallow and deep strata around the 70 m depth mark.

Habitat or substratum type was also influential in discriminating between the assemblages observed in the Flinders CMR shelf. Despite preferentially sampling clusters of sites surrounding our initial mixed/reef sites, a large number of all sites surveyed contained only sediments, at least within the field of view of the BRUVs. This occurred because low-profile shelf reefs in the CMR are small scale features (tens of meters) and their distribution is highly patchy. Sediment sites were generally less species rich than those found on mixed or reef substrata. However, the composition of these sites varied enough to form three distinct assemblage groups, primarily on the basis of their depth distribution. The lower diversity of sediment assemblages has been attributed to

lower habitat complexity with fewer niches to support co-existing species [49]. Still, sediment-associated assemblages contained some species with affinity for reef habitats. This may be because reef habitat was present outside the BRUVs field of view but within the swimming distance of fishes resulting in a 'halo' effect where species with an affinity for reef and sediments co-occur [49]. In subtropical shallow reefs this halo can extend 200 m from the reef edge [49]. The importance of local substratum in structuring fish assemblages highlights the need to develop comprehensive, high resolution habitat maps for the reserve. Assembling these habitat maps for marine reserves will require a commitment to strategic mapping of the seabed using multibeam sonar, so that over time complete coverage is achieved. For the continental shelf, this mapping could take years to decades [50]. In the meantime, management will need to adopt spatially representative sampling such as presented in this study to provide habitat and biodiversity inventories and to underpin monitoring programs.

Geomorphic features of the seafloor, such as submarine canyons, can substantially alter water flows, enhance local productivity [51] [52] and influence fish assemblage patterns [53]. Canyons are a conservation feature of the Flinders CMR and several canyons intersect the shelf edge. Canyons were represented in our sampling by a cluster of sites near a canyon head in the north of the survey area. Assemblages adjacent to canyon heads were not however, unique to canyon heads and were also found in other deeper sections of the shelf. Neither were these assemblages particularly species rich. Despite our preliminary findings, a better understanding of the ecological significance of canyon heads in this system will require a more targeted and intensive sampling strategy within and around the canyon head systems, e.g. [51].

Implications for Monitoring and Management

An objective of the Flinders CMR − and the larger Southeast Marine Reserve Network of which it is part - is to protect the biodiversity, natural and cultural values of the region. These include "representative examples of the ecosystems, communities and habitats" associated with the Tasmanian Shelf and Southeast Shelf Transition Provinces [21]. While these values are defined in broad terms from the national bioregionalisation, these have yet to be translated into inventories of conservation values, with specific metrics for monitoring their status. This study is a first step towards developing both an inventory of biological values, and elements of an effective biological monitoring program, including relevant environmental influences, for fish communities represented in the shelf component of the Flinders CMR. We suggest that representative demersal fish communities include those found on dynamic, low- profile reefs in both shallow (<70 m) and deep (70–160 m) areas of the shelf as well as those found on sediments in the same depth strata, and here we have characterised each of these assemblages. Several prevalent commercial and recreational species, including striped trumpeter and jackass morwong, which are likely to be responsive to management actions may be good candidates as indicator species and this work will provide the basis for further research into deriving specific monitoring targets.

We used a non-standard sampling design, GRTS, to survey fish assemblages and to satisfy the objectives of the broader survey of the Flinders CMR. Due to the cost and logistics associated with sampling in offshore and/or remote environments, the use of survey designs that are efficient, representative and flexible enough to accommodate multiple objectives is likely to become increasingly important. This is particularly true when simultaneously considering the inventory and monitoring needs of multiple reserves within networks. In the present study, the GRTS sampling ensured that clusters of reef and sediment sites sampled with BRUVs are well spread out across the reserve (within the distribution of each habitat type) and consequently encompass broad spatial and environmental gradients. This in turn provides confidence in the patterns observed and provides a better basis for additional or repeated sampling in a monitoring context. The site-ordered sampling of GRTS has the potential disadvantage of using ship time inefficiently if there is a large cumulative distance between deployments. This would have been the case in our study, but was circumvented by clustering samples around a subset of original sites. Conversely, the site ordering of GRTS allows dynamic altering of samples according to situations in the field, while maintaining spatial balance within strata.

The GRTS methodology will also be useful when choosing samples for future sampling, typically as part of an adaptive monitoring program. The master list of samples, that encompasses all possible sampling locations [28], can be subset in different ways by varying the inclusion probabilities for factors such as habitat types or depth, to accommodate new research questions or changing management objectives. While such changes lead to higher or lower sampling rates for different habitats or strata, the set of sites selected always maintains spatial balance within strata. Inclusion probabilities are taken into account when deriving representative estimates of the occurrence, proportions or abundance of metrics or indicators across the entire sampling area [18]. Once suitable indicator species and metrics have been selected, the GRTS master sample can provide the basis for rotating panel designs [28] for repeat sampling that aims to detect changes in the abundance of these species, and possibly also major shifts in the distribution of assemblage types. Another advantage of GRTS in this respect, is its use of a local neighbourhood variance estimator, which is generally unbiased and more precise than variance estimators from other types of sampling designs [26]. Future sampling within the context of monitoring however, will need to incorporate comparable areas outside of the multiple use management zone of the Flinders CMR. In addition sufficient sample sizes will be needed to provide enough power to distinguish between natural variability and biological trends in order to be able to assess whether current management controls are effective in achieving their stated objectives.

The use of baited systems to survey fish communities has sometimes been criticised because of the bias towards sampling scavenging and predatory species [54]. However, all fish sampling methods have selectivity biases, and alternatives for sampling deep continental shelf fishes (e.g. trawls, meshnets and traps [41]) are extractive (and some destructive to the habitats the reserve is intended to protect). We recognise that by using BRUVs we have preferentially observed a subset of species that is attracted to, or undeterred by, baits. However, BRUVs data have been shown to clearly discriminate between fish assemblages in a variety of environmental settings [43] and have better statistical power to detect spatial and temporal changes in assemblage structure and abundances of individuals than unbaited methods [54]. In the

future, we may complement BRUVs with other non-extractive video techniques (e.g. forward looking cameras on remotely or autonomously operated vehicles) to capture other components of the community. Never-the-less, in our present study BRUVs successfully observed a broad range of species, including key ecological species and potential indicator species. Standardised methodologies, in combination with robust sampling designs, will be essential for ongoing monitoring and assessment across multiple reserves in Australia's reserve network. Finally, the challenges we outline for the inventory and long-term monitoring of large and remote marine reserves are not unique to the Australian reserve network, nor exclusively to marine reserves. Therefore the approach that we have demonstrated may be usefully applied in other systems and under other management scenarios.

Supporting Information

Figure S1 The location of the forty GRTS sites sampled across the Flinders CMR shelf in phase one. Sites are coloured according to the broad habitat type: sediment (yellow); mixed, low-profile reef and sediments (red); canyon head (blue) recorded during phase one of sampling. Sites are labelled with their GRTS site number which, when sampled in order, represents a spatially balanced sample. The table shows the subset of sites sampled as the basis of clusters in the phase two sampling with BRUVs. These sites are the first three GRTS sites classified as sediment and the first eight classified as mixed reef.

Data S1 Matrix of the species observed at each BRUV deployment. Environmental variables used in analyses are also included.

Data S2 Attributes of species used in analysis, including trophic level, habitat preference and endemicity. Summary statistics on the prevalence and abundance of species across the survey area are also given.

Acknowledgments

The authors would like to thank the crew of *Challenger*, Matt Francis and Tim Green, for their expertise and dedication during fieldwork, Richard Zavalas for assistance with fieldwork and video scoring, Johnathan Kool for assistance with fieldwork, and Vanessa Lucieer for assistance in the planning phase. SN publishes with permission of the Chief Executive Officer of Geoscience Australia.

Author Contributions

Conceived and designed the experiments: NAH NB EL JMD SN KRH. Performed the experiments: NAH NB JH SN. Analyzed the data: NAH NB EL JH KRH. Contributed reagents/materials/analysis tools: NAH NB EL JH SN AW KRH. Wrote the paper: NAH NB EL JMD SN AW KRH.

References

1. Halpern BS, Walbridge S, Selkoe KA, Kappel CV, Micheli F, et al. (2008) A global map of human impact on marine ecosystems. Science 319: 948–952.
2. Halpern BS, Longo C, Hardy D, McLeod KL, Samhouri JF, et al. (2012) An index to assess the health and benefits of the global ocean. Nature; 615–620.
3. Day J, Dudley N, Hockings M, Holmes G, Laffoley D, et al. (2012) Guidelines for applying the IUCN Protected Area Management Categories to Marine Protected Areas. Gland, Switzerland: IUCN. 36 p.
4. Kemp J, Gibson R, Atkinson R, Gordon J, Hughes R, et al. (2012) Measuring the performance of spatial management in marine protected areas. Oceanography and Marine Biology: An Annual Review 50: 287–314.
5. Gubbay S, editor (1995) Marine Protected Areas: principles and techniques for management. London: Chapman and Hall.
6. CBD (1992) Convention on Biological Diversity. 5 June 1992 Rio de Janerio (Brazil).
7. CBD (1995) The Jakarta Mandate on Marine and Coastal Biological Diversity. Decisions of the Second Meeting of the Conference of the Parties to the Convention on Biological Diversity. Jakarta, Indonesia, 6–17 November 1995: UNEP.
8. Spalding MD, Meliane IN, Milam A, Fitzgerald C, Hale LZ (2013) Protecting marine spaces: Global targets and changing approaches. Ocean Yearbook Online 27: 213–248.
9. Edgar GJ, Stuart-Smith RD, Willis TJ, Kininmonth S, Baker SC, et al. (2014) Global conservation outcomes depend on marine protected areas with five key features. Nature advance online publication.

10. Barr LM, Possingham HP (2013) Are outcomes matching policy commitments in Australian marine conservation planning? Marine Policy 42: 39–48.

11. ANZECC (1998) Guidelines for establishing the National Representative System of Marine Protected Areas. Canberra: Australian and New Zealand Environment and Conservation Council (ANZECC) Task Force on Marine Protected Areas.

12. Commonwealth of Australia (2006) A Guide to the Integrated Marine and Coastal Regionalisation of Australia Version 4.0. In: Heritage DotEa, editor. Canberra, Australia.

13. Lucieer V, Lamarche G (2011) Unsupervised fuzzy classification and object-based image analysis of multibeam data to map deep water substrates, Cook Strait, New Zealand. Continental Shelf Research 31: 1236–1247.

14. Bailey DM, King NJ, Priede IM (2007) Cameras and carcasses: historical and current methods for using artificial food falls to study deep-water animals. Marine Ecology Progress Series 350: 179–191.

15. Willis TJ, Babcock RC (2000) A baited underwater video system for the determination of relative density of carnivorous reef fish. Marine and Freshwater Research 51: 755–763.

16. Langlois TJ, Harvey ES, Meeuwig JJ (2012) Strong direct and inconsistent indirect effects of fishing found using stereo-video: Testing indicators from fisheries closures. Ecological Indicators 23: 524–534.

17. Harvey ES, Cappo M, Kendrick GA, McLean DL (2013) Coastal fish assemblages reflect geological and oceanographic gradients within an Australian zootone. Plos One 8.

18. Stevens DL, Olsen AR (2004) Spatially balanced sampling of natural resources. Journal of the American Statistical Association 99: 262–278.

19. Olsen AR, Kincaid TM, Payton Q (2012) Spatially balanced survey designs for natural resources. In: Gitzen RA, Millspaugh JJ, Cooper AB, Licht DS, editors. Design and Analysis of Long-term Ecological Monitoring Studies. Cambridge, England: Cambridge University Press. pp. 126–150.

20. Commonwealth of Australia (2006) A Guide to the Integrated Marine and Coastal Regionalisation of Australia Version 4.0. Canberra, Australia: Department of the Environment and Heritage. 16 p.

21. Director of National Parks (2013) South-east Commonwealth Marine Reserves Network management plan 2013–23. Director of National Parks, Canberra.

22. Fandry C (1983) Model for the three-dimensional structure of wind-driven and tidal circulation in Bass Strait. Marine and Freshwater Research 34: 121–141.

23. National Tidal Centre (2014) Tide predictions for Australia, South Pacific and Antarctica.

24. Harris G, Nilsson C, Clementson L, Thomas D (1987) The water masses of the East Coast of Tasmania: Seasonal and interannual variability and the influence on phytoplankton biomass and productivity. Australian Journal of Marine and Freshwater Research 38: 569–590.

25. Whiteway TG (2009) Australian Bathymetry and Topography Grid. Canberra, Australia: Geoscience Australia. 46 p.

26. Stevens DL, Olsen A (2003) Variance estimation for spatially balanced samples of environmental resources. Environmetrics 14: 593–610.

27. Dambacher JM, Jones KK, Larsen DP (2009) Landscape-level sampling for status review of great basin redband trout. North American Journal of Fisheries Management 29: 1091–1105.

28. Larsen D, Olsen A, Stevens D (2008) Using a Master Sample to Integrate Stream Monitoring Programs. Journal of Agricultural, Biological, and Environmental Statistics 13: 243–254.

29. Harvey ES, Newman SJ, McLean DL, Cappo M, Meeuwig JJ, et al. (2012) Comparison of the relative efficiencies of stereo-BRUVs and traps for sampling tropical continental shelf demersal fishes. Fisheries Research 125–126: 108–120.

30. Cappo M, De'ath G, Speare P (2007) Inter-reef vertebrate communities of the Great Barrier Reef Marine Park determined by baited remote underwater video stations. Marine Ecology Progress Series 350: 209–221.

31. Seager J (2014) Transect Measure. 2.30 ed: SeaGIS.

32. Maechler M, Rousseeuw P, Struyf A, Hubert M, Hornik K (2013) cluster: Cluster Analysis Basics and Extensions. R package. 1.14.4 ed.

33. R Development Core Team (2008) R: A language and environment for statistical computing. Vienna, Austria: R Foundation for Statistical Computing.

34. Anderson MJ, Gorley RN, Clarke KR (2008) PERMANOVA+ for PRIMER: Guide to software and statistical methods. Plymouth, UK: PRIMER-E. 214 p.

35. Edgar GJ (2008) Australian Marine Life: the plants and animals of temperate waters. New Holland, Sydney. Sydney: Reed New Holland.

36. Gomon MF, Bray DJ, Kuiter RH (2008) Fishes of Australia's Southern Coast. Chatswood, Australia: Reed New Holland.

37. (2014) FishBase. In: Froese R, Pauly D, editors. Available: www.fishbase.org: World Wide Web electronic publication.

38. Venables WN, Ripley BD (2002) Modern Applied Statistics with S. New York: Springer.

39. Last PR, White WT, Gledhill DC, Pogonoski JJ, Lyne V, et al. (2011) Biogeographical structure and affinities of the marine demersal ichthyofauna of Australia. Journal of Biogeography 38: 1484–1496.

40. Phillips J (2001) Marine macroalgal biodiversity hotspots: why is there high species richness and endemism in southern Australian marine benthic flora? Biodiversity and Conservation 10: 1555–1577.

41. Williams A, Bax N (2001) Delineating fish-habitat associations for spatially based management: an example from the south-eastern Australian continental shelf. Marine and Freshwater Research 52: 513–536.

42. Fitzpatrick BM, Harvey ES, Heyward AJ, Twiggs EJ, Colquhoun J (2012) Habitat specialization in tropical continental shelf demersal fish assemblages. PLoS ONE 7: e39634.

43. Zintzen V, Anderson MJ, Roberts CD, Harvey ES, Stewart AL, et al. (2012) Diversity and composition of demersal fishes along a depth gradient assessed by Baited Remote Underwater Stereo-Video. PLoS ONE 7: e48522.

44. Chatfield BS, Van Niel KP, Kendrick GA, Harvey ES (2010) Combining environmental gradients to explain and predict the structure of demersal fish distributions. Journal of Biogeography 37: 593–605.

45. Gratwicke B, Speight MR (2005) The relationship between fish species richness, abundance and habitat complexity in a range of shallow tropical marine habitats. Journal of Fish Biology 66: 650–667.

46. Fulton CJ, Bellwood DR (2004) Wave exposure, swimming performance, and the structure of tropical and temperate reef fish assemblages. Marine Biology 144: 429–437.

47. Bruce BD, Evans K, Sutton CA, Young JW, Furlani DM (2001) Influence of mesoscale oceanographic processes on larval distribution and stock structure in jackass morwaong (Nemodactlus macropertus: Cheilodactylidae). ICES Journal of Marine Science 58: 1072–1080.

48. Bax NJ, Burford M, Clementson L, Davenport S (2001) Phytoplankton blooms and production sources on the south-east Australian continental shelf. Marine and Freshwater Research 52: 451–462.

49. Schultz AL, Malcolm HA, Bucher DJ, Smith SDA (2012) Effects of reef proximity on the structure of fish assemblages of unconsolidated substrata. Plos One 7.

50. Kloser RJ, Williams A, Butler A (2006) Exploring surveys of seabed habitats in Australia's deep ocean using remote sensing – needs and realities. In: Todd BJ, Greene HG, editors. Mapping the Seafloor for Habitat Characterization. pp. 93–109.

51. Currie DR, McClatchie S, Middleton JF, Nayar S (2012) Biophysical factors affecting the distribution of demersal fish around the head of a submarine canyon off the Bonney Coast, South Australia. Plos One 7.

52. Vetter EW, Smith CR, De Leo FC (2010) Hawaiian hotspots: enhanced megafaunal abundance and diversity in submarine canyons on the oceanic islands of Hawaii. Marine Ecology 31: 183–199.

53. Leathwick JR, Elith J, Francis MP, Hastie T, Taylor P (2006) Variation in demersal fish species richness in the oceans surrounding New Zealand: an analysis using boosted regression trees. Marine Ecology Progress Series 321: 267–281.

54. Harvey E, Cappo M, Butle RJ, Hall N, Kendrick G (2007) Bait attraction affects the performance of remote underwater video stations in assessment of demersal fish community structure. Marine Ecology Progress Series 350: 245–254.

10

A High-Resolution Chronology of Rapid Forest Transitions following Polynesian Arrival in New Zealand

David B. McWethy[1]*, **Janet M. Wilmshurst**[2,4], **Cathy Whitlock**[1,3], **Jamie R. Wood**[2], **Matt S. McGlone**[2]

1 Department of Earth Sciences, Montana State University, Bozeman, Montana, United States of America, **2** Landcare Research, Lincoln, New Zealand, **3** Institute on Ecosystems, Montana State University, Bozeman, Montana, United States of America, **4** School of Environment, University of Auckland, Auckland, New Zealand

Abstract

Human-caused forest transitions are documented worldwide, especially during periods when land use by dense agriculturally-based populations intensified. However, the rate at which prehistoric human activities led to permanent deforestation is poorly resolved. In the South Island, New Zealand, the arrival of Polynesians c. 750 years ago resulted in dramatic forest loss and conversion of nearly half of native forests to open vegetation. This transformation, termed the Initial Burning Period, is documented in pollen and charcoal records, but its speed has been poorly constrained. High-resolution chronologies developed with a series of AMS radiocarbon dates from two lake sediment cores suggest the shift from forest to shrubland occurred within decades rather than centuries at drier sites. We examine two sites representing extreme examples of the magnitude of human impacts: a drier site that was inherently more vulnerable to human-set fires and a wetter, less burnable site. The astonishing rate of deforestation at the hands of small transient populations resulted from the intrinsic vulnerability of the native flora to fire and from positive feedbacks in post-fire vegetation recovery that increased landscape flammability. Spatially targeting burning in highly-flammable seral vegetation in forests rarely experiencing fire was sufficient to create an alternate fire-prone stable state. The New Zealand example illustrates how seemingly stable forest ecosystems can experience rapid and permanent conversions. Forest loss in New Zealand is among the fastest ecological transitions documented in the Holocene; yet equally rapid transitions can be expected in present-day regions wherever positive feedbacks support alternate fire-inhibiting, fire-prone stable states.

Editor: Christopher Carcaillet, Ecole Pratique des Hautes Etudes, France

Funding: This research was supported in part by National Science Foundation Grants OISE-0966472 (DBM, CW), and BCS-1024413 (DBM, CW), New Zealand Foundation for Research Science and Technology funding to JMW, JRW and MSM, and Royal Society of New Zealand Marsden funding to MSM, JMW, and CW. The funders had no role in study design, data collection and analysis, decision to publish, or preparation of the manuscript.

Competing Interests: The authors have declared that no competing interests exist.

* Email: dmcwethy@montana.edu

Introduction

The composition and structure of forests worldwide are being altered by intensification and expanding land-use, changing climatic conditions and increasing incidence of large fires [1,2]. Sudden transitions between forest, savanna and grassland biomes can be linked to feedbacks between land-use practices, climate change and disturbances such as fire [3]. Research suggests that the extent of forest cover reinforces conditions that maintain or inhibit fire (e.g., high forest cover often promotes greater fuel moisture), and consequently, the stability of forested, savanna or grassland vegetation [4]. Recent deforestation has been shown to initiate positive feedbacks replacing forested landscapes that rarely burned with fire-prone shrubland, savanna and grasslands [5]. Human-mediated forest transitions are not a new phenomenon, however. Humans have effected forest transitions for millennia through their use of fire, and widespread rapid transitions in the past are often linked to expanding human settlements and cultivation [6,7].

New Zealand provides a prime example of a relatively recent and major prehistoric forest loss. Here, the arrival of humans c.

750 years ago [8] instigated a period of deliberate burning, termed the Initial Burning Period (IBP), that led to the loss of nearly 50% of New Zealand's forests [9,10] and conversion of podocarp dominated forests to open shrubland vegetation. The conversion of forests in the South Island, presents a striking paradox, that extremely small populations (by any estimate – human mtDNA sequences suggest an initial founding population of between 50–100 females and a low population growth rate of c. 1% [11]) were able to instigate widespread and permanent forest transitions. Paleoenvironmental records from New Zealand suggest that ecosystem feedbacks during the initial burning period accelerated the rapid rate of forest demise and help explain this paradox [5,9]. In this paper, we examine the rate at which relatively stable forest biomes in New Zealand were converted to open shrublands in the 13th century, and how forest transitions following early human-set fires in the South Island of New Zealand compares with other regions in the world currently experiencing deforestation.

During the IBP, nearly all pollen and charcoal records derived from the drier parts of New Zealand show a distinct decline in pollen from native podocarp tree taxa (and to a lesser extent beech forest trees), an increase in pollen from seral taxa particularly bracken fern (*Pteridium esculentum*) and grasses (Poaceae), and an increase in charcoal (e.g., [9,12,13]. Previous results from radiocarbon dated high-resolution microscopic charcoal analyses from a network of sites across the central South Island showed a distinct increase in charcoal accumulation rates during the IBP within 200 years following Polynesian (Māori) settlement, although the increase is diachronous among sites [9]. At all but the most remote and wet sites, paleofire reconstructions suggest that repeated fires (2 or more) were responsible for the watershed conversion of native forest to a fern/scrub or grassland. [9,10,14]. However, the age-depth chronologies described in [9,14] were based on a small number of radiocarbon dates selected prior to and after the IBP, and consequently, the rate of forest loss remains poorly constrained.

To precisely determine the rate at which prehistoric forest transitions occurred following initial human arrival in the South Island, New Zealand, we developed high-resolution reconstructions of vegetation (pollen) and fire (macroscopic charcoal) from radiocarbon dated lake-sediment records from two small, closed-basin lakes, Lake Kirkpatrick and Dukes Tarn (Fig. 1). We specifically choose two sites that represent different vulnerabilities to human-set fires: one drier lowland site and a second, wetter high-elevation site. The chronology of these reconstructions and rates-of-change calculations are based on a stratigraphic series of Accelerator Mass Spectrometry (AMS) radiocarbon dates from the sediment cores (Table S1).

Results

Chronology model results were highly convergent, suggesting forest transitions occurred within decades of the first human-set fires (Figs. S1–S5). Significant (>15% decline in native taxa) forest loss occurred within 17 yrs (SD = 7) at Lake Kirkpatrick and 48 yrs (SD = 19) at Dukes Tarn (Fig. 2). At Lake Kirkpatrick, the IBP was punctuated by two fire episodes at c. AD 1367 and AD 1391 (Fig. 3, Bchron chronology). Following the initial increase in charcoal accumulation rates (CHAR; particles cm^{-2} yr^{-1}), fire episodes occurred periodically (every 50–100 yrs) until c. AD 1600 when fire activity decreased. A second increase in fire activity coincided with European arrival ca. AD 1800. Vegetation assemblages changed dramatically following the first fires associated with human arrival. Native trees (e.g., *Nothofagus menziesii*, Nothofagus spp., *Podocarpus* spp. and *Prumnopitys* spp.) declined from 99 to 47% of the total terrestrial pollen percentage from 1331 to 1391 yr AD, whereas disturbance related-taxa (e.g. Poaceae and *Pteridium*) increased from up to 25 and 27%, respectively, over the same period (see Fig. S6 for additional pollen information). Pollen of native trees increased slightly to 64% at c. 1642 yr AD at the expense of grass pollen. Tree pollen percentages then declined, recovered again at c. 1792 yr AD before declining to <30% in the late 20th century. Pollen of exotic taxa (Pinaceae, *Rumex* spp. and *Taraxacum* spp.) reached 2% by the mid 19th century and increased to >10% at c. 1947 yr AD.

At Dukes Tarn, charcoal data register initial fire events at c. AD 1337, and 1387, and the pollen data indicate that forests recovered within decades even though fire events occurred periodically (once every 50–100 yrs) from the IBP until the last century (Fig. 3). Pollen of some native trees (*Nothofagus fuscospora* type) collectively declined from 97 to 83% whereas disturbance-related taxa, notably Poaceae and *Pteridium*, increased from 1 to 9% and 0 to 7% respectively during the interval from c. AD 1355 to 1414 (see Fig. S7 and [9] for additional pollen information). In contrast to Lake Kirkpatrick, pollen of native trees increased to 94% by c. AD 1502 at the expense of Poaceae, then fluctuated between 85–94%,

and accounted for 97% at c. 1694 yr AD (back to pre-deforestation levels). Tree percentages decreased to 83% in the last century. Non-native plant taxa (Pinaceae, *Rumex* spp. and *Taraxacum* spp.) account for >1% of the pollen at c. 1928 yr AD and 5% at c. 1980 yr AD before declining to <1% in recent decades. Fires that were recorded during the same time period at both sites (Fig. 4) likely represent a pattern of burning associated with humans moving through the landscape to enhance resource conditions and facilitate travel to nearby greenstone quarries [15] and could have been facilitated by extreme fire weather (e.g., hot, dry, windy conditions). Compared to Lake Kirkpatrick where forest transitions led to open vegetation that persists to the present, native vegetation at Duke's Tarn experienced relatively minor modification and showed partial recovery within several decades of the first IBP fires and full recovery 350 years after the IBP. Post IBP fires at Duke's Tarn seem to have had less of a lasting impact on the long-term structure and composition of vegetation.

Statistical analyses of rates of change reflect the rapid transition from native trees to grasses and bracken, evident in changes in pollen percentages (Fig. 3). Sørensen's index quantifying dissimilarity in pollen taxa prior to and during the IBP shows sharp increases in the dissimilarity in vegetation composition and dominance, with transitions in vegetation taxa peaking during the IBP at both Lake Kirkpatrick and Dukes Tarn. The Sørensen's index (dissimilarity between each sample and the previous sample) was greatest at c. 1363 yr AD (10^{-2}) at Lake Kirkpatrick and at c. 1375 AD (10^{-2}) at Dukes Tarn, coinciding with the first IBP fires at both sites. Dissimilarity between pre-human vegetation assemblages and post-IBP vegetation taxa increased from present to the pre-IBP period at both sites but was much more pronounced at Lake Kirkpatrick where dramatic vegetation transitions from forest to open shrublands persisted until present. Changes in vegetation assemblages at Dukes Tarn are characterized by an initial forest loss, a late recovery of native trees, followed by an increase in grasses and the presence of introduced taxa associated with European colonization.

Discussion

Human-caused forest transitions are documented worldwide, especially during periods when land use by dense agriculturally-based populations intensified [6,16,17]. However, the rate at which prehistoric human activities led to a persistent biome shift is commonly poorly resolved [18]. In the South Island of New Zealand, the arrival of Polynesians resulted in dramatic forest loss, with conversion of nearly half of native forests to open vegetation by a small number of sparsely distributed populations [10]. Paleoenvironmental records suggest that this conversion occurred within decades to centuries of human arrival but the speed at which site conversion took place once burning commenced has been unclear until now. Chronological models (Bchron, Oxcal and MCAgeDepth) that explicitly estimate joint uncertainty of all radiocarbon calibration dates suggest that the forest loss during IBP was rapid, occurring in 10–24 years at the drier site, Lake Kirkpatrick (median = 17) and 29–67 years at the wetter site, Dukes Tarn (median = 48). The speed at which vegetation conversion occurred is likely even more rapid than estimates based on age-depth chronologies alone because they incorporate joint uncertainty originating from both AMS radiocarbon dating and radiocarbon calibration distributions. Despite inherent limitations in calculating the rate of past vegetation change from lake-sediment records, forest loss in New Zealand was rapid by any measure. These rates of deforestation parallel those made for the eastern North Island where dry, lowland podocarp forests

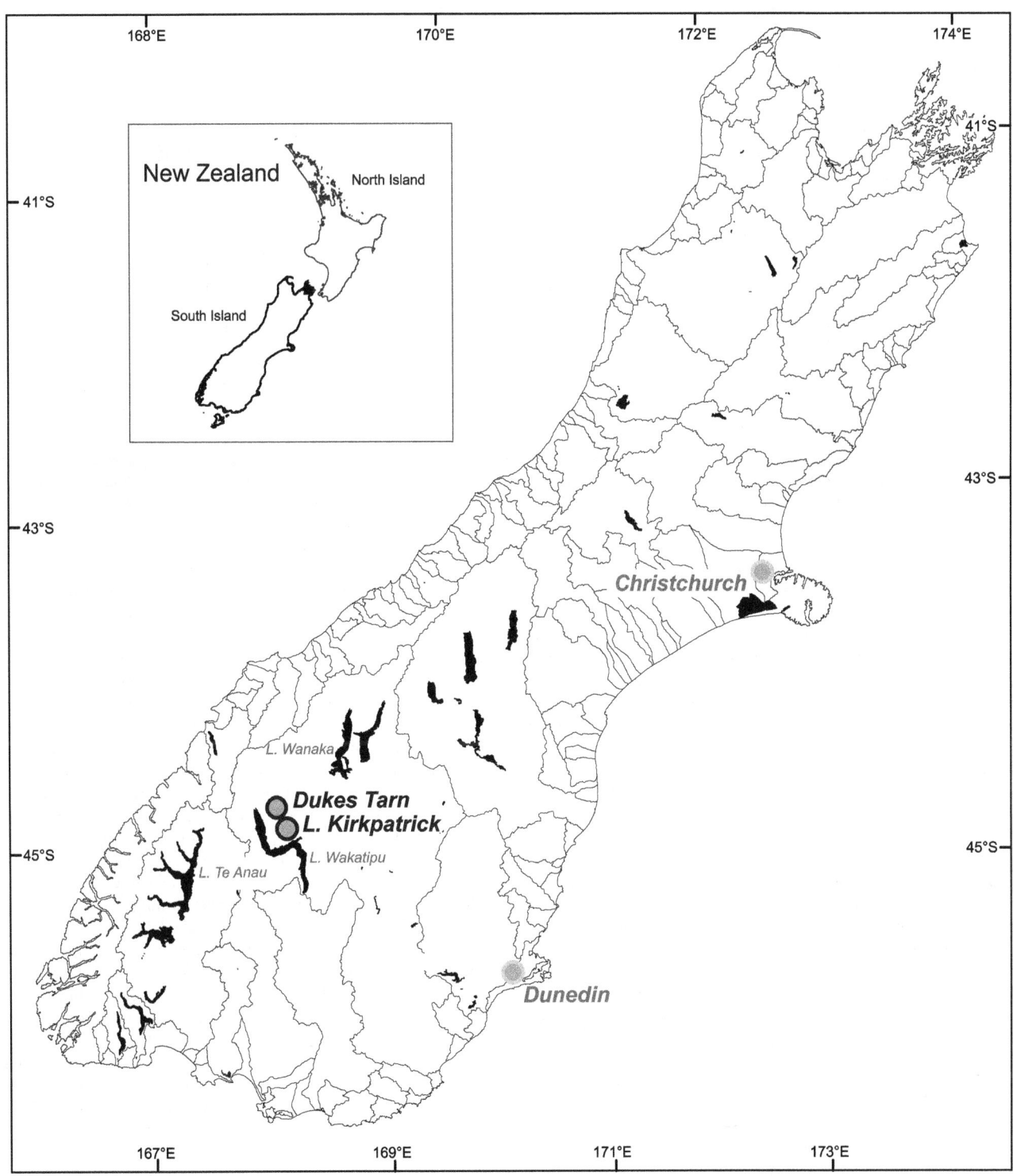

Figure 1. Location of study sites, South Island, New Zealand.

transitioned to bracken fern shrubland within decades of the first human-set fires whereas loss of more mesic, upland forests took a century or longer [19].

For most of the South Island, fire impacts on forested ecosystems were ecologically insignificant throughout the Holocene, in that changes in the structure and composition of vegetation were poorly related to the occurrence of fire [20].

The widespread and mostly one-way shift from forest to open shrubland that occurred with the arrival of Polynesians, and then following European arrival, was unprecedented in the Holocene, even after extreme disturbance events (e.g., large volcanic eruptions). This was the case for Lake Kirkpatrick and many other sites throughout similarly dry forests of eastern New Zealand including the eastern North Island [21]. The higher-elevation site,

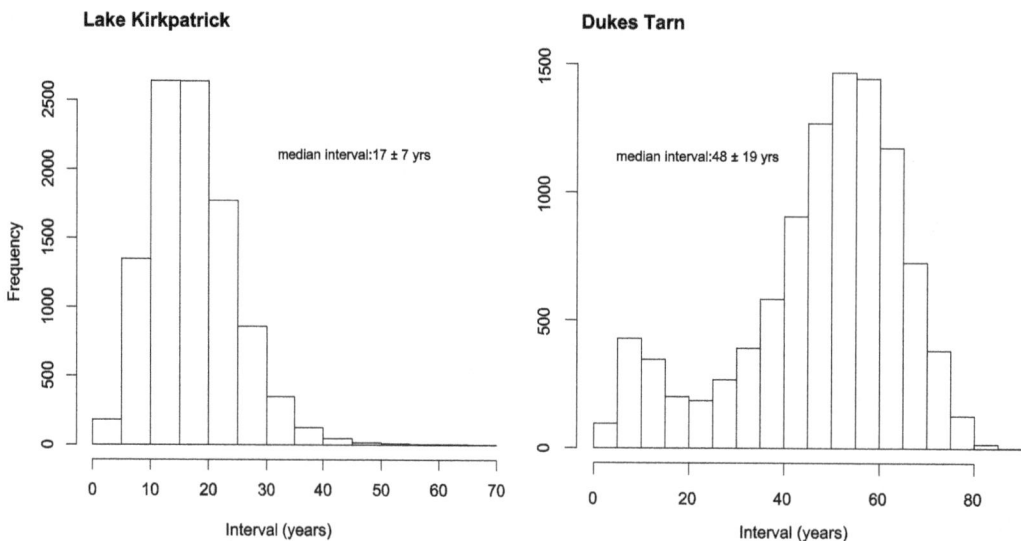

Figure 2. Bchron posterior frequency distribution of the interval (yrs) associated with forest transitions at Lake Kirkpatrick and Dukes Tarn, South Island, New Zealand. The interval associated with forest transitions was identified using two criteria: 1) visible change in core lithology, and 2) a rapid decline in pollen percentages of native forest taxa and a coincident increase in pollen percentages of open vegetation taxa.

Dukes Tarn, experienced equally rapid vegetation change, but declines in forest taxa at this site were less pronounced. This is most likely because this site is wetter and because the deeply-incised forested drainage basin south of the tarn has provided an ongoing refuge from fire, contributing significant wind-transported pollen to the tarn sediments. Differences in the extent to which fire activity and vegetation changed following human arrival at Lake Kirkpatrick and Dukes Tarn may be explained by variation in the intensity of human activity at these two sites; the most parsimonious explanation, however, centers on differences in rainfall and inferred fuel moisture at these two sites and is supported by previous results comparing a network of sites from the South Island ([9]). Paleofire records at even wetter sites (>1600 mm/yr) near or west of the Southern Alps of the South Island indicate only limited fire activity and minor associated vegetation change [9,22]. Similarly, along the western coast of the North Island moist conditions limited the extent to which Māori were able to burn forests [23]. Here, Māori gardens were established by burning small patches of forests but the scale of forest clearing was insignificant compared to deforestation that occurred in drier regions of the North and South Island sites [23]. What is surprising is the extent and rapid rate of forest clearance that occurred in the South Island where prehistoric human populations were at their lowest compared to the North Island, and cultivation was limited to a few coastal sites north of latitude 44° South.

Dynamic Global Vegetation Models that estimate rates of prehistoric deforestation at a regional to global scale suggest that the extent and pattern of preindustrial deforestation in most regions was strongly linked to population densities, inherent landscape productivity and lifestyles and technological innovation [16,17]. Estimates of population density and reconstructions of environmental change in these models are derived from archeological and paleoenvironmental records, and in most regions, rates of deforestation are poorly resolved. Our high-resolution chronology of environmental change from the South Island, New Zealand indicates rapid forest transitions occurred during the IBP during a time of exceedingly low population numbers and limited use of technological innovation (i.e., intensive cultivation). Why

then were New Zealand's forests particularly vulnerable to fire? The lowland evergreen conifer/angiosperm forests were characterized as being both fire sensitive and ignition limited [20]. Low- to mid-elevation forests experiencing moderate to low levels of annual rainfall (300–1600 mm/yr) persisted for millennia simply because ignitions were rare and fuels were typically wet [20,24]. It was only when human-set fires expanded the proportion of flammable early-seral vegetation on the landscape that widespread forest transitions were possible [5]. Positive feedbacks between the fire sensitivity of the vegetation, introduction of human-set ignitions, and changes in landscape flammability with the spread of early-seral vegetation offer an explanation for these rapid transitions. Perry et al. [5,22] provide empirical support for these feedbacks, showing that dry, low- to mid-elevation evergreen conifer/angiosperm forests in the South Island New Zealand were vulnerable to fire, and that targeted burning created positive feedbacks where burning of forest resulted in more flammable seral vegetation which was more vulnerable to ignitions. The model estimated that conversion of the forest to shrubland could have occurred in 50–200 years if fires were targeted both spatially (in highly-flammable early-seral vegetation) and temporally (every few decades to maintain early-seral vegetation).

Forest transitions occurring at present in a number of regions around the world bring new attention to conditions and feedbacks that drive rapid transitions between seemingly stable states [25–28]. Analyzing global tree cover and fire activity data, Staver et al. [4] show that mesic landscapes (>2500 mm/y rainfall) with >60% tree cover are resistant to changes in fire activity and biome shifts, whereas areas with intermediate levels of rainfall (1000–2500 mm/y) and <50% tree cover are vulnerable to rapid transitions to savanna. Hirota et al. [29] suggest that landscapes with intermediate levels of tree cover are extremely rare because the interaction between climate, edaphic conditions and fire activity interact to attract vegetation to relatively stable states of higher (forest) and lower (savanna) levels of tree cover. A number of mechanisms create positive feedbacks that trigger and maintain these multiple stable states. Landscapes with high canopy cover of mature forests are resilient to increased fire activity because they maintain fuel moisture levels that suppress fire and often lack

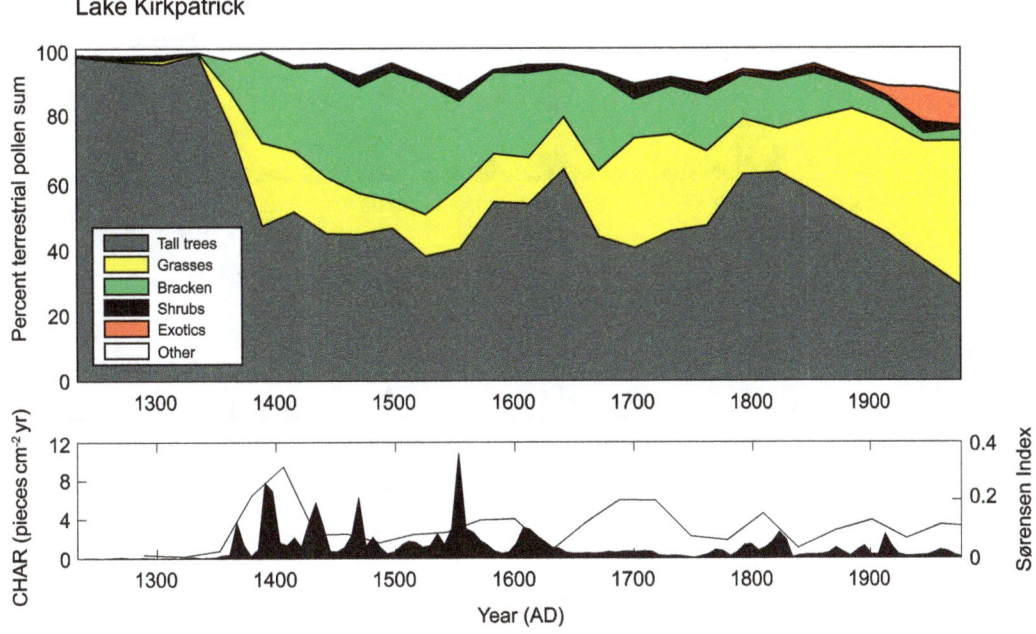

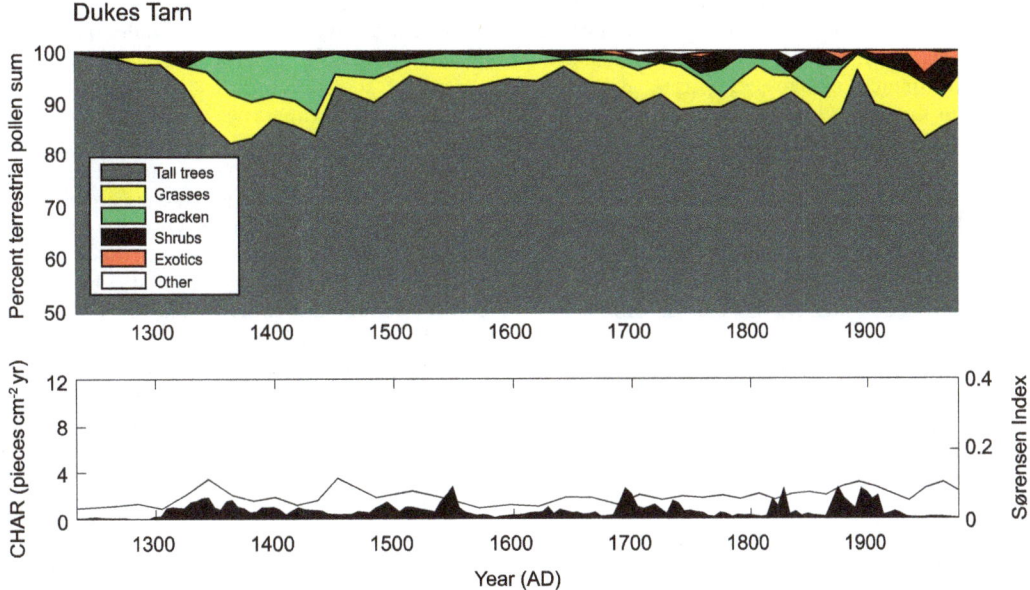

Figure 3. Statistical analysis of vegetation change and rate of vegetation change (based on Sørensen's similarity index) and changes in fire activity (based on CHAR, charcoal particles cm^{-2} yr^{-1}, and fire event determination), Lake Kirkpatrick and Dukes Tarn. Colored panels show change in percent of total terrestrial pollen percentages for native trees and disturbance-associated taxa (e.g., Poaceae, *Pteridium*) and non-native taxa (Pinaceae, *Rumex*, *Taraxacum*-type) introduced by Europeans for Lake Kirkpatrick (top) and Dukes Tarn (bottom). Black and white panels show charcoal accumulation rates and Sørensen's distance between each pollen sample and the next oldest pollen sample.

continuous understory fuels for carrying fire [3]. Dry landscapes reinforce open conditions by supporting low fuel moisture and species with plant functional traits that promote fire.

A consequence of these attracting states is the potential for human activities to reduce forest density and/or canopy cover past a tipping point where they become increasingly vulnerable to rapid transitions to more fire-prone states [30]. This has been shown to be particularly true for forests where flammability decreases with forest age. Kitzberger et al. [3] demonstrate that

even small increases in the frequency of ignitions and/or climate variability in these forests can result in dramatic shifts in fire regimes at large scales. The New Zealand example suggests that targeted ignitions alone where sufficient to initiate feedbacks promoting high landscape flammability and increased fire activity.

Today, the vulnerability of temperate forests to rapid transitions is evident in a number of regions. For example, alterations to fuel and microclimate conditions following logging activities in evergreen forests of the northwestern US increase post-fire

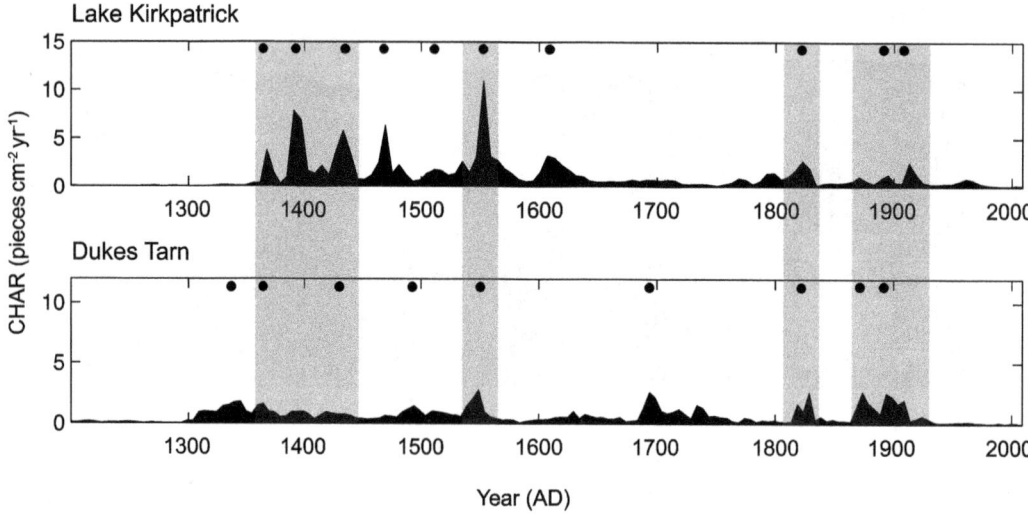

Figure 4. Variation in CHAR (charcoal particles cm^{-2} yr^{-1}), and peak identification for Lake Kirkpatrick and Dukes Tarn. Black circles identify peaks in CHAR that are statistically significant from background variation in CHAR and are thus likely to be local fires (within 1–3 kms of lake). Shaded areas highlight time periods when fire events were recorded at both lakes.

flammability, ultimately increasing fire severity such that forest recovery is delayed or non-existent [31,32]. Lindenmayer et al. [27] describe a similar feedback in mountain ash forests in Victoria, Australia where logging activity results in an abundance of fine fuels, slash and more flammable early-seral vegetation. While fire occurred in these forests in the past, it was typically infrequent, and eventually followed by conditions that promoted robust recruitment of mountain ash seedlings. In recent decades, the increase in fine fuels, slash and persistently drier microclimatic conditions following logging activity and subsequent increased fire activity interact to create these new fire-prone 'landscape traps' [27]. Likewise, Cochrane et al. [33] demonstrate that previously burned forests in Amazonia are highly vulnerable to subsequent fires that further promote more open, highly-flammable vegetation. In southern South America, the combination of increased anthropogenic fires and herbivory of native seedlings by introduced species is rapidly facilitating the replacement of mesic *Nothofagus* forests with fire-prone shrubs [34]. On Great Barrier Island North Island, New Zealand, Perry et al. [35] found that conversion of mesic broadleaf forests to shrublands initiated positive feedbacks in which an increase in fire activity not only reinforced the spread and persistence of highly flammable early-seral woody vegetation but also promoted flammable invasive species and degraded soil conditions, especially on drier aspects. Lacking fire, these forests would likely recover but Perry et al. [35] observe that, along with a continued loss of topsoil, seed trees and dispersal agents (e.g. avian frugivores) are becoming increasingly rare such that forest recovery becomes ever more improbable. These recent forest transitions demonstrate the strong influence of human-initiated positive feedbacks that increase landscape flammability and highlight the inherent vulnerability of seemingly stable temperate forests to rapid biome and fire regime shifts. Widespread forest transitions are predicted for the future in regions with similar age-flammability and low-ignition characteristics [36].

Conclusions

The timing of an abrupt change from forest to open shrubland following human-set fires at two sites in the South Island of New

Zealand is well-constrained in sediment cores using high-resolution chronology models, suggesting widespread deforestation occurred within decades of the first anthropogenic fires in the most vulnerable drier regions. The fire history of New Zealand is an example of how a seemingly stable ecosystem can pass natural tipping points and experience persistent transformations. The rate at which anthropogenic forest loss occurred was rapid, taking place within decades following the initial arrival of a small founding population. The rapidity of the response was partly facilitated by positive feedbacks created by the introduction of a new and frequent ignition source by humans. In addition, the dominant forest taxa lacked natural adaptations to fire (e.g., serotiny, resprouting) which made them highly vulnerable to increased ignitions [24]. The newly created fire-prone landscapes persisted even when fire activity decreased after the IBP. These shrublands will likely be maintained for centuries if not millennia unless fire is removed from the system for long periods of time (>100 years). As with temperate forests experiencing similar transitions today, simply removing fire from New Zealand landscapes could eventually lead to forest recovery. However, recent examples suggest that a confluence of factors that degrade biophysical conditions, impede seedling regeneration, remove source populations and dispersal agents and introduce fire-adapted weedy species will continue to hinder native forest recovery in New Zealand and elsewhere. Even in the absence of fire, management of landscapes to promote open vegetation, agriculture, and exotic forestry plantations promises to reinforce conditions that prevent regeneration of native species. The rapid rate of forest transitions witnessed in New Zealand 750 years ago portends potentially rapid transitions in biomass-rich regions experiencing similar land-use-disturbance interactions and feedbacks and highlights challenges to native forest recovery.

Materials and Methods

Study Sites

The study regions lie in the southeastern hill country and southern alpine region of the South Island of New Zealand, an area that ranges between 250 and 1900 m elevation (Fig. 1). The climate is cool year-round and moderately wet with most of the precipitation occurring in winter. There is a strong steep east/west gradient both in precipitation and elevation. During the Holocene the vegetation of the region was dominated by closed-canopy broadleaf evergreen forests [20]. Natural fires occurred in areas of forest in the dry southeastern region of New Zealand but were characterized by long intervals between fires [37]. The original forest extent and composition are inferred from pollen records, isolated surviving forest stands and remnant wood [38,39]. Permission to conduct fieldwork at the two study sites was granted by the New Zealand Department of Conservation (DOC).

Lake Kirkpatrick (lat. 45.03°S, long. 168.57°E; 567 masl). The Lake Wakatipu catchment is a mid-elevation (570 masl) site receiving moderate levels of rainfall (1077 mm/yr) and was part of a trade route for highly prized greenstone that the indigenous population (Māori) used and traded [15]. Lake Kirkpatrick lies within the Lake Wakatipu catchment. The small lake (~3.0 ha) is surrounded by pastureland with browntop (*Agrostis capillaris*), Poa spp. and other introduced grasses, clover (*Trifolium* spp.), *Hypericum* spp.; pine plantations (*Pinus radiata*); and small areas of native silver beech forest (*Nothofagus menziesii*).

Dukes Tarn (lat. 44.96°S, long. 168.49°E; 825 masl). Dukes Tarn is approximately 9.5 km northwest of Lake Kirkpatrick at 830 masl and also located within the Lake Wakatipu catchment. Dukes Tarn (~1.5 ha) receives more annual rainfall (1340 mm/yr) than Lake Kirkpatrick and vegetation surrounding the lake consists of patches of black beech (*Nothofagus solandri*), small trees and shrubs (e.g., *Coprosma* spp. *Phyllocladus, Dracophyllum longifolium, Hebe* spp., *Gaultheria* spp.), native tussock (*Chionochloa* spp.) and introduced pasture grasses (*Festuca* spp.). Deeply-incised drainages supporting native forest remnants border Dukes Tarn to the south whereas open vegetation surrounds the tarn on rolling terrain to the north.

Lake Sediment Cores

Lake sediment cores were obtained from Lake Kirkpatrick and Dukes Tarn using a polycarbonate tube (Klien core). Cores were split at University of Minnesota's LacCore facility for charcoal and pollen analysis. A 195 cm sediment core was retrieved from Lake Kirkpatrick in 2009 and a 173 cm sediment core was retrieved from Dukes Tarn in 2008. Core lithology for Lake Kirkpatrick consisted of light green gyttja 0–27 cm (depth below surface), gray silty clay 27–29 cm, dark green gyttja 29–105 cm, light gray silty clay 105–115 cm, and light green gyttja 115–195 cm. Core lithology for Dukes Tarn consisted of light green gyttja 0–42 cm, silty clay 42–43 cm, light green gyttja 43–128 cm, dark silty clay with charcoal 128–136 cm, and light green gyttja 138–173 cm.

Reconstruction of Vegetation and Fire Histories

Pollen analysis followed the preparation methods of Moore et al. [40]. Pollen counts exceeded 250 terrestrial pollen grains and bracken (*Pteridium*) spores. Percentage calculations were based on this terrestrial sum (excluding other ground ferns and tree fern spores as they tend to be over-represented). We followed the protocol of McGlone and Wilmshurst [10] to detect initial human impact on the vegetation and grouped pollen taxa into tall forest taxa (predominantly *Fuscospora* spp. and podocarps); disturbance-related taxa including ferns (*Pteridium esculentum*), *Coriaria* spp. and grasses (Poaceae); and introduced taxa, including Pinaceae and *Rumex* spp. Remaining taxa, mostly herbaceous species, were labeled "other". These groupings were used to identify the transition from closed forest to open shrubland and grassland.

High-resolution charcoal analysis followed methods of Whitlock and Larsen [41]. Charcoal particles greater (>125 μm in diameter) were examined to reconstruct local fire events based on changes in charcoal accumulation rates (CHAR; particles cm^{-2} yr^{-1}). Decomposition of the charcoal time series employed a Gaussian mixture model to identify the mean and variance of the background CHAR (BCHAR) distribution [42]. The 99th percentile of this distribution is defined as the threshold value separating peak fire events from "noise". Significant charcoal peaks, determined by CHAR values greater than a locally-defined threshold value, identified specific fire events. Because each 1-cm vertical interval in the lakes encompasses 1–10 years at both sites, a charcoal peak is assumed to represent one or more fires occurring within the time span of a charcoal peak. The statistical significance of each peak was evaluated by comparing the original charcoal counts against the values in samples occurring 35 years before the peak. If the maximum count of a peak had a >5% chance of coming from the same Poisson-distributed population as the minimum charcoal count within the preceding 35 years, then a "peak" was not identified. Fire size and/or intensity were inferred from the magnitude of individual charcoal peaks (particles cm^{-2} yr^{-1}) [43].

Chronology Models

We obtained 22 radiocarbon dates at two sites to develop high-resolution chronologies (Table S1). Radiocarbon dates were obtained from macrofossils found during the interval of the first IBP fires where changes in charcoal accumulation rates and pollen percentages indicate dynamic changes in fire activity and vegetation. The chronology of these reconstructions and rates-of-change calculations are based on a stratigraphic series of AMS radiocarbon dates from the sediment cores. We incorporated this suite of radiocarbon dates into three readily accessible chronology models designed to estimate lake-sediment age-depth chronologies and chronological uncertainty: Bchron [44], Oxcal [45] and McAgeDepth [43]. Model outputs were highly convergent for all three models (Figs. S1–S5). We highlight results from Bchron because the chronology model has an environment event analysis function which explicitly estimates the duration of the interval at which significant environmental events occurred (Fig. 2).

Accelerator mass spectrometry (AMS) ^{14}C dates obtained on twig charcoal and terrestrial plant macrofossils, provided the chronology for precisely estimating rates of past environmental change. Thirteen radiocarbon dates were obtained from Lake Kirkpatrick and nine from Dukes Tarn. Eleven radiocarbon dates were obtained from across 20 centimeters of sediment where charcoal and pollen percentages identify the IBP at Lake Kirkpatrick and six radiocarbon dates were obtained across the comparable 10-centimeter interval at Dukes Tarn. We used this set of radiocarbon dates in a suite of age-depth chronology models to estimate the duration at which forest transitions occurred.

To derive age-depth chronologies we used three statistical age-depth chronology models: Bchron [44], Oxcal [45] and McAge-Depth [43] which all address two primary types of uncertainty: uncertainty in radiocarbon dates and their calibration, and variability in the sedimentation process itself. These models share a number of features that are desirable for developing a continuous age-depth chronology for sediment cores including, monotonicity (deeper sediments are likely older), variable sedimentation (rates are assumed to be variable), probabilistic interpolation based on uncertainty associated with calibrated probability distributions (posterior samples or chronologies are based on joint estimates of uncertainty of all dates), and increased uncertainty at depths far from radiocarbon-dated material [46]. Each model uses a Markov-Chain Monte-Carlo algorithm to estimate model parameters, which provides a distribution of the chronologies that are most likely given all radiocarbon-dated probability distributions. Most importantly, they explicitly account for and estimate joint uncertainty of all of the radiocarbon dates and consider the probability density functions of calibrated age ranges in chronology interpolation. All high-resolution chronology models achieved convergence and identified no outliers at Dukes Tarn, but two outliers at Lake Kirkpatrick (macrofossil dated material at depths 110 and 111 cm), which were subsequently left out of modeled chronologies. The radiocarbon material at these two depths were identified as having low mass during AMS analyses (KCCAMS lab), which likely increased their age uncertainty and inaccuracy.

Rate of vegetation change

We used the Sørensen's similarity statistic [47] to assess the rate of vegetation change between pollen samples. Sørensen's index, similar to the Bray-Curtis dissimilarity [48] is a similarity index that measures compositional similarity between two sites, or in this case, compositional similarity in pollen taxa at two depths where pollen percentages are measured. Sørensen's distance statistic is calculated between each pollen sample and the preceding pollen sample providing a statistical measure of vegetation change over time.

Supporting Information

Figure S1 Comparison of chronology model estimates (circles and bars represent median and 95% Confidence Intervals) for the interval associated with forest transitions that followed the first human-set fires, Lake Kirkpatrick and Dukes Tarn, South Island, New Zealand. Median (and 95% confidence estimate ranges) for the interval associated with forest transitions for Lake Kirkpatrick were 17 yrs (3–13, Bchron), 27 yrs (7–47, MCAgeDepth), and 31 yrs (16–46, Oxcal). Median and 95% confidence estimate ranges (in parentheses) for the interval associated with forest transitions for Dukes Tarn were 47 yrs (35–59, Bchron), 44 yrs (24–64, MCAgeDepth), and 25 yrs (14–36, Oxcal).

Figure S2 MCAgeDepth age depth model for Lake Kirkpatrick with sedimentation rates and sample resolution. Chronologies developed using a weighted cubic smoothing spline which takes into account the number and uncertainty of age estimates. Gray shading represent 95% confidence intervals reflect the combined uncertainty of all age estimates in a model and are derived from 1000 bootstrapped chronologies.

Figure S3 MCAgeDepth age depth model for Dukes Tarn with sedimentation rates and sample resolution. Chronologies developed using a weighted cubic smoothing spline

which takes into account the number and uncertainty of age estimates. Gray shading represent 95% confidence intervals reflect the combined uncertainty of all age estimates in a model and are derived from 1000 bootstrapped chronologies.

Figure S4 Oxcal age-depth model for Lake Kirkpatrick using the P_Sequence deposition model which assumes deposition to be random (k parameter set at 1 – which assumes a postulated event spacing of approximately 1 cm). Light grey distributions represent likelihoods for single calibrated dates and in darker grey shading represents the marginal posterior distributions which take into account the depth model; the depth model curves show envelopes for the 95% and 68% highest probability density (HPD) ranges.

Figure S5 Oxcal age-depth model for Dukes Tarn using the P_Sequence deposition model which assumes deposition to be random (k parameter set at 1 – which assumes a postulated event spacing of approximately 1 cm). Light grey distributions represent likelihoods for single calibrated dates and in darker grey shading represents the marginal posterior distributions which take into account the depth model; the depth model curves show envelopes for the 95% and 68% highest probability density (HPD) ranges.

Figure S6 Pollen diagram for Lake Kirkpatrick, South Island, New Zealand.

Figure S7 Pollen diagram for Dukes Tarn, South Island, New Zealand.

Table S1 AMS radiocarbon dating determined age information for Lake Kirkpatrick and Dukes Tarn, New Zealand.

Acknowledgments

We thank Mairie Fromont for conducting pollen analyses for Dukes Tarn, John Southon (University of California-Irvine, Keck-Carbon Cycle AMS Facility) for radiocarbon dates and Philip Higuera for help with Figure 3 and the application of Sørensen's similarity index to pollen data.

Author Contributions

Conceived and designed the experiments: DBM CW JMW. Performed the experiments: DBM CW JMW JRW. Analyzed the data: DBM CW JMW JRW. Contributed reagents/materials/analysis tools: DBM CW JMW JRW. Wrote the paper: DBM CW JMW JRW MSM.

References

1. Nepstad DC, Stickler CM, Soares B, Merry F (2008) Interactions among Amazon land use, forests and climate: prospects for a near-term forest tipping point. Philosophical Transactions of the Royal Society B-Biological Sciences 363: 1737–1746.
2. Mayer AL, Khalyani AH (2011) Grass Trumps Trees with Fire. Science 334: 188–189.
3. Kitzberger T, Araoz E, Gowda J, Mermoz M, Morales J (2011) Decreases in fire spread probability with forest age promotes alternative community states, reduced resilience to climate variability and large fire regime shifts. Ecosystems: 1–16.
4. Staver AC, Archibald S, Levin SA (2011) The global extent and determinants of savanna and forest as alternative biome states. Science 334: 230–232.
5. Perry GLW, Wilmshurst JM, McGlone MS, McWethy DB, Whitlock C (2012) Explaining fire-driven landscape transformation during the Initial Burning Period of New Zealand's prehistory. Global Change Biology 18: 1609–1621.

6. Pongratz J, Reick C, Raddatz T, Claussen M (2008) A reconstruction of global agricultural areas and land cover for the last millennium. Global Biogeochem Cycles 22: GB3018.
7. Kaplan JO, Krumhardt KM, Ellis EC, Ruddiman WF, Lemmen C, et al. (2011) Holocene carbon emissions as a result of anthropogenic land cover change. The Holocene 21: 775–791.
8. Wilmshurst JM, Anderson AJ, Higham TFG, Worthy TH (2008) Dating the late prehistoric dispersal of Polynesians to New Zealand using the commensal Pacific rat. Proceedings of the National Academy of Sciences 105: 7676–7680.
9. McWethy DB, Whitlock C, Wilmshurst JM, McGlone MS, Fromont M, et al. (2010) Rapid landscape transformation in South Island, New Zealand, following initial Polynesian settlement. Proceedings of the National Academy of Sciences 107: 21343–21348.
10. McGlone MS, Wilmshurst JM (1999) Dating initial Maori environmental impact in New Zealand. Quaternary International 59: 5–16.

11. Murray-McIntosh RP, Scrimshaw BJ, Hatfield PJ, Penny D (1998) Testing migration patterns and estimating founding population size in Polynesia by using human mtDNA sequences. Proceedings of the National Academy of Sciences of the United States of America 95: 9047–9052.

12. Newnham R, Lowe DJ, McGlone MS, Wilmshurst JM, Higham TFG (1998) The Kaharoa Tephra as a critical datum for earliest human impact in northern New Zealand. Journal of Archaeological Science 25: 533–544.

13. Horrocks M, Deng Y, Ogden J, Alloway B, Nichol S, et al. (2001) High spatial resolution of pollen and charcoal in relation to the c. 600 yr BP Kaharoa Tephra: implications for Polynesian settlement of Great Barrier Island, northern New Zealand. Journal of Archaeological Science 28: 153–168.

14. McWethy DB, Whitlock C, Wilmshurst JM, McGlone MS, Li X (2009) Rapid deforestation of South Island, New Zealand by early Polynesian fires. The Holocene 19: 883–897.

15. CINZAS (2008) Central Index of New Zealand Archaeological Sites. nzarchaeologyorg.

16. Kaplan JO, Krumhardt KM, Zimmermann N (2009) The prehistoric and preindustrial deforestation of Europe. Quaternary Science Reviews 28: 3016–3034.

17. Pfeiffer M, Spessa A, Kaplan JO (2013) A model for global biomass burning in preindustrial time: LPJ-LMfire (v1.0). Geosci Model Dev\ 6\: 643\-685\.

18. Williams M (2000) Dark ages and dark areas: global deforestation in the deep past. Journal of Historical Geography 26: 28–46.

19. Wilmshurst JM, McGlone MS, Partridge TR (1997) A late Holocene history of natural disturbance in lowland podocarp/hardwood forest, Hawke's Bay, New Zealand. New Zealand Journal of Botany 35: 79–96.

20. Ogden J, Basher L, McGlone M (1998) Fire, forest regeneration and links with early human habitation: Evidence from New Zealand. Annals of Botany 81: 687–696.

21. Wilmshurst JM (1997) The impact of human settlement on vegetation and soil stability in Hawke's Bay, New Zealand. New Zealand Journal of Botany 35: 97–111.

22. Perry GLW, Wilmshurst JM, McGlone MS, Napier A (2012) Reconstructing spatial vulnerability to forest loss by fire in pre-historic New Zealand. Global Ecology and Biogeography: no-no.

23. Wilmshurst JM, Higham TFG, Allen H, Johns D, Phillips C (2004) Early Maori settlement impacts in northern coastal Taranaki, New Zealand. New Zealand Journal of Ecology 28: 167–179.

24. Perry G, Wilmshurst J, McGlone M (2014) The ecology and long-term history of fire in New Zealand's ecosystems. New Zealand Journal of Ecology 38.

25. Kitzberger T, Brown PM, Heyerdahl EK, Swetnam TW, Veblen TT (2007) Contingent Pacific-Atlantic ocean influence on multi-century wildfire synchrony over western North America. Proceedings of the National Academy of Sciences 104: 543–548.

26. Field RD, Van der Werf G, Shen SSP (2009) Human amplification of drought-induced biomass burning in Indonesia since 1960. Nature Geoscience 2: 185–188.

27. Lindenmayer DB, Hobbs RJ, Likens GE, Krebs CJ, Banks SC (2011) Newly discovered landscape traps produce regime shifts in wet forests. Proceedings of the National Academy of Sciences 108: 15887–15891.

28. Odion DC, Moritz MA, DellaSala DA (2010) Alternative community states maintained by fire in the Klamath Mountains, USA. Journal Of Ecology 98: 96–105.

29. Hirota M, Holmgren M, Van Nes EH, Scheffer M (2011) Global Resilience of Tropical Forest and Savanna to Critical Transitions. Science 334: 232–235.

30. McWethy DB, Higuera PE, Whitlock C, Veblen TT, Bowman DMJS, et al. (2013) A conceptual framework for predicting temperate ecosystem sensitivity to human impacts on fire regimes. Global Ecology and Biogeography 22: 900–912.

31. Thompson JR, Spies TA, Ganio LM (2007) Reburn severity in managed and unmanaged vegetation in a large wildfire. Proceedings of the National Academy of Sciences 104: 10743–10748.

32. Odion DC, Frost EJ, Strittholt JR, Jiang H, Dellasala DA, et al. (2004) Patterns of fire severity and forest conditions in the western Klamath Mountains, California. Conservation Biology 18: 927–936.

33. Cochrane MA, Alencar A, Schulze MD, Souza CM Jr., Nepstad DC, et al. (1999) Positive Feedbacks in the Fire Dynamic of Closed Canopy Tropical Forests. Science 284: 1832–1835.

34. Veblen TT, Holz A, Paritsis J, Raffaele E, Kitzberger T, et al. (2011) Adapting to global environmental change in Patagonia: What role for disturbance ecology? Austral Ecology: no-no.

35. Perry GLW, Ogden J, Enright NJ, Davy LV (2010) Vegetation patterns and trajectories in disturbed landscapes, Great Barrier Island, northern New Zealand. New Zealand Journal of Ecology: 311–323.

36. Moritz MA, Parisien M-A, Batllori E, Krawchuk MA, Van Dorn J, et al. (2012) Climate change and disruptions to global fire activity. Ecosphere 3: art49.

37. McGlone MS, Wilmshurst JM, Leach HM (2005) An ecological and historical review of bracken (Pteridium esculentum) in New Zealand, and its cultural significance. New Zealand Journal of Ecology 29: 165–184.

38. McGlone MS (1995) Lateglacial landscape and vegetation change and the Younger Dryas climatic oscillation in New Zealand. Quaternary Science Reviews 14: 867–881.

39. Wardle P (2001) Distribution of native forest in the upper Clutha district, Otago, New Zealand. New Zealand Journal of Botany 39: 435–446.

40. Moore PD, Webb JA, Collinson ME (1991) Pollen Analysis. Oxford, UK: Blackwell Scientific.

41. Whitlock C, Larsen C. (2001) Charcoal as a Fire Proxy. In: Smol JP, Birks, H.J.B, Last, W M., editor.Tracking Environmental Change Using Lake Sediments: Volume 3 Terrestrial, Algal, and Siliceous indicators.Dordrecht: Kluwer Academic Publishers. pp. 75–97.

42. Higuera PE, Peters ME, Brubaker LB, Gavin DG (2007) Understanding the origin and analysis of sediment-charcoal records with a simulation model. Quaternary Science Reviews 26: 1790–1809.

43. Higuera PE, Brubaker LB, Anderson PM, Hu FS, Brown TA (2009) Vegetation mediated the impacts of postglacial climate change on fire regimes in the south-central Brooks Range, Alaska. Ecological Monographs 79: 201–219.

44. Haslett J, Parnell A (2008) A simple monotone process with application to radiocarbon-dated depth chronologies. Journal of the Royal Statistical Society: Series C (Applied Statistics) 57: 399–418.

45. Bronk Ramsey C (2008) Deposition models for chronological records. Quaternary Science reviews 27: 42–60.

46. Parnell AC, Buck CE, Doan TK (2011) A review of statistical chronology models for high-resolution, proxy-based Holocene palaeoenvironmental reconstruction. Quaternary Science Reviews 30: 2948–2960.

47. Sørensen T (1948) A method of establishing groups of equal amplitude in plant sociology based on similarity of species and its application to analyses of the vegetation on Danish commons. Kongelige Danske Videnskabernes Selskab 5: 1–34.

48. Bray JR, Curtis JT (1957) An ordination of upland forest communities of southern Wisconsin. Ecological Monographs 27: 325–349.

11

Spatial and Temporal Variations of Crop Fertilization and Soil Fertility in the Loess Plateau in China from the 1970s to the 2000s

Xiaoying Wang[1,2], Yanan Tong[1,2]*, Yimin Gao[1], Pengcheng Gao[1], Fen Liu[1], Zuoping Zhao[1], Yan Pang[1]

1 College of Natural Resources and Environment, Northwest A&F University, Yangling, China, 2 Key Laboratory of Plant Nutrition and the Agri-environment in Northwest China, Ministry of Agriculture, Yangling, China

Abstract

Increased fertilizer input in agricultural systems during the last few decades has resulted in large yield increases, but also in environmental problems. We used data from published papers and a soil testing and fertilization project in Shaanxi province during the years 2005 to 2009 to analyze chemical fertilizer inputs and yields of wheat (*Triticum aestivum* L.) and maize (*Zea mays* L.) on the farmers' level, and soil fertility change from the 1970s to the 2000s in the Loess Plateau in China. The results showed that in different regions of the province, chemical fertilizer NPK inputs and yields of wheat and maize increased. With regard to soil nutrient balance, N and P gradually changed from deficit to surplus levels, while K deficiency became more severe. In addition, soil organic matter, total nitrogen, alkali-hydrolysis nitrogen, available phosphorus and available potassium increased during the same period. The PFP of N, NP and NPK on wheat and maize all decreased from the 1970s to the 2000s as a whole. With the increase in N fertilizer inputs, both soil total nitrogen and alkali-hydrolysis nitrogen increased; P fertilizer increased soil available phosphorus and K fertilizer increased soil available potassium. At the same time, soil organic matter, total nitrogen, alkali-hydrolysis nitrogen, available phosphorus and available potassium all had positive impacts on crop yields. In order to promote food safety and environmental protection, fertilizer requirements should be assessed at the farmers' level. In many cases, farmers should be encouraged to reduce nitrogen and phosphate fertilizer inputs significantly, but increase potassium fertilizer and organic manure on cereal crops as a whole.

Editor: Cheng-Sen Li, Institute of Botany, China

Funding: The authors thank the Special Fund for Agro-scientific Research in the Public Interest of China (201103003) and the Soil Quality Foundation of China (2012BAD05B03) for their financial support. The funders had no role in study design, data collection and analysis, decision to publish, or preparation of the manuscript.

Competing Interests: The authors have declared that no competing interests exist.

* Email: tongyanan@nwsuaf.edu.cn

Introduction

China has only 9% of the world's arable land and feeds nearly 22% of the world population [1–2]. This depends heavily on increasing grain production with the use of chemical fertilizers. Before the 1970s, farmers maintained the original agricultural practices, such as crop rotation, diversified plantation, manure application and legume crop integration, for soil fertility maintenance and pest and disease control. Since the late 1980s, the practice of applying organic manure in arable cropping systems has nearly come to an end [2–6]. From then on, almost all available organic manure has been used on vegetables and fruit trees, while the nutrients for cereal crops have been mainly in the form of chemical fertilizers. From 1970 to 2010, total annual grain production in China increased from 240 to 546 million tons (a 128% increase). However, inorganic fertilizer application increased from 3.51 to 55.62 million tons (a 1485% increase) over the same period [7].

Soil quality indicators are measurable soil properties that benefit food production or other specific functions, including physical, chemical and biological characteristics [8]. The increase or decrease in single soil index values, such as soil organic matter, total nitrogen and available nutrients, amplitude of variation and variation in time, can be used as a monitoring index for agricultural land management [9–11]. Given the spatial and temporal variation in characteristics of soil quality, it is necessary to compare or analyze two or more phase changes to understand the nature and mechanisms of soil quality [12].

Farmland fertilization is one of the most effective ways to maintain soil fertility and increase crop yields [13–15]. For this reason, information on household fertilization levels is of great value. In addition, wheat and maize are two of the most important food crops throughout the world, and they account for 51.7% of the total area for food crops and 53.5% of the total food production in 2010 in China [7]. Chemical fertilizer consumption data from official Chinese statistics do not contain information on usage for each kind of crop. It is imprecise to analyze and evaluate fertilizer efficiency using total amounts, because the distribution and application of fertilizer on specific crops are ambiguous [16].

Thus, the objectives of this study were to: (1) reveal the spatial and temporal variations of chemical fertilization and yields of wheat and maize at the farmers' level from the 1970s to the 2000s in the Loess Plateau in China; (2) reveal the spatial and temporal variations of soil fertility over the same period; and (3) reveal the relationships among fertilizer inputs, crop yields and soil fertility.

Materials and Methods

Ethics Statement

This study has been approved by the Agricultural Technology Extension Center of Shaanxi province, which is responsible for fertilization and soil fertility in Shaanxi province. All data in this study can be published and shared.

Study area

Shaanxi province (Figure 1) is located in the middle reaches of the Yellow River and the upper reaches of the Yangtze River of the eastern part of northwest China, and it falls between latitudes 31°42' and 39°35'N, and longitudes 105°29' and 111°15'E. The area is 2.058×10^5 km^2, extending about 880 km from north to south and 160 to 490 km from east to west. The whole province from north to south can be divided into four agro-ecological zones, which include the Loess Plateau area of northern Shaanxi, the Weibei dry plateau, the Guanzhong irrigated area and the Qin-Ba mountain area of southern Shaanxi; the previous three regions belong to the Loess Plateau and in this study they are abbreviated as North, Weibei, and Guanzhong, respectively. The Loess Plateau region in China, covers five provinces (including Shaanxi province), stretches over an area of 0.62 million km^2, and consists of typical semiarid and arid areas with rainfed farming [17–18]. Winter wheat is planted in the regions of Weibei and Guanzhong, while summer maize is planted in the Guanzhong region and spring maize in the North and Weibei regions. Main soil types and climatic conditions in the different regions are shown in Table 1.

Data sources

The data from the 1970s to the 1990s was extracted from 380 published papers reporting household fertilization and soil fertility in the study area; the screening process and results are shown in Figure 2. Data from the 2000s was collected from the project "soil testing and formulated fertilization in Shaanxi province during the years 2005 to 2009."

Statistics

The data were analyzed by EXCEL software. In this study, we used the following equations to analyze the soil nutrient balance and partial factor productivity (PFP) of fertilizer:

$$Soil\ nutrient\ balance = nutrient\ input\ rate$$
$$- nutrient\ output\ rate \qquad (Eq\ 1)$$

where the nutrient input rate represents chemical fertilizer input, and the nutrient output rate represents amounts extracted in crop products and above ground biomass;

$$PFP = Y/F \qquad (Eq\ 2)$$

where Y represents crop yields, and F represents chemical fertilizer input.

Results

Spatial and temporal variations of chemical fertilization and yields of wheat and maize at the farmers' level in different regions of Shaanxi province

The average chemical fertilizer NPK inputs for both wheat and maize at the farmers' level increased for decades in the different regions (Figure 3). In the Weibei and Guanzhong regions,

chemical fertilizer N inputs for wheat in the 1970s were 45 kg ha^{-1} and 52 kg ha^{-1}, respectively, and in the 2000s they increased to 185 kg ha^{-1} and 195 kg ha^{-1}, respectively. In these two regions, chemical fertilizer P$_2$O$_5$ inputs were 45 kg ha^{-1} and 46 kg ha^{-1} in the 1970s and they increased to 112 kg ha^{-1} and 115 kg ha^{-1} in the 2000s. In the 1980s, farmers started to use the chemical fertilizer K$_2$O for wheat, which was increased from 0.5 kg ha^{-1} and 2.3 kg ha^{-1} to 22.8 kg ha^{-1} and 22.5 kg ha^{-1}, respectively, during the 1980s to the 2000s in the two regions. For maize in the North, Weibei and Guanzhong regions, chemical fertilizer N inputs were 48 kg ha^{-1}, 89 kg ha^{-1} and 36 kg ha^{-1} and they increased to 237 kg ha^{-1}, 223 kg ha^{-1} and 244 kg ha^{-1}, respectively, from the 1970s to the 2000s. Unlike wheat, from the 1980s onward farmers were awarded for using the chemical fertilizers P$_2$O$_5$ and K$_2$O for maize, and their use has increased greatly.

In accordance with increased chemical fertilizer NPK inputs (Figure 3), the average yields of wheat and maize showed increasing trends in the different regions over the four decades (Figure 4). In the Weibei and Guanzhong regions, from the 1970s to the 2000s, yields of wheat changed from 1883 kg ha^{-1} and 3377 kg ha^{-1} to 4269 kg ha^{-1} and 6437 kg ha^{-1}, with increase rates of 127% and 91%, respectively. In the North, Weibei and Guanzhong regions, yields of maize changed from 3636 kg ha^{-1}, 2519 kg ha^{-1} and 4232 kg ha^{-1} to 7867 kg ha^{-1}, 7077 kg ha^{-1} and 6886 kg ha^{-1}, with increase rates of 116%, 181% and 63%, respectively, for the same period.

Spatial and temporal variations of soil nutrient balance from the inputs and uptake on wheat and maize plots in different regions of Shaanxi province

Because the farmers tended not to use organic manure for cereal crops, especially from the 1980s onward, the soil nutrient inputs only include chemical fertilizers, and the nutrient uptakes include those extracted in crop products and above ground biomass. The nutrient balance was calculated as the difference between the average input and uptake (Eq. 1). Other losses, from leakage and gaseous loss, were not included in these calculations. In the 1970s, N was deficient on wheat and maize plots in the different regions (except for maize plots in the Weibei region). Then from the 1980s N was consistently at surplus levels, and it displayed an upward trend with time. In the 2000s, N surpluses on wheat plots were 74 kg ha^{-1} and 29 kg ha^{-1} in the Weibei and Guanzhong regions, respectively; meanwhile N surpluses on maize plots were 64 kg ha^{-1}, 67 kg ha^{-1} and 93 kg ha^{-1} in the North, Weibei and Guanzhong regions, respectively (Figure 5).

In the Weibei and Guanzhong regions, the amount of surplus P$_2$O$_5$ on wheat plots increased each year from the 1970s to the 2000s, and surplus amounts increased from 24 kg ha^{-1} and 9 kg ha^{-1} to 65 kg ha^{-1} and 44 kg ha^{-1}, respectively. In the North, Weibei and Guanzhong regions, P$_2$O$_5$ was deficient on maize plots in the 1980s; then it gradually reached surplus levels until the 2000s with the increased application of chemical fertilizer phosphorus. The balance of P$_2$O$_5$ on maize plots increased from −34 kg ha^{-1}, −7 kg ha^{-1} and −35 kg ha^{-1} to 29 kg ha^{-1}, 28 kg ha^{-1} and −11 kg ha^{-1}, respectively, in the three regions from the 1980s to the 2000s. It is worth noting, that winter wheat and summer maize were in a rotation system in the Guanzhong region, so total P$_2$O$_5$ was in surplus in this region in the 2000s and the amount was 33 kg ha^{-1} (Figure 5).

Although farmers have been awarded for using K$_2$O chemical fertilizer in recent years, the amount used was still small (Figure 3), and it was usually from compound fertilizers. So K$_2$O deficiency has become more serious (Figure 5). In the 2000s, K$_2$O deficiency

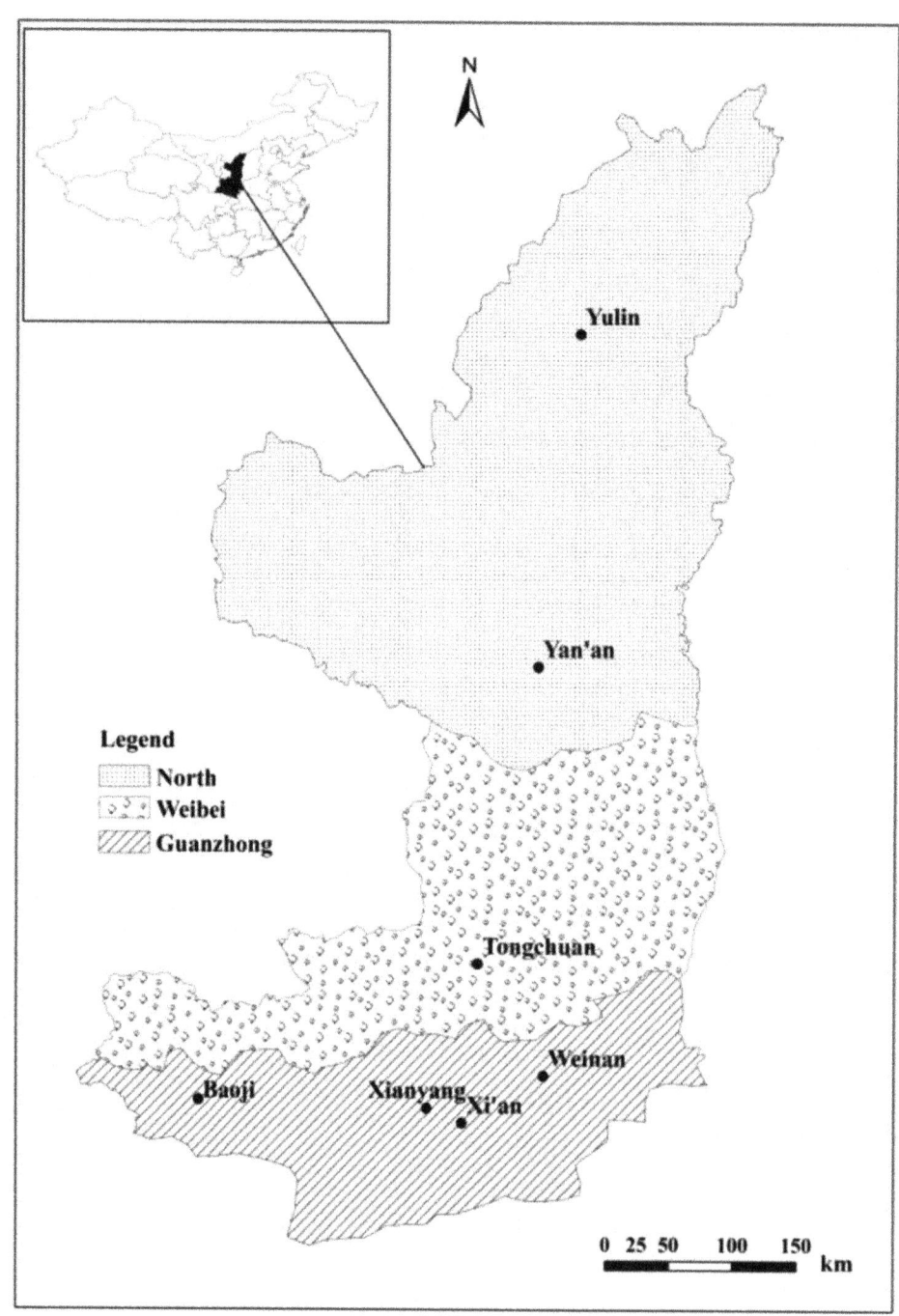

Figure 1. Map of the study area.

Table 1. Main soil types and climatic conditions in the different regions.

Region	Main soil types	Annual mean temperature (°C)	Annual precipitation (mm)
North	Castanozems, Sierozems, Loess soils	8~11	275~590
Weibei	Black loess soils, Loess soils	9~13	530~630
Guanzhong	Cinnamon soils	10~14	600~720

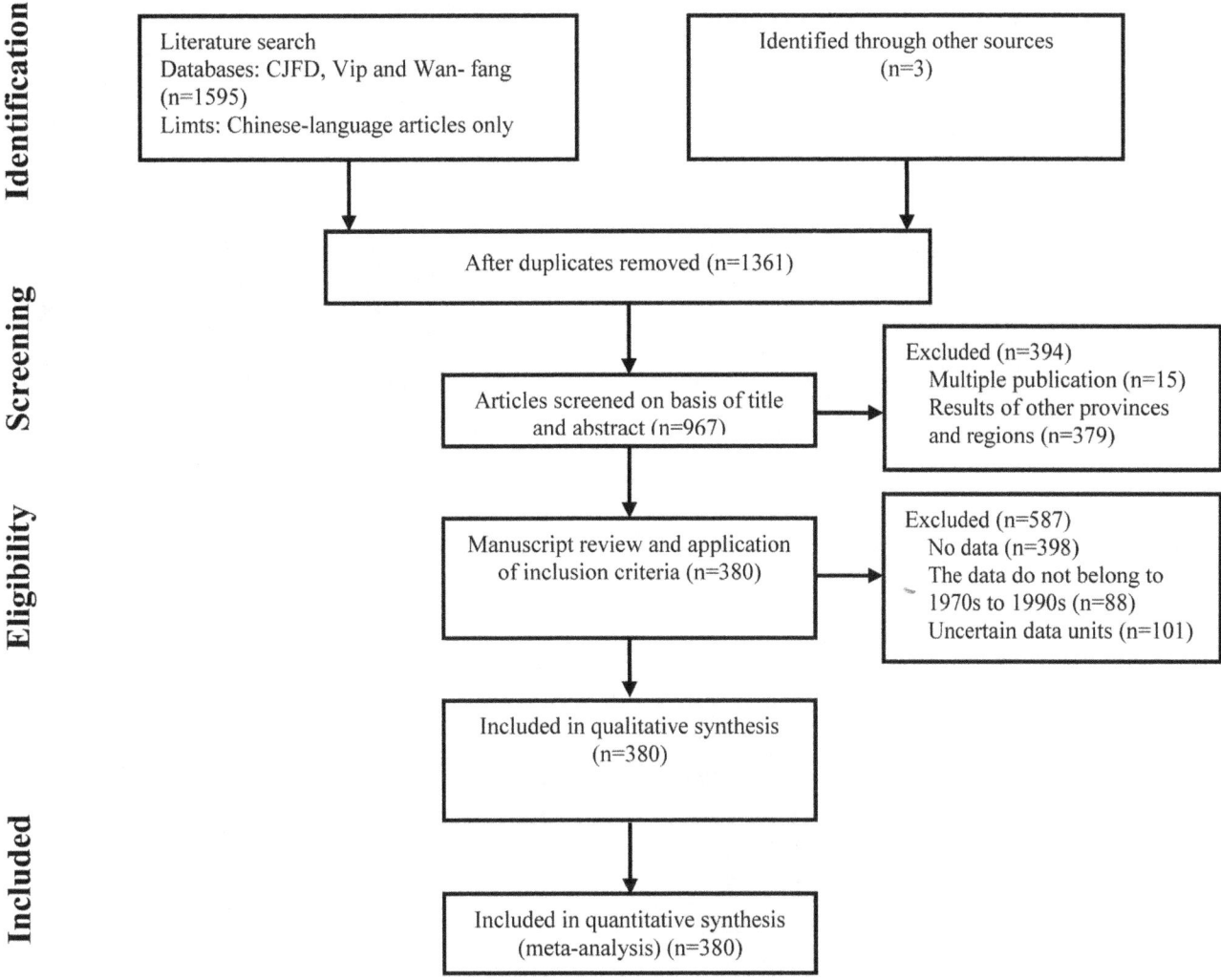

Figure 2. The screening process and results for literature from the 1970s to the 1990s.

levels on wheat plots were -102 kg ha^{-1} and -165 kg ha^{-1} in the Weibei and Guanzhong regions, respectively; meanwhile K$_2$O deficiency levels on maize plots were -179 kg ha^{-1}, -137 kg ha^{-1} and -147 kg ha^{-1} in the North, Weibei and Guanzhong regions, respectively.

Spatial and temporal variations of soil fertility in different regions of Shaanxi province

In the different regions of Shaanxi province, soil fertility indexes, including organic matter, total nitrogen, alkali-hydrolysis nitrogen, available phosphorus and available potassium, all increased from the 1970s to the 2000s. Simultaneously, each of these five indicators increased from the north to the south during the same period (North<Weibei<Guanzhong) (Figure 6). In the North, Weibei and Guanzhong regions from the 1970s to the 2000s, organic matter varied from 0.57%, 1.01% and 1.12% to 0.83%, 1.26% and 1.50%, with increase rates of 46%, 26% and 43%, respectively; total nitrogen varied from 0.04%, 0.07% and 0.07% to 0.05%, 0.08% and 0.09%, with increase rates of 42%, 9% and 14%, respectively; alkali-hydrolysis nitrogen varied from 29.95 mg kg^{-1}, 20.43 mg kg^{-1} and 30.81 mg kg^{-1} to 35.20 mg kg^{-1}, 58.70 mg kg^{-1} and 68.40 mg kg^{-1}, with increase rates of 18%, 187% and 122%, respectively; available phosphorus varied

from 4.98 mg kg^{-1}, 7.13 mg kg^{-1} and 9.90 mg kg^{-1} to 8.10 mg kg^{-1}, 14.60 mg kg^{-1} and 26.40 mg kg^{-1}, with increase rates of 63%, 105% and 167%, respectively; available potassium varied from 85.60 mg kg^{-1}, 56.78 mg kg^{-1} and 111.75 mg kg^{-1} to 99.60 mg kg^{-1}, 160.70 mg kg^{-1} and 170.40 mg kg^{-1}, with increase rates of 16%, 183% and 52%, respectively.

Relationships among fertilizer inputs, crop yields and soil fertility in different regions of Shaanxi province

Because farmers used little P and K fertilizers in the 1970s and 1980s (Figure 3), only PFP of N, NP and NPK were calculated in the study (Eq. 2). The PFP of N, NP and NPK on wheat and maize decreased from the 1970s to the 2000s as a whole in the different regions (Table 2). The PFP of N on wheat in the Weibei and Guanzhong regions were 42 kg kg^{-1} and 65 kg kg^{-1}, respectively, in the 1970s, which decreased to 23 kg kg^{-1} and 33 kg kg^{-1}, respectively, in the 2000s. Meanwhile the PFP of N on maize in the North and Guanzhong regions were 76 kg kg^{-1} and 118 kg kg^{-1}, respectively, and they decreased to 33 kg kg^{-1} and 28 kg kg^{-1} from the 1970s to the 2000s. In the Weibei region, the PFP of N on maize changed slightly from 28 kg kg^{-1} to 32 kg kg^{-1}, which resulted from the use of high N inputs relative to the other two regions (up to 89 kg ha^{-1}) in the 1970s (Figure 3). This led to

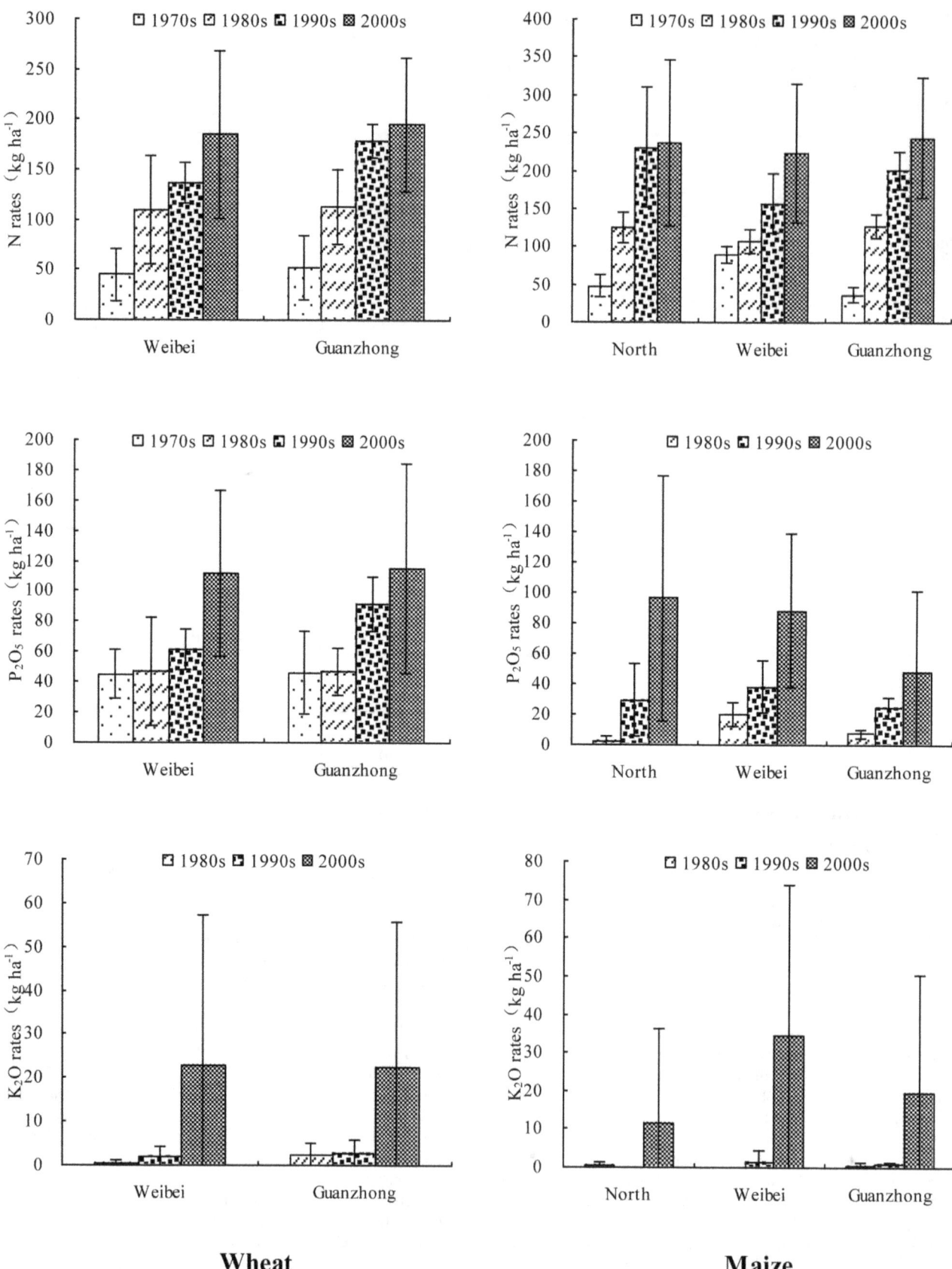

Wheat

Maize

Figure 3. Variations of chemical fertilization for wheat and maize at the farmers' level in different regions of Shaanxi province (error bars show standard deviations).

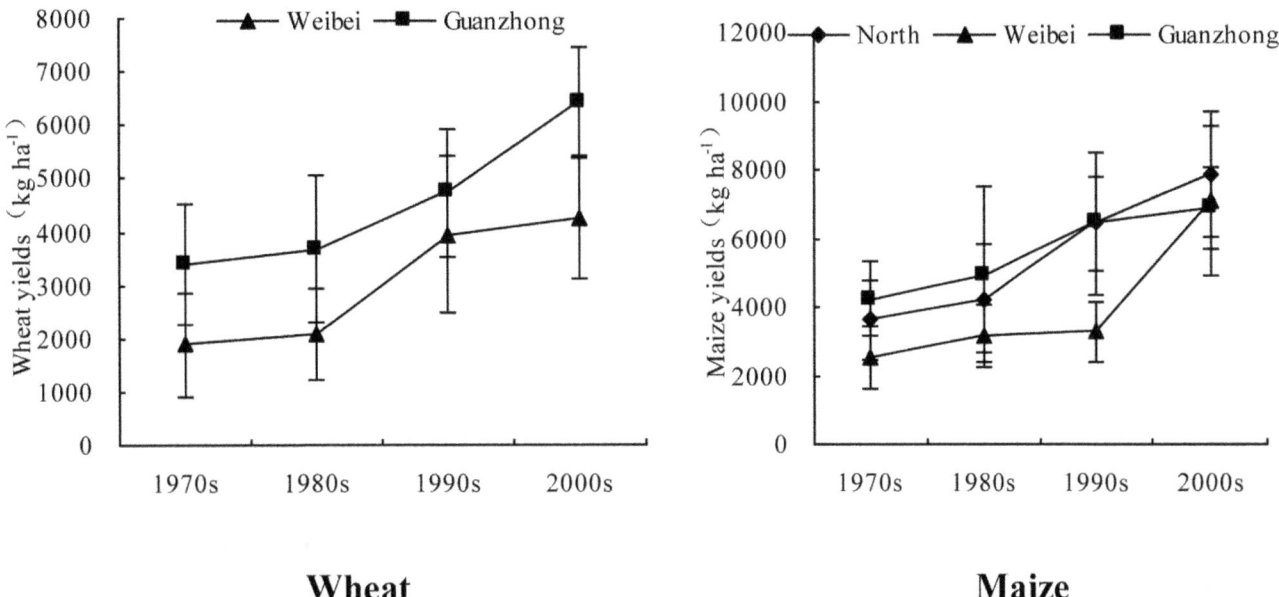

Figure 4. Variations of yields for wheat and maize at the farmers' level in different regions of Shaanxi province (error bars show standard deviations).

a low PFP of N in that period. The PFP of NP on wheat decreased to 14 kg kg^{-1} and 21 kg kg^{-1} in the Weibei and Guanzhong regions, respectively; in maize it decreased to 24 kg kg^{-1}, 23 kg kg^{-1} and 24 kg kg^{-1} in the North, Weibei and Guanzhong regions, respectively. Similar to N and NP, the PFP of NPK on wheat decreased to 13 kg kg^{-1} and 19 kg kg^{-1} in the Weibei and Guanzhong regions, respectively; in maize it decreased to 23 kg kg^{-1}, 20 kg kg^{-1} and 22 kg kg^{-1} in the North, Weibei and Guanzhong regions, respectively (Table 2).

In order to find relationships among soil fertility, crop yields and fertilizer rates, we used the Weibei region as an example. The values of fertilization, crop yields and soil fertility did not have one to one correspondence from the 1970s to the 1990s, so their mean value from each period was examined (Figures 7 and 8). Although the sample size was small and some relationships did not reach significant levels, with the increase in N fertilizer inputs, soil total nitrogen and alkali-hydrolysis nitrogen both increased. P fertilizer increased soil available phosphorus and K fertilizer increased soil available potassium significantly (Figure 7). At the same time, soil organic matter, total nitrogen, alkali-hydrolysis nitrogen, available phosphorus and available potassium all had positive impacts on wheat yields (Figure 8).

Discussion

Fertilizer use efficiency of both wheat and maize decreased from the 1970s to the 2000s as a whole in the Loess Plateau of Shaanxi (Table 2), which was consistent with national trends. Nitrogen fertilizer, phosphorus fertilizer and potassium fertilizer use efficiencies were 30–35%, 15–20% and 35–50%, respectively, from 1981 to 1983, and the average values decreased to 28%, 12% and 32% on cereal crops by 2001 to 2005 in China [19]. This suggested that the effect of chemical fertilizers on increasing grain production had diminished. The PFP of N on wheat in the Weibei and Guanzhong regions decreased to 23 kg kg^{-1} and 33 kg kg^{-1}, respectively, and the PFP of N on maize in the North, Weibei and Guanzhong regions were 33 kg kg^{-1}, 32 kg kg^{-1} and 28 kg kg^{-1}, respectively, in the 2000s (Table 2). Zhang et al. [19] reported

average PFP values of N for wheat and maize of 43 kg kg^{-1} and 52 kg kg^{-1}, respectively, in China. Dobermann and Cassman [20] reported a global average PFP of N for cereals of 44 kg kg^{-1}. This indicated that nitrogen use efficiency on wheat and maize in the Loess Plateau of Shaanxi was much lower than the current national and global levels. Excessive fertilization has been the main reason for low fertilizer use efficiency in China [19]. In addition, Liu et al. [21] reported that in agro-ecosystems, surplus N increased from 1978 to 2005 throughout the country, and our findings on the Loess Plateau were consistent with this trend. For example, in the 2000s, chemical fertilizer N inputs on maize were 237 kg ha^{-1}, 223 kg ha^{-1} and 244 kg ha^{-1} in the North, Weibei and Guanzhong regions, respectively (Figure 3); meanwhile N surpluses on maize plots were 64 kg ha^{-1}, 67 kg ha^{-1} and 93 kg ha^{-1}, respectively, in the three regions (Figure 5). This indicated that excessive N fertilization was a serious problem in the Loess Plateau, and the same phenomenon has been reported many times in China, for example, in Beijing [16,22], Shandong [1,23–25], and Jiangsu [26–27]. Excessive N fertilization not only wastes resources, but also leads to many serious environmental problems [28–31] including nitrate pollution of groundwater [32–37], eutrophication of surface water [38–39], greenhouse gas emissions and other forms of air pollution [40–42], acid rain [43–46], soil acidification [36,47–50] and so on. On the other hand, a lower fertilization rate does not necessarily reduce crop yields [51]. Many studies have shown that reducing the current N application rates by 30 to 60% could increase N fertilizer efficiency, while still maintaining crop yields and substantially reducing N losses to the environment [31,52–53].

Like nitrogen, phosphate fertilizer inputs (Figure 3), P surpluses (Figure 5) and soil available phosphorus levels (Figure 6) all increased in the last 40 years on the Loess Plateau in Shaanxi. Similar results have been noted in north China and all over the country [25,54]. Yang et al. [55] reported that maintaining soil available phosphorus at a relatively high level requires a P application rate of about 80 kg ha^{-1} yr^{-1} in winter wheat/summer maize rotation systems in the Guanzhong region. Our

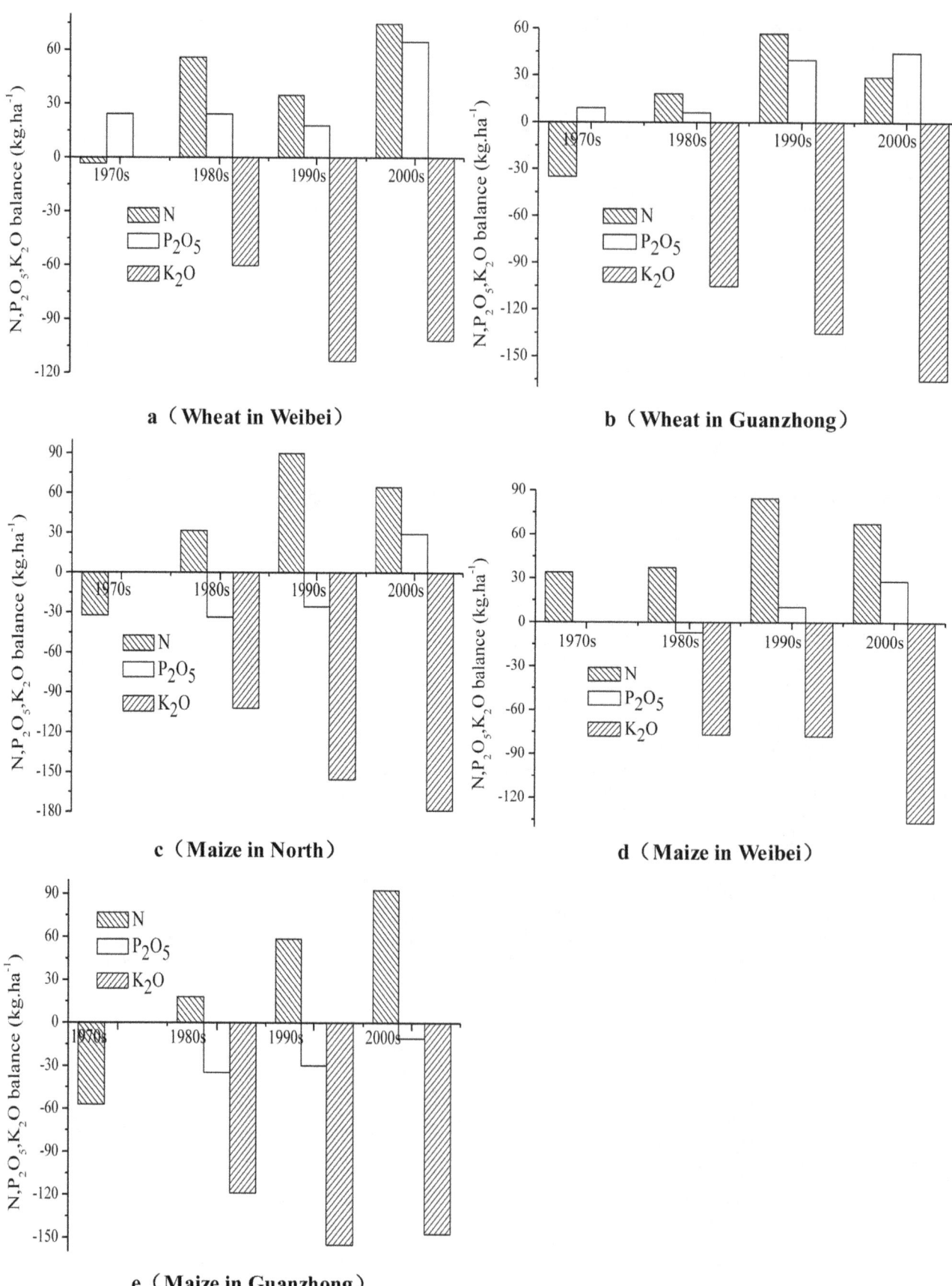

Figure 5. Variations of soil nutrient balance on wheat and maize plots in different regions of Shaanxi province.

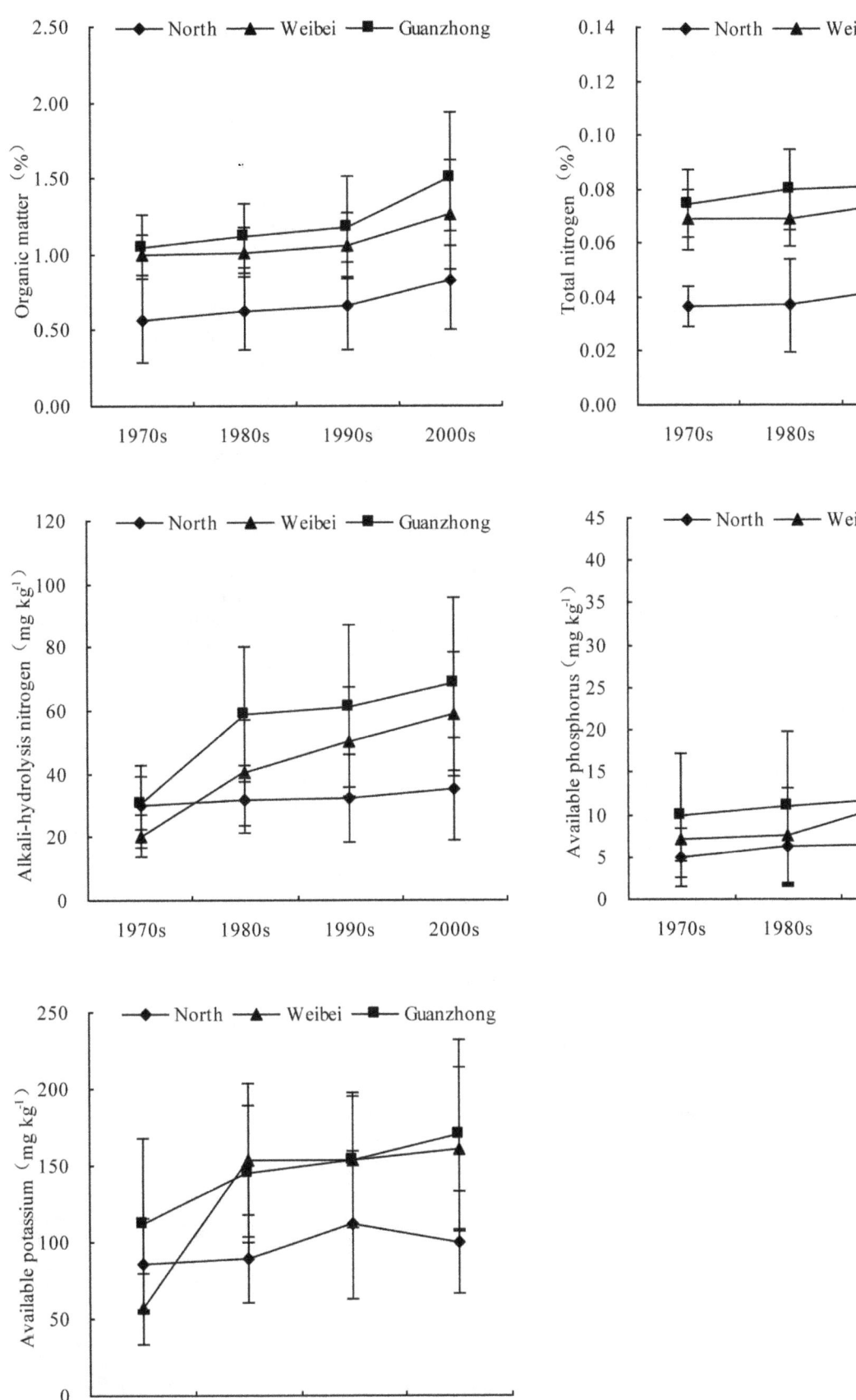

Figure 6. Variations of soil organic matter, total nitrogen, alkali-hydrolysis nitrogen, available phosphorus and available potassium in different regions of Shaanxi province (error bars show standard deviations).

results showed phosphate fertilizer inputs of up to 163 kg ha^{-1} in winter wheat/summer maize rotation systems in this region in the 2000s (Figure 3). This indicated that P fertilization was also excessive, which not only wasted resources but also led to many serious environmental problems [28–31]. Phosphate fertilizer production consumes more than 80% of the phosphate rock resources [56], but phosphate rock resources are limited and high grade material is in short supply [57]. In addition, the phosphate fertilization utilization ratio of the main crops ranged from 7% to 20%. It averages 12% in China [19], which has led to phosphorus accumulation in the soil, increasing the risk of non-point source pollution from surface runoff [58]. Agricultural non-point source pollution has become an increasingly serious problem in China, primarily because it leads to eutrophication.

In spite of increased K fertilizer inputs on wheat and maize in recent years (Figure 3), the soil K balance has become increasingly negative (Figure 5) and soil available potassium has increased (Figure 6) in the last 40 years. This phenomenon was previously reported in northwest and north China [55,59–60]. Evidently, K fertilizer application was not the only source of K absorbed by crops. The primary sources of K for crops were weathering of parent materials [60–61], release of K into the soil from increased soil organic matter and changes in soil pH [61]. Yang et al. [55] found that soil organic matter content in all treatments (including those without fertilizer) significantly increased over time and soil pH dropped from the initial value of 8.65 to 8.58 from 1991 to 2010 during long-term field trials in the Guanzhong region. Our results showed that in the North, Weibei and Guanzhong regions soil organic matter increased from 0.57%, 1.01% and 1.12% to 0.83%, 1.26% and 1.50%, respectively, from the 1970s to the 2000s (Figure 6). The average soil pH has declined 0.5 units with the overuse of N fertilizer in the past two decades in China [62]. Li et al. [63] reported that the soil pH decreased from the initial value of 8.76 to 8.56 from 1992 to 2008 during long-term field trials in

the North region. There may be other mechanisms involved, for example, crops might draw on K in the deeper soil layers or from the non-exchangeable pool. The contribution of K from the subsoil could be considerable [64]. Witter and Johansson [65] found that 41–47% of the K was from the subsoil for green manure crops. Many studies have shown that crops use non-exchangeable K [66–67]. Decreases in the abundance of non-exchangeable K with simultaneous increases in exchangeable and water-soluble K concentrations suggest that much of the K taken up by crops comes from non-exchangeable species via solution and exchangeable phases in a way that establishes and maintains the equilibrium between various forms of K in the soil [66].

Fertilizer rates had a large effect on soil fertility. With the increase in N fertilizer inputs, both soil total nitrogen and alkali-hydrolysis nitrogen increased; P fertilizer increased soil available phosphorus and K fertilizer increased soil available potassium significantly in the Weibei region (Figure 7). It has been reported that after 25 years of N fertilization, soil organic carbon and total nitrogen had increased by 18% and 26%, respectively, from 1984 to 2009 in the Weibei region [18]. Cai and Hao [68] also found that accumulation of soil nitrogen initially increased and then decreased with increasing nitrogen, and total nitrogen and alkali-hydrolysis nitrogen content reached the highest value or the second highest value of 135 kg ha^{-1} on wheat plots in the Weibei region, which was in accordance with findings in northwest and north China by Li et al. [63] and Lin et al. [69]. Through long-term field experimentation on the Loess Plateau in Shaanxi, Li et al. [63] and Hao et al. [70] found that with increases in P fertilizer inputs, soil available P increased significantly. Similar results have been obtained in northeast and northwest China by Geng et al. [71] and Zhao et al. [72], and also in America by Griffin et al. [73]. In addition, Li et al. [74] found that with increased K fertilizer inputs, soil available K increased significantly in a long-term field experiment on the Loess Plateau. Further-

Table 2. Variations of PFP of fertilizer on wheat and maize in the different regions (kg kg^{-1}).

Crop	Fertilizer type	Region	1970s	1980s	1990s	2000s
Wheat	N	Weibei	42	19	29	23
		Guanzhong	65	33	26	33
	N+P$_2$O$_5$	Weibei	21	13	20	14
		Guanzhong	34	23	17	21
	N+P$_2$O$_5$+K$_2$O	Weibei	21	13	20	13
		Guanzhong	34	23	17	19
Maize	N	North	76	34	28	33
		Weibei	28	30	21	32
		Guanzhong	118	39	32	28
	N+P$_2$O$_5$	North	76	33	25	24
		Weibei	28	25	17	23
		Guanzhong	118	37	29	24
	N+P$_2$O$_5$+K$_2$O	North	76	33	25	23
		Weibei	28	25	17	20
		Guanzhong	118	37	29	22

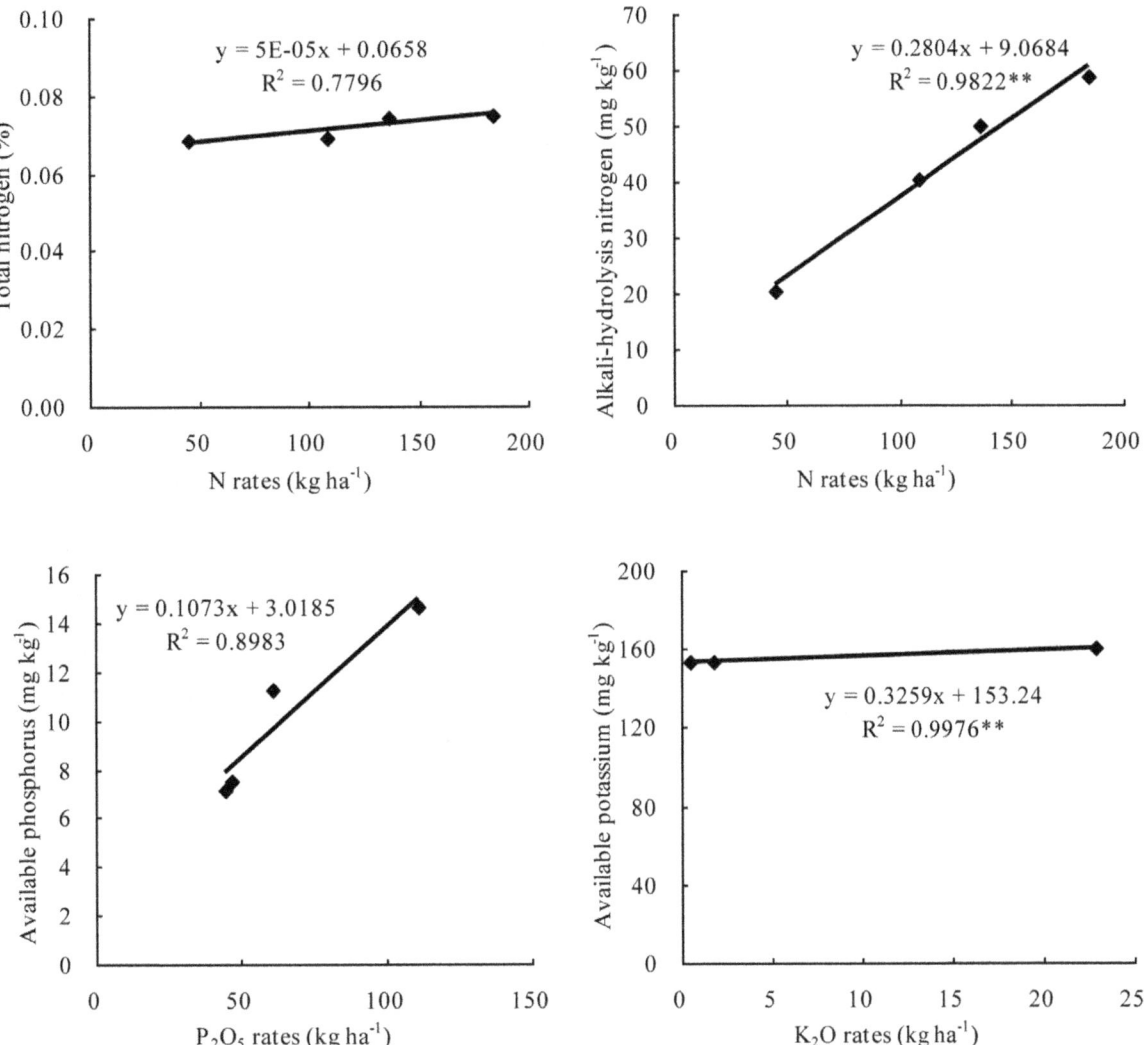

Figure 7. Relationships between N rates and total nitrogen, N rates and alkali-hydrolysis nitrogen, P₂O₅ rates and available phosphorus and K₂O rates and available potassium on wheat plots in the Weibei region of Shaanxi province. **Significance level: P< 0.01.

more, many studies in this area have shown that on the basis of N and P fertilizer application, long-term K fertilizer application can increase soil available K and grain yields [75–76].

Our research also found that soil fertility had a positive impact on crop yields (Figure 8). Zhou et al. [77] revealed that soil organic carbon and total nitrogen concentrations had a significant effect on crop yields in the semi-arid Loess Plateau by long-term experimentation. Higher yields without fertilizer were generally obtained in soils with higher average soil organic matter concentrations. For example, yields without fertilizer <4000 kg ha^{-1} were obtained with average soil organic matter concentrations of 1.41% for winter wheat and 1.46 for summer maize. In contrast, average soil organic matter concentrations were 1.69% for winter wheat and 1.61% for summer maize for plots with yields>6000 kg ha^{-1} without fertilizer in north China [78]. Gong et al. [79] also found that the contribution percentage of basic soil productivity to wheat yield was significantly correlated with soil organic carbon, total nitrogen, available nitrogen, available phosphorus and available potassium in long-term soil fertility experiments in north China. Similar results have been obtained in

other parts of mainland China [80], indicating that inherent soil productivity contributed to the substantial increase in China's crop yields.

In addition, although the use of chemical fertilizers to supplement NPK nutrients in the soil is important, many researchers at home and abroad reported that the application of chemical fertilizer in combination with organic manure is helpful in maintaining soil fertility (especially soil organic carbon) and buffering capacity, and in reducing NO₃-N accumulation in the soil, while maintaining high soil productivity [4,81–87].

Conclusions

From the 1970s to the 2000s in the North, Weibei and Guanzhong regions of the Loess Plateau in Shaanxi province, chemical fertilizer NPK inputs and yields of wheat and maize increased at the farmers' level. In the 1970s, N was deficient on wheat and maize plots in the different regions; thereafter N was in surplus. In the same way, P gradually changed from deficit to surplus levels. In addition, soil organic matter, total nitrogen, alkali-hydrolysis nitrogen, available phosphorus and available

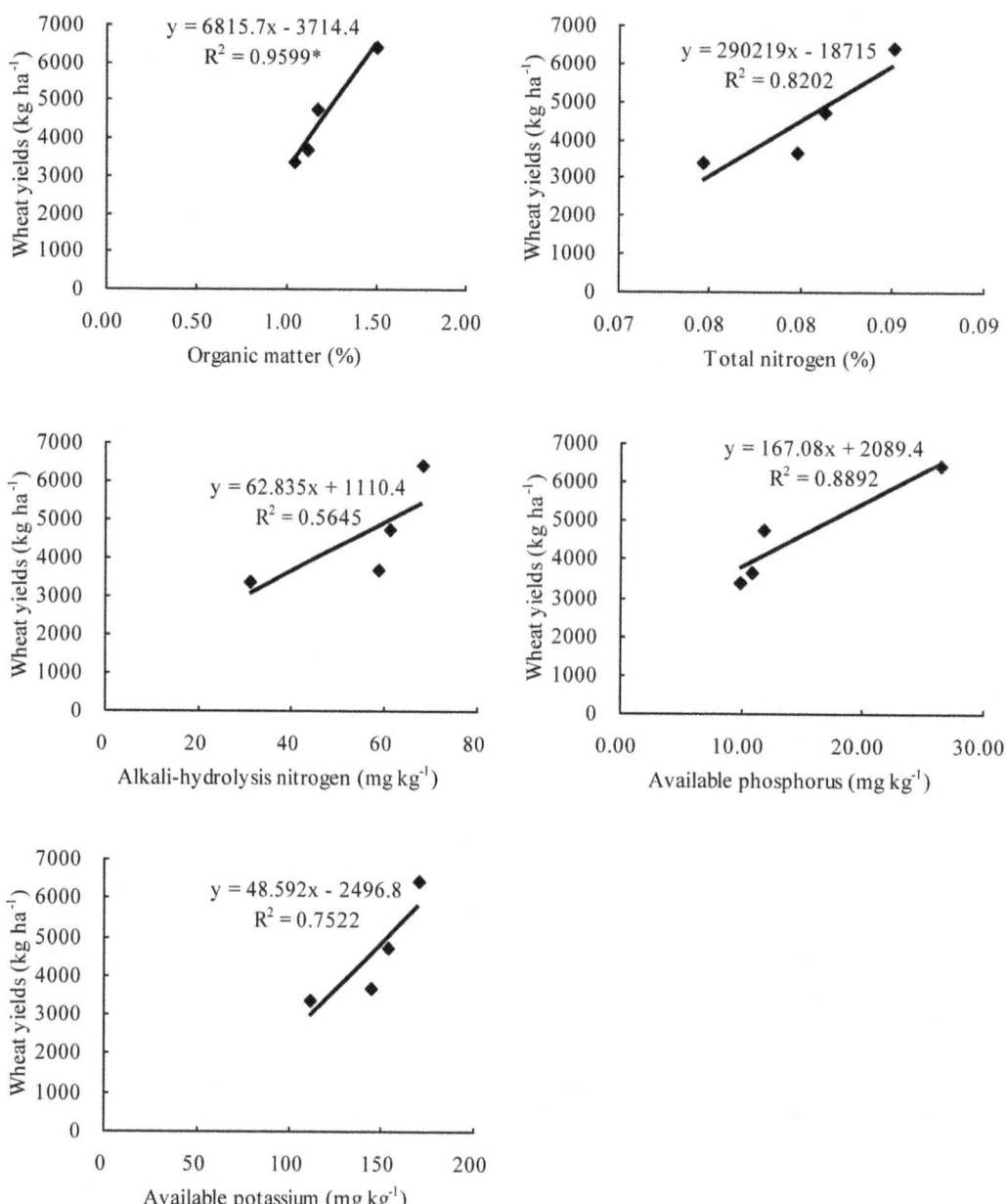

Figure 8. Relationships between wheat yield and soil organic matter, total nitrogen, alkali-hydrolysis nitrogen, available phosphorus and available potassium in the Weibei region of Shaanxi province. *Significance level: P<0.05.

potassium increased over the same period. However, K deficiencies became more and more severe. The PFP of N, NP and NPK on wheat and maize all decreased from the 1970s to the 2000s as a whole. With the increase in N fertilizer inputs, both soil total nitrogen and alkali-hydrolysis nitrogen increased; P fertilizer increased soil available phosphorus and K fertilizer increased soil available potassium significantly. At the same time, soil organic matter, total nitrogen, alkali-hydrolysis nitrogen, available phosphorus and available potassium all had positive impacts on crop yields. In order to promote food safety and environmental protection, farmers should be encouraged to assess their fertilizer needs carefully. Many can reduce nitrogen and phosphate fertilizer inputs significantly and increase potassium fertilizer and organic manure on cereal crops.

Acknowledgments

We are grateful to Harald Grip and Lars Lövdahl for their help in writing this paper. The authors would also like to thank the Agricultural Technology Extension Center of Shaanxi province for the help with data collection.

Author Contributions

Conceived and designed the experiments: YT PG. Analyzed the data: XW YG. Contributed reagents/materials/analysis tools: YT YG PG. Wrote the paper: XW. Collected the data: XW FL ZZ YP.

References

1. Cui ZL, Chen XP, Zhang FS (2010) Current nitrogen management status and measures to improve the intensive wheat–maize system in China. AMBIO 39: 376–384.

2. Gao C, Sun B, Zhang TL (2006) Sustainable nutrient management in Chinese agriculture: challenges and perspective. Pedosphere 16(2): 253–263.

3. Zhu ZL, Chen DL (2002) Nitrogen fertilizer use in China-Contributions to food production, impacts on the environment and best management strategies. Nutrient Cycling in Agroecosystems 63: 117–127.

4. Jiang D, Hengsdijk H, Dai TB, de Boer W, Qi J, et al. (2006) Long-term effects of manure and inorganic fertilizers on yield and soil fertility for a winter wheat-maize system in Jiangsu, China. Pedosphere 16(1): 25–32.

5. Luo SM (2007) To discover the secret of traditional agriculture and serve the modern ecoagriculture. Geographical Research 26(3): 609–615. (in Chinese).

6. Zhang F, Qiao Y, Wang F, Zhang W (2007) A perspective on organic agriculture in China: Opportunities and challenges. Proceedings of 9th German Scientific Conference on Organic Agriculture.

7. Department of Rural Surveys, National Bureau of Statistics (1971–2011) China Rural Statistical Yearbook. China Statistics Press. Beijing, China. (in Chinese).

8. Karlen DL, Mausbach MJ, Doran JW, Cline RG, Harris RF, et al. (1997) Soil quality: A concept, definition and framework for evaluation. Soil Science Society of America Journal 61: 4–10.

9. Wang XJ, Gong ZT (1998) Assessment and analysis of soil quality changes after eleven years of reclamation in subtropical China. Geoderma 81: 339–355.

10. Arshad MA, Martin S (2002) Identifying critical limits for soil quality indicators in agro-ecosystems. Agriculture, Ecosystems & Environment 88: 153–160.

11. Huang B, Sun WX, Zhao YC, Zhu J, Yang RQ, et al. (2007) Temporal and spatial variability of soil organic matter and total nitrogen in an agricultural ecosystem as affected by farming practices. Geoderma 139: 336–345.

12. Hoosbeek MR, Bryant RB (1992) Towards the quantitative modeling of pedogenesis: A review. Geoderma 55: 183–210.

13. Smil V (2001) Enriching the Earth: Fritz Haber, Carl Bosch, and the Transformation of World Food Production. MIT Press. Cambridge, UK.

14. Gong W, Yan XY, Wang JY (2011) Effect of long-term fertilization on soil fertility. Soils 43: 336–342. (in Chinese).

15. Bierman PM, Rosen CJ, Venterea RT, Lamb JA (2012) Survey of nitrogen fertilizer use on corn in Minnesota. Agricultural Systems 109: 43–52.

16. Wang SR (2002) Current status and evaluation of crop fertilization in Shaanxi province and Beijing city, Ph.D. thesis, China Agricultural University, Beijing, China. (in Chinese).

17. Liu GB (1999) Soil conservation and sustainable agriculture on the Loess Plateau: Challenges and prospects. AMBIO 28: 663–668.

18. Guo SL, Zhu HH, Dang TH, Wu JS, Liu WZ, et al. (2012) Winter wheat grain yield associated with precipitation distribution under long-term nitrogen fertilization in the semiarid Loess Plateau in China. Geoderma 189: 442–450.

19. Zhang FS, Wang JQ, Zhang WF, Cui ZL, Ma WQ, et al. (2008) Nutrient use efficiencies of major cereal crops in China and measures for improvement. Acta Pedologica Sinica 45: 915–924. (in Chinese).

20. Dobermann A, Cassman KG (2005) Cereal area and nitrogen use efficiency are drivers of future nitrogen fertilizer consumption. Science in China (Series C: Life Sciences) 48: 745–758.

21. Liu Z, Li BG, Fu J (2009) Nitrogen balance in agro-ecosystem in China from 1978 to 2005 based on DSS. Transactions of the CSAE 25(4): 168–175. (in Chinese).

22. Zhao JR, Guo Q, Guo JL, Wei DM, Wang CW, et al. (1997) The chemical fertilizer inputs and yields of grain fields in the suburbs of Beijing. Beijing Agricultural Sciences 15(2): 36–38. (in Chinese).

23. Ma WQ (1999) Current status and evaluation of crop fertilization in Shandong province, Ph.D. thesis, China Agricultural University, Beijing, China. (in Chinese).

24. Li JL, Cui DJ, Meng XX, Li XL, Zhang FS (2002) The study of fertilization condition and question in protectorate vegetable in Shouguang Shandong. Chinese Journal of Soil Science 33: 126–128. (in Chinese).

25. Zhen L, Zoebisch MA, Chen GB, Feng ZM (2006) Sustainability of farmers' soil fertility management practices: A case study in the North China Plain. Journal of Environmental Management 79: 409–419.

26. Richter J, Roelcke M (2000) The N-cycle as determined by intensive agriculture–examples from central Europe and China. Nutrient Cycling in Agroecosystems 57: 33–46.

27. Ma LH, Zhang Y, Sui B, Liu CL, Wang P, et al. (2011) The impact factors of excessive fertilization in Jiangsu province. Journal of Yangzhou University 32(2): 48–52, 80. (in Chinese).

28. Gao XZ, Ma WQ, Du S, Zhang FS, Mao DR (2001) Current status and problems of fertilization in China. Chinese Journal of Soil Science 32(6): 258–261. (in Chinese).

29. Cui ZL, Chen XP, Miao YX, Zhang FS, Sun QP, et al. (2008) On-farm evaluation of the improved soil N_{min}-based nitrogen management for summer maize in North China Plain. Agronomy Journal 100: 517–525.

30. Cui ZL, Zhang FS, Chen XP, Miao YX, Li JL, et al. (2008) On-farm evaluation of an in-season nitrogen management strategy based on soil N_{min} test. Field Crops Research 105: 48–55.

31. Ju XT, Xing GX, Chen XP, Zhang SL, Zhang LJ, et al. (2009) Reducing environmental risk by improving N management in intensive Chinese agricultural systems. Proceedings of the National Academy of Sciences 106: 3041–3046.

32. Tong YA, Emteryd O, Lu DQ, Grip H (1997) Effect of organic manure and chemical fertilizer on nitrogen uptake and nitrate leaching in a Eum-orthic anthrosols profile. Nutrient Cycling in Agroecosystems 48: 225–229.

33. Ju XT, Kou CL, Zhang FS, Christie P (2006) Nitrogen balance and groundwater nitrate contamination: Comparison among three intensive cropping systems on the North China Plain. Environmental Pollution 143: 117–125.

34. Ju XT, Liu XJ, Zhang FS, Roelcke M (2004) Nitrogen fertilization, soil nitrate accumulation, and policy recommendations in several agricultural regions of China. AMBIO 33: 300–305.

35. Yan X, Jin JY, He P, Liang MZ (2008) Recent advances on the technologies to increase fertilizer use efficiency. Agricultural Sciences in China 7(4): 469–479.

36. Guo SL, Wu JS, Dang TH, Liu WZ, Li Y, et al. (2010) Impacts of fertilizer practices on environmental risk of nitrate in semiarid farmlands in the Loess Plateau of China. Plant and Soil 330: 1–13.

37. Gao Y, Yu G, Luo C, Zhou P (2012) Groundwater nitrogen pollution and assessment of its health risks: A case study of a typical village in rural-urban continuum, China. PLoS ONE 7(4): e33982.

38. Tilman D, Fargione J, Wolff B, D'Antonio C, Dobson A, et al. (2001) Forecasting agriculturally driven global environmental change. Science 292: 281–284.

39. Huang GQ, Wang XX, Qian HY, Zhang TL, Zhao QG (2004) Negative impact of inorganic fertilizer application on agricultural environment and its countermeasures. Ecology and Environment 13(4): 656–660. (in Chinese).

40. Mosier AR, Duxbury JM, Freney JR, Heinemeyer O, Minami K (1996) Nitrous oxide emissions from agricultural fields: Assessment, measurement and mitigation. Plant and Soil 181: 95–108.

41. Zhang JF, Han XG (2008) N_2O emission from the semi-arid ecosystem under mineral fertilizer (urea and superphosphate) and increased precipitation in northern China. Atmospheric Environment 42: 291–302.

42. Li H, Qiu JJ, Wang LG, Tang HJ, Li CS, et al. (2010) Modelling impacts of alternative farming management practices on greenhouse gas emissions from a winter wheat–maize rotation system in China. Agriculture, Ecosystems and Environment 135: 24–33.

43. Krusche AV, de Camargo PB, Cerri CE, Ballester MV, Lara LBLS, et al. (2003) Acid rain and nitrogen deposition in a sub-tropical watershed (Piracicaba): ecosystem consequences. Environmental Pollution 121: 389–399.

44. Menz FC, Seip HM (2004) Acid rain in Europe and the United States: an update. Environmental Science & Policy 7: 253–265.

45. Wu D, Wang SG, Shang KZ (2006) Progress in research of acid rain in China. Arid Meteorology 24(2): 70–77. (in Chinese).

46. Huang DY, Xu YG, Peng PA, Zhang HH, Lan JB (2009) Chemical composition and seasonal variation of acid deposition in Guangzhou, South China: Comparison with precipitation in other major Chinese cities. Environmental Pollution 157: 35–41.

47. Dai ZH, Liu YX, Wang XJ, Zhao DW (1998) Changes in pH, CEC, and exchangeable acidity of some forest soils in southern China during the last 32–35 years. Water, Air, and Soil Pollution 108: 377–390.

48. Zhang HM, Wang BR, Xu MG, Fan TL (2009) Crop yield and soil responses to long-term fertilization on a red soil in southern China. Pedosphere 19(2): 199–207.

49. Zhao X, Xing GX (2009) Variation in the relationship between nitrification and acidification of subtropical soils as affected by the addition of urea or ammonium sulfate. Soil Biology & Biochemistry 41: 2584–2587.

50. Huang S, Zhang WJ, Yu XC, Huang QR (2010) Effects of long-term fertilization on corn productivity and its sustainability in an Ultisol of southern China. Agriculture, Ecosystems and Environment 138: 44–50.

51. Ma WQ, István S (2008) Can sharp decrease of fertilizer input lead obvious reduction of crop yield?. Ecology and Environment 17: 1296–1301. (in Chinese).

52. Peng SB, Buresh RJ, Huang JL, Yang JC, Zou YB, et al. (2006) Strategies for overcoming low agronomic nitrogen use efficiency in irrigated rice systems in China. Field Crops Research 96: 37–47.

53. Yi Q, Zhang XZ, He P, Yang L, Xiong GY (2010) Effects of reducing N application on crop N uptake, utilization, and soil N balance in rice-wheat rotation system. Plant Nutrition and Fertilizer Science 16: 1069–1077. (in Chinese).

54. Cao N, Zhang YB, Chen XP (2009) Spatial-temporal change of phosphorus balance and the driving factors for agroecosystems in China. Chinese Agricultural Science Bulletin 25: 220–225. (in Chinese).

55. Yang XY, Sun BH, Zhang SL (2014) Trends of yield and soil fertility in a long-term wheat-maize system. Journal of Integrative Agriculture 13: 402–414.

56. Zhang WX (2011) Development and utilization trend of phosphate resources in China. Journal of Wuhan Institute of Technology 33: 1–5. (in Chinese).

57. Zhang WF, Ma WQ, Zhang FS, Ma J (2005) Comparative analysis of the superiority of China's phosphate rock and development strategies with that of the United States and Morocco. Journal of Natural Resources 20: 378–386. (in Chinese).

58. van Bochove E, Thériault G, Dechmi F, Leclerc ML, Goussard N (2007) Indicator of risk of water contamination by phosphorus: Temporal trends for the Province of Quebec from 1981 to 2001. Canadian Journal of Soil Science 87: 121–128.

59. Liu EK, Yan CR, Mei XR, He WQ, Bing SH, et al. (2010) Long-term effect of chemical fertilizer, straw, and manure on soil chemical and biological properties in northwest China. Geoderma 158: 173–180.

60. Tan DS, Jin J Y, Jiang LH, Huang SW, Liu ZH (2012) Potassium assessment of grain producing soils in North China. Agriculture, Ecosystems and Environment 148: 65–71.

61. Munson RD (1985) Potassium in Agriculture. Soil Science Society of America Madison, Wisconsin, USA.

62. Guo JH, Liu XJ, Zhang Y, Shen JL, Han WX, et al. (2010) Significant acidification in major Chinese croplands. Science 327: 1008–1010.

63. Li Q, Xu MX, Liu GB, ZhaoYG, Tuo DF (2013) Cumulative effects of a 17-year chemical fertilization on the soil quality of cropping system in the Loess Hilly Region, China. Journal of Plant Nutrition and Soil Science 176: 249–259.

64. Kautz T, Amelung W, Ewert F, Gaiser T, Horn R, et al. (2013) Nutrient acquisition from arable subsoils in temperate climates: A review. Soil Biology & Biochemistry 57: 1003–1022.

65. Witter E, Johansson G (2001) Potassium uptake from the subsoil by green manure crops. Biological Agriculture & Horticulture 19: 127–141.

66. Singh M, Singh VP, Reddy DD (2002) Potassium balance and release kinetics under continuous rice-wheat cropping system in Vertisol. Field Crops Research 77: 81–91.

67. Sharma A, Jalali VK, Arora S (2010) Non-exchangeable potassium release and its removal in foot-hill soils of North-west Himalayas. Catena 82: 112–117.

68. Cai Y, Hao MD (2013) Effects of long-term nitrogen fertilization on wheat in Loess Plateau. Journal of Triticeae Crops 33: 983–987. (in Chinese).

69. Lin ZA, Zhao BQ, Yuan L, Bing-So H (2009) Effects of organic manure and fertilizers long-term located application on soil fertility and crop yield. Scientia Agricultura Sinica 42: 2809–2819. (in Chinese).

70. Hao MD, Fan J, Wei XR, Pen LF, Lu L (2005) Effect of fertilization on soil fertility and wheat yield of dryland in the Loess Plateau. Pedosphere 15(2): 189–195.

71. Geng YH, Cao GJ, Ye Q, Qi QG, Wu P, et al. (2013) Effects of different phosphorus applications on soil available phosphorus, phosphorus absorption and yield of spring maize. Journal of South China Agricultural University 34: 470–474. (in Chinese).

72. Zhao J, Hou ZA, Li SX, Liu LP, Huang T, et al. (2014) Effects of P rate on soil available P, yield and nutrient uptake of maize. Journal of Maize Sciences 22: 123–128. (in Chinese).

73. Griffin TS, Honeycutt CW, He Z (2003) Changes in soil phosphorus from manure application. Soil Science Society of America Journal 67: 645–653.

74. Li LF, Hao MD, Li YM, Gao CQ (2009) Research on characteristics of spatial distribution and availability of soil potassium forms under long-term fertilization in the dryland of the Loess Plateau. Agricultural Research in the Arid Areas 27: 127–131, 142. (in Chinese).

75. Wang HT, Jin JY, Wang B, Zhao PP (2010) Effects of long-term potassium application and wheat straw return to cinnamon soil on wheat yields and soil potassium balance in Shanxi. Plant Nutrition and Fertilizer Science 16: 801–808. (in Chinese).

76. Zhang YL, Lu JL, Jin JY, Li ST, Chen ZQ, et al. (2012) Effects of chemical fertilizer and straw return on soil fertility and spring wheat quality. Plant Nutrition and Fertilizer Science 18: 307–314. (in Chinese).

77. Zhou ZC, Gan ZT, Shangguan ZP, Zhang FP (2013) Effects of long-term repeated mineral and organic fertilizer applications on soil organic carbon and total nitrogen in a semi-arid cropland. European Journal of Agronomy 45: 20–26.

78. Fan MS, Lai R, Cao J, Qiao L, Su YS, et al. (2013) Plant-based assessment of inherent soil productivity and contributions to China's cereal crop yield increase since 1980. PloS ONE, 8(9): e74617.

79. Gong FF, Zha Y, Wu XP, Huang SM, Xu MG, et al. (2013) Analysis on basic soil productivity change of winter wheat in fluvo-aquic soil under long-term fertilization. Transactions of the Chinese Society of Agricultural Engineering 29(12): 120–129. (in Chinese).

80. Tang YH, Huang Y (2009) Spatial distribution characteristics of the percentage of soil fertility contribution and its associated basic crop yield in mainland China. Journal of Agro-Environment Science 28: 1070–1078. (in Chinese).

81. Gami SK, Ladha JK, Pathak H, Shah MP, Pasuquin E, et al. (2001) Long-term changes in yield and soil fertility in a twenty-year rice-wheat experiment in Nepal. Biology and Fertility of Soils 34: 73–78.

82. Yang SM, Li FM, Malhi SS, Wang P, Suo DR, et al. (2004) Long-term fertilization effects on crop yield and nitrate nitrogen accumulation in soil in northwestern China. Agronomy Journal 96: 1039–1049.

83. Mando A, Ouattara B, Somado AE, Wopereis MCS, Stroosnijder L, et al. (2005) Long-term effects of fallow, tillage and manure application on soil organic matter and nitrogen fractions and on sorghum yield under Sudano-Sahelian conditions. Soil Use and Management 21: 25–31.

84. Li J, Zhao BQ, Li XY, Jiang RB, Bing SH (2008) Effects of long-term combined application of organic and mineral fertilizers on microbial biomass, soil enzyme activities and soil fertility. Agricultural Sciences in China 7(3): 336–343.

85. Banger K, Kukal SS, Toor G, Sudhir K, Hanumanthraju TH (2009) Impact of long-term additions of chemical fertilizers and farmyard manure on carbon and nitrogen sequestration under rice–cowpea cropping system in semi-arid tropics. Plant and Soil 318: 27–35.

86. Majumder B, Mandal B, Bandyopadhyay PK (2008) Soil organic carbon pools and productivity in relation to nutrient management in a 20-year-old rice-berseem agroecosystem. Biology and Fertility of Soils 44: 451–561.

87. Moharana PC, Sharma BM, Biswas DR, Dwivedi BS, Singh RV (2012) Long-term effect of nutrient management on soil fertility and soil organic carbon pools under a 6-year-old pearl millet–wheat cropping system in an Inceptisol of subtropical India. Field Crops Research 136: 32–41.

12

Interactions between Benthic Copepods, Bacteria and Diatoms Promote Nitrogen Retention in Intertidal Marine Sediments

Willem Stock[1]*, Kim Heylen[2], Koen Sabbe[1], Anne Willems[2], Marleen De Troch[1]

1 Department of Biology, Ghent University, Ghent, Belgium, 2 Department of Biochemistry and Microbiology, Ghent University, Ghent, Belgium

Abstract

The present study aims at evaluating the impact of diatoms and copepods on microbial processes mediating nitrate removal in fine-grained intertidal sediments. More specifically, we studied the interactions between copepods, diatoms and bacteria in relation to their effects on nitrate reduction and denitrification. Microcosms containing defaunated marine sediments were subjected to different treatments: an excess of nitrate, copepods, diatoms (*Navicula* sp.), a combination of copepods and diatoms, and spent medium from copepods. The microcosms were incubated for seven and a half days, after which nutrient concentrations and denitrification potential were measured. Ammonium concentrations were highest in the treatments with copepods or their spent medium, whilst denitrification potential was lowest in these treatments, suggesting that copepods enhance dissimilatory nitrate reduction to ammonium over denitrification. We hypothesize that this is an indirect effect, by providing extra carbon for the bacterial community through the copepods' excretion products, thus changing the C/N ratio in favour of dissimilatory nitrate reduction. Diatoms alone had no effect on the nitrogen fluxes, but they did enhance the effect of copepods, possibly by influencing the quantity and quality of the copepods' excretion products. Our results show that small-scale biological interactions between bacteria, copepods and diatoms can have an important impact on denitrification and hence sediment nitrogen fluxes.

Editor: Candida Savage, University of Otago, New Zealand

Funding: This research was conducted within the frame of research project GOA 01GA1911W of the Special Research Fund at Ghent University (BOF-UGent) on "Understanding biodiversity effects on the functioning of marine benthic ecosystems". M. De Troch is a postdoctoral researcher financed by the same project. Kim Heylen is supported by the Flemish Fund for Scientific Research (FWO11/PDO/084). The funders had no role in study design, data collection and analysis, decision to publish, or preparation of the manuscript.

Competing Interests: The authors have declared that no competing interests exist.

* Email: Willem.Stock@Ugent.be

Introduction

Over the past century anthropogenic activities have dramatically increased the amount of reactive nitrogen on Earth [1]. It has been estimated that nitrogen inputs have increased as much as tenfold in coastal ecosystems [2,3]. As a result, these often nitrogen-limited areas [4] have experienced severe eutrophication, resulting in anoxia and changes in community structure [5]. Denitrification and anaerobic ammonium oxidation (anammox) are capable of countering eutrophication by removing reactive nitrogen from the ecosystem as nitrous oxide (N_2O) or nitrogen gas (N_2) [6]. In contrast, during dissimilatory nitrate reduction to ammonium (DNRA), nitrate (NO_3^-) and nitrite (NO_2^-) are reduced to ammonium (NH_4^+), preserving reactive nitrogen in the system. In coastal environments, anammox, denitrification and DNRA are all catalysed in the anoxic sediment, but by different microbial assemblages [7].

Denitrification and DNRA are carried out by a different but diverse range of mostly heterotrophic microorganisms [8] and are assumed to be *in situ* mutually exclusive processes determined by the C/N ratio of the system [9]. DNRA is thought to be favoured in nitrate-limited environments rich in labile carbon [9], since the energy yield per nitrate reduced is higher for nitrate ammonification than for denitrification (the reduction of nitrate to ammonia consumes eight electrons rather than five in denitrification, thus more carbon can be oxidised per nitrate reduced; [10]). Anammox is apparently only conducted by members of the *Planctomycetes* group [11] and probably is a less important nitrogen sink than denitrification in nutrient-loaded coastal areas [12].

In the past decades, strong efforts have been made to unravel which benthic organisms affect nitrogen cycling in intertidal sediments and how they do this (e.g. [13,14]). Macrofauna for example, has been shown to impact DNRA and denitrification by turbating the sediment (e.g. [15,16]). Other studies focussed on the impact of microphytobenthos (e.g. [17,18]) on denitrification. The effect of meiofauna (e.g. nematodes and copepods), the intermediate trophic level, on nitrogen fluxes has to date been almost completely neglected [19]. Although the effects of the meiofaunal bioturbation – confined to the superficial sediments – will be far less pronounced than those of macrofaunal bioturbation [20], these organisms can potentially impact benthic nitrate reduction in other ways. Meiofauna is, for instance, capable of eating its body weight equivalent in microorganisms each day [21]. By grazing on microphytobenthos and bacteria, meiofauna will not only counteract the effects of the microphytobenthos and bacteria on nitrate reduction, but also release high amounts of organic nitrogen and carbon [22] into the interstitial environment, thus potentially impacting the C/N ratio. We hypothesized that the meiofauna can impact nitrogen reduction in marine sediments through their grazing activity. The aim of this study was therefore to investigate the impact of meiofauna and its interactions with its food sources, diatoms and bacteria, on denitrification in marine sediments. For this purpose an experiment was setup in which all possible combinations of meiofauna, diatoms and bacteria were included. Harpactecoid copepods were used as meiofauna representative since they occur in high densities at the study site (230 ± 194 ind. 10 cm^{-2}, [23]) and have been well-studied in terms of both composition [23] and feeding ecology (e.g. [24–26]) in this tidal flat.

In the experiment, both nitrate reduction (the combined activity of denitrification, DNRA and anammox) and denitrification as such were measured as these biochemical reactions are relevant and important ecosystem functions in coastal sediments. Furthermore, nitrate reduction and denitrification can serve as proxies for the overall functioning of the benthic microbial community. In microcosm experiments with sediment, harpacticoid copepods (Crustacea, *Copepoda*) and diatoms from an intertidal flat (Paulina Polder, Westerschelde estuary, The Netherlands), nutrient dynamics and the potential for nitrate reduction and denitrification were monitored. Nitrate reduction and denitrification rates could not be measured *in situ*, as they are largely anaerobic processes whilst copepods are strictly aerobic. Both rates were therefore measured indirectly by making a subsample of the homogenised microcosm anaerobic and measuring the potential rates under non-carbon or -nitrogen limiting conditions.

Methodology

Field sampling

Silty sediment was collected from the intertidal mudflat Paulina (Westerschelde estuary, The Netherlands; 51°20′ N, 3°43′ E) in February 2013 by scraping the top layer (0–3 cm) of the sediment at low tide. Seawater (salinity: 19.3; 1.85 ± 1.11 µM NO$_2$; 122.02 ± 8.17 µM NO$_3^-$; 2.50 ± 1.70 µM NH$_4^+$; 0.73 ± 0.60 µM PO$_4^{3-}$; 88.89 ± 0.16 µM Si^{4+}; N = 3) was collected from the same site and was filtered over a 0.22 µm filter (Corning 500 mL Bottle Top Vacuum Filter) and stored in the dark at 4°C (filtered seawater: FSW). No permits were required for the sampling nor were there any endangered or protected species involved.

Experimental setup

Collected sediment was washed over a 250 µm sieve to remove all benthic fauna. The sieved sediment (average median grain size: 56.89 ± 0.25 µm; N = 3) was divided in equal aliquots of 80 g in polyethylene containers (microcosms). The microcosms with ± 2.5 cm of sieved sediment were stored frozen (-20°C) to kill all the remaining fauna. A microcosm was defrosted two days before the start of a treatment. After adding 60 ml of filtered seawater (FSW), the thawed sediment was thoroughly mixed. Right before the start of a treatment, the FSW was drained off.

The experimental design included a blank: the defaunated sediment in which only bacteria were present. To verify the effect of copepods and diatoms, independently of one another, on the bacteria, they were added separately to the microcosm. To cover the interaction effects between copepods and diatoms, both of them were added to the microcosm. In order to discriminate between the effects of the activity of the copepods themselves and the waste products that they produce, a treatment was included in which the spent medium from copepods was added to the microcosm.

Together with the positive control (increased NO$_3^-$), this resulted in a total of six different treatments.

Each treatment starting with a microcosm containing 80 g of defaunated and thawed sediment: (1) Blank: +70 ml FSW; (2) Increased NO$_3^-$: +0.1 mmol KNO$_3$ in 70 ml FSW; (3) Copepod: + 200 phototactic copepods collected from Paulina Polder in 70 ml FSW; (4) Diatom: +4×10^5 cells of *Navicula* sp. (36.27 ± 2.30 µm; isolated from the study site in 2012) in 70 ml FSW (corresponds to chlorophyll a concentrations observed in the study site.); (5) Copepod+diatom: +4×10^5 cells of *Navicula* sp. and 200 copepods (as above) in 70 ml FSW; (6) Spent medium:+70 ml of FSW, in which 200 copepods were fed 4×10^5 *Navicula* cells over a period of one week. A visual (microscopic) screening of the treatments revealed that most diatoms had been eaten by the copepods after one week. Prior to the start of the treatment this spent medium was stored at -20°C after manually removing the copepods.

One hour after the start of the experiment, 10 ml of water was extracted for nutrient analysis (initial nutrient concentration) from each microcosm. All microcosms were then incubated for 7.5 days, at 15°C.

To study the effects of the photosynthetic activity of the diatoms on the nitrogen fluxes, all treatments were run (1) under a diurnal (12 h/12 h) light regime and (2) in the dark. Cold-white fluorescent lamps provided the necessary light at a rate of 20–25 µmol photons m^{-2} s^{-1}. Four replicates were used for each treatment under each light condition. To avoid depletion of active nitrogen in the microcosms, half of the SW was renewed on day 5 and 6 of the experiment. At the end of the incubation period, 10 ml of SW was stored for nutrient analysis (final nutrient concentration).

An additional experiment was setup in which the blank and copepod+diatom treatment were repeated to verify the effects of copepods and diatoms on the oxygen pentration depths (Text S1).

Denitrification rates were measured using the so-called acetylene inhibition method [27][16] (cf. Fig S1 which illustrates the design of the experiment). In the presence of acetylene the final reaction of denitrification, in which N$_2$O is converted to N$_2$, is inhibited, causing N$_2$O to accumulate. The easily quantifiable N$_2$O can then be measured as a proxy for denitrification. The rate at which NO$_3^-$ is consumed is a good proxy for the combined activity of all three reduction pathways.

At the end of each treatment, a serum vial was filled with 30 g (wet weight) sediment and 20 ml incubation water (collected from the treatments; Fig S1). To prevent nitrogen and carbon limitation, the water was supplemented with 0.5 mmol KNO$_3$ and 1 mmol α-D-glucose. After vigorous shaking, 1 ml was extracted to determine the initial NO$_3^-$/NO$_2^-$ concentration (t$_0$). The vials were hermetically sealed and flushed five times with helium to remove oxygen. After adding 10% acetylene, the vials were incubated at 25°C under a constant stirring rate of 90 rpm. The N$_2$O concentrations were measured every two hours by injecting 1 ml of headspace in a GC-TCD (Gas Chromatography-Thermal Conductivity Detector; MICRO E-0391, Interscience; LOD 13.55 ppm N$_2$O). This was done at four time points (t$_1$–t$_4$) for each serum vial (Fig. S1). N$_2$O concentrations were corrected for headspace volume changes, pressure and dissolving of the gas into the liquid phase. To ascertain potential side-effects of acetylene [28], the process was repeated with a technical replicate without acetylene (data not shown). At the second (t$_2$) and final sampling event (t$_4$) from the replicate without acetylene, 800 µl of fluid was extracted for later NO$_3^-$/NO$_2^-$ determination (indicative for the NO$_3^-$ reduction activity; Fig. S1).

Nutrient analysis

Nutrient concentrations (NO_3^-, NO_2^-, NH_4^+ and PO_4^-) of the samples collected at the start and the end of the incubation period (initial and final nutrient concentration) were analysed with an automatic chain (SAN [plus] segmented flow analyser, SKALAR) according to Beyst et al. [29].

Samples extracted from the serum vials for NO_3^-/NO_2^- determination were analysed differently because the above used method required higher sample volumes. The samples were centrifuged (14000 rpm, 5 min) and the supernatants were stored frozen ($-20°C$) prior to analysis. Analysis of NO_3^- and NO_2^- was based on a colorimetric method as described by Cataldo et al. [30], based on Griess [31] with adjustments from Navarro-Gonzálvez et al. [32].

Data analysis

The software package R 2.15.0. was used for data analysis. Differences in initial and final nutrient concentrations between the treatments and light conditions were detected using a two-way ANOVA on the rank transformed concentrations, performed in the software package R 2.15.0. Pairwise differences were unravelled using Dunnett's Modified Tukey-Kramer Pairwise Multiple Comparison Test (DTK) [33] using 95% confidence limits.

N_2O production rates and NO_3^- reduction rates were calculated according to both Magalhaes et al. [34], assuming no bacterial growth between the N_2O, respectively NO_3^-/NO_2^-, samplings (i.e. between t_0 and t_4), and Stenström et al. [35], assuming that bacterial growth does occur. Magalhaes et al. [34] obtained the N_2O production rates by dividing the N_2O concentration at t_4 by t_4. Stenström et al. [35], however, propose an exponential regression to accommodate for the increasing gas production rate between samplings: $p(t) = p_0 + \frac{r_{N2O}}{\mu}(e^{\mu t} - 1)$, with p = the amount of gas at time t, p_0 = the amount of product at t_0, r_{N2O} = the N_2O production rate (see below) and μ = the specific growth rate constant. Since the serum vials were flushed with helium, there was no N_2O at the start of the incubation and p_0 was set to zero. This function was adapted to fit the data for the NO_3^- reduction (in the serum vials) to $s(t) = s_0 - \frac{r_{NO3}}{\mu}(e^{\mu t} - 1)$, with s = the amount of substrate (NO_3^-) at time t, s_0 = the amount of substrate at t_0 and r_{NO3} = the initial NO_3^- reduction rate, further referred to as "NO_3^- reduction rate".

The N_2O production rates (r_{N2O}) obtained from the regression analysis are the production rates of N_2O at the start (t_0) of the denitrification potential experiment, i.e. after the 7.5 days incubation period (Fig. S1). Likewise, the initial NO_3^- reduction rates (r_{NO3}) express the NO_3^- reduction rate at the start of the denitrification potential experiment. Thus the rates are corrected for any microbial growth occurring after the NO_3^- and glucose addition.

Differences in the obtained N_2O production and the NO_3^- reduction rates between the different treatments and light conditions were analysed using a permutation-based two-way ANOVA [36] since the data were not normally distributed. Pairwise differences were analyzed using Wilcoxon rank-sum post-hoc test using 95% confidence limits, with Bonferroni correction.

Results

Viability

At the end of the incubation, viability of copepods and diatoms were checked in all treatments. Both microscopic observations (after collecting the cells with the lens-tissue method) and pulse-amplitude modulation (PAM; Maxi-Imaging PAM M-series, Walz) showed healthy and active diatom cells. Visual observations showed active copepods in all microcosms of the copepod and copepod + diatom treatment.

Nutrient levels

Initial nutrient concentrations. Except for the spent medium and the increased NO_3^- treatment, the initial nutrient concentrations in the microcosms did not differ among treatments (ANOVA; $p < 0.05$; Table 1). The nitrate concentration in the spent medium was significantly lower than in the other treatments, while in the increased NO_3^- treatment, the nitrate concentration was, as expected, significantly higher. Nitrite was significantly lower in the spent medium and in the increased NO_3^- treatment. The initial ammonium concentration was highest in the copepod + diatom treatment, but the difference was only significant compared to the increased NO_3^- treatment. In contrast, the phosphate concentration in the spent medium treatment was 2.5–4 times higher than in the other treatments.

Final nutrient concentrations. After 7.5 days of incubation, the nutrient concentrations did not differ significantly between light conditions (two-way ANOVA; $p > 0.05$).

At the end of the incubation period, nitrate and nitrate were almost completely depleted in all treatments (on average 1.12 ± 0.03 µM and 0.30 ± 0.01 µM, respectively; Table 1), despite renewal of half of the SW on days 5 and 6. The ammonium concentration more or less quadrupled towards the end of the experiment (average 384.08 ± 3.04 µM). The final ammonium concentration in the copepod treatment was significantly higher than in the diatom and the blank treatment. The final ammonium concentration in the copepod + diatom treatment was the highest of all treatments, although it was not significantly different from the other treatments due to considerable variation between the replicates.

The same pattern was observed in the final phosphate concentrations. The phosphate concentrations strongly increased during the incubation period as the final average phosphate concentration (18.60 ± 0.21 µM) was almost twenty times higher than the initial one.

Potential for nitrate reduction and denitrification

During the measurement of the denitrification potential, nitrate was consumed, while nitrite and nitrous oxide were produced (see Fig. S2 for the average NO_3^-, NO_2^- and N_2O concentrations during the measurement of the denitrification potential). Since the exponential function proposed by Stenström et al. [35] had a significantly better fit than the linear regression proposed by Magalhaes et al. [34], rates were calculated with the exponential function.

The NO_3^- reduction and N_2O production rates did not differ between light conditions. In contrast to the NO_3^- reduction rate, which did not differ between treatments (Fig 1, black bars), the N_2O production rate did significantly differ between the treatments ($p < 0.01$; Fig. 1, white bars). The N_2O production rate was significantly lower for the copepod + diatom treatment compared to the blank ($p < 0.001$), diatom ($p < 0.001$), increased NO_3^- ($p < 0.001$) and spent medium ($p < 0.001$) treatments.

The correlation between N_2O production and NO_3^- reduction rates was weak but significantly positive (Pearson's $r = 0.30$, $p < 0.05$).

Table 1. Initial and final nutrient concentrations of the different treatments.

Nutrients	NO$_3^-$, mean ±SE (µM)		NO$_2^-$, mean ±SE (µM)		NH$_4^+$, mean ±SE (µM)		PO$_4^{3-}$, mean ±SE (µM)	
Treatments	Initial***	Final	Initial***	Final	Initial*	Final**	Initial***	Final***
Blank	125.93±5.00[a]	2.24±0.79	4.35±0.20[a]	0.23±0.10	99.67±6.54[ab]	360.03±24.42[a]	0.85±0.13[a]	18.90±1.90[ab]
Increased NO$_3^-$	1135.29±258.98[b]	1.29±0.36	3.43±0.21[b]	0.25±0.03	105.56±5.06[a]	400.71±22.01[ab]	0.75±0.11[a]	15.30±1.74[a]
Copepod	424.17±265.21[a]	0.83±0.24	4.48±0.21[a]	0.32±0.06	89.13±6.27[ab]	467.19±9.70[b]	0.98±0.16[a]	23.27±2.36[b]
Diatom	130.86±4.04[a]	0.76±0.21	4.17±0.14[a]	0.20±0.03	88.97±6.75[ab]	372.18±21.35[a]	0.94±0.17[a]	13.79±1.83[a]
Copepod+diatom	131.17±2.95[a]	1.43±0.58	4.39±0.27[a]	0.65±0.18	78.16±2.25[b]	485.69±34.52[ab]	0.66±0.09[a]	25.98±2.61[b]
Spent medium	34.33±10.63[c]	0.65±0.31	1.42±0.32[c]	0.17±0.05	94.43±6.97[ab]	425.93±16.24[ab]	2.44±0.30[b]	26.29±2.53[b]

Significant differences of the ANOVAs indicated with symbols (*** $\leq$0.001<** $\leq$0.01<* $\leq$0.05< <0.1). The different superscripted letters indicate significant differences (P<0.05; DTK) between the treatments. Light conditions (not shown) did not alter the outcome of treatments (see text).

Discussion

Effect of copepods

The presence of copepods or their spent medium resulted in elevated phosphate and ammonium concentrations in the microcosms. The copepods' outfluxes (including excretion products, moults, remnants of sloppy feeding) therefore proved to be an important source of both N and P which are, in coastal areas, potentially limiting nutrients [37].

Contrary to macrofauna, not much is known about the impact of meiofauna on nitrogen fluxes. Macrofauna seems to have its biggest impact through bioturbation (e.g. [15]). However, the oxygen penetration depth did not increase in the presence of copepods (Text S1), and consequently the effects of copepods are not related to oxygen. Since the N$_2$O production rates of the copepod and their spent medium treatments were low, the copepods appeared to negatively impact denitrification, at least partially via their outfluxes. The outfluxes of the copepods are, apart from being an important nutrient source, a source of organic compounds and hence a substrate for bacterial growth [38]. Since organic matter loading may be one of the most important variables controlling denitrification in aquatic ecosystems [39], the impact of these carbon outfluxes should not be underestimated. The copepod outfluxes contain high amounts of labile carbon [22], which are known to stimulate DNRA over denitrification and anammox (Fig. 2, dashed arrow 1; [9]). Copepods can also indirectly stimulate labile carbon production by mechanically breaking down detrital particles [40,41]. In addition, more organic matter results in a higher sulphate reduction rate (Fig. 2, dashed arrow 2; [42]). The main product of sulphate reduction is hydrogen sulphide, which inhibits denitrification and nitrification, but not DNRA [7]. These findings are supported by the higher final ammonium concentrations in the treatments with copepods or their spent medium. Furthermore, the NO$_3^-$ reduction rate did not differ significantly between the treatments, suggesting that the reduced denitrification activity in the treatments with copepods or their spent medium was compensated by another nitrate reducing process.

Effects of diatoms

Since there were no differences between the final nitrate, nitrite and ammonium concentrations of the blank and the diatom treatment, diatoms seemed to have no net effect on the nitrogen fluxes in the microcosm. Likewise, N$_2$O production rates did not differ between the blank and the diatom treatment either. It thus appears that diatoms had no or very little impact on denitrification in the microcosms. This is inconsistent with previous reports, where the presence of benthic microalgae generally had a negative impact on denitrification rates [43] as microalgae outcompeted the bacteria for nitrogen [18]. The opposite has however also been observed (e.g. [44]) were microalgae enhanced denitrification. It is unlikely that the inoculation concentration (2×10^5 diatoms/cm^2) used in this study was too low to have a significant effect on the denitrification rate as it was comparable to the diatom concentrations in other studies were diatoms did have an effect on the nitrogen fluxes (e.g. [18]). It is, however, possible that positive and negative effects of the diatoms cancelled each other out or that the overall effect was too small to be detected. However, one should bear in mind that the used incubation time might be too short to obtain strong effects caused by the diatoms [18,45].

Effects of diatoms + copepods

The negative effect of the copepods on denitrification was most pronounced and only significant when diatoms were added to the

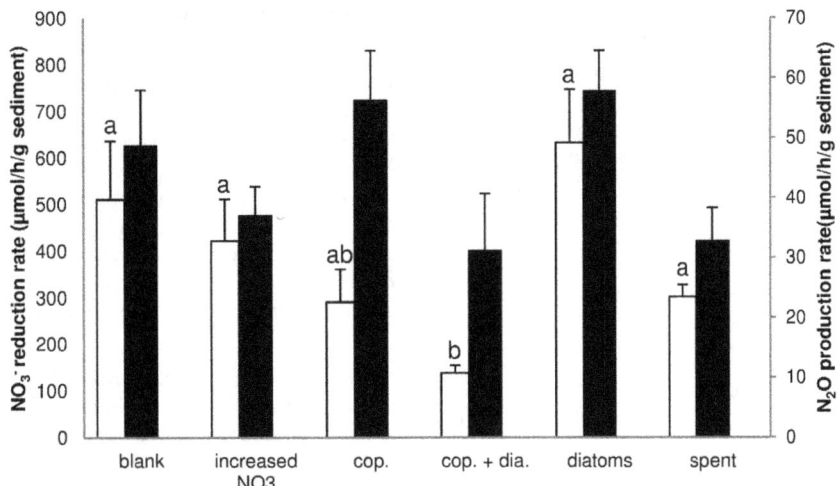

Figure 1. NO$_3^-$ reduction and N$_2$O production rates. NO$_3^-$ reduction rate (black bars; left y-axis) and N$_2$O production rates (white bars; right y-axis) during the measurement of the denitrification potential for the different treatments (mean ± SE; cop. = copepods; cop. + dia. = copepod+ diatom). The different letters above the bar indicate significant differences (P<0.05; DTK) between the treatments. Light conditions (not shown) did not affect the outcome of treatments.

system. This suggests an important interaction effect between diatoms and copepods. Since the diatoms themselves had no effect on denitrification and nitrate reduction, it is unlikely that they were directly responsible for the difference between the copepods + diatoms and the copepods treatments. The diatoms can, however, influence the composition of the copepod outfluxes

(Fig. 2, dashed arrow 4), which depends on the food type [22]. As diatoms are thought to be a better food source for copepods than bacteria (e.g. [26,46]), they might also enhance the survival of the copepods and, accordingly, their activity and the quantity of the excretion products (Fig. 2, dashed arrows 3–4).

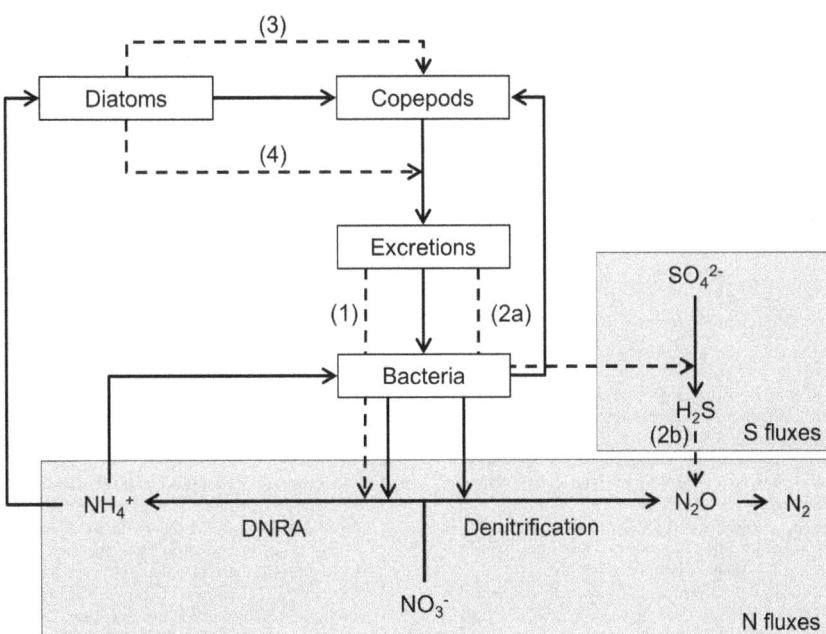

Figure 2. Summary of the assumed interactions to explain the observed differences in N$_2$O production rates. The assumed interactions which affect denitrification are indicated with dashed arrows. Bacteria mediated relevant reduction reactions of the nitrogen pathway and sulfur pathway are enclosed by grey boxes indicated with respectively 'N fluxes' and 'S fluxes'. Copepods feed on both diatoms and bacteria, and produce excretion producs (excretions). Bacteria feed on the excretion produces and are also responsible for the reduction of SO$_4^{2-}$ to H$_2$S and of NO$_3^-$ to NH$_4^+$ (DNRA) and N$_2$O+N$_2$ (denitrification) in the microcosm. The produced NH$_4^+$ is assimilated by both bacteria and diatoms. Copepods affect the N$_2$O production rate by producing excretion products which provide an extra carbon source, of which mainly the DNRA bacteria can take advantage (1) and also enhances SO$_4^{2-}$ reduction (2a), which results in more H$_2$S. The increased H$_2$S inhibits denitrification (2b). Diatoms have no direct effect on the N$_2$O production rate, but do have an indirect effect by enhancing the survival of the copepods (3) and influencing the quantity and composition of the copepods' excretion products (4).

Effects of the nitrate addition

The N_2O production rate for the increased NO_3^- treatment was unexpectedly similar to the blank. In general, the denitrification rate is positively influenced by the nitrate load (e.g. [47,48]), but such a relation was not observed here. However, a preliminary test in which we used a shorter, three and a half day long incubation period showed that the N_2O production rate was almost twice as high for the increased NO_3^- treatment compared to the blank (data not shown). This indicates that the supplemented nitrate was depleted within the first few days and that the denitrifying community changed accordingly.

Final considerations

Our findings have potentially important implications for our understanding of nutrient fluxes in marine sediment and the role of meiofauna. It was already known that meiofauna facilitates biomineralization of organic material and enhances nutrient regeneration [40]. This is an indirect process, by stimulating the bacterial community [40] through the production of excretion products [49] and bioturbation [40]. The present study suggests that these processes may also negatively impact denitrification. Consequently, more active nitrogen will be preserved in the ecosystem in the presence of meiofauna. Furthermore, they also increase the freely available phosphorus and carbon (this study, [22]). These elevated nutrient levels will benefit both bacteria and microphytobenthos. Our observations should, however, be interpreted with care as they were obtained from a short-term microcosm experiment. They might therefore not be representative for the highly dynamic estuarine sediments these organisms where obtained from. Our results do, however, prove that the interactions between meiofauna, diatoms and bacteria can potentially impact the nitrogen fluxes and that they should therefore not be neglected in future research. The time-dependent effect of the increased NO_3^- treatment clearly illustrates the importance of the temporal scale in this setup. It might therefore prove useful for further research to investigate the effects of the different treatments at different time intervals. Furthermore, additional experiments (for instance relying on the ^{15}N technique; [50]) will be necessary to fully unravel these fine-scale interactions.

Supporting Information

Figure S1   Design of the denitrification potential experiment. Incubation steps are indicated by dotted arrows. Manipulations are indicated by a dashed arrow. Each microcosm (here represented as a rectangle) was incubated for 7.5 days, after which water (blue) and sediment (brown) were transferred from the microcosm to the serum vials (represented as trapezoids). Glucose and potassium nitrate were added to the vial after which it was thoroughly homogenized. The headspace was flushed with helium to remove oxygen, after which acetylene was injected. The vials were sampled four times for N_2O determination, every two hours.

Figure S2   NO3-, NO2- and N2O concentrations in the serum vials after the addition of 0.5 mmol KNO3 during the measurement of the denitrification potential (from t0 to t4). Amounts (µmol/h/g sediment) are averaged over all samples (all treatments), as is the time at with the sample was taken (Time; h:min) ± SE. N2O concentration(blue) plotted on the right y axis; NO3- (red) and NO2- (green) concentrations on the left y axis.

Text S1   Measurement of oxygen penetration depth. To verify the effects of copepods and diatoms on the oxygen pentration depths, the experiment as described in the method section was repeated for the blank, diatom and copepod+diatom treatment. The methodology and results of the additional experiment are shown here.

Acknowledgments

Evie De Brandt, Bram Vekeman, Sven Hoefman and Dirk Van Gansbeke are acknowledged for their assistance with the lab analyses. Frederik De Laender is acknowledged for his help with the regressions and improving the R-script. The authors thank the 2 reviewers for their comments that helped improve and clarify this manuscript.

Author Contributions

Conceived and designed the experiments: WS KH KS AW MDT. Performed the experiments: WS. Analyzed the data: WS. Contributed reagents/materials/analysis tools: MDT AW KS. Wrote the paper: WS KH KS AW MDT.

References

1. Leach AM, Galloway JN, Bleeker A, Erisman JW, Kohn R, et al. (2012) A nitrogen footprint model to help consumers understand their role in nitrogen losses to the environment. Environmental Development 1: 40–66.

2. Paerl HW (2006) Assessing and managing nutrient-enhanced eutrophication in estuarine and coastal waters: Interactive effects of human and climatic perturbations. Ecological Engineering 26: 40–54.

3. Pätsch J, Serna A, Dähnke K, Schlarbaum T, Johannsen A, et al. (2010) Nitrogen cycling in the German Bight (SE North Sea)—Clues from modelling stable nitrogen isotopes. Continental Shelf Research 30: 203–213.

4. Howarth RW, Marino R (2006) Nitrogen as the limiting nutrient for eutrophication in coastal marine ecosystems: evolving views over three decades. Limnology and Oceanography 51: 364–376.

5. Koop-Jakobsen K, Giblin AE (2010) The effect of increased nitrate loading on nitrate reduction via denitrification and DNRA in salt marsh sediments. Limnology and Oceanography 55: 789.

6. Schreiber F, Wunderlin P, Udert KM, Wells GF (2012) Nitric oxide and nitrous oxide turnover in natural and engineered microbial communities: biological pathways, chemical reactions, and novel technologies. Front Microbiol 3: 372.

7. An S, Gardner WS (2002) Dissimilatory nitrate reduction to ammonium (DNRA) as a nitrogen link, versus denitrification as a sink in a shallow estuary (Laguna Madre/Baffin Bay, Texas). Marine Ecology Progress Series 237.

8. Herbert R (1999) Nitrogen cycling in coastal marine ecosystems. FEMS microbiology reviews 23: 563–590.

9. Burgin AJ, Hamilton SK (2007) Have we overemphasized the role of denitrification in aquatic ecosystems? A review of nitrate removal pathways. Frontiers in Ecology and the Environment 5: 89–96.

10. Strohm TO, Griffin B, Zumft WG, Schink B (2007) Growth yields in bacterial denitrification and nitrate ammonification. Applied and environmental microbiology 73: 1420–1424.

11. Jetten MS, Wagner M, Fuerst J, van Loosdrecht M, Kuenen G, et al. (2001) Microbiology and application of the anaerobic ammonium oxidation ('anammox') process. Current opinion in biotechnology 12: 283–288.

12. Teixeira C, Magalhaes C, Joye SB, Bordalo AA (2012) Potential rates and environmental controls of anaerobic ammonium oxidation in estuarine sediments. Aquatic Microbial Ecology 66: 23–32.

13. Risgaard-Petersen N (2003) Coupled nitrification-denitrification in autotrophic and heterotrophic estuarine sediments: On the influence of benthic microalgae. Limnology and Oceanography 48: 93–105.

14. Ferguson A, Eyre B (2013) Interaction of benthic microalgae and macrofauna in the control of benthic metabolism, nutrient fluxes and denitrification in a shallow sub-tropical coastal embayment (western Moreton Bay, Australia). Biogeochemistry 112: 423–440.

15. Braeckman U, Provoost P, Gribsholt B, Van Gansbeke D, Middelburg JJ, et al. (2009) Role of macrofauna functional traits and density in biogeochemical fluxes and bioturbation. Marine Ecology Progress Series 399: 173.

16. Binnerup SJ, Jensen K, Revsbech NP, Jensen MH, Sørensen J (1992) Denitrification, dissimilatory reduction of nitrate to ammonium, and nitrification

in a bioturbated estuarine sediment as measured with 15N and microsensor techniques. Applied and environmental microbiology 58: 303–313.

17. Christensen PB, Nielsen LP, Sørensen J, Revsbech NP (1990) Denitrification in nitrate-rich streams: diurnal and seasonal variation related to benthic oxygen metabolism. Limnology and Oceanography 35: 640–651.

18. Risgaard-Petersen N, Nicolaisen MH, Revsbech NP, Lomstein BA (2004) Competition between ammonia-oxidizing bacteria and benthic microalgae. Appl Environ Microbiol 70: 5528–5537.

19. Parent S, Morin A (1999) Role of copepod-dominated meiofauna in the nitrification process of a cold marine mesocosm. Canadian Journal of Fisheries and Aquatic Sciences 56: 1639–1648.

20. Martin P, Boes X, Goddeeris B, Fagel N (2005) A qualitative assessment of the influence of bioturbation in Lake Baikal sediments. Global and Planetary Change 46: 87–99.

21. Montagna PA (1984) In situ measurement of meiobenthic grazing rates on sediment bacteria and edaphic diatoms. Marine ecology progress series Oldendorf 18: 119–130.

22. Frangoulis C, Christou ED, Hecq JH (2005) Comparison of marine copepod outfluxes: nature, rate, fate and role in the carbon and nitrogen cycles. Adv Mar Biol 47: 253–309.

23. Cnudde C, De Troch M. unpubl. data.

24. De Troch M, Houthoofd L, Chepurnov V, Vanreusel A (2006) Does sediment grain size affect diatom grazing by harpacticoid copepods? Mar Environ Res 61: 265–277.

25. De Troch M, Vergaerde I, Cnudde C, Vanormelingen P, Vyverman W, et al. (2012) The taste of diatoms: the role of diatom growth phase characteristics and associated bacteria for benthic copepod grazing. Aquatic Microbial Ecology 67: 47–58.

26. Cnudde C, Moens T, Hoste B, Willems A, De Troch M (2013) Limited feeding on bacteria by two intertidal benthic copepod species as revealed by trophic biomarkers. Environ Microbiol Rep 5: 301–309.

27. Sørensen J (1978) Capacity for denitrification and reduction of nitrate to ammonia in a coastal marine sediment. Applied and Environmental Microbiology 35: 301–305.

28. Groffman PM, Altabet MA, Böhlke J, Butterbach-Bahl K, David MB, et al. (2006) Methods for measuring denitrification: diverse approaches to a difficult problem. Ecological Applications 16: 2091–2122.

29. Beyst B, Hostens K, Mees J (2001) Factors influencing fish and macrocrustacean communities in the surf zone of sandy beaches in Belgium: temporal variation. Journal of Sea Research 46: 281–294.

30. Cataldo D, Maroon M, Schrader L, Youngs V (1975) Rapid colorimetric determination of nitrate in plant tissue by nitration of salicylic acid 1. Communications in Soil Science & Plant Analysis 6: 71–80.

31. Griess P (1879) Bemerkungen zu der Abhandlung der HH. Weselsky und Benedikt "Ueber einige Azoverbindungen" . Berichte der deutschen chemischen Gesellschaft 12: 426–428.

32. Navarro-Gonzálvez JA, García-Benayas C, Arenas J (1998) Semiautomated measurement of nitrate in biological fluids. Clinical chemistry 44: 679–681.

33. Lau M (2009) DTK: Dunnett–Tukey–Kramer pairwise multiple comparison test adjusted for unequal variances and unequal sample sizes. R package version 2.

34. Magalhaes CM, Machado A, Matos P, Bordalo AA (2011) Impact of copper on the diversity, abundance and transcription of nitrite and nitrous oxide reductase genes in an urban European estuary. FEMS Microbiol Ecol 77: 274–284.

35. Stenström J, Hansen A, Svensson B (1991) Kinetics of microbial growth-associated product formation. Swedish Journal of Agricultural Research (Sweden).

36. Wheeler B (2010) lmPerm: Permutation tests for linear models. R package version 1.1–2.

37. Vadstein O, Andersen T, Reinertsen HR, Olsen Y (2012) Carbon, nitrogen and phosphorus resource supply and utilisation for coastal planktonic heterotrophic bacteria in a gradient of nutrient loading. Mar Ecol Prog Ser 447: 55–75.

38. De Troch M, Cnudde C, Willems A, Moens T, Vanreusel A (2010) Bacterial colonization on fecal pellets of harpacticoid copepods and on their diatom food. Microbial ecology 60: 581–591.

39. Cornwell JC, Kemp WM, Kana TM (1999) Denitrification in coastal ecosystems: methods, environmental controls, and ecosystem level controls, a review. Aquatic Ecology 33: 41–54.

40. Coull BC (1999) Role of meiofauna in estuarine soft-bottom habitats*. Australian Journal of Ecology 24: 327–343.

41. Nascimento FJ, Näslund J, Elmgren R (2012) Meiofauna enhances organic matter mineralization in soft sediment ecosystems. Limnology and Oceanography 57: 338.

42. Berner RA, Westrich JT (1985) Bioturbation and the early diagenesis of carbon and sulfur. American Journal of Science 285: 193–206.

43. Sundbäck K, Miles A, Linares F (2006) Nitrogen dynamics in nontidal littoral sediments: Role of microphytobenthos and denitrification. Estuaries and coasts 29: 1196–1211.

44. An S, Joye SB (2001) Enhancement of coupled nitrification-denitrification by benthic photosynthesis in shallow estuarine sediments. Limnology and Oceanography 46: 62–74.

45. Nilsson P, Jonsson B, Swanberg IL, Sundback K (1991) Response of a marine shallow-water sediment system to an increased load of inorganic nutrients. Marine Ecology Progress Series MESEDT 71.

46. Sundbäck K, Nilsson P, Nilsson C, Jönsson B (1996) Balance between autotrophic and heterotrophic components and processes in microbenthic communities of sandy sediments: a field study. Estuarine, Coastal and Shelf Science 43: 689–706.

47. Bartoli M, Castaldelli G, Nizzoli D, Viaroli P (2012) Benthic primary production and bacterial denitrification in a Mediterranean eutrophic coastal lagoon. Journal of Experimental Marine Biology and Ecology 438: 41–51.

48. Koch M, Maltby E, Oliver G, Bakker S (1992) Factors controlling denitrification rates of tidal mudflats and fringing salt marshes in south-west England. Estuarine, Coastal and Shelf Science 34: 471–485.

49. De Troch M, Steinarsdóttir MB, Chepurnov V, Ólafsson E (2005) Grazing on diatoms by harpacticoid copepods: species-specific density-dependent uptake and microbial gardening. Aquatic microbial ecology 39: 135–144.

50. Giblin AE, Tobias CR, Song B, Weston N, Banta GT, et al. (2013) The importance of dissimilatory nitrate reduction to ammonium (DNRA) in the nitrogen cycle of coastal ecosystems.

13

The Relationship between the Distribution of Common Carp and Their Environmental DNA in a Small Lake

Jessica J. Eichmiller*, Przemyslaw G. Bajer, Peter W. Sorensen

Department of Fisheries, Wildlife, and Conservation Biology, Minnesota Aquatic Invasive Species Research Center, University of Minnesota, Twin Cities, St. Paul, Minnesota, United States of America

Abstract

Although environmental DNA (eDNA) has been used to infer the presence of rare aquatic species, many facets of this technique remain unresolved. In particular, the relationship between eDNA and fish distribution is not known. We examined the relationship between the distribution of fish and their eDNA (detection rate and concentration) in a lake. A quantitative PCR (qPCR) assay for a region within the cytochrome b gene of the common carp (*Cyprinus carpio* or 'carp'), an ubiquitous invasive fish, was developed and used to measure eDNA in Lake Staring (MN, USA), in which both the density of carp and their distribution have been closely monitored for several years. Surface water, sub-surface water, and sediment were sampled from 22 locations in the lake, including areas frequently used by carp. In water, areas of high carp use had a higher rate of detection and concentration of eDNA, but there was no effect of fish use on sediment eDNA. The detection rate and concentration of eDNA in surface and sub-surface water were not significantly different ($p \geq 0.5$), indicating that eDNA did not accumulate in surface water. The detection rate followed the trend: high-use water > low-use water > sediment. The concentration of eDNA in sediment samples that were above the limit of detection were several orders of magnitude greater than water on a per mass basis, but a poor limit of detection led to low detection rates. The patchy distribution of eDNA in the water of our study lake suggests that the mechanisms that remove eDNA from the water column, such as decay and sedimentation, are rapid. Taken together, these results indicate that effective eDNA sampling methods should be informed by fish distribution, as eDNA concentration was shown to vary dramatically between samples taken less than 100 m apart.

Editor: Arga Chandrashekar Anil, CSIR- National institute of oceanography, India

Funding: Funding for this project was provided by the Minnesota Environment and Natural Resources Trust Fund as recommended by the Legislative-Citizen Commission on Minnesota Resources (LCCMR). The funders had no role in study design, data collection and analysis, decision to publish, or preparation of the manuscript.

Competing Interests: The authors have declared that no competing interests exist.

* Email: eich0146@umn.edu

Introduction

Methods to quantify the abundance of fish populations, such as mark-recapture and electrofishing, are costly and time-consuming. In addition, fish are often difficult to capture and detect at low densities, and capture methods themselves can lead to behavioral changes of the target species [1–3]. Molecular methods to detect the DNA released by aquatic organisms into their environment are non-invasive, rapid, and potentially more sensitive than traditional census techniques [4–6]. This environmental DNA (eDNA) is released through processes such as cell sloughage, mucus excretions, and defecation [7]. Notably, eDNA is currently used to monitor the presence of invasive Bigheaded carps (often called 'Asian carps') (*Hypophthalmichthys* spp.) in the Chicago Area Waterway System and the Mississippi River [8]. Although initially developed as a detection tool, molecular techniques that utilize eDNA are evolving to answer more complex questions. For example, several studies have established relationships between eDNA concentration and biomass in aquatic habitats [9–11]. Next-generation sequencing approaches have successfully identified multiple species simultaneously [11,12].

Despite the immense potential for eDNA technology to revolutionize monitoring programs for fish and other aquatic species, little is known about the production, fate, and distribution of eDNA in the natural environment. The distribution of eDNA is of particular importance for development of effective monitoring methods [6]. Surprisingly, Pilliod et al. [9] found that time of day, sampling location, and distance from the target organism (salamanders) had no apparent effect on eDNA concentration in small streams. In contrast, eDNA from snails was more abundant in the middle of a river channel relative to the channel margins [13]. Surface water samples are widely used for eDNA studies [8,9,14]. The rationale for this approach has only been confirmed in one study done in experimental ponds [15]. The possibility that eDNA concentration within a water body may be influenced by fish distribution was initially posed by Takahara et al. [10]. In a lagoon in winter, the concentration of eDNA from common carp (*Cyprinus carpio*, hereafter 'carp') was positively correlated with water temperature and was spatially heterogeneous. The cause of this pattern, and in particular whether it was due to the distribution of carp or higher metabolic activity of fish in warmer waters, was not examined as the distribution of carp was not measured. From the few studies that address the question of eDNA distribution in water, it is clear that it varies among habitat types, and more conclusive explanations of eDNA distribution patterns are needed. Also of interest is the distribution of eDNA in sediments, as sediments likely retain eDNA for long periods of time [16].

To determine whether fish distribution affects eDNA concentration and detection rate in lake water and sediment, we examined the distribution of carp eDNA in a small, shallow lake and compared it to known patterns of carp distribution, which had been monitored for several years. We were interested both in detection rate (percentage of samples in which eDNA levels were present above detection threshold) as well as concentration, because the former is commonly used to assess the likely distribution of invasive Bigheaded carps while the latter measure, if understood, might add more resolution and value to the technique. First, a qPCR assay specific for *C. carpio* eDNA was developed and validated in the lab. Next, since eDNA is often assumed to accumulate in surface water and sediment, surface, sub-surface, and sediment samples were taken throughout the lake. Finally, the concentration and detection rate of eDNA was compared between areas of low- and high-fish use identified from radiotelemetry data. Results of this study provide insights into optimal eDNA sampling methods for small lakes as well as information on how eDNA is distributed in aquatic systems in relation to the distribution of target organisms.

Materials and Methods

Quantitative PCR marker development and validation

Although two *C. carpio* qPCR assays had been developed prior to this study [10,17], a screen against the NCBI database indicated potential non-specific amplification of non-target fish species (Table S1). Therefore, a qPCR assay was developed for the current study. Four genes were considered in the development of a novel qPCR marker specific to the common carp: (1) mitochondrial gene cytochrome *b*, (2) mitochondrial gene cytochrome *c* oxidase subunit 1, (3) mitochondrial gene control (D-loop) region, and (4) the nuclear gene recombination activating gene 1 (RAG1). Candidate primer sets were identified by NCBI Primer-BLAST using sequences under GenBank accession number X61010.1 [18] for mtDNA and EF458304.1 [19] for the RAG1 gene. Specificity was initially screened against the BLASTn database sequences for 15 fish species (Table S1). Minor groove binder (MGB) probes were manually designed using the Primer Probe Test Tool in Primer Express Software v3.0.1 (Life Technologies, Grand Island, NY). Assays with amplification efficiency outside the range of acceptable values of 90–110% or a limit of detection above 300 copies per reaction were not considered. We defined the limit of detection (LOD) as the lowest value at which three replicate reactions would successfully amplify with a quantification cycle (Cq) value of less than 40 cycles within the linear range of the standard curve.

Candidate markers were screened for specificity for carp by testing for amplification of 15 ng of fin clip DNA from carp and 34 native and non-native fish species (Table S1). Fin clip samples for genetic marker specificity testing were extracted using the DNeasy Blood and Tissue Kit (Qiagen, Hilden, Germany) and assayed as described below.

Next, we tested markers using aqueous samples. Three 340 L flow-through tanks were set up to confirm the ability of the marker to detect carp eDNA. Prior to this experiment, all tanks were treated with 10% bleach for 30 minutes to remove all traces of DNA. The flow through rate was set at 600 mL/min, and

temperature was maintained at 18°C. The first tank was stocked with 10 carp (35 g), and the second tank was stocked with 10 goldfish (*Carassius auratus*) (50 g), while the third tank was stocked with five fish of both species. These stocking levels corresponded to a biomass of 0, 438, and 875 mg/L of carp. Fish were fed once daily *ad libitum* a combination of flake feed (Color Tropical Marine Flake, Pentair Aquatic Eco-systems, Inc., Apopka, FL) and 2.5 mm pellet feed (Oncor Fry, Skretting USA, Tooele, UT) that did not contain target genetic markers. After 6 days, 4 1 L water samples were collected from each tank, immediately stored at 4°C, and filtered within 4 h. Molecular analyses followed protocols described below. This study was carried out in strict accordance with the recommendations in the Guide for the Care and Use of Laboratory Animals of the National Institutes of Health. The protocol for care and holding of laboratory fish was approved by the University of Minnesota's Institutional Animal Care and Use Committee (IACUC) (Protocol: 1407-31659A). No anesthesia or euthanasia was required as part of this study.

Study site

The study site was Lake Staring, a small freshwater lake located in the Upper Mississippi River Basin (44°50'14" N, −93°27'18" W). Lake Staring is a small, shallow lake that experiences frequent mixing due to wind and is typical of high carp density lakes in this region [20]. The surface area of the lake is 65.7 ha, consisting mostly of littoral zone with a depth of less than 2 m. The maximum depth is 4.8 m, and the lake bed is composed of fine sediment. Due to high carp density, the lake lacks aquatic vegetation except for white water lily (*Nymphaea odorata*), which covers less than 10% of the lake area.

Carp population abundance in Lake Staring was estimated in 2011 using a mark-recapture analysis [20]. This analysis showed that the lake was inhabited by approximately 26,000 carp, 95% CI [21,000, 31,000], or approximately 400 carp/ha. The mean body length of carp was 444 mm, indicating that the population was primarily composed of adults [20]. Approximately 14,000 fish were removed from the lake in the winters of 2012 and 2013, and a new population estimate was generated in the fall of 2013 by conducting a mark-recapture analysis. For the mark-recapture analysis, 46 carp were marked with individually numbered tags in November 2013, of which 22 were recaptured during the following four months among 5,457 carp that were captured and examined for marks. Using these data we estimated that Lake Staring was inhabited by 11,153 carp, 95% CI [7,972, 14,334] in the fall of 2013. The biomass decreased only slightly, from 490 kg/ha in 2011 to 397 kg/ha in 2013 because the mean body length of carp increased to 559 mm over this time frame. The biomass at the time of this study was approximately 20 mg/L, assuming an average lake depth of 2 m.

Since 2011, the distribution of carp has been regularly assessed by locating 10–20 carp equipped with internal radio tags (F 1850, Advanced Telemetry Systems, Isanti, MN). Carp location was determined by identification of signal directionality (bearing) with a hand-held antenna and a compass while positioned within 200 m of the radiotagged fish. Two bearings were measured for each fish, each from a different location, and their intersection calculated (LOAS, Ecological Software Solutions, CA) to estimate fish location. Mean measurement error (30 m) was estimated using dummy tags.

Using previously determined carp locations, the pattern of carp habitat usage was examined for the warm season (June–October) when carp maintain stable summertime distributions [21]. A total of 12 radiotelemetry surveys within the 2011–2013 time frame were conducted. Individual locations (N = 135) were pulled across fish and years. Areas of high carp use were estimated by calculating kernel density (search radius = 35 m, approximately one SE of carp location estimate; output cell size = 4.2 m) using spatial analyst in ArcMap (10.0, Esri, Redlands, CA).

Carp showed well-defined areas of habitat usage in Lake Staring (Figure 1A). A density of 800 radiotagged carp/km^2 was used as the cut-off between high- and low-use areas. As such, areas with lower values were considered to be low-use areas, whereas areas with higher densities were classified as high-use areas. The value of 800 radiotagged carp/km^2 corresponded to approximately 1,248 carp/ha. This cut-off value was chosen because lower densities were associated with relatively isolated radiotagged carp observations. The high-use areas were located within or near the patches of lilies, which the carp most likely use for cover as the lake lacks other physical structure. In addition, all high-use areas were within 200 m of the shoreline and less than 2 m deep.

Field sampling

Field sampling took place on 8 October 2013. Average wind speed was 20 km/h from the S [22], while air and water temperatures were 17.8°C and 19.8°C, respectively. Water and sediment samples were collected from 24 locations within the lake. Samples were taken at 18 points at 4 to 5 locations along three N to S transects of the lake and at one location at the E and W ends. Six additional sampling points were added within three patches (two samples per patch) of *N. odorata* where we knew carp were generally found. At each sampling site, surface water, sub-surface (0.5 m depth) water, and sediment were sampled. A surface water sample was also taken in both the inflow and outflow of the lake.

Water samples were collected in 1-L HDPE bottles (Nalgene, Rochester, NY) that had been previously soaked in 10% bleach for at least 30 min to remove all traces of DNA. Bottles were subsequently rinsed with distilled water to remove residual bleach. Surface water samples were taken by partially submerging a

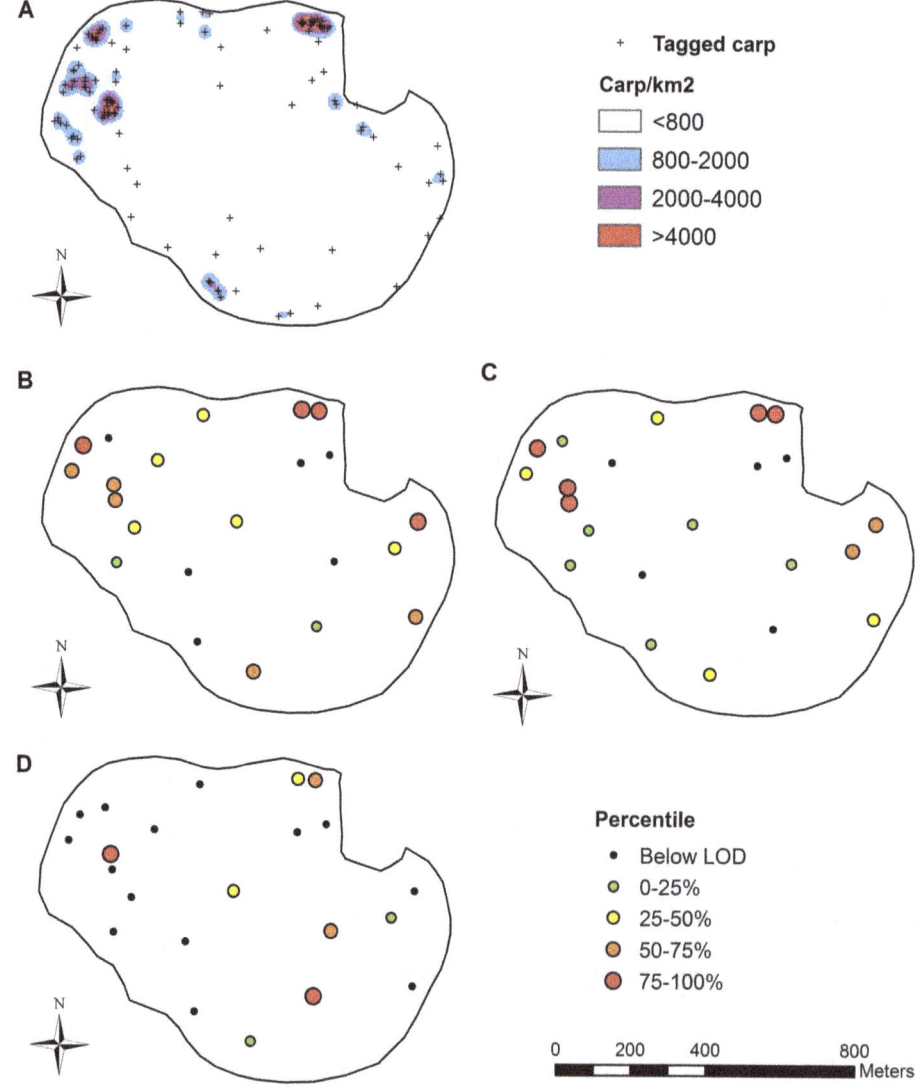

Figure 1. Carp use and distribution of eDNA in Lake Staring. Panel A shows locations of radiotagged carp and high- and low-use areas. Density categories represent the average number of locations of radiotagged carp/km^2. The high- and low-use area cut-off value of 800 radiotagged carp/km^2 corresponded to approximately 1,248 carp/ha. Panels B–D show the pattern of eDNA detection and concentration in surface water (B), sub-surface water (C), and sediment (D). All figures have the same scale. The symbol legend in the upper right refers to panel A, whereas lower right refers to panels B–D.

sample bottle to collect water from the top few cm. Sub-surface samples were taken using a stainless steel Van Dorn sampler (Wildlife Supply Company, Yulee, FL). Sediment samples were collected using a stainless steel Petite Ponar grab sampler (Wildlife Supply Company, Yulee, FL). Sediment was transferred to a sterile Whirl-Pak bag (Nasco, Fort Atkinson, WI) using a sterile polystyrene spatula (Bel-Art, Wayne, NJ). Once collected, samples were immediately placed on ice. Water and sediment samples were stored at 4°C and were filtered within 24 h. No specific permissions were required for access to the study site or collection of samples as part of this study. No animals were collected as part of this study.

Molecular Analyses

Water samples were filtered through Whatman 934-AH 1.5 μm glass microfiber filters (GE Whatman, Fairfield, CT) using a polyphenylsulfone filter funnel (Pall Corporation, Port Washington, NY). Filter funnels and forceps were soaked in 10% bleach and rinsed in distilled water prior to use and between samples. For tank samples, 1 L of water was filtered per sample. For field samples, only 200 mL could be filtered per sample due to clogging from the high amount of suspended solids. Filters were stored at −80°C until DNA extraction.

Sediment samples were homogenized, and a 0.1 g subsample was stored at −80°C for DNA extraction. Preliminary experiments showed that extraction of greater than 0.1 g of sediment lead to reduction in eDNA yield and inhibition of qPCR, regardless of post-extraction inhibitor removal protocols or inclusion of PCR adjuvants. For moisture content analysis, a 10 g subsample of sediment was weighed and then dried at 100°C for 24 h.

DNA was extracted using the QIAamp DNA Stool Mini Kit (Qiagen, Hilden, Germany) using the human DNA analysis protocol. Frozen filters were sliced into 1 mm×5 mm fragments with a sterile razor blade and then transferred to extraction tubes. For sediment samples, extraction buffer was directly added to frozen sediment. Before extraction, 50 ng of UltraPure Salmon Sperm Solution (Life Technologies, Grand Island, NY) was added to adjust for extraction efficiency of DNA as previously described [23]. DNA was eluted in a final volume of 50 μL. To further remove potential inhibitors, all DNA extracts were processed with the Wizard Genomic DNA Purification Kit (Promega, Madison, WI).

A multiplex qPCR assay was designed to amplify both the CarpCytb and the extraction control targets, and the oligonucleotide concentrations were optimized for this study (Table 1). CarpCytb standard was created by cloning PCR product amplified from carp fin clip DNA for the CarpCytb genetic marker (Table 1) using the StrataClone PCR kit (Stratagene, Santa Clara, CA). Purified plasmid DNA was quantified by using a QuantiFluor-ST Fluorometer (Promega, Madison, WI). For the extraction control, standards were created by diluting UltraPure Salmon Sperm Solution (Life Technologies, Grand Island, NY). CarpCytb and extraction control standards were combined prior to preparation of five qPCR standards, ranging from 50 to 300,000 CarpCytb copies and 1.6 to 10,000 pg control DNA per 5 μL.

The assay used iTaq Universal Probes Supermix (Bio-Rad, Hercules, CA). Reactions contained 12.5 μL mastermix, 10 μg bovine serum albumin (New England Biolabs Inc., Ipswich, MA) primers and probe, and water or sediment DNA in a final reaction volume of 25 μL. The volume of water and sediment DNA added to the qPCR reaction was adjusted by testing a dilution series of a subset of 5 samples from water and sediment to confirm that inhibition was not present. For sediment samples, 2.5 μL of DNA

extract was added to the reaction, and for water samples 5 μL of DNA extract was added. Reaction conditions consisted of an initial denaturation at 95°C for 3 min, followed by 40 cycles of denaturation at 95°C for 15 s and an annealing and extension step at 60°C for 1 min. Each qPCR run contained triplicate reactions of standards, non-transcript controls, and samples. Amplifications were performed using the StepOnePlus Real-Time PCR System (Life Technologies, Grand Island, NY), and Cq values were automatically determined using the system software. Sample marker concentrations were calculated on a per-run basis. All sediment values are reported per dry g.

Statistical analyses

Detection rate of eDNA was defined as the proportion of samples that were above the qPCR assay LOD. To analyze the effects of water sample depth (surface, sub-surface), carp usage (low-use, high-use), and matrix type (water, sediment) on eDNA detection rate, the number of detections and non-detections were statistically compared using Fisher's exact test. Fisher's exact test was used due to low expected values (<5) in some cells.

Concentration of eDNA in water was analyzed using a three-way Analysis of Variance (ANOVA). Main effects and 2-way interactions of carp usage (low-use, high-use), water sample depth (surface, sub-surface), and lake depth (m) were examined. Lake depth was included to determine the potential for suspended sediment to affect water column eDNA concentration, with shallower depths more likely to be affected by sediment mixing into the water column. The 3-way ANOVA was restricted to sites greater than 0.5 m and less than 2 m depth (i.e. omitting 2 shallower and 7 deeper sampling sites) because of a partial confound (high-use areas were not found at depths greater than 2 m, and no low-use areas were sampled at depths less than 0.5 m). Finally, student's t-test was used to determine whether there was a significant difference between eDNA concentration of sites included in the ANOVA and those excluded for the low-use areas. Since only 2 sites within the high-use area were excluded, no statistical comparison was done.

For all parametric descriptive analyses and statistical tests, eDNA concentrations were log_{10} transformed to achieve normal data distribution. Values below the LOD were given a value of half the LOD prior to analysis in order to reduce skewing of data. For graphical representation of eDNA concentrations across sampling points, data above the LOD were divided into four equal percentile categories. Percentiles were determined independently for the two sample types: (1) surface and sub-surface water and (2) sediment samples. All statistical tests were conducted in JMP, Version 10 (SAS Institute Inc., Cary, NC).

Results

Carp genetic marker development and laboratory validation

A carp-specific genetic marker (CarpCytb) was developed for a 149 bp region in the cytochrome b gene (Table 1). This assay had an R^2 of over 0.99 and an average PCR efficiency of 92%. The lowest copy number that all three replicate reactions reliably and successfully amplified was determined to be the assay LOD. The assay LOD was 50 copies per reaction, which corresponded to $2.0×10^4$ copies/L for water samples and $1.0×10^5$ copies/g for sediment samples (calculated from the amount of DNA extract analyzed and the volume filtered or weight extracted). The sample LOD varied slightly for individual samples depending on extraction efficiency. Further details regarding qPCR calibration curves can be found in Table S2. The average extraction efficiency

Table 1. Primers and probes used for multiplex quantitative PCR.

Assay	Target	Locus	Primer/Probe	Sequence (5' to 3')[a]	Conc. (nM)	Ref.
CarpCyt*b*	*Cyprinus carpio*	Cytochrome *b*	CCcytbF	CTAGCACTATTCTCCCCTAACTTAC	200	This study
			CCcytbR	ACACCTCCGAGTTTGTTTGGA	200	
			CCcytbP	(6FAM) CCCTCTAGTTACACCACC (MGBNFQ)	200	
Extraction control	*Oncorhynchus keta*	ITS[b] region 2	SketaF2	GGTTTCCGCAGCTGGG	200	[36]
			SketaR3	CCGAGCCGTCCTGGTCTA	200	
			SketaP2	(JOE) AGTCGCAGGCGGCCACCGT (BHQ-1)	100	

[a]BHQ1, black hole quencher-1; 6FAM, 6-carboxyfluorescein; JOE, 6-carboxy-4',5'-dichloro-2',7'-dimethoxyfluorescein.
[b]ITS, internal transcribed spacer.

was 11% for water samples and 6% for sediment samples. The assay reliably quantified up to 3.0×10^5 copies per reaction, the highest standard tested. No amplification of non-transcript controls was observed. The assay did not amplify DNA from a selection of 34 native and non-native fish species, including Bigheaded carps and other related Cyprinids (Table S1).

In validation tests using laboratory tanks which were stocked with combinations of carp and goldfish, no CarpCyt*b* markers were detected in the tank that contained only goldfish. In the tank with 10 carp, the concentration of markers averaged 1.3×10^7 copies/L 95% CI $[1.0 \times 10^7, 1.6 \times 10^7]$. In the tank with 5 each of carp and goldfish, the concentration of markers was 4.5×10^6 copies/L, 95% CI $[3.1 \times 10^6, 6.7 \times 10^6]$ which was significantly lower ($p = 0.007$, Student's t-test) than that of carp alone, indicating that the presence of more individuals yielded more eDNA. On average, individual carp contributed 3.1×10^{11} copies of CarpCyt*b* in the mixed species tank and 4.4×10^{11} copies per individual in the carp only tank.

Detection rates of eDNA in Lake Staring

The overall detection rate of CarpCyt*b* in water samples was 75% (Table 2). The distribution of eDNA was patchy, with carp eDNA detected tens of meters from sites where eDNA was not detected (Fig. 1B, C). The detection rate was not statistically different between surface and sub-surface samples ($p = 1.00$, Fisher's exact test). However, detection rate was significantly higher in water samples collected in high-use areas ($p = 0.009$, Fisher's exact test), a difference of nearly 40% (Table 2).

The overall detection rate of CarpCyt*b* in sediment was only 36% (Table 3). Similar to the water samples, detection pattern was patchy (Fig. 1D). The detection rate of CarpCyt*b* in sediment was slightly higher in high-use areas relative to low use areas, however, the difference between these use areas was not significant ($p = 1.00$, Fisher's exact test). The detection rate of CarpCyt*b* in low-use areas in the sediment was approximately 30% less than for the average value for water samples, however, there was no difference between the detection rate ($p = 0.11$, Fisher's exact test). The detection rate of eDNA in sediment within high-use areas was nearly 60% less than water within high-use areas, and the difference was statistically significant ($p = 0.005$, Fisher's exact test).

Concentration of eDNA in Lake Staring

CarpCyt*b* concentration in water ranged from below the LOD (2.0×10^4 copies/L) to 1.7×10^6 copies/L (Table S3). The mean CarpCyt*b* concentration across all water samples was 5.7×10^4 copies/L, 95% CI $[3.9 \times 10^4, 8.3 \times 10^4]$ (Table 2). Most samples

(84%) had less than 3.0×10^5 copies/L. Only three samples had a marker concentration above 5.0×10^5 copies/L, Surface water samples from the inflow and outflow streams had CarpCyt*b* concentrations of 6.0×10^4 and 3.4×10^4 copies/L, respectively.

Three-way ANOVA showed that carp use pattern had a significant (main) effect on eDNA concentration (Table 4). There was no significant effect of lake depth or sample depth on eDNA, and no 2-way interactions were significant (Table 4). For the subset of sampling locations considered in the ANOVA (0.5 to 2 m lake depth), in low-use areas the average CarpCyt*b* concentration was 3.3×10^4 copies/L, 95% CI $[1.8 \times 10^4, 5.9 \times 10^4]$, and in high use areas CarpCyt*b* concentration averaged 1.6×10^5 copies/L, 95% CI $[1.1 \times 10^5, 2.4 \times 10^5]$. There was no significant difference in eDNA concentration of the shallow water sites included in the ANOVA and those excluded for the low carp use areas ($p = 0.75$).

For sediment samples, the concentration of eDNA ranged from below the LOD (1.0×10^5 copies/g) to 5.4×10^5 copies/g (Table S3). For sediment samples above the LOD, the concentration of CarpCyt*b* was slightly higher in high-use areas (Table 3), but the difference was not significant ($p = 0.3$, Student's t-test). On a per mass basis, the lowest measureable sediment concentration of CarpCyt*b* was nearly two orders of magnitude greater than the water sample with the highest concentration of CarpCyt*b*.

Discussion

This study found that both the detection rate and concentration of carp eDNA strongly correlated with the distribution of carp in lake water. In water, the concentration of the carp genetic marker CarpCyt*b* was over 7 times greater in high-use areas as opposed to low-use areas, and detection rate rose from 63% to 100%. The detection rate and concentration of eDNA did not differ between surface and sub-surface water samples. Detection rate was comparably low in sediment, at 36%. The distribution of eDNA is fundamentally important in the design of eDNA sampling schemes and accurate interpretation of eDNA data [6]. Thus, we have shown that the distribution of a target organism must be carefully considered in the design of eDNA sampling schemes and accurate interpretation of eDNA data. Specifically, eDNA is patchily distributed in the environment, and the probability of detecting a target organism may drastically decline tens of meters from areas that are frequently inhabited. As part of this study, a highly-specific qPCR assay for common carp, an invasive and broadly distributed fish, was developed and validated.

The patchiness of eDNA distribution within water and sediment samples taken from Lake Staring was unexpected, given that the lake is small, shallow, and has a high biomass of carp. However,

Table 2. Concentration and detection rates of CarpCytb in water.

Carp usage	Surface			Sub-surface			Total		
	Mean (copies/L) [95% CI]	Detection rate (%)	N	Mean (copies/L) [95% CI]	Detection rate (%)	N	Mean (copies/L) [95% CI]	Detection rate (%)	N
Low-use	3.8×10^4 [2.2×10^4, 6.8×10^4]	60	15	2.7×10^4 [1.8×10^4, 4.2×10^4]	67	15	3.1×10^4 [2.1×10^4, 4.5×10^4]	63	30
High-use	2.1×10^5 [9.8×10^4, 3.7×10^5]	100	7	2.6×10^5 [1.3×10^5, 3.7×10^5]	100	7	2.4×10^5 [1.4×10^5, 3.8×10^5]	100	14
Total	6.6×10^4 [3.7×10^4, 1.2×10^5]	73	22	5.5×10^4 [3.1×10^4, 9.9×10^4]	77	22	5.7×10^4 [3.9×10^4, 8.3×10^4]	75	44

Pilliod et al. [9] also observed high variation in amphibian eDNA concentration of replicate water samples in small freshwater streams. The authors hypothesized that variation was due to downstream pulses of eDNA due to activity of the target organism or variation in the cell type or form (free or cellular) of eDNA. An alternative, and perhaps complimentary explanation, is that mechanisms of eDNA removal from the water column, such as sedimentation and decay, are rapid. Lake Staring is a eutrophic lake, with high productivity and turbid, nutrient rich water. Carp eDNA is primarily contained in the particle size fraction ranging from 1.0–10 μm [24]. Particulate eDNA is continuously settling into lake sediments, but suspended particles are also hot spots of microbial degradation in aquatic systems [25]. Although decay of eDNA was not measured in this study, microcosm studies suggest that eDNA decays rapidly in the environment, and eDNA degrades nearly 90% within several days in water [12,26,27,28]. Taken together, it is likely that rapid removal of eDNA from the water column through processes of decay and sedimentation prevented its accumulation and diffusion from release points in Lake Staring, leading to significantly higher concentrations of eDNA in areas where carp were present.

Regardless of the cause of eDNA's patchy distribution, it has implications for the optimal sampling of eDNA. In low-use areas, the detection rate of eDNA was 40% lower in low-use areas as opposed to high-use areas. Distances between high-use and low-use sampling sites were tens or hundreds of meters apart, and that small distance, in some instances, affected whether eDNA was detected or not. Differences on a fine spatial scale have been observed in streams [9,29], experimental ponds [15], and lakes [10]. Therefore, we conclude that eDNA sampling should be conducted on small spatial scales. Only after extensive testing and consideration should sampling intervals of greater distances be used.

To the authors' knowledge, only one other study has attempted to correlate fish eDNA distribution within a lentic system. A positive relationship between carp eDNA concentration and temperature was noted by Takahara et al. [10] within a Japanese lagoon using a different marker on a slightly coarser scale without explicit information of carp distribution. The authors posited that the distribution corresponded to temperature because carp prefer warmer water; however, the data were not compared to the actual distribution of carp, as in the present study. Therefore other factors, such as higher metabolic activity of fish within warmer waters, could not be ruled out. Nevertheless, there are concordances with patterns observed between these studies. For example, excluding the lagoon channel, areas with higher eDNA appeared to have been located near the shoreline [10]. Similarly, hot spots of eDNA near shore was observed within Lake Staring, due to carp aggregation.

The overall detection rate of carp eDNA in Lake Staring water was 77%, similar to other studies conducted in areas of high fish abundance. For example, the eDNA of Bigheaded carps in the surface waters of a reach of the Mississippi River with a high target population was detected in 64% of samples [30]. Similarly, a 90% of samples were positive for common carp eDNA in a lagoon used for breeding purposes by Takahara et al. [10].

Despite the assumption of eDNA accumulation in surface waters, water depth did not affect concentration or detection rate of carp eDNA in the study lake. Therefore, both surface and sub-surface samples were equally effective for eDNA sampling in the present study. Although most eDNA samples are taken from the water's surface [11,14,26,31], only one study, done in experimental ponds, has confirmed that eDNA is most frequently detected in surface waters [15]. The convention of surface water sampling

Table 3. Concentration and detection rates of CarpCytb in sediment.

	Conc. of samples above LOD			Detection rate	
	Mean (copies/g)		N	(%)	N
Carp usage	[95% CI]				
Low-use	1.2×10^5		5	33	15
	$[7.5 \times 10^4, 1.8 \times 10^5]$				
High-use	2.3×10^5		3	43	7
	$[1.1 \times 10^5, 4.8 \times 10^5]$				
Total	1.5×10^5		8	36	22
	$[1.1 \times 10^5, 2.1 \times 10^5]$				

may be a holdover from early methods, wherein eDNA was used to detect floating feces of large marine mammals [32]. Our results indicate that the level of eDNA accumulation in surface water may differ among species, and accumulation likely depends on the relative proportion of eDNA sources, buoyancy of fecal material, and site-specific factors.

Although eDNA is hypothesized to accumulate in sediment, the detection rate was unexpectedly low, at 36%. This is likely due to the high LOD of the CarpCytb marker in sediment. The LOD in sediment was 100,000 copies/g, nearly 4 orders of magnitude higher than for water samples on a per g basis. The high LOD is likely partly due to the limited amount of sediment that could be extracted. DNA extracts prepared with more than 0.1 g of sediment were observed to inhibit the qPCR reaction; therefore, greater amounts of sediment were not capable of being processed. Although the extraction efficiency of sediment was within the range previously observed for commercial DNA extraction kits [33], it was approximately half that of water samples. We do not know the cause of this discrepancy, but we hypothesize that sample chemistry can differentially affect extraction efficiency. Due to the difficulty of DNA recovery from sediment and the potential for qPCR inhibition, water sampling was more efficient for detection of carp in the study lake.

Regardless of the low detection rate of eDNA in sediment, its importance as a reservoir of carp eDNA cannot be disregarded. The concentration of CarpCytb was high in sediment locations where eDNA was detected, but as the majority of the samples (63%) were below the LOD, we were unable to reliably calculate a mean sediment eDNA concentration. The accumulation of eDNA

in sediment has been suspected based on the high concentration of microbial DNA in sediment [34,35], but measurements of fish eDNA in sediment have not been previously published. Hot spots of eDNA in sediment did not correlate with carp use. Therefore, there is a need for future studies to measure factors that may control eDNA distribution in sediment, such as deposition, resuspension, and degradation rates.

Conclusions

Sampling design has been previously identified as one of the four critical aspects that must be optimized in a DNA-based monitoring program [6]. The present study showed that common carp distribution led to spatial patterns in both eDNA concentration and detection rate in a small, shallow lake. Our results show that while eDNA is relatively evenly distributed in the water column, eDNA is patchily distributed horizontally. The large variation of eDNA on a small spatial scale, of tens to hundreds of meters, indicates that sampling for aquatic species using eDNA should use a similarly fine scale, at least for initial surveys. The results of this study also indirectly suggest that mechanisms of eDNA removal from the water column are rapid and may partially control eDNA distribution. Although the observations of the current study may not be universally applicable to all species and habitats, our results indicate that eDNA sampling schemes should be critically evaluated for the specific organism and the type of aquatic environment they inhabit. Future research is needed to examine the role of decay, sediment re-suspension, and eDNA release on eDNA distribution in aquatic habitats.

Supporting Information

Table S1 List of fish species tested for marker specificity.

Table S2 Quantitative PCR calibration data.

Table S3 Coordinates, lake depth, fish use, and eDNA concentration at sampling sites.

Acknowledgments

The authors thank Sendréa Best for field and laboratory assistance. Fish telemetry data was collected by Mary Headrick, Joseph Lechelt, Brett Miller, Robert Mollenheuer, and Tracy Szela. Erik Smith and Justine Koch provided assistance with ArcGIS. Mark Hove graciously provided fin

Table 4. Results of a 3-way ANOVA for CarpCytb marker in water samples.

Effects	df	F ratio	P value
Sample depth (surface, sub-surface)	1	0.07	0.79
Lake depth (m)	1	0.08	0.78
Carp use (low-use, high-use)	1	5.77	0.03*
Sample depth × Lake depth	1	0.18	0.67
Lake depth × Carp use	1	0.62	0.44
Sample depth × Carp use	1	0.26	0.62
Error	19		

clips for use in specificity testing. Commercial fisher Tim Adams and crew also provided valuable assistance obtaining fin clips for specificity testing from the Mississippi River.

Author Contributions

Conceived and designed the experiments: JJE PGB PWS. Performed the experiments: JJE PGB. Analyzed the data: JJE PGB PWS. Contributed reagents/materials/analysis tools: JJE PGB PWS. Wrote the paper: JJE PGB PWS.

References

1. Mesa MG, Schreck CB (1989) Electrofishing mark-recapture and depletion methodologies evoke behavioral and physiological changes in cutthroat trout. Trans Amer Fish Soc 118: 644–658.
2. Cross DG, Stott B (1975) The effect of electric fishing on the subsequent capture of fish. J Fish Biol 7: 349–357.
3. Bayley PB, Austen DJ (2002) Capture efficiency of a boat electrofisher. Trans Amer Fish Soc 131: 435–451.
4. Lodge DM, Turner CR, Jerde CL, Barnes MA, Chadderton L, et al. (2012) Conservation in a cup of water: estimating biodiversity and population abundance from environmental DNA. Mol Ecol 21: 2555–2558.
5. Olson ZH, Briggler JT, Williams RN (2012) An eDNA approach to detect eastern hellbenders (*Cryptobranchus a. alleganiensis*) using samples of water. Wildl Res 39: 629–636.
6. Darling JA, Mahon AR (2011) From molecules to management: adopting DNA-based methods for monitoring biological invasions in aquatic environments. Env Res 111: 978–988.
7. Ficetola GF, Miaud C, Pompanon F, Taberlet P (2008) Species detection using environmental DNA from water samples. Biol Lett 4: 423–425.
8. Jerde CL, Mahon AR, Chadderton WL, Lodge DM (2011) "Sight-unseen" detection of rare aquatic species using environmental DNA. Conserv Lett 4: 150–157.
9. Pilliod DS, Goldberg CS, Arkle RS, Waits LP (2013) Estimating occupancy and abundance of stream amphibians using environmental DNA from filtered water samples. Can J Fish Aquat Sci 70: 1123–1130.
10. Takahara T, Minamoto T, Yamanaka H, Doi H, Kawabata Z (2012) Estimation of fish biomass using environmental DNA. PLoS One 7: e35868.
11. Thomsen PF, Kielgast J, Iversen LL, Wiuf C, Rasmussen M, et al. (2012) Monitoring endangered freshwater biodiversity using environmental DNA. Mol Ecol 21: 2555–2558.
12. Thomsen PF, Kielgast J, Iversen LL, Møller PR, Rasmussen M, et al. (2012) Detection of a diverse marine fish fauna using environmental DNA from seawater samples. PLoS One 7: e41732.
13. Goldberg CS, Sepulveda A, Ray A, Baumgardt J, Waits LP (2013) Environmental DNA as a new method for early detection of New Zealand mudsnails (*Potamopyrgus antipodarum*). Freshw Sci 32: 2555–2558.
14. U.S. Fish and Wildlife Service (2013) Quality assurance project plan (QAPP): eDNA monitoring of bighead and silver carps. 89 pp.
15. Moyer GR, Díaz-Ferguson E, Hill JE, Shea C (2014) Assessing environmental DNA detection in controlled lentic systems. PLoS One 9: e103767.
16. Bohmann K, Evans A, Gilbert TP, Carvalho GR, Creer S, et al. (2014) DNA for wildlife biology and biodiversity monitoring. Trends Ecol Evol 29: 358–357.
17. Mahon AR, Jerde CL, Galaska M, Bergner JL, Chadderton WL, et al. (2013) Validation of eDNA surveillance sensitivity for setection of Asian carps in controlled and field experiments. PLoS One 8: e58316.
18. Chang Y, Huang F, Lo UT (1994) The complete nucleotide sequence and gene organization of carp (*Cyprinus carpio*) mitochondrial genome. J Mol Evol 38: 138–155.
19. Mayden RL, Tang KL, Conway KW, Freyhof J, Sudkamp M, et al. (2007) Phylogenetic relationships of *Danio* within the order Cypriniformes: A framework for comparative and evolutionary studies of a model species. J Exp Zoo B Mol Dev Evol 308: 642–654.
20. Bajer PG, Sorensen PW (2012) Using boat electrofishing to estimate the abundance of invasive common carp in small midwestern lakes. N Amer J Fish Manag 32: 817–822.
21. Bajer PG, Chizinski CJ, Sorensen PW (2011) Using the Judas technique to locate and remove wintertime aggregations of invasive common carp. Fish Manag Ecol 18: 497–505.
22. (NOAA) National Oceanic and Atmospheric Administration Global Historical Climate Network. Available: http://www.ncdc.noaa.gov/cdo-web/datasets#GHCND. Accessed 11 February 2014.
23. Haugland RA, Siefring SC, Wymer LJ, Brenner KP, Dufour AP (2005) Comparison of *Enterococcus* measurements in freshwater at two recreational beaches by quantitative polymerase chain reaction and membrane filter culture analysis. Water Res 39: 559–568.
24. Turner CR, Barnes MA, Xu CCY, Jones SE, Jerde CL, et al. (2014) Particle size distribution and optimal capture of aqueous macrobial Edna. Method Ecol Evol doi: 10.1111/2041-210X.12206.
25. Simon M, Grossart H-P, Schweitzer B, Ploug H (2002) Microbial ecology of organic aggregates in aquatic ecosystems. Aquat Microb Ecol 28: 175–211.
26. Pilliod DS, Goldberg CS, Arkle RS, Waits LP (2014) Factors influencing detection of eDNA from a stream-dwelling amphibian. Mol Ecol Res 14: 109–116.
27. Dejean T, Valentini A, Duparc A, Pellier-Cuit S, Pompanon F, et al. (2011) Persistence of environmental DNA in freshwater ecosystems. PLoS One 6: e23398.
28. Barnes MA, Turner CR, Jerde CL, Renshaw MA, Chadderton WL, et al. (2014) Environmental conditions influence eDNA persistence in aquatic systems. Environ Sci Technol 48: 1819–1827.
29. Jane SF, Wilcox TM, McKelvey KS, Young MK, Schwartz MK, et al. (2014) Distance, flow, and PCR inhibition: eDNA dynamics in two headwater streams. Mol Ecol Res doi: 10.1111/1755-0998.12285.
30. Amberg JJ, McCalla SG, Miller L, Sorensen P, Gaikowski MP (2013) Detection of environmental DNA of Bigheaded carps in samples collected from selected locations in the St. Croix River and in the Mississippi River: U.S. Geological Survey Open-File Report 2013-1080, 44p.
31. Takahara T, Minamoto T, Doi H (2013) Using environmental DNA to estimate the distribution of an invasive fish species in ponds. PLoS One 8: e56584.
32. Tikel D, Blair D, Marsh HD (1996) Marine mammal faeces as a source of DNA. Mol Ecol 5: 456–457.
33. Mumy KL, Findlay RH (2004) Convenient determination of DNA extraction efficiency using an external DNA recovery standard and quantitative-competitive PCR. J Microbiol Meth 57: 259–268.
34. Anno AD, Corinaldesi C (2004) Degradation and turnover of extracellular DNA in marine sediments: ecological and methodological considerations. Appl Environ Microbiol 70: 4384–4386.
35. Pietramellara G, Ascher J, Borgogni F, Ceccherini MT, Guerri G, et al. (2008) Extracellular DNA in soil and sediment: fate and ecological relevance. Biol Fertil Soils 45: 219–235.
36. Domanico MJ, Phillips RB, Oakley TH (1997) Phylogenetic analysis of Pacific salmon (genus *Oncorhynchus*) using nuclear and mitochondrial DNA sequences. Can J Fish Aquat Sci 54: 1865–1872.

14

Seed Dormancy, Seedling Establishment and Dynamics of the Soil Seed Bank of *Stipa bungeana* (Poaceae) on the Loess Plateau of Northwestern China

Xiao Wen Hu[1]*, Yan Pei Wu[1], Xing Yu Ding[1], Rui Zhang[1], Yan Rong Wang[1]*, Jerry M. Baskin[2], Carol C. Baskin[2,3]

1 State Key Laboratory of Grassland Agro-ecosystems, College of Pastoral Agriculture Science and Technology, Lanzhou University, Lanzhou, 730020, China, **2** Department of Biology, University of Kentucky, Lexington, Kentucky 40506-0225, United States of America, **3** Department of Plant and Soil Sciences, University of Kentucky, Lexington, Kentucky 40546-0312, United States of America

Abstract

Studying seed dormancy and its consequent effect can provide important information for vegetation restoration and management. The present study investigated seed dormancy, seedling emergence and seed survival in the soil seed bank of *Stipa bungeana*, a grass species used in restoration of degraded land on the Loess Plateau in northwest China. Dormancy of fresh seeds was determined by incubation of seeds over a range of temperatures in both light and dark. Seed germination was evaluated after mechanical removal of palea and lemma (hulls), chemical scarification and dry storage. Fresh and one-year-stored seeds were sown in the field, and seedling emergence was monitored weekly for 8 weeks. Furthermore, seeds were buried at different soil depths, and then retrieved every 1 or 2 months to determine seed dormancy and seed viability in the laboratory. Fresh seeds (caryopses enclosed by palea and lemma) had non-deep physiological dormancy. Removal of palea and lemma, chemical scarification, dry storage (afterripening), gibberellin (GA$_3$) and potassium nitrate (KNO$_3$) significantly improved germination. Dormancy was completely released by removal of the hulls, but seeds on which hulls were put back to their original position germinated to only 46%. Pretreatment of seeds with a 30% NaOH solution for 60 min increased germination from 25% to 82%. Speed of seedling emergence from fresh seeds was significantly lower than that of seeds stored for 1 year. However, final percentage of seedling emergence did not differ significantly for seeds sown at depths of 0 and 1 cm. Most fresh seeds of *S. bungeana* buried in the field in early July either had germinated or lost viability by September. All seeds buried at a depth of 5 cm had lost viability after 5 months, whereas 12% and 4% seeds of those sown on the soil surface were viable after 5 and 12 months, respectively.

Editor: Jian Liu, Shandong University, China

Funding: This study was supported by Program for Changjiang Scholars and Innovative Research Team in University (IRT13019), the National Key Technology Research and Development Program (2011BAD17B02) and the Gansu Provincial Key Grant Project (1203FKDA035). The funders had no role in study design, data collection and analysis, decision to publish, or preparation of the manuscript.

Competing Interests: The authors have declared that no competing interests exist.

* Email: huxw@lzu.edu.cn (XWH); yrwang@lzu.edu.cn (YRW)

Introduction

Soil erosion is the main cause of land degradation in arid and semiarid regions, and it is a widespread problem on the Chinese Loess Plateau [1]. One way to restore degraded soils and reduce soil erosion is by revegetation [2,3]. The first step in any program of rehabilitation of soils degraded by erosion is to select the most suitable species to use for revegetation, based on the capacity of the seeds to germinate and the seedlings to become established [4,5].

Stipa bungeana Trin. (Poaceae) is a perennial grass that mainly occurs in semi-arid areas of the temperate steppe zone in Eurasia. The species is widely distributed on the Loess Plateau and other areas of western China. It is the main wild forage species in natural grasslands of northwestern China and also plays important roles in protecting the soil from erosion and reducing water loss by runoff. Due to its environmental benefits and economic value, Cheng *et*

al. [6] and Hu *et al.* [7] suggested that *S. bungeana* is a potential key species for revegetation of degraded land on the Loess Plateau.

Seed dormancy is the failure of viable seeds to germinate in a specified period of time under conditions suitable for their germination after they become nondormant [8,9]. Dormancy could prevent or delay germination even under favorable conditions, thus enabling seeds to accumulate in the soil seed bank and preventing plants from expending their entire reproductive outputs at a given time [10]. As such, then, seed dormancy is expected to be important in optimizing the timing of germination to maximize seedling establishment. Although *S. bungeana* produces up to 1430 seeds m^{-2} in typical *S. bungeana*-dominated rangeland, only a low portion of them germinate in the field; thus, few seedlings became established following seed dispersal [7]. Two reasons may contribute to low seedling establishment: 1) fresh seeds of *S. bungeana* exhibit primary dormancy, which prevents seed germination immediately after dispersal; and 2) environmental conditions during the dispersal season prevent seeds from germinating. In the first case, no seeds, or only a small portion of them, germinate even under otherwise favorable conditions until dormancy release. Thus, we expected that primary dormancy plays a role in regulating the time of seed germination and seedling recruitment in the field. Hu et al. [7] showed that germination of S. bungeana seeds was inhibited by light and sensitive to water stress, implying that most seeds would germinate slowly or not at all on the soil surface. Thus, seed burial in the soil may play a key role in determining whether they can germinate after dispersal.

The effects of temperature [11], light [11], water stress [7] and burial depth [7] on germination and of fungicide pretreatment on seed survival in the field have been determined for seeds of S. bungeana stored (afterripened) in the lab for 1 year. However, no studies have been done to test for seed dormancy in S. bungeana and its underlying mechanism, and consequently its effect on seedling emergence and seed survival in the soil. Moreover, seeds stored in the lab for 1 year were used in previous studies. As reported by Baskin and Baskin [12], results from studies initiated after seeds have been stored dry may have little ecological relevance. That is, the germination responses of seeds may have changed through time, and thus interpretation of results obtained using stored seeds may differ from fresh seeds dispersed in the natural environment.

Thus, the aims of this study were to determine: 1) whether fresh seeds are dormant and if so why; 2) the effect of storage condition on seed dormancy of fresh seeds; 3) the effect of seed dormancy on seedling emergence in relation to sowing depth; and 4) seed dormancy and survival of fresh seeds in relation to burial depth and duration in the field.

Materials and Methods

Seed collection

Stipa bungeana flowers in the early May, and seeds mature and are dispersed in late June in the study area. The dispersal unit of S. bungeana is a caryopsis tightly enclosed by the palea and lemma. It is 5–6 mm in length and 0.7–1.0 mm in width. Hereafter, the dispersal unit of S. bungeana will be referred to as a seed.

S. bungeana seeds were collected on 23 June 2012 and 28 June 2013 from a field on the Yuzhong Campus of Lanzhou University, Gansu Province (35°57′N, 104°10′E). The mean annual temperatures is 6.7°C and mean annual rainfall 350 mm, most of which falls from July to September. The soils consist of silt (66.9%), clay (20.8%) and sand (12.3%), and natural vegetation is dominated by S. bungeana. Other species growing at this site include Achnatherum inebrians, Artemisia spp., Glycyrrhiza spp. and Lespedeza davurica. Infructescences with ripe seeds were collected from several hundred plants and taken to the laboratory, where the seeds were separated from them, cleaned and dried at room temperature for one week (RH, 20–35%, 18–25°C) and stored at 4°C until used in experiments. The experiments were conducted within two weeks after seed collection except for the one on seedling emergence.

Seed viability

Viability of fresh seeds collected in 2012 and 2013 was determined. Seeds were soaked in distilled water for 12 h, after which hulls and half of endosperm were removed. Then the remaining part of the seed containing the embryo was soaked in 1% tetrazolium phosphate-buffer solution for 6–8 hours at 30°C in the dark. Seeds with embryos that stained red were considered to be viable and those with unstained embryos nonviable. For fresh

seeds collected in 2012 and 2013, four replicates of 50 seeds were tested.

Effect of light and temperature on germination

The aim of this experiment was to determine whether fresh seeds are dormant. Fresh seeds were tested at four constant (10°C, 15°C, 20°C and 25°C) and three alternating (10/20°C, 15/25°C and 20/30°C) temperature regimes (12 h/12 h). At each temperature, seeds were incubated at a 12 h/12 h daily photoperiod (hereafter light) or in continuous darkness. For treatments in light, seeds were exposed to light produced by white fluorescent tubes with a photon irradiance of 60 $\mu mol \cdot m^{-2} \cdot s^{-1}$ (400–700 nm). For continuous darkness, Petri dishes were covered with two layers of aluminum foil, and seeds were monitored for germination (root emergence) daily under a LED green safe light (520 nm ± 10 nm, Sanpai, Shanghai, China). Photon irradiance at Petri dishes level was 10 $\mu mol \cdot m^{-2} \cdot s^{-1}$, as determined by use of a quantum sensor (LI-190SA) connected to a LI-6400 portable photosynthesis system (LI-COR, USA). For each treatment, four replicates of 50 seeds each were placed in 11-cm-diameter Petri dishes on two sheets of filter paper (Shuangquan, Hangzhou) moistened with 8 mL of distilled water. Seeds incubated in both light and dark were examined for germination daily for 14 days, and any seedlings present were removed from the Petri dishes.

Effect of hulls, half-endosperm removal and scarification with NaOH on germination

The aim of this experiment was to determine the role of hulls and endosperm in controlling seed dormancy and effect of NaOH on breaking dormancy. The mechanical removal experiment consisted of a control (caryopsis with intact lemma and palea) and three treatments: 1) hulls (lemma and palea) removed; 2) hulls removed and put back in their original position enclosing the caryopsis; and 3) half of endosperm (and enclosing portions of hulls) removed from the endosperm-end of the seed without damaging the embryo. Before performing the removal treatments, seeds were immersed in distilled water for 12 h, and then the hulls were removed from the caryopses using forceps. For treatment 2, the hulls were removed from the seed and then re-attached loosely to their original position starting at the embryo end. For treatment 3, half of the endosperm (along with enclosing portions of hulls) was removed from the endosperm-end of the seed with a scalpel.

For scarification with NaOH, seeds (with hulls) were soaked in 50 mL of a 30% NaOH solution at 20°C for 20, 40 or 60 min. Then the seeds were rinsed thoroughly with tap water five times and allowed to dry on filter paper for 48 h on the laboratory bench. For each treatment, four replicates of 50 seeds were used for testing germination at 20°C in dark. Seeds without pretreatment were used as a control. Germination was monitored daily for 14 days as described above.

Effect of dry storage on seed dormancy

To determine the effect of storage duration and temperature on seed dormancy, fresh seeds were placed in a paper bag and stored in darkness at 5°C and at 20°C for 1, 3 and 6 months. After each treatment, germination was tested in light and in darkness at 20°C. There were four replicates of 50 seeds per treatment. Germination was monitored daily for 14 days as described above.

Effect of fluridone, GA3 and KNO3 on seed germination

To determine the role of plant growth regulator and potassium nitrate (KNO3) in controlling seed dormancy of S. bungeana, seeds were treated with gibberellic acid (GA3, Sigma, China);

fluridone (FLU, Sigma, China), an inhibitor of abscisic acid (ABA) biosynthesis, or KNO_3. Fluridone and GA_3 each were dissolved in 2 mL ethanol prior to dilution in water, and the final concentration of FLU and GA_3 was 200 μM. A preliminary experiment showed that a low concentration of ethanol did not affect germination of *S. bungeana* seeds. The potassium nitrate solution was 1 mM. Seeds incubated in distilled water were used as control. For each treatment, four replicates of 50 seeds were incubated at 20°C in dark. Germination was monitored daily for 14 days as described above.

Seedling emergence in field

To determine the effect of seed dormancy on seedling emergence, the seedling emergence experiment was conducted at the Yuzhong Campus from 16 July 2013 to 12 September 2013. Seeds that had been stored dry at 20°C for 1 year and fresh seeds collected on 28 June 2013 were used in this experiment. There were two seed lots (fresh, afterripened) × three burial depths (0, 1, 5 cm) in a completely randomized design. Seeds were sown on 16 July directly in PVC pots (15 cm in diameter, 11 cm in height) that were buried in the field with the rim 5 cm above the soil surface. Soil level in the pot was even with the soil surface. To avoid seed contamination, soil was passed through a 0.5 mm-mesh wire sieve to remove any *S. bungeana* seeds present in it. Then, the soil was placed in the pots to the desired depth, and seeds were placed at this depth and covered with soil. The pots were covered with nylon mesh to prevent animal predation and contamination by extraneous seeds of *S. bungeana*. Ten replicates of 50 seeds each were used for each treatment. Seedling emergence was monitored weekly for 8 weeks, and any seedlings present were counted and removed. The vegetation surrounding the pots was removed by hand every week. The speed of seedling emergence was calculated using the emergence index (*EI*):

$$EI = \sum \left(\frac{Et}{Tt} \right)$$

where Et is the number of seeds emerged on tth week, and Tt is the weeks of seedling emergence from sowing [13].

Seed burial experiment

To determine the effect of burial depth and burial duration on seed survival and seed dormancy of *S. bungeana*, seeds collected on 23 June 2012 were put into 96 15 cm×10 cm nylon mesh bags and buried at the Yuzhong Campus of Lanzhou University on 5 July 2012. The permeable nylon fabrics allowed movement of water, air and microbes between inside and outside of bags. The burial site is about 300 m from the seed collection site and has a sparse vegetation cover. The vegetation within and surrounding burial site was removed before burial. Forty-eight bags with 50 seeds each were placed on the soil surface (0 cm) and 48 at a soil depth of 5 cm. They were arranged in a randomized complete block design. For 0 cm burial depth, the bags were fixed to the ground by iron nails so that each bag was in contact with the soil and won't move to the other soil profile. Physical removal of weeds within burial site was applied every week during experimental period. Six bags each were retrieved from 0 cm and 5 cm burial depths on 28 July, 30 August, 30 September, 31 October and 30 December 2012 and on 2 March, 9 May and 5 July 2013. For each of these dates, the seeds in each of 12 bags were put into one 11-cm-diameter Petri dish with two layers of filter paper moistened with 8 mL of distilled water and incubated in darkness at 20°C. Thus, there were six replications each for the 0 cm and 5 cm

burial depth treatments. The number of germinated seeds was counted after 14 days. Germination of 6 replicates of 50 seeds was tested before burial as described above. For all germination tests, seeds failed to germinate were tested for viability.

Temporal changes in soil seed bank size

To determine soil seed bank size in the field, soil samples were taken on 28 July and 28 August, 2012 and 2 March and on 20 May, 2013 on the Yuzhong Campus of Lanzhou University. The sampling site is 2000 m^2 and dominant plants were *S. bungeana* and *Medicago sativa*. Other species at the site included *Achnatherum inebrians* and *Artemisia* spp. Sixteen 1 m×1 m quadrats were haphazardly established at the study site for each sampling time. In each quadrat, five soil subsamples 0–5 cm deep were collected using a 10 cm×10 cm×5 cm soil sampler, mixed together, air dried and sieved through a 0.5 mm sieve. Seeds of *S. bungeana* were separated from the litter and incubated in 11-cm-diameter Petri dishes with two layers of filter paper moistened with 8 mL of distilled water in darkness at 20°C. Germination percentages were determined after 14 days of incubation, and seeds that failed to germinate were tested for viability.

Statistical analysis

A two way ANOVA at a significance level of $P<0.05$ was used to analyze the effect of light, temperature and their interaction on seed germination and of burial time, burial depth and their interaction on seed viability and seed dormancy. Duncan's multiple range tests was used to compare means of germination percentage between treatments when significant differences were found. Independent t-test was used to compare the mean viability of seeds collected in 2012 and 2013. Germination percentage data were arcsine transformed to increase homogeneity of variance prior to analysis, but nontransformed data are shown in all figures and in tables. All analyses were conducted in SPSS 15.0 software.

Results

Seed viability

Fresh seeds collected in 2012 and 2013 showed no significant difference in terms of seed viability which was 85±4.6% and 87±3.7%, respectively.

Effect of light and temperature on germination

Light, temperature and their interaction had significant effects on germination (Fig. 1, Table 1). Light inhibited germination at the three temperatures at which germination occurred. The highest germination was 25%, in darkness at 20°C, and only 7% of the seed germinated in light. Seeds germinated to significantly lower percentage at 15/25°C and 20/30°C in both darkness and light. No seeds germinated at 10, 15, 25 or 10/20°C in darkness or in light.

Effect of hulls, half-endosperm removal and scarification with NaOH on seed germination

Hulls, endosperm and scarification with NaOH had a significant effect on release of seed dormancy ($P<0.05$, Fig. 2). Twenty-five percent of fresh seeds (with hulls) without pretreatment germinated, and dormancy was completely released when the hulls were removed. However, seeds in which the hulls were put back to their original position germinated to a significantly lower percentage (46) than those hulls removed (94). Removal of half of the endosperm increased germination from 25% to 45%

Table 1. Two way ANOVA of the effects of temperature, light and their interaction on germination of fresh seeds of *Stipa bungeana* (n = 4).

Source	Sum of Squares	df	F	P-value
Temperature(T)	1790	6	39.2	.000
Light(L)	330	1	43.4	.000
T * L	623	6	13.6	.000

(P<0.05). Seeds treated with NaOH for 20, 40 and 60 minutes germinated to 45, 63 and 82%, respectively (Fig. 2).

Effect of dry storage on dormancy break

Storage at 5°C and 20°C significantly increased germination percentages in light and in dark. However, there was little difference in afterripening at the two storage temperatures, although seeds afterripened of 20°C for 3 months germinated to significantly higher percentage in light and dark than those afterripened at 5°C, and seeds stored at 20°C for 6 months germinated to a significantly higher percentage in light than those afterripened at 5°C (Fig. 3).

Effect of fluridone, GA$_3$ and KNO$_3$ on germination

GA$_3$ and KNO$_3$ significantly increased germination from 25% to 36% and 44%, respectively (P<0.05), but FLU had no effect. Germination percentage was significantly higher for seeds treated with KNO$_3$ than for those treated with GA$_3$ (Fig. 4).

Effect of dry storage and burial depth on seedling emergence

Percentage and speed of seedling emergence in the field varied with burial depth and seed lot (fresh vs. one-year-stored seeds) (Fig. 5, Table 2). At 0, 1 and 5 cm burial depths, seeds stored one year had a significantly higher emergence speed (higher emergence index) than fresh seeds. Most seeds stored for 1 year germinated within 3 weeks after sowing. However, fresh seeds had just begun to germinate by the third week, and they continued to do so for another 4–5 weeks. On the other hand, there was no significant difference in final seedling emergence between stored and fresh seeds sown at 0 cm and 1 cm (Table 2). In contrast, seedling emergence percentage at 5 cm was significantly higher for stored seeds than for fresh seeds, but emergence percentages were <10% for both burial depths. The highest final seedling emergence percentage was for 1 cm burial depth, with 33% for one year stored seeds and 27% for fresh seeds.

Effect of burial on seed viability and dormancy

Burial depth, burial duration and their interaction had a significant effect on seed viability (germinated seeds in the lab +

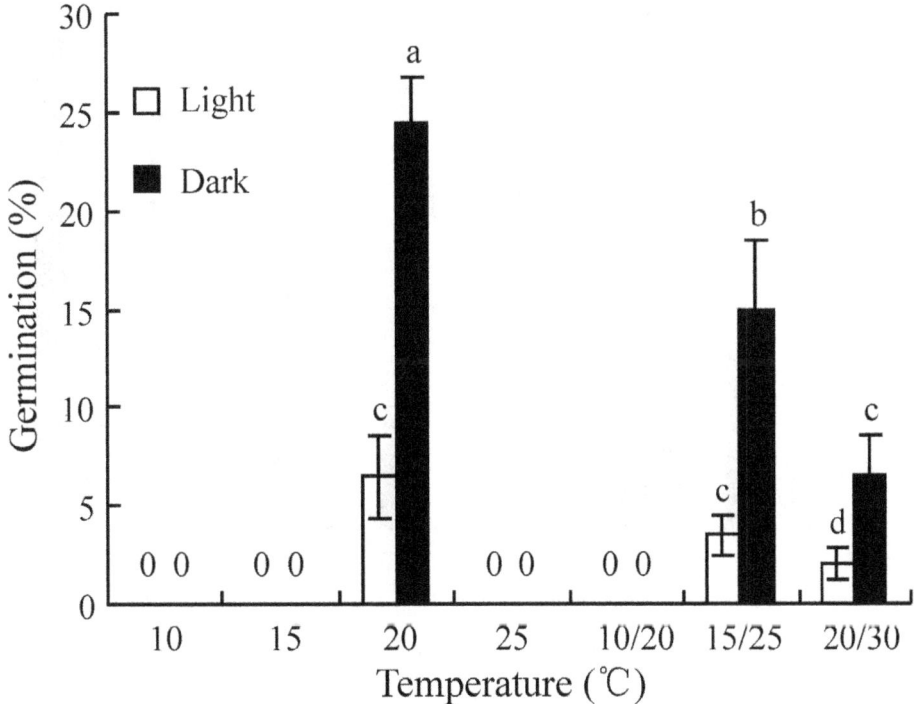

Figure 1. Effect of temperature on germination of fresh seeds of *Stipa bungeana* **in a 12 h/12 h photoperiod and in dark.** Different letters indicate significant difference (P<0.05) among all treatments (n = 4).

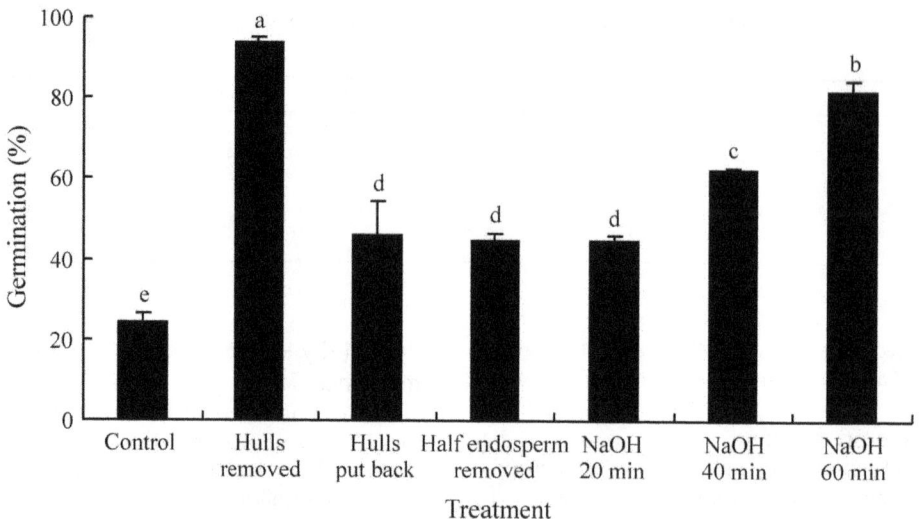

Figure 2. Effect of hulls, half-endosperm removal and NaOH scarification on germination of fresh seeds of *Stipa bungeana* **at 20°C in dark.** Different letters indicate significant difference (P<0.05) among all treatments (n = 4).

dormant seeds) and on seed dormancy (Fig. 6, Table 3). After five months of burial at 5 cm, all seeds had lost viability, whereas 12% and 4% of those on the soil surface were viable after 5 and 12 months, respectively. Further, seeds buried at 5 cm lost dormancy more quickly than those sown on the soil surface. For example, almost 99% of seeds buried at 5 cm depth had lost dormancy after 3 months, while 87% seeds of those on the soil surface had done so.

Temporal changes in soil seed bank size

The size of the soil seed bank of *S. bungeana* declined with time and significantly so between August and March. The highest density of seeds in the seed bank was 869 m^{-2}, in July, and the lowest density was 31 m^{-2}, in May (Fig. 7).

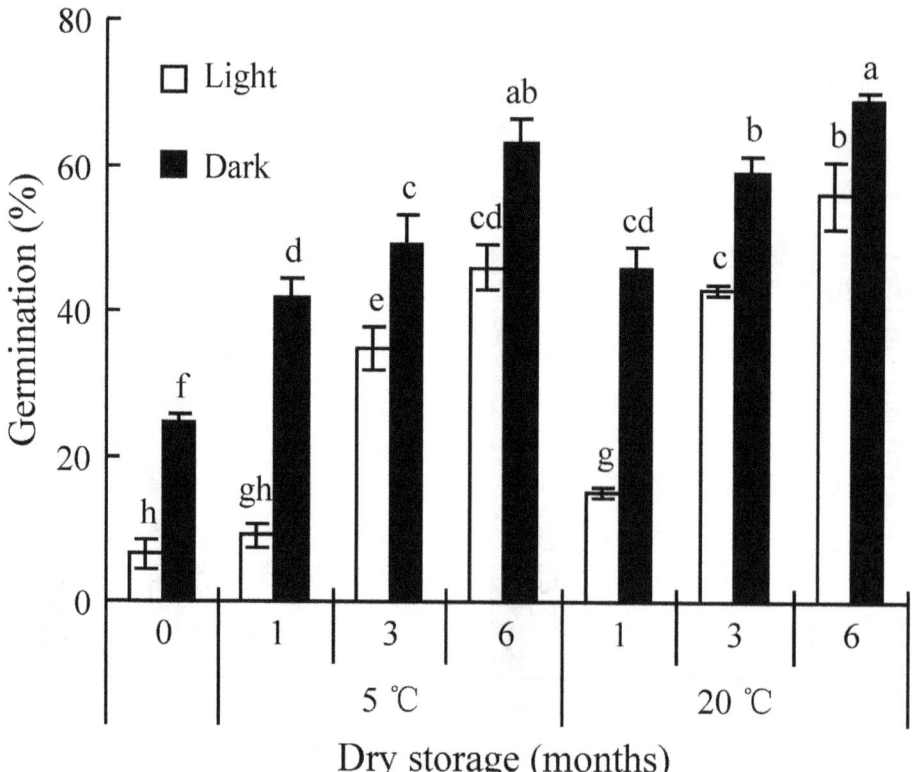

Figure 3. Effect of dry storage (afterripening) on germination of seeds of *Stipa bungeana* **in a 12 h/12 h photoperiod and in dark at 20°C.** Different letters indicate significant difference (P<0.05) among all treatments (n = 4).

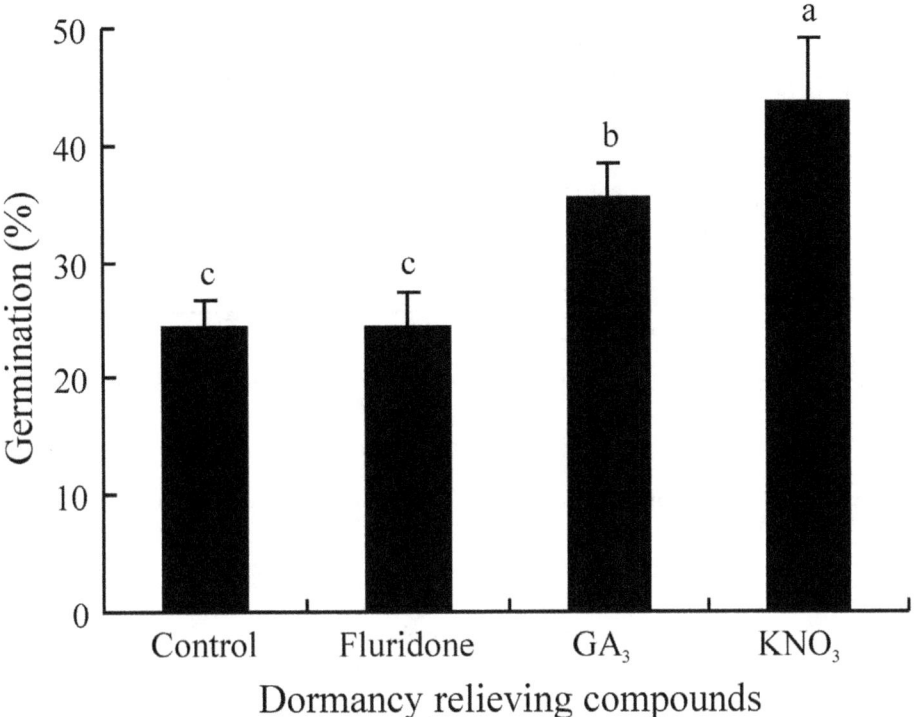

Figure 4. Effect of fluridone, GA₃ and KNO₃ on germination of fresh seeds of *Stipa bungeana* **in dark at 20°C.** Different letters indicate significant difference (P<0.05) among all treatments (n = 4).

Discussion

Seed dormancy and its underlying mechanism

Seed dormancy prevents or delays germination even under favorable conditions and spreads the risk of recruitment over time. It generally is accepted that seed dormancy plays a significant role in ensuring seed germination at the right time and in the proper sites to maximize the probability of successful seedling establishment [9,14]. It is clear from the present study that *S. bungeana* seeds exhibit primary dormancy, since germination percentages were low at various combinations of light and temperature, and they were increased significantly by several dormancy breaking treatments. According to the seed dormancy classification system of Baskin & Baskin [8,9], physiological dormancy can be caused by the tissues surrounding the embryo, by low growth potential of the embryo or by a combination of the two. The palea and lemma clearly play an important role in regulating germination of fresh *S.*

bungeana seeds since their removal released dormancy completely. This is consistent with other studies that reported hulls imposed dormancy in other grass species, for example, *Stipa viridula* [15], *S. tenacissima* [16], *Leymus secalinus* [17], *L. chinensis* [18–20], *Hordeum spontaneum* [21,22] and *H. vulgare* [23].

When hulls were removed from the dispersal unit of *S. bungeana* and then loosely reattached in their original position, seeds germinated to significantly lower percentage than those with hulls removed and not reattached. This suggests the possibility of the presence of germination inhibitors in the hulls, since the loosely-attached hulls should not have inhibited germination via mechanical restriction. Chemical germination inhibitors have been isolated from the hulls of grass seeds [24,25]. ABA in hulls of *Leymus chinensis* inhibited germination of this species in vitro [26,27]. Soluble germination inhibitors in the hulls of *Aegilops geniculata* [28,29] may regulate germination response to amount of rain [30]. However, seed dormancy of *Stipa viridula* was not

Table 2. Seedling emergence percentages and index for fresh one-year-stored seeds of *Stipa bungeana* at sowing depths of 0, 1 and 5 cm.

Seed lots	Sowing depth (cm)	Emergence percentage	Emergence index
Fresh seed	0	18±2.9c	4.1±1.2d
	1	27±2.6ab	6.2±1.4c
	5	1±0.8e	0.2±0.2e
Stored seed	0	22±2.6bc	9.5±1.5b
	1	33±2.9a	14.2±1.2a
	5	8±1.5d	2.8±0.5d

Different letters within a column indicate significant differences (P<0.05, n = 10).

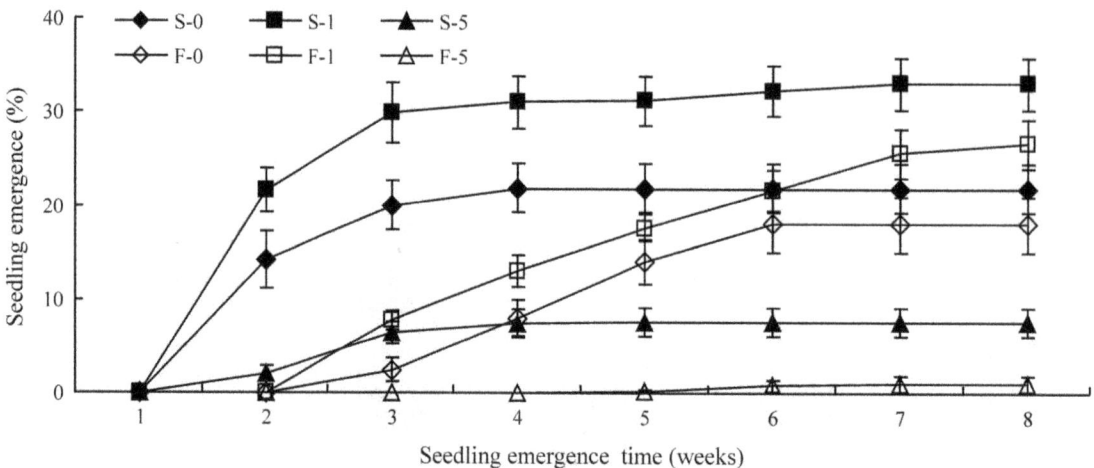

Figure 5. Seedling emergence from fresh (F) and one-year-stored (S) seeds of *Stipa bungeana* at 0, 1 and 5 cm sowing depths. F-0, F-1 and F-5 indicate fresh seeds sown at a depths of 0, 1 and 5 cm, respectively (n = 10). S-0, S-1 and S-5 indicate seeds stored for one year and then sown at depths of 0, 1 and 5 cm, respectively (n = 10).

caused by chemical inhibitors in the palea and lemma since seeds with palea and lemma clipped on both ends without damage to the enclosed caryopsis and seeds with the lemma and palea removed germinated equally well [15]. The presence vs. absence of chemical inhibition of hulls on germination may vary among and within species [16,24]. Huang *et al.* [31], Ma *et al.* [18], and He [32] suggested that the endosperm was responsible for seed dormancy in *Leymus racemosus* and *L. chinensis*, respectively. However, this obviously was not the case in *S. bungeana* since almost all fresh intact caryopses (with hulls removed) germinated.

Scarification with acids or alkalis has been shown to be effective in breaking seed dormancy in grasses [24,25], and scarification with NaOH significantly increased germination of *S. bungeana* seeds. This may be due to damage to the hulls, thus decreasing their mechanical resistance to germination. Also, an increase in permeability of embryo covering tissues after treatment with NaOH may favor leaching of germination inhibitors from them. A combination of NaOH soaking and exogenous GA$_3$ completely broke seed dormancy in *Leymus chinensis* [32].

ABA plays a role in the induction and maintenance of seed dormancy, whereas gibberellins (GAs) are associated with dormancy breaking and germination [33–35]. GA$_3$ has been reported to stimulate germination of seeds of many grass species [36], and it does so by increasing the growth potential of the

embryo [25,33]. The small but significant increase in germination percentage of *S. bungeana* seeds by GA$_3$ indicates that this plant growth regulator increased the growth potential of embryos in only some of the seeds to the point where the embryo overcame the mechanical restriction of its covering layers. The failure of fluridone, to promote germination suggests that ABA biosynthesis during imbibition may not be the primary cause of dormancy in *S. bungeana* seeds. Gianinetti & Vernieri [37] concluded that ABA is not the primary mediator of dormancy in imbibed rice seeds. However, a correlation between embryonic ABA level, sensitivity to ABA and hull imposed dormancy was found in barley [23].

Potassium nitrate is used extensively to break grass seed dormancy under laboratory conditions [38]. It alleviated the light inhibition of germination of *S. bungeana* seeds [27] and significantly increased germination of fresh seeds of this species (this study). Nitrate is one of the most ubiquitous inorganic ions in soils [39], and it can release seed dormancy and stimulate germination [9,25,36]. The much better regeneration of *Plantago lanceolata* in gaps than in closed vegetation may be attributed to higher nitrate level in gaps (0.2–1.1 mM) than that in closed vegetation (0.1 mM) [40]. Peaks in seedling emergence of *Capsella bursa-pastoris* in summer were correlated with increases in nitrate level in the soil [41]. Sensitivity of *Arabidopsis thaliana* seed to nitrate level at 20°C was highest in summer-early autumn when

Table 3. Two way ANOVA of the effects of burial time, burial depth and their interaction on seed viability and seed dormancy of *Stipa bungeana* (n = 6).

Source	Sum of Squares	df	F	P-value
Seed viability				
Burial time (BT)	24348	7	131.4	.000
Burial depth (BD)	1226	1	46.3	.000
BT * BD	638	7	3.4	.003
Seed dormancy				
Burial time (BT)	800	7	28.8	.000
Burial depth (BD)	224	1	56.3	.000
BT * BD	234	7	8.4	.001

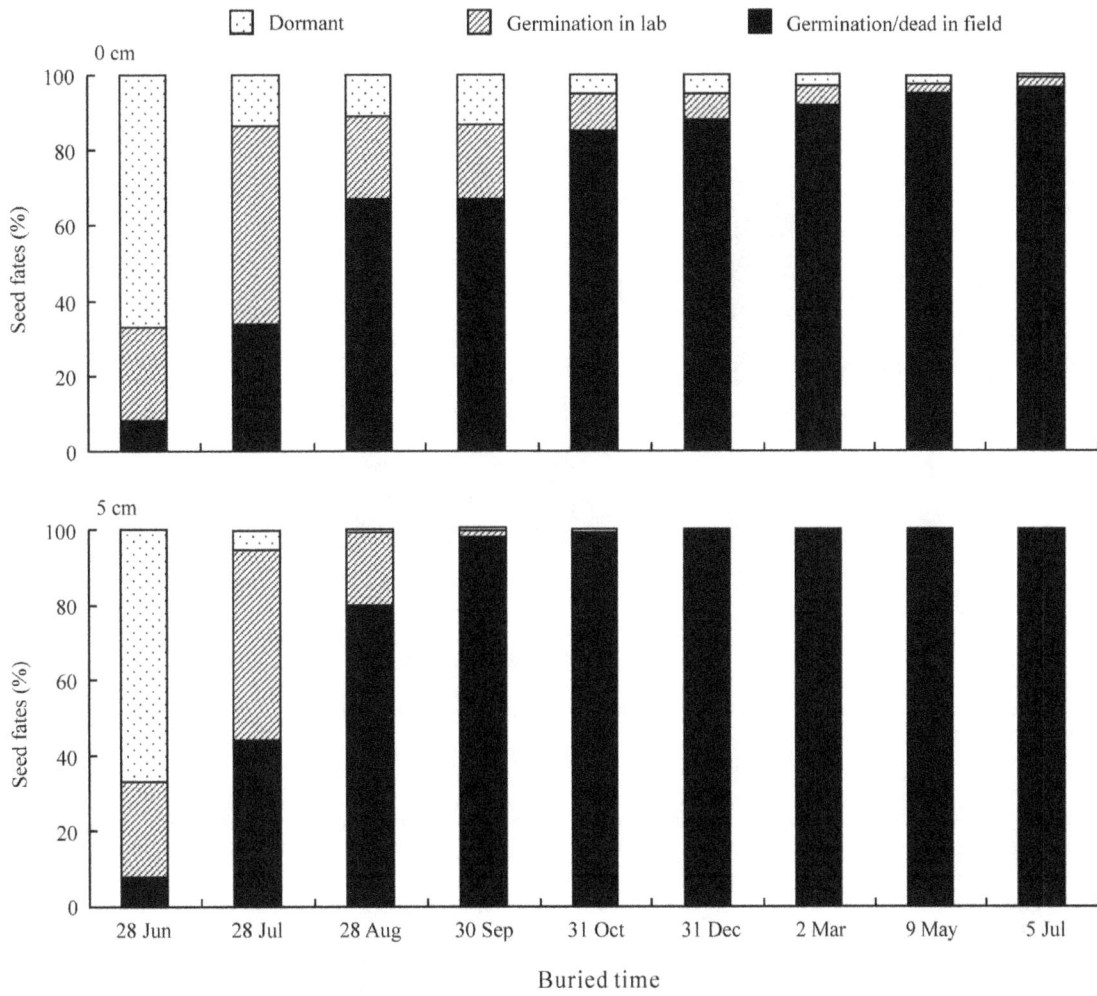

Figure 6. Fates of seeds in the field at 0 (surface) and 5 cm soil depths (n = 6).

dormancy level in the seeds was low and expression of nitrate transporter 1.1 (*NRT1.1*) and nitrate reductase 1 (*NR1*) genes was highest [42].

Seed afterripening can be characterized by dormancy release and an increase of germination speed during dry storage [9,43,44]. Seeds of many grass species come out of dormancy during dry storage [16,25,36]. Dry storage for 4 months significantly increased germination percentage of *Stipa tenacissima* seeds [16], and the germination percentage of *S. bungeana* seeds increased significantly during dry storage for 1, 3 and 6 months at 5°C and 20°C. Effect of storage temperature on germination varies with the species [45]. A positive relationship between dormancy release rate and afterripening temperature has been found in seeds of some grass species [46–48]. Rate of afterripening of *Bromus tectorum* seeds was approximately the same at 20°C and 30°C [49]. Overall, however, storage temperature had only a small effect on germination of *S. bungeana* seeds, which germinated to 63% and 69% at 20°C in dark after dry storage for 6 months at 5°C and 20°C, respectively.

In sum, fresh seeds of *S. bungeana* have hull-imposed dormancy, and it can be released completely by removal of the hulls and in part by GA_3, KNO_3, NaOH scarification and dry storage. These results suggest that seeds of *S. bungeana* have non-deep physiological dormancy [8], as have been reported for seeds of many other species of grasses [9].

Effect of seed dormancy on seedling emergence

On the Chinese Loess Plateau, caryopses of *S. bungeana* normally mature and are dispersed by the end of June (before rainy season). It is expected that seeds will germinate and seedlings become established in summer and autumn due to suitable temperature and precipitation for them to do so from July to September. Indeed, seedlings recruit mainly from July to September [7,50]. Although they were dormant at the time of burial, and most seeds of *S. bungeana* that germinated from July to September (Fig. 6), and most of them were depleted from the soil during August and September (Fig. 7). There was no significant difference in the final percentage of seedling emergence in the field between stored and fresh seeds sown at 0 cm and those buried 1 cm (Table 2).

Although in final emergence percentage of stored and fresh seeds sown in the field did not differ, fresh seeds germinated much slower than stored seeds, indicating that primary dormancy delayed germination. The timing of germination has pronounced effects on subsequent survival and phenology of seedling, and, in turn, affect the reproductive output of adult plants [9,14]. Water stress is one of the most important factors limiting seedling establishment in arid and semi-arid area. The long term (1957–2009) data showed that the humid index of study area ranked as September > August > July. Thus, seeds germinated in August or September will have an advantage for seedling growth and

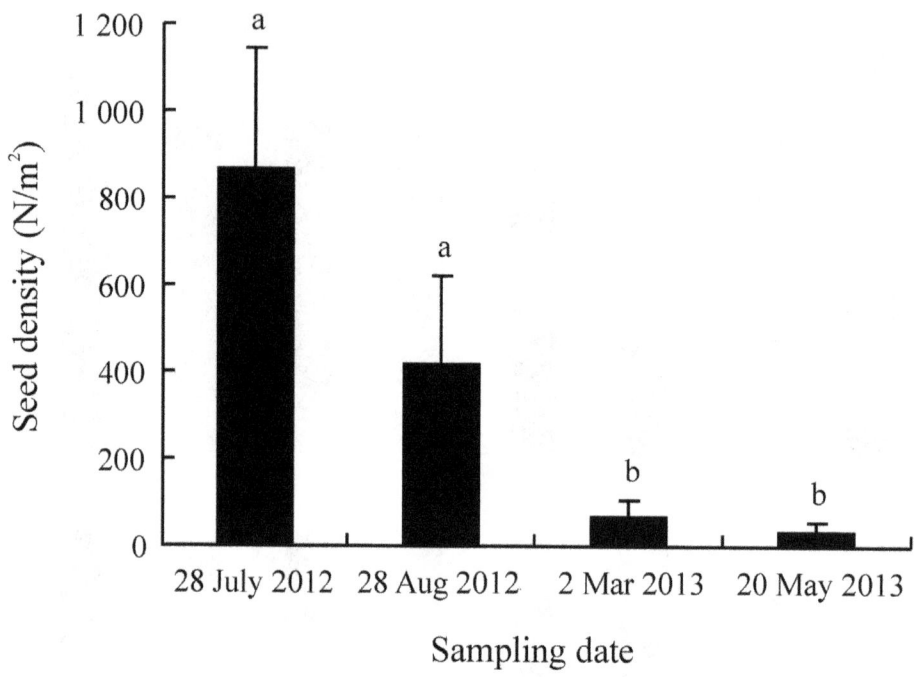

Figure 7. Seed density on different sampling dates in a *S. bungeana*-dominated grassland. Different letters indicate significant difference (P<0.05) among sampling dates (n = 16).

development with less water stress. However, severe winter conditions can over-ride this advantage through increasing seedling mortality. In present study, fresh seeds sown at the soil surface mostly emerged during 4–6 weeks after sown which corresponding the late July to mid-August after seed dispersal. However, stored seeds mostly emerged during 2–3 weeks after sown which corresponding the mid-July after seed dispersal. Thus, the delay in germination by seed dormancy of *S. bungeana* seems to be a compromise between water stress avoidance and seedling overwinter. This may provide an adaptive advantage of *S. bungeana* in arid environment.

Timing of seed germination largely depends on dormancy release which regulated by various environmental factors [9]. Dormancy of fresh seeds of *S. bungeana* buried in the soil was released more quickly than those on the soil surface (Fig. 6). There are at least three possible reasons for this difference. One is that nitrate in the soil may promote dormancy release (see above). Second, seeds were exposed to a higher level of hydration than those on the soil surface, which via microbial decay activity might have decreased mechanical resistance of the hull to embryo growth and thus germination. Third, seeds had high moisture content, which favored seed afterripening. Increasing the moisture content of rice seeds from 8% to 11% resulted in a 2.5-fold reduction in the storage period required for a given level of germination [51]. Seed dormancy release speed of *Lolium rigidum* increased as seed water content increased from 6% to 18% [48]. Increased dormancy release by burial may have ecological significance in that the seeds moisture conditions beneath the soil surface are more suitable for germination and subsequent seedling establishment than they are on the surface [7,52].

Effect of seed burial depth on seedling emergence and soil seed bank

The vertical distribution of a seed in the soil plays an important role in determining whether it remains dormant, germinates or

dies [14]. Emergence percentage and speed (Table 2) were significantly higher for both fresh and stored seeds of *S. bungeana* buried 1 cm deep than they were for seeds sown on the soil surface (Table 2), probably due to the negative effect of light and water stress on germination [7]. Further, 16% seeds on the soil surface were dormant after one month of burial, but in only 5% of those buried at 5 cm (Fig. 6). Thus, buried seeds germinated to a higher percentage and speed than those on the soil surface.

All seeds buried 5 cm in soil and 88% of those on the soil surface on 28 June had either germinated or died by 31 December (Fig. 6). Further, the number of seeds in the soil seed bank decreased from 869 m^{-2} to 31 seeds m^{-2} between July and May (Fig. 7). Thus, our study indicates that only a small number of seeds produced by *S. bungeana* may have the potential to form a persistent seed bank, i.e. remain viable for ≥1 year in/on soil. Hou [50] reported that *S. bungeana* had a transient seed bank when seeds were hand-buried in nylon bags. However, *S. bungeana* formed a natural persistent seed bank in a typical prairie [53] and on eroded slopes [54], respectively, on the Loess Plateau.

Practical implications

Fresh *S. bungeana* seeds exhibit primary dormancy, and thus seed pretreatment to release dormancy is important for quick establishment of *S. bungeana* plants. Although most seeds will germinate in 4–6 weeks after dispersal if soil moisture is suitable for them to do so, we recommended sowing one-year-stored (after-ripened) seeds 1 cm deep to attain a uniform stand. An option is to pretreat fresh seeds with a 30% NaOH solution for 60 min before sowing them. Trampling by livestock during a short-term grazing period in early July immediately following seed dispersal may promote seed burial in the soil, which promotes germination. However, disturbance, such as by grazing, during the stand establishment period should be avoided in order to prevent seedlings from being injured or destroyed.

Author Contributions

Conceived and designed the experiments: XWH YRW. Performed the experiments: YPW XYD RZ. Analyzed the data: YPW XWH JMB CCB. Contributed reagents/materials/analysis tools: YRW XWH. Contributed to the writing of the manuscript: XWH JMB CCB.

References

1. Zheng FL (2006) Effect of vegetation changes on soil erosion on the Loess Plateau. Pedosphere 16: 420–427.
2. Mensching HG (1986) Desertification in Europe? A critical comment with examples from Mediterranean Europe. In: Fantechi R, NS Margaris NS, editors. Desertification in Europe. Dordrecht, Holland: Reidel Publishing Company. 3–8 p.
3. Francis CF, Thornes JB (1990) Matorral: erosion and reclamation. In: Albaladejo J, Stocking MA, Díaz E (editors) Soil degradation and rehabilitation in Mediterranean environmental conditions. Murcia: Consejo Superior de Investigaciones Científicas. 87–116 p.
4. Morgan RPC, Rickson RJ, Wright E (1990) Regeneration of degraded soils. In: Albaladejo J, Stocking MA, Díaz E, editors. Soil degradation and rehabilitation in Mediterranean environmental conditions. Murcia: Consejo Superior de Investigaciones Científicas. 69–85 p.
5. Albaladejo J, Castillo V, Roldán A (1996) Rehabilitation of degraded soils by water erosion in semiarid environments. In: Rubio JL, Calvo A, editor. Soil degradation and desertification in Mediterranean environments. Logrono, Spain: Geoforma Ediciones. 265–278 p.
6. Cheng J, Hu TM, Cheng JM, Wu GL (2010) Distribution of biomass and diversity of Stipa bungeana community to climatic factors in the Loess Plateau of northwestern China. Afr J Biotechnol 9: 6733–6737.
7. Hu XW, Zhou ZQ, Li TS, Wu YP, Wang YR (2013) Environmental factors controlling seed germination and seedling recruitment of Stipa bungeana on the Loess Plateau of northwestern China. Ecol Res 28: 801–809.
8. Baskin JM, Baskin CC (2004) A classification system for seed dormancy. Seed Sci Res 14: 1–16.
9. Baskin CC, Baskin JM (2014) Seeds: ecology, biogeography and evolution of dormancy and germination. Second edition. San Diego: Elsevier/Academic Press.
10. Koller D (1969) The physiology of dormancy and survival of plants in desert environments. In: Woolhouse HW, editor. Dormancy and Survival. Symposium of the Society for Experimental Biology 23: 449–469.
11. Zhou ZQ, Li TS, Wu YP, Hu XW (2013) A study of optimum germination condition of Stipa bungeana seeds. Pratacultural Science, 30: 218–222.
12. Baskin CC, Thompson K, Baskin JM (2006) Mistakes in germination ecology and how to avoid them. Seed Sci Res 16: 165–168.
13. Wang YR, Zhang JQ, Liu HX, Hu XW (2004). Physiological and ecological responses of lucerne and milkvetch seed to PEG priming. Acta Ecologia Sinica 24: 402–408.
14. Fenner M, Thompson K (2005) The ecology of seeds. Cambridge, UK: Cambridge University Press.
15. Fulbright TE, Redente EF, Wilson AM (1983) Germination requirements of green needle-grass (Stipa viridula Trin.). J Range Manage 36: 390–394.
16. Gasque M, Garcia-Fayos P (2003) Seed dormancy and longevity in Stipa tenacissima (L.) Poaceae. Plant Ecol 168: 279–290.
17. Zhu YJ, Dong M, Huang ZY (2007) Caryopsis germination and seedling emergence in an inland dune dominant grass Leymus secalinus. Flora 202: 249–257.
18. Ma HY, Liang ZW, Wang ZC (2008) Lemmas and endosperms significantly inhibited germination of Leymus chinensis (Trin.) Tzvel. (Poaceae). J Arid Environ 72: 573–578.
19. Ma HY, Liang ZW, Wu HT, Huang LH, Wang ZC (2010) Role of endogenous hormones, glumes, endosperm and temperature on germination of Leymus chinensis (Poaceae) seeds during development. J Plant Ecol 3: 269–277.
20. He XQ, Hu XW, Wang YR (2010) Study on seed dormancy mechanism and breaking technique of Leymus chinensis. Acta Botanica Boreali -Occidentalia Sinica 30: 120–125.
21. Gutterman Y, Corbineau F, Côme D (1996) Dormancy of Hordeum spontaneum caryopses from a population on the Negev Desert Highlands. J Arid Environ 3: 337–345.
22. Zhang FC, Gutterman Y (2003) The trade-off between breaking of dormancy of caryopses and revival ability of young seedlings of wild barley (Hordeum spontaneum). Can J Bot 81: 375–382.
23. Benech-Arnold RL, Giallorenzi MC, Frank J, Rodriguez V (1999) Termination of hull-imposed dormancy in developing barley grains is correlated with changes in embryonic ABA levels and sensitivity. Seed Sci Res 9: 39–47.
24. Simpson GM (1990) Seed dormancy in grass. Cambridge: Cambridge University Press.
25. Adkins SW, Bellairs SM, Loch DS (2002) Seed dormancy mechanisms in warm season grass species. Euphytica 126: 13–20.
26. Yi J, Li QF, Tian RH (1997) Seed dormancy and hormone control of germination in Leymus. Acta Agrestia Sinica 5: 93–100.
27. Hu XW, Huang XW, Wang YR (2012) Hormonal and temperature regulation of seed dormancy and germination in Leymus chinensis. Plant Growth Regul 67: 199–207.
28. Lavie D, Levy EC, Cohen A, Evenari M, Gutterman Y (1974) New germination inhibitor from Aegilops ovata L. Nature 249: 388.
29. Gutterman Y, Evenari M, Cooper R, Levy EC, Lavie D (1980) Germination inhibition activity of a naturally occurring lignin from Aegilops ovata L. in green and infrared light. Experientia 36: 662–663.
30. Gutterman Y (2002) Survival strategies of annual desert plants. Berlin: Springer-Verlag.
31. Huang ZY, Dong M, Gutterman Y (2004) Caryopses dormancy, germination and seedling emergence in sand, of Leymus racemosus (Poaceae), a perennial sand dune grass inhabiting the Junggar Basin of Xinjiang, China. Aust J Bot 52: 519–528.
32. He XQ (2011) Seed coat permeability and location of semipermeable layer in seeds of several grass species. Ph.D. Thesis, Lanzhou University, Lanzhou.
33. Kucera B, Cohn MA, Leubner-Metzger G (2005) Plant hormone interactions during seed dormancy release and germination. Seed Sci Res 15: 281–307.
34. Graeber K, Nakabayashi K, Miatton E, Leubner-Metzger G, Soppe WJJ (2012) Molecular mechanisms of seed dormancy. Plant Cell Environ 35: 1769–1786.
35. Linkies A, Leubner-Metzger G (2012) Beyond gibberellins and abscisic acid: how ethylene and jasmonates control seed germination. Plant Cell Rep 31: 253–270.
36. Baskin CC, Baskin JM (1998) Ecology of seed dormancy and germination in grasses. In: Cheplick GP, editor. Population biology of grasses. Cambridge, UK: Cambridge University Press. 30–83 p.
37. Gianinetti A, Vernieri P (2007) On the role of abscisic acid in seed dormancy of red rice. J Exp Bot 58: 3449–3462.
38. International Seed Testing Association (2012) International rules for seed testing. Seed Sci Technol 27 (Supplement).
39. Brady NC, Weil RR (2008) The nature and properties of soils. 14th Edition. Upper Saddle River: Prentice Hall.
40. Pons TL (1989) Breaking of seed dormancy by nitrate as a gap detection mechanism. Ann Bot 63: 139–143.
41. Popay AI, Roberts EH (1970) Ecology of Capsella bursa-pastoris (L.) Medik and Senecio vulgaris L. in relation to germination behavior. J Ecol 58: 123–139.
42. Footitt S, Douterelo-Soler I, Clay H, Finch-Savage WE (2011) Dormancy cycling in Arabidopsis seeds is controlled by seasonally distinct hormone-signaling pathways. PNAS 108: 20236–20241.
43. Finch-Savage Leubner-Metzger (2006) Seed dormancy and the control of germination. New Phytol 171: 501–23.
44. Finkelstein R, Reeves W, Ariizumi T, Steber C (2008) Molecular aspects of seed dormancy. Ann Rev Plant Biol 59: 387–415.
45. Liu K, Baskin JM, Baskin CC, Bu HY, Liu MX, et al (2010) Effect of storage conditions on germination of seeds of 489 species from high elevation grasslands of the eastern Tibet Plateau and some implications for climate change. Am J Bot 98: 12–19.
46. Roberts EH (1962) Dormancy in rice seed. III. The influence of temperature, moisture and gaseous environment. J Exp Bot 13: 75–94. editor.
47. Probert RJ (2000) The role of temperature in the regulation of seed dormancy and germination. In: Fenner M, editor. Seeds: the ecology of regeneration in plant communities. Wallingford: CAB International. 261–292 p.
48. Steadman KJ, Crawford AD, Gallagher RS (2003) Dormancy release in Lolium rigidum seeds is a function of thermal after-ripening time and seed water content. Funct Plant Biol 30: 345–352.
49. Bair NB, Meyer SE, Allen PS (2006) A hydrothermal after-ripening time model for seed dormancy loss in Bromus tectorum L. Seed Sci Res 16: 17–28.
50. Hou JW (2009) Effects of fungicide seed treatment on soil seed bank under various conditions. Ph.D. Thesis, Lanzhou University, Lanzhou.
51. Ellis RH, Hong TD, Roberts EH (1983) Procedure for the safe removal of dormancy from rice seed. Seed Sci Technol 11: 77–112.
52. Thanos CA, Georghiou K, Douma DJ, Marangaki CJ (1991) Photoinhibition of seed germination in Mediterranean maritime plants. Ann Bot 68: 469–475.
53. Zhao LH, Cheng JM, Wan HE (2008) Dynamic analysis of the soil seed bank in typical prairie on the Loess Plateau. Bulletin of Soil and Water Conservation 28: 60–66.
54. Wang N, Jiao JY, Jia YF, Wang DL (2011) Seed persistence in the soil on eroded slopes in the hilly-gullied Loess Plateau region, China. Seed Sci Res 21: 295–304.

15

Coryphoid Palm Leaf Fossils from the Maastrichtian–Danian of Central India with Remarks on Phytogeography of the Coryphoideae (Arecaceae)

Rashmi Srivastava[1]*, Gaurav Srivastava[1], David L. Dilcher[2]

1 Cenozoic Palaeoflorist Laboratory, Birbal Sahni Institute of Palaeobotany, 53 University Road, Lucknow- 226 007, Uttar Pradesh, India, **2** Department of Geology, Indiana University, 1001 E. Tenth St. Bloomington- 47405, Indiana, United States of America

Abstract

Premise of research: A large number of fossil coryphoid palm wood and fruits have been reported from the Deccan Intertrappean beds of India. We document the oldest well-preserved and very rare costapalmate palm leaves and inflorescence like structures from the same horizon.

Methodology: A number of specimens were collected from Maastrichtian–Danian sediments of the Deccan Intertrappean beds, Ghughua, near Umaria, Dindori District, Madhya Pradesh, India. The specimens are compared with modern and fossil taxa of the family Arecaceae.

Pivotal results: *Sabalites dindoriensis* sp. nov. is described based on fossil leaf specimens including basal to apical parts. These are the oldest coryphoid fossil palm leaves from India as well as, at the time of deposition, from the Gondwana-derived continents.

Conclusions: The fossil record of coryphoid palm leaves presented here and reported from the Eurasian localities suggests that this is the oldest record of coryphoid palm leaves from India and also from the Gondwana- derived continents suggesting that the coryphoid palms were well established and wide spread on both northern and southern hemispheres by the Maastrichtian–Danian. The coryphoid palms probably dispersed into India from Europe via Africa during the latest Cretaceous long before the Indian Plate collided with the Eurasian Plate.

Editor: Qi Wang, Institute of Botany, China

Funding: These authors have no support or funding to report.

Competing Interests: The authors have declared that no competing interests exist.

* Email: rashmi57.bsip@gmail.com

Introduction

Palms (Arecaceae/Palmae) are considered an important and characteristic component of tropical rainforest ecosystems having a pantropical distribution [1]. The family has been placed within the commelinid clade of the monocotyledons [2,3], and is composed of five subfamilies: Arecoideae, Calamoideae, Ceroxyloideae, Coryphoideae and Nypoideae [4,5]. The family comprises 188 genera and about 2600 species [5,6,7]. Palm species richness is the highest in tropical Asia (>1200 species) and the higher in the Americas (730 species) than in Africa (only 65 species) [5]. It has been suggested that the low diversity of palms in Africa in contrast to Asia and America is due to Neogene aridification in Africa [8]. However, recent studies suggest *in situ* diversification in other regions like Asia and America etc [9,10]. In Indian subcontinent, palms consist of 20 genera and 88 species [5] with 24 species belonging to 9 genera endemic [11]. Among the five subfamilies of the Arecaceae, Coryphoideae is sister to a clade comprising Arecoideae and Ceroxyloideae. Asmussen et al. [12]

considered Coryphoideae as one of the earliest diverging members of Arecaceae from which both pinnate and palmate leaves may have evolved. However, Baker and Couvreur [9,10] on the basis of molecular data suggest that the divergence of Coryphoideae occurred at about 87 Ma (95% HPD 86–88) in Laurasia in which *Sabalites carolinensis* Berry described from the late Coniacian–early Santonian (85.8–83.5 Ma) of South Carolina, USA was used as a calibration point [13]. Kvaček and Herman [14] recorded *S. longirachis* Kvaček and Herman from the early Campanian of Austria. A large number of fossil records attributed to Coryphoideae in the form of fruit and wood are also reported from the Deccan Intertrappean sediments. These are: *Hyphaeneocarpon indicum* Bande, Prakash and Ambwani [15], *Palmocarpon coryphoidium* Shete and Kulkarni [16], *Palmocaulon costapalmatum* Kulkarni and Patil [17], *P. hyphaeneoides* Shete and Kulkarni [18], *Palmoxylon coryphoides* Ambwani and Mehrotra [19] and *P. hyphaeneoides* Rao and Shete [20].

Here we report very rare and well-preserved costapalmate palm leaves under the organ genus *Sabalites* (*S. dindoriensis* sp. nov.)

from the Deccan Intertrappean sediments (Maastrichtian–Danian) of Central India. This is the oldest fossil record of costapalmate palm leaves from India and the Gondwana- derived continents. The locality bearing the fossils was situated in a low palaeolatitude ~18.09° S near the equator (Fig. 1) [21] when the leaves were deposited. Attempts have also been made to discuss the origin and phytogeography of the subfamily Coryphoideae in Indian context.

The Deccan Traps: a brief review

The Deccan Traps (Continental Flood Basalt) are one of the largest igneous provinces of the world. The area occupied by the Deccan Traps today is about 500,000 sq km in peninsular India which includes Andhra Pradesh, Gujarat, Karnataka, Madhya Pradesh and Maharashtra (Fig. 2A). The original stretch may have been over 1.5 million sq km including sediments found in the Arabian Sea to the west of Mumbai [22]. The outpouring of magma/lava was associated with the northward voyage of the Indian Plate after it was separated from Gondwana during the Early Cretaceous and moved over the Reunion Hot Spot situated east of Madagascar in the Indian Ocean [23,24]. The extensive volcanic eruptions with associated magma and lava outpouring that formed the Deccan Traps and associated sedimentary beds has been difficult to date and thus is an active topic of discussion among geologists and palaeontologists. Recent studies based on $^{40}Ar/^{39}Ar$ dating indicate that the duration of the volcanism extended from 69–61 Ma and the major eruptions took place between 67–65 Ma [25,26] rather than a short duration of only one million years [27].

The sedimentary sequences between two successive magma/lava flows were deposited in lacustrine, fluviatile and palustrine environments during quiescent (inactive) phases of volcanic activity mainly while the Indian Plate was still an isolated land mass moving toward Asia. These are repeated episodic events resulting in the multiple sequences of fossiliferous beds and basalts. The fossiliferous sediments of the intertrappean beds are exposed mainly in Central India, western India and to the south in parts of Andhra Pradesh and Karnataka, including Rajahmundry (Fig. 2A). The age of the Deccan Intertrappean beds was previously thought to be early Palaeogene due to the abundance of angiospermous remains [28,29]. However, microfloral studies

and faunal assemblages suggest a Maastrichtian age for most of the intertrappean exposures but there are a few Palaeocene indicators [30–33]. Currently, based on radiometric dating and planktonic foraminifera, the age of the intertrappean sediments is considered to be upper Maastrichtian–Danian [34–37].

Recent studies of the sedimentary sequences associated with the Deccan Traps (both Infratrappean and Intertrappean) have been conducted to resolve their role in mass extinction at the $K–Pg$ boundary. Cripps et al. [38] working on Mumbai Intertrappeans concluded that volcanic activity had hardly any effect on the floristic elements. However, pollen analysis shows distinct floral changes at different stratigraphic levels [32,33].

The flora reported from Deccan Intertrappean beds is unique and one of the richest fossil plant assemblages in India. The fossil plant assemblage includes all the plant groups ranging from algae to angiosperms [39]. Most of the fossil flora (mainly angiosperms) from intertrappean beds is reported from Central India (Madhya Pradesh and Maharashtra) with only a few elements of the flora reported from western India [40–42]. The majority of the plant macrofossils reported from the Deccan Traps are permineralized woods, fruits with only a few leaf impressions [39], however, microfossils have also been recorded [33].

Materials and Methods

The fossil palm leaves were collected from Umaria near Ghughua (23° 7' N; 80° 37' E), in the premises of Ghughua Fossil National Park, Dindori District, Madhya Pradesh. The fossil site is situated about 76 km east of Jabalpur and spreads over an area of 27.34 ha in Ghughua and Umaria villages (Fig. 2B). The locality is very rich in permineralized angiospermous woods (both palms and eudicots), but leaves and other plant organs are rarely preserved and thus very rarely found. The studied fossil leaf specimens were first cleaned with a chisel and hammer and then photographed in natural low angled light using a 10 megapixel digital camera (Canon SX110). All the figured fossil specimens (Specimen nos 40073–40077) are housed in the repository of Birbal Sahni Institute of Palaeobotany, Lucknow, India. The fossil leaves were compared with the nearest living relatives in the herbaria of the Central National Herbarium, Howrah, Forest Research Institute, Dehradun, National Botanical Research

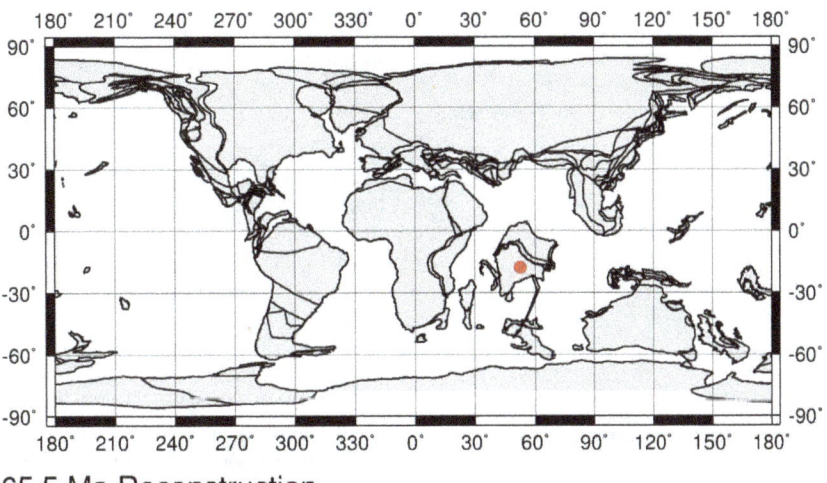

65.5 Ma Reconstruction

Figure 1. Palaeocontinental map showing the position of India and fossil locality (red dot) at 65.5 Ma [21].

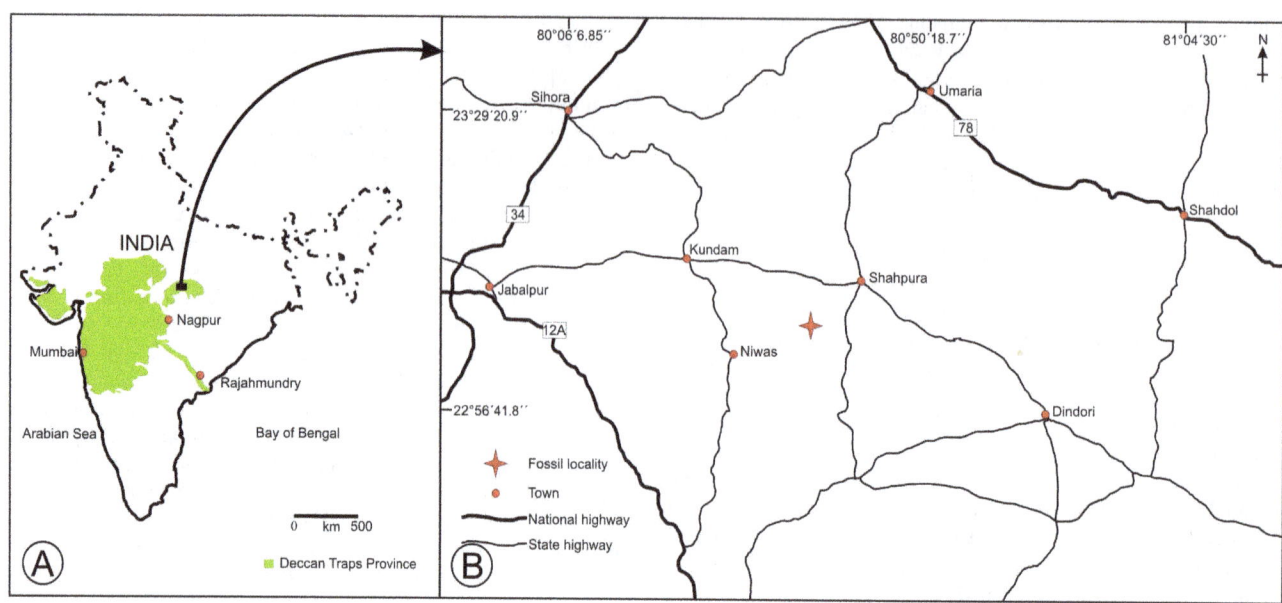

Figure 2. Map of India showing fossil locality. A. Map of India showing extent of Deccan traps. B. High resolution map showing the fossil locality (marked by asterisk) [95].

Institute, Lucknow and the website of Royal Botanic Gardens, Kew. Attempts were made to extract pollen from the floral axis but it could not be recovered. The Director of Birbal Sahni Institute of Palaeobotany, Lucknow has permitted to publish the present work (Ref. No. BSIP/RDCC/Publication no. 22).

Read and Hickey [43] gave five basic characters of palm leaves that can be used alone, or in various combinations to differentiate fossil palm leaves. We followed their classification and placed our specimens in the genus *Sabalites* G. Saporta [44] emended Read and Hickey [43], which they proposed for costapalmate fossil palm leaves.

Results

Family. Arecaceae Schultz Sch.

Subfamily. Coryphoideae Burnett

Genus. *Sabalites* G. Saporta emended Read and Hickey

Species. *Sabalites dindoriensis* R. Srivastava, G. Srivastava and D. L. Dilcher, sp. nov.

Etymology. The specific epithet is named after the fossil locality.

Holotype. BSIP Museum No. 40073, Fig. 3A; designated here.

Paratypes. BSIP Museum nos 40074, 40075, 40076, 40077.

Horizon. Deccan Intertrappean Beds.

Type locality. Umaria near Ghughua Fossil National Park, Dindori District, Madhya Pradesh, India.

Age. Maastrichtian–Danian.

Diagnosis

Leaves costapalmate. Costa/petiole very thick at the basal portion and gradually tapers towards apex, petiole robust, unarmed; a number of longitudinal fibre like structures seen on the petiole/costa. Leaf segments plicate, emerging at an acute angle from costa; fused at emerging point. Mid-veins of each segment thick; two orders of veins on either side of mid-vein,

segments near the petiole narrow becoming broader away from the petiole; transverse veins rarely preserved, very fine, perpendicular or obliquely oriented to parallel veins.

Description. The species is described based on the five specimens shown in Figures 3–5. One is the basal part having a thick petiole (Fig. 3A), two specimens (Figs 3C, 4C) are the middle-upper part. Apical portions of two specimens (Figs 4A, 5A) have faint impressions of axis bearing flower. The leaf segments are preserved only near the costa, where they are attached, so the complete size and shape of an entire leaf is uncertain.

Basal part: Holotype- BSIP Museum No. 40073, Fig. 3A, 3B.

The preserved length of the specimen is about 45 cm and width about 13.5 cm, petiole with costa is about 26 cm long and 4.2 cm broad at the base that gradually tapers towards the apex and shows attached plicate leaf segments, petiole armature not seen. There are numerous longitudinal fibre like structures present on the petiole and costa. About 15 leaf segments arise from the distal portion of the costa and are crowded together. Leaf segments broaden away from the costa, measuring about 1.0–2.0 cm in width. The mid-veins of leaf segments are about 1–1.5 mm thick and two vein orders present on either side of the mid vein. The higher order venation is not preserved due to the coarse matrix. Transverse veins are rarely preserved but wherever visible, they are very fine and are oriented perpendicularly or obliquely to the parallel veins.

Middle part: Paratypes- BSIP Museum nos 40074 and 40075, Figs 3C, 4C.

The specimen in Figure 3C shows the middle portion with a preserved length of about 42.5 cm and a width of 14.5 cm. The costa is about 3.3 cm broad at the base and tapers to 0.7 cm distally. Leaf segments are preserved along both sides of costa but their terminal portions are broken. The segments are inserted on the costa and their maximum length is about 20 cm. The specimen illustrated in Figure 4C contains only plicate leaf segments with a few segments flattened on the rock surface. The detached segments have a maximum width of about 5.4 cm.

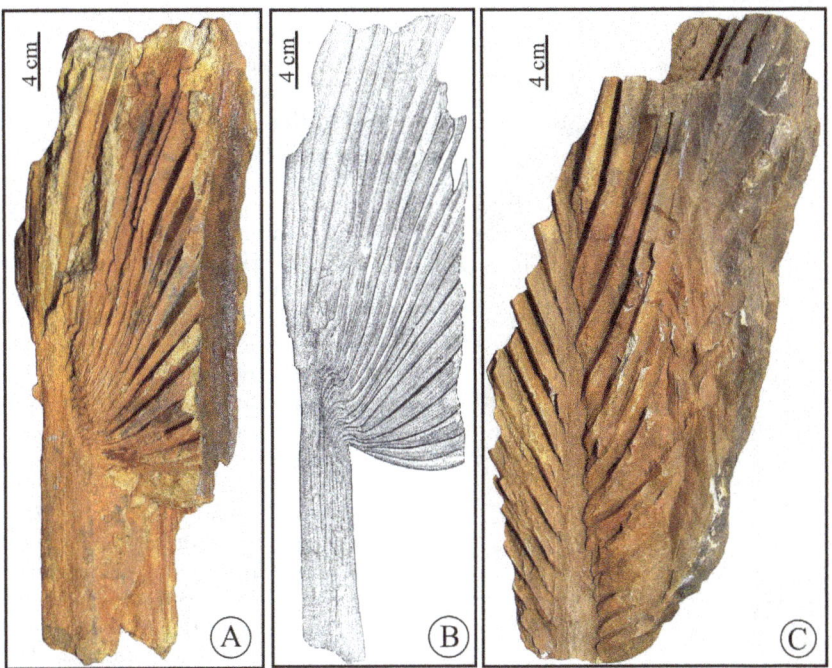

Figure 3. *Sabalites dindoriensis* **sp. nov.** A. Basal portion of *Sabalites dindoriensis* sp. nov. showing thick costa. B. Drawing of the same fossil. C Middle portion of the fossil leaf showing leaf segments attached to costa.

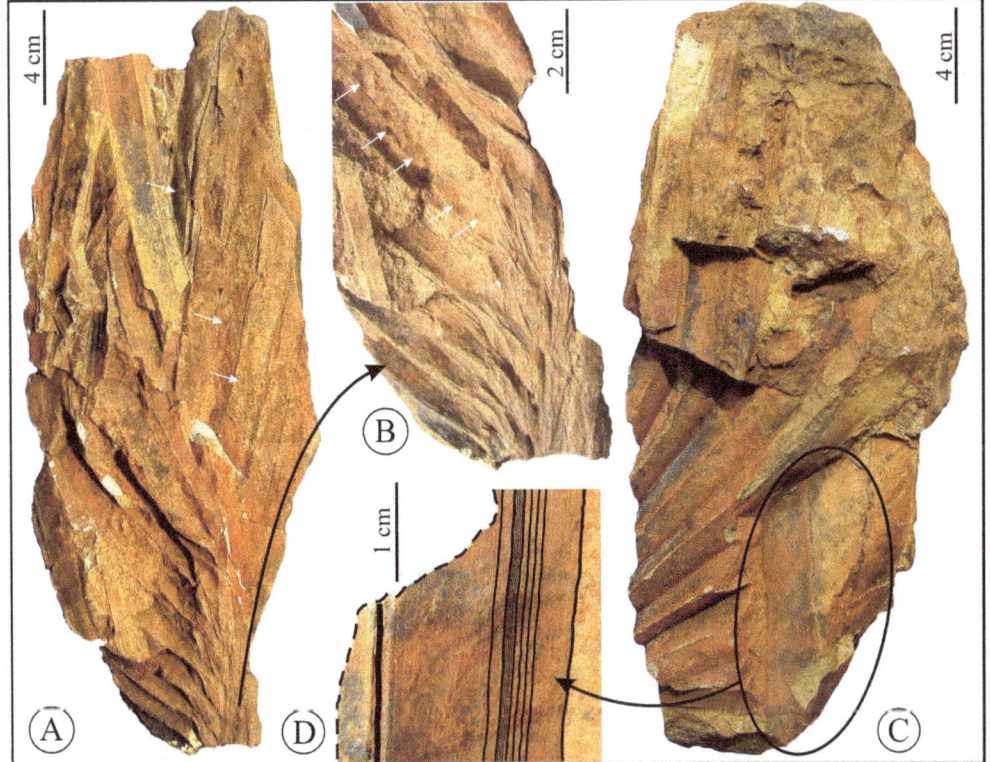

Figure 4. *Sabalites dindoriensis* **sp. nov.** A. Specimen seems to be of apical portion showing faint impressions of rachilla like structure (white arrows). B. Enlarged portion of the same specimen showing rachilla like structure (white arrows). C. Specimen seems to be of middle portion. D. Enlarged portion showing high order venation.

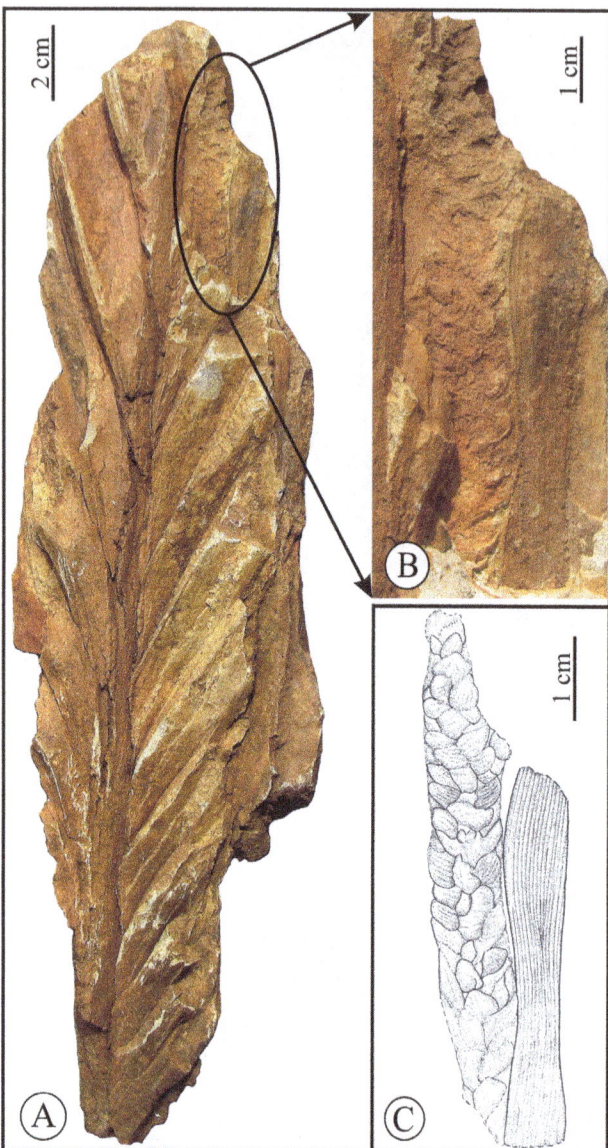

Figure 5. *Sabalites dindoriensis* **sp. nov.** A. Apical portion of the fossil leaf showing tapering costa with leaf segments having rachilla like structure. B. Enlarged portion of the axis bearing flower showing impression of spirally arranged abscised flowers, bract and spatulate rachillar bract. C. Drawing of the axis bearing flower with bract.

These leaf segments clearly show a distinct mid-vein with about 16 major parallel veins on either side of mid-vein, each with a minor vein between them (Fig. 4D).

Apical part: Paratypes: BSIP Museum nos 40076 and 40077, Figs 4A, 5A.

Specimens (BSIP No. 40076 and 40077) are incomplete and broken. The preserved length is about 30 cm and the width is 7 cm each. The costa tapers gradually from 0.9 cm to 0.3 cm along the apex (Fig. 5A). All the leaf segments are preserved incompletely near the costa; numerous 2° veins run parallel on either side of the midrib.

Also present in this specimen is a poorly preserved impression of a rachilla- like structure (Figs 5A–C) with a preserved length of 9 cm and a width of 1.4 cm. This axis is reminiscent of an

inflorescence after the spirally arranged flowers were abscised with striated bracts probably adanate to axis. A spatulate rachillar bract of preserved length 6.6 cm and width 1.4 cm is present adjacent to the floral axis having numerous parallel veins (Figs 5A–C).

Affinities. The diagnostic features of the fossil leaves include: palmate, plicate leaves with long costa (costapalmate) and unarmed petiole. These characters suggest that the fossil leaves have affinities with the subfamily Coryphoideae in the Arecaceae [5]. A number of palm taxa were examined at the Central National Herbarium, Howrah, Forest Research Institute, Dehradun, National Botanical Research Institute, Lucknow and the website of Royal Botanic Gardens, Kew [45] to find species with similar characters. The fossil leaves show resemblance with a number of coryphoid palms with costapalmate leaves in gross morphology such as *Bismarckia nobilis* Hildebr. & Wendl., *Borassus aethiopum* Mart., *B. flabellifer* L., *Corypha taliera* Roxb., *Hyphaene coriacea* Gaertn. (Fig. 6), *H. dichotoma* Furtado, *H. thebaica* Mart., *Livistona australis* Mart., *L. boninensis* Nakai, *L. carinensis* Dransf. and Uhl, *Sabal bermudana* Bailey and *Trachycarpus martianus* H. Wendle. Except *Bismarckia* and *Sabal* all taxa have armed petiole different from our fossil. The leaves of *Trachycarpus* H. Wendl. are non-costate which differentiates it from the present fossil. The inflorescence like structure of the fossil shows a close resemblance with the *Hyphaene* Gaertn. by having characteristic shape, striated bracteoles and spatulate large bract associated with floral axis which also gets support from the previous fossil records of *Hyphaene* from the same horizon [15,18,20]. However, due to the lack of spines on the petiole of fossil it cannot be assigned exactly to the modern taxa. As Read and Hickey [43] stated that "Since it is very difficult to identify specimens of modern palms accurately from their leaves alone, no attempt should be made to place fossil palm fragments in genera of modern palms unless unquestionably identifiable with them". Under these circumstances the fossils are placed in the organ genus *Sabalites* G. Saporta [44] proposed for costapalmate fossil leaves.

A number of palm leaves have been described from Upper Cretaceous–Neogene deposits of India under various fossil taxa (table 1) [46–71]. The species attributed to *Amesoneuron* (Goeppert) Read and Hickey [47–53] cannot be compared with the leaves under consideration as they are isolated fragments of lamina with parallel veins but the fragments are not attached to main rachis making it impossible to determine whether they belong to pinnate or palmate leaves. *Malpophyllum dakshinens* [58] is based on anatomical features and *Malpophyllum* sp. [58] is based on a very fragmentary specimen in which costa and other characteristic features are not preserved.

Phoenicites Brongniart species (*P. lakhanpalii* Guleria and Mehrotra [48]; *P. indica* Guleria et al. [51]; *Phoenicites* sp. [63,64]; *P. siwalikensis* Bonde [65]), palm leaf [55], *Sabalites* sp. [66,67], *Zalaccites jaintiensis* Barman and Duara [71] and the leaf of cf. *Iguanura wallichiana* Srivastava, Mehrotra and Bauer [57] are pinnate leaves and therefore differ from *Sabalites dindoriensis*.

Palmate leaf fossil taxa reported from Indian Upper Cretaceous–Neogene sediments includes *Palmacites*, a taxon for leaves lacking primary costa. Costapalmate forms referred to *Sabalites* and *Sabalophyllum* and two specimens placed in the modern genera, *Livistona* and *Trachycarpus*. The *Palmacites* species (*Palmacites* sp. [50], *P. makumensis* Srivastava, Mehrotra and Bauer [57], *P. khariensis* Lakhanpal and Guleria [61], *P. tsokarensis* Paul et al. [62]) lack a costa and therefore differ from *Sabalites dindoriensis*. *Sabalophyllum livistonoides* Bonde [68] is based on a petrified leaf with anatomical features and therefore it could not be compared with *S. dindoriensis*. *Sabalites microphylla*

Figure 6. Modern leaf of *Hyphaene coriacea* (modified after http://specimens.kew.org/herbarium/K000462899 [45].

Sahni [66] and *Sabalites* sp. [67] are fragmentary palmate leaves not attached to petiole, so it is not clear whether they are costapalmate. *Trachycarpus ladakhensis* Lakhanpal et al. [69,70] lacks primary costa. *Livistona wadiai* Lakhanpal et al. [56] is the only costapalmate leaf reported from late Eocene–Oligocene

sediments of Ladakh Himalaya, it is a smaller leaf with thin costa and narrower leaf segments.

In a detailed work on coryphoid palms from the Eocene of Texas, Daghlian [72] reported a number of fossil palm leaves under various fossil taxa, namely, *Costapalma*, *Palmacites*, *Palustrapalma*, *Sabal* and *Sabalites* based on cuticular features. But none of them show morphological resemblance with the leaves of *Sabalites dindoriensis*. Recently, Zhou et al. [73] reported many coryphoid palm leaves from the Eocene of southern China. But all the specimens are much smaller in size, with intact cuticular structures and short costa. Besides, the fossil records of palms are abundant worldwide and it is not possible to explore all of them here. Therefore, we compare *Sabalites dindoriensis* with those fossil costapalmate leaves that are most similar. *Sabal chinensis* (Endo) Huzioka and Takahashi [74] from the Eocene of Northeast China, Japan and Russia is similar to the basal part of one specimen (Fig. 3A, BSIP Museum no. 40073) in having a thick stout unarmed petiole extending into the lamina (costapalmate) and leaf segments fused at their emerging point. It differs from the present species by having a shorter costa with narrower segments. The specimens of the middle and apical portions of *Sabilites dindoriensis* (Figs 3C, 4A–D and 5A BSIP Museum nos 40074–40077) show a close resemblance to *S. longirachis* [14] reported from the early Campanian of Austria and Maastrichtian of the Pyrénées [75] in having a thick long costa and segments with a similar angle of attachment. However, this species is based on cuticular features and fine venation, none of which are preserved in *S. dindoriensis* leaves. Therefore, in the absence of any similar leaves we propose the new species, *Sabalites dindoriensis* R. Srivastava, G. Srivastava and D. L. Dilcher sp. nov.

Discussion

Modern distribution of palms and their ecology

Palms are largely distributed and diversified in tropical areas [5,76] with 90% of the species diversity restricted to tropical rainforest [1]. They are much less prominent and diverse in temperate regions [5,77,78], thus showing very restricted frost tolerance [77]. The low frost tolerance of palms is considered to be an evolutionarily conserved trait. Their architecture and more notably a crown composed of large evergreen leaves [79], which has limited frost resistance [80] and unique stem physiology doesn't allow dormancy [81]. Palms also exhibit a strong latitudinal diversity gradient [82] and need water accessibility for their survival [5,83]. Palms grow mainly under the top canopy of tropical rain forests along low hills and streams in warm and humid conditions, while a few grow in open areas. They are also dominant in coastal swamps and mangrove forests [8,84]. Studies of new world palms indicate that solar radiation as related to absolute latitude and water is the main factor that determines the richness of palms species [85]. However, the subfamily Coryphoideae is distributed in a wider range of habitats such as pantropical to warm temperate areas of the world (Fig. 7). It is also found in climatic extremes such as cold and arid regions [5].

Origin and possible migratory path of the Coryphoideae from the Northern Hemisphere to Indian subcontinent

Phylogenetic and molecular clock studies indicate that the palms originated and diversified in Laurasia around 100 Ma [1,10] while the coryphoids diverged at about 87 Ma (95% HPD 86–88) which was constrained by the calibration point from a *Sabalites* fossil [9,10] and diversified during the Late Cretaceous to Cenozoic in boreotropical regions [5,86]. The subfamily Coryphoideae includes four major clades including (1) New world thatch palm

Table 1. Fossil palm leaves from Upper Cretaceous–Neogene sediments of India.

Fossil species/References	Locality/Horizon	Age
Fossil palm leaf and stem [46]	Polgaon, DIB Nagpur	Maastrichtian
Amesoneuron borassoides Bonde [47]	Chhindwara, DIB	Maastrichtian–Danian
A. deccanensis Guleria and Mehrotra [48,50]	Seoni and Dindori, DIB; East Garo Hills; Tura Fm.	Maastrichtian–Danian; Upper Palaeocene
A. ladakhensis Mehrotra et al. [49]	Hemis Conglomerate Horizon, Ladakh	Late Eocene–Oligocene
A. lakhanpalii Mehrotra [50]	East Garo Hills; Tura Fm.	Upper Palaeocene
A. manipurensis Guleria et al. [51]	Imphal	Late Eocene
sahnii Guleria et al. [52]	Solan; Kasauli Fm.	Lower Miocene
A. siwalicus Prasad [53]	Jawalamukhi, Siwalik	Middle Miocene
Borassiod palm leaf [54]	Chhindwara, DIB	Maastrichtian–Danian
Palm leaves [55]	East Godavari, east of Rajahmundry	Late Tertiary
Livistona wadiai Lakhanpal et al. [56]	Hemis Conglomerate Horizon, Ladakh	Late Eocene–Oligocene
cf. *Iguanura wallichiana* Srivastava et al. [57]	Tinsukia, Makum; Tikak Parbat Fm.	Late Oligocene
Malpophyllum dakshinens Kumaran [58]	Chhindwara, DIB	Maastrichtian–Danian
Malpophyllum sp. [58]	Solan; Kasauli Fm.	Lower Miocene
*Nypa fruticans*Wurmb [59]	Tinsukia, Makum; Tikak Parbat Fm.	Late Oligocene
Palmacites sp. [50]	Barail; Tirap	Oligocene
Palmacites sp. [60]	Kangra, Ranital; Lower Siwalik	Middle Miocene
P. khariensis Lakhanpal and Guleria [61]	Kutch, Khari Series	Miocene
P. makumensis Srivastava et al. [57]	Tinsukia, Makum; Tikak Parbat Fm.	Late Oligocene
P. tsokarensis Paul et al. [62]	Hemis Conglomerate Horizon, Ladakh	Late Eocene–Oligocene
Phoenicites sp. [63,64]	Tura Fm., Garo Hills Laisong Fm.	Eocene Late Eocene–Oligocene
P. indica Guleria et al. [51]	Imphal	Late Eocene
P. siwalikensis Bonde [65]	Darjeeling; Middle Siwalik	Miocene
P. lakhanpalii Guleria and Mehrotra [48]	Seoni; DIB	Maastrichtian–Danian
Sabalites sp. [66,67]	Near Chakoti river, Jhelum, Kashmir;	Miocene
	Solan, Kasauli Fm.	Lower Miocene
S. microphylla Sahni [66]	Solan, Kasauli Fm.	Lower Miocene
Sabalophyllum livistonoides Bonde [68]	Nawargaon, DIB	Maastrichtian–Danian
Trachycarpus ladakhensis Lakhanpal et al. [69,70]	Ladakh, Liyan Fm.; Lower Siwalik	Miocene; Early Miocene;
Zalaccites jaintiensis Burman and Daura [71]	Khasi and Jaintia Hills	Palaeocene

DIB: Deccan Intertrappean Beds; Fm.: Formation.

clade consisting of tribes Sabaleae and Cryosophileae; (2) Syncarpous clade consisting of tribes Chuniophoeniceae, Caryoteae, Coryphae and Borasseae; (3) tribe Phoeniceae and (4) tribe Trachycarpeae [5,87]. Based on molecular phylogenetic analysis, Baker and Couvreur [9] suggested that the New world thatch palm clade diverged at 55 Ma (95% HPD 39–72) in North America. The Syncarpous clade diverged in Eurasia at 66 Ma (95% HPD 51–80). The tribe Phoeniaceae diverged from Trachycarpeae around 49 Ma (95% HPD 33–65) in Eurasia. Out of the four aforesaid clades, the syncarpous clade is the earliest diverging clade (66 Ma) that also corresponds to the age of Deccan Intertrappean beds to which our fossils belong. In syncarpous clade, the Caryoteae can be differentiated from the present fossil by having pinnate or bipinnate leaves while amongst Chuniophoeniceae, Coryphae and Borasseae the fossil probably shows near resemblance with floral axis of *Hyphaene* (Borasseae) by having the characteristic shape and striate bractioles which also corroborate with the previous fossil records of *Hyphaene* from the

same horizon [15,18,20]. In the subsequent study Baker and Couvreur [10] suggested that only one dispersal event occurred from Indian Ocean into India (including Sri Lanka) during the Miocene but the palm fossils reported from the Maastrichtian–Danian sediments of Deccan Intertrappean beds [39] opens a new dispersal route (Fig. 8).

The oldest fossil records of Coryphoideae are reported from the Northern Hemisphere, such as: *Sabalites carolinensis* [13] from late Coniacian–early Santonian of South Carolina and *Sabal bigbendense* Manchester, Lehman and Wheeler [88] from Maastrichtian of Texas, USA. *Sabalites longirhachis* [14] were reported from the lower Campanian of Austria and from the Maastrichtian of the Pyrénées [75].

The fossil records show that the continent of Africa and India possessed much richer palm flora in the past than at present. In Africa, there are definite evidences of palm pollen from the Campanian (83.5–70.6 Ma) and they became much abundant and more diverse during the Maastrichtian (70–65.5 Ma). This period

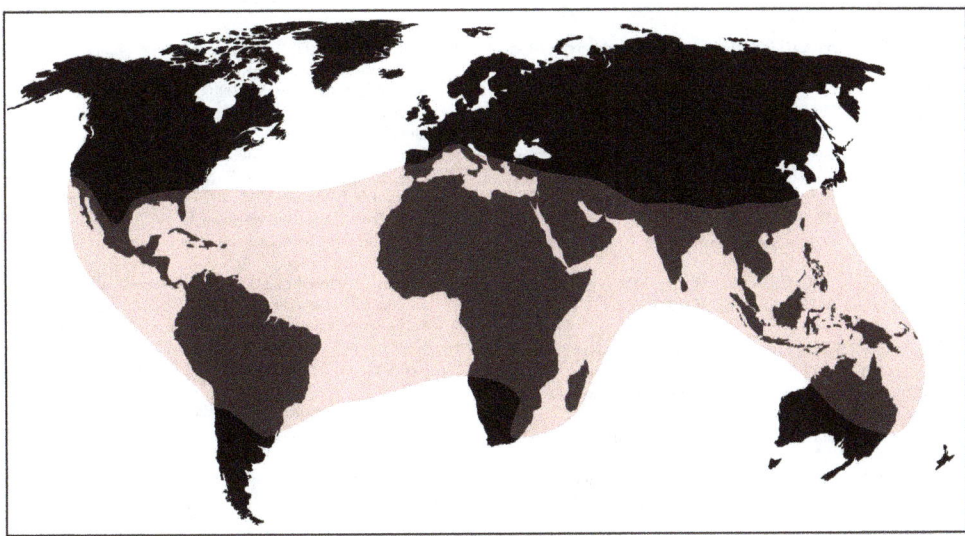

Figure 7. World map showing modern distribution of Coryphoideae [5].

is referred as 'Late Cretaceous Palm Province' [8,89]. The subfamilies such as Nypoideae and Calamoideae have been recorded from the Maastrichtian [90,91], the Coryphoideae in the form of seed has been recorded from the Danian (65.5–61.7 Ma) sediments of Egypt [92]. A large number of fossils attributed to Coryphoideae have been reported from the Deccan Intertrappean sediments such as: *Hyphaeneocarpon indicum* [15], *Palmocarpon coryphoidium* [16], *Palmocaulon costapalmatum* [17], *P. hyphae-neoides* [18], *Palmoxylon coryphoides* [19] and *P. hyphaeneoides*

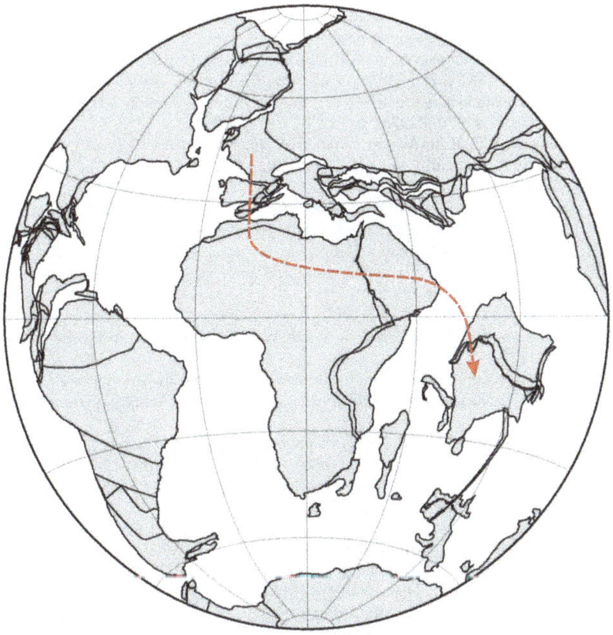

[20]. It is interesting to note that several of the *Palmoxylon* species reported from the Upper Cretaceous sediments of the Indian subcontinent [66] have also been reported from the late Eocene to early Miocene sediments of Egypt [93].

All the above fossil records and the similarity in coryphoid fossil palm records between India and Africa suggest that the coryphoid palms probably dispersed into India from Europe via Africa. During the Early Cretaceous the Indian plate was separated from the other Gondwana continents and moved northward. It collided with the Kohistan-Ladakh arc at ~85 Ma (an island like structure) re-establishing the land connection between India and Africa ~70 Ma [24]. This facilitated the interchange of various plants and also several Maastrichtian dinosaurs [24]. During the Campanian–Maastrichtian, Africa was also connected with Europe by land [94], which most likely facilitated the entry of coryphoid palms from Europe to Africa. Thus, the migration model of coryphoid palms we propose (Fig. 8) fits well with the plate tectonic models. However, with the aforesaid model, long distance oceanic dispersal cannot be ruled out [9,10]. In future better preserved palm fossils assignable to modern genera are needed to further strengthen the proposed model.

Acknowledgments

We are thankful to Prof. Sunil Bajpai, Director, Birbal Sahni Institute of Palaeobotany, Lucknow for permission to publish the paper. We are also indebted to the Directors, Botanical Survey of India, Kolkata, Forest Research Institute, Dehradun and National Botanical Research Institute, Lucknow for permission to consult their herbaria. RS thanks Dr. J. S. Guleria (Retired Scientist) of Birbal Sahni Institute of Palaeobotany, Lucknow for his help during the field work. Thanks are also due to Mr. Pawan Kumar for the line drawings. We are extremely grateful to the two anonymous reviewers and Prof. Qi Wang (Academic Editor) for their invaluable constructive comments for improving the manuscript.

Author Contributions

Conceived and designed the experiments: RS GS DLD. Performed the experiments: RS GS DLD. Analyzed the data: RS GS DLD. Contributed reagents/materials/analysis tools: RS. Wrote the paper: RS GS DLD.

Figure 8. Palaeogeographic map at 65.5 Ma [21] showing possible dispersal path of Coryphoideae from Europe to India via Africa (red broken line).

References

1. Couvreur TLP, Forest F, Baker WJ (2011) Origin and global diversification patterns of tropical rain forests: inferences from a complete genus-level phylogeny of palms. BMC Biology 9: 44.

2. Chase MW, Fay MF, Devey DS, Maurin O, Ronsted N, et al. (2006) Multigene analyses of monocot relationships: A summary. Aliso 22: 63–75.

3. Davis JI, Petersen G, Seberg O, Stevenson DW, Hardy CR, et al. (2006) Are mitochondrial genes useful for the analysis of monocot relationships?. Taxon 55: 857–870.

4. Dransfield J, Uhl NW, Asmussen CB, Baker WJ, Harley MM, et al. (2005) A new phylogenetic classification of the palm family, Arecaceae. Kew Bulletin 60: 559–569.

5. Dransfield J, Uhl NW, Asmussen CB, Baker WJ, Harley MM, et al. (2008) Genera Palmarum: The evolution and classification of palms. Kew, UK, Royal Botanic Gardens.

6. Govaerts R, Dransfield J (2005) World checklist of palms. UK, Kew: Royal Botanic Gardens.

7. Mabberley DJ (2005) The plant book, a portable dictionary of the vascular plants. Cambridge: Cambridge University Press.

8. Morley RJ (2000) Origin and evolution of tropical rain forests. Chichester: John Wiley and Sons, Ltd.

9. Baker WJ, Couvreur TLP (2013) Global biogeography and diversification of palms sheds light on the evolution of tropical lineages. I. Historical biogeography. Journal of Biogeography 40: 274–285.

10. Baker WJ, Couvreur TLP (2013) Global biogeography and diversification of palms sheds light on the evolution of tropical lineages. II. Diversification history and origin of regional assemblages. Journal of Biogeography 40: 286–298.

11. Kulkarni AR, Mulani RM (2004) Indigenous palms of India. Current Science 86 (12): 1598–1603.

12. Asmussen CB, Baker WJ, Dransfield J (2000) Phylogeny of the palm family (Arecaceae) based on *rps*16 intron and *trn*L-*trn*F plastid DNA sequences. In: Wilson K, Morrison D, (Eds), Proceedings of II International Conference on Comparitive Biology of Monocotyledons. Australia, Sydney: CSIRO. pp. 525–535.

13. Berry EW (1914) The Upper Cretaceous and Eocene floras of South Carolina and Georgia. US Geological Survey professional paper 84: 5–200.

14. Kvaček J, Herman AB (2004) Monocotyledons from Early Campanian (Cretaceous) of Grunbach, Lower Austria. Review of Palaeobotany and Palynology 128: 323–353.

15. Bande MB, Prakash U, Ambwani K (1982) A fossil palm fruit *Hyphaeneocarpon indicum* gen. et sp. nov. from the Deccan Intertrappean beds of India. Palaeobotanist 30(3): 303–309.

16. Shete RH, Kulkarni AR (1985) *Palmocarpon coryphoidium* sp. nov., a coryphoid palm fruit from the Deccan Intertrappean beds of Wardha District, Maharashtra. Journal of the Indian Botanical Society 64: 45–50.

17. Kulkarni AR, Patil KS (1977) *Palmocaulon costapalmatum*, a petrified palm leaf axis from the Deccan Intertrappean beds of Wardha District, Maharashtra. Geophytology 7(2): 208–213.

18. Shete RH, Kulkarni AR (1980) *Palmocaulon hyphaeneoides* sp. nov. from the Deccan Intertrappean beds of Wardha District, Maharashtra, India. Palaeontographica B172: 117–124.

19. Ambwani K, Mehrotra RC (1990) A new fossil palm wood from the Deccan Intertrappean bed of Shahpura, Mandla District, Madhya Pradesh. Geophytology 19(1): 70–75.

20. Rao GV, Shete RR (1989) *Palmoxylon hyphaeneoides* sp. nov. from the Deccan Intertrappean beds of Wardha District, Maharashtra. In: N V . Biradar, Proceedings of Special Indian Geophytological Conference Poona. pp. 123–128.

21. http://www.odsn.de/odsn/index.html (accessed 2013 May 19).

22. Jay AE, Widdowson M (2008) Stratigraphy, structure and volcanology of the SE Deccan continental flood basalt province: implications for eruptive extant and volumes. Journal of the Geological Society, London 165: 177–188.

23. Smith AG, Smith DG, Funnell M (1994) Atlas of Mesozoic and Cenozoic coastlines. Cambridge: Cambridge University Press.

24. Chatterjee S, Goswami A, Scotese CR (2013) The longest voyage: Tectonic, magmatic, and palaeoclimatic evolution of Indian plate during its northward flight from Gondwana to Asia. Gondwana Research 23: 238–267.

25. Sheth HC, Pande K, Bhutani R (2001) ^{40}Ar/^{39}Ar ages of Bombay trachytes: evidence for a Palaeocene phase of Deccan volcanism. Geophysical Research Letters 28: 3513–3516.

26. Chenet AL, Courtillot V, Fluteau F, Gérard M, Quidelleur X, et al. (2009) Determination of rapid Deccan eruptions across the Cretaceous–Tertiary boundary using paleomagnetic secular variation: 2. Constraints from analysis of eight new sections and synthesis for a 3500-m-thick composite section. Journal of Geophysical Research 114: B06103.

27. Hofmann C, Feraund G, Courtillot V (2000) ^{40}Ar/Ar^{40} dating of mineral separates and whole rocks from the Western Ghats lava pile: further constraints on duration and age of the Deccan traps. Earth and Planetary Science Letters 180: 13–27.

28. Sahni B (1934) The Deccan Traps: Are they Cretaceous or Tertiary? Current Science 3: 134–136.

29. Bande MB (1992) The Palaeogene vegetation of Peninsular India (megafossil evidences). Palaeobotanist 40: 275–284.

30. Kar RK, Srinivasan S (1988) Late Cretaceous palynofossils from the Deccan Intertrappean beds of Mohgaon-Kalan, Chhindwara District, Madhya Pradesh. Geophytology 27: 17–22.

31. Khosla A, Sahni A (2003) Biodiversity during the Deccan volcanic eruptive episode. Journal of Asian Earth Sciences 21: 895–908.

32. Samant B, Mohabey DM (2005) Response of flora to Deccan volcanism: A case study from Nand-Dongargaon basin of Maharashtra, implications to environment and climate. Gondwana Geological Magazine Special Publication 8: 151–164

33. Samant B, Mohabey DM (2009) Palynoflora from Deccan volcano-sedimentary sequence (Cretaceous-Palaeogene transition) of central India: implications for spatio-temporal correlation. Journal of Biosciences 34 (5): 811–823

34. Venkatesan TR, Pande K, Gopalan V (1993) Did Deccan volcanism predate the Cretaceous-Tertiary transition? Earth and Planetary Science Letters 119: 181–189.

35. Shukla PN, Shukla AD, Bhandari N (1997) Geochemical characterization of the Cretaceous-Tertiary sediments at Anjar, India. Palaeobotanist 46(1–2): 127–132.

36. Khosla SC (1999) *Costabuntonia*, a new genus of ostracoda from the Intertrappean beds (Paleocene) of east coast of India. Micropaleontology 45: 319–323.

37. Keller G, Adatte T, Bajpai S, Mohabey DM, Widdowson M, et al. (2009) K-T Transition in Deccan Traps of central India marks major marine seaway across India. Earth and Planetary Science Letters 282: 10–23.

38. Cripps JA, Widdowson M, Spicer RA, Jolly DW (2005) Coastal ecosystem response to late stage Deccan Trap volcanism: the post K-T boundary (Danian) palynofacies of Mumbai (Bombay), west India. Paleogeography, Paleoclimatology, Paleoecology 216(1–4): 303–332.

39. Srivastava R (2011) Indian Upper Cretaceous-Tertiary flora before collision of Indian Plate: A reappraisal of central and western Indian flora. Memoir of the Geological Society of India 77: 281–292.

40. Lakhanpal RN, Maheshwari HK, Awasthi N (1976) A catalogue of Indian fossil plants. Lucknow: Birbal Sahni Institute of Palaeobotany.

41. Srivastava R (1991) A catalogue of fossil plants from India–4. Cenozoic (Tertiary) megafossils. Lucknow: Birbal Sahni Institute of Palaeobotany.

42. Srivastava R, Guleria JS (2005) A catalogue of Cenozoic (Tertiary) plant megafossils from India (1989–2005). Lucknow: Birbal Sahni Institute of Palaeobotany.

43. Read RW, Hickey LJ (1972) A revised classification of fossil palm and palm-like leaves. Taxon 21: 129–137.

44. Saporta G (1865) Études sur la vegetation du sud-est de la France a l époque tertiare. Annales des Sciences Naturelles (Botanique) 5(3): 5–152.

45. The Herbarium Catalogue, Royal Botanic Garden, Kew. Published on internet http://www.kew.org/herbcat (accessed 2014 August 18).

46. Mohabey DM (1986) Depositional environment of Lameta Formation (Late Cretaceous) of Nand-Dongargaon inland basin, Maharashtra. The fossil and lithological evidences. Memoir of the Geological Society of India 37: 363–386.

47. Bonde SD (1986) *Amesoneuron borassoides* sp. nov., a borassoid palm leaf from Deccan Intertrappean beds at Mohgaonkalan, India. Biovigyanam 12: 89–91.

48. Guleria JS, Mehrotra RC (1999) On some plant remains from Deccan Intertrappean localities of Seoni and Mandla districts of Madhya Pradesh, India. Palaeobotanist 47: 68–87.

49. Mehrotra RC, Ram-Awatar, Sharma A, Phartiyal B (2007) A new palm leaf from the Indus suture zone, Ladakh Himalayas, India. Journal of Palaeontological Society of India 52: 159–162.

50. Mehrotra RC (2000) Study of plant megafossils from the Tura Formation of Nangwalbibra, Garo Hills, Meghalaya, India. Palaeobotanist 49(2): 225–237.

51. Guleria JS, Singh Hemanta RK, Mehrotra RC, Soibam I, Kishor R (2005) Palaeogene plant fossils of Manipur and their palaeoecological significance. Palaeobotanist 54: 61–77.

52. Guleria JS, Srivastava R, Prasad M (2000) Some fossil leaves from the Kasauli Formation of Himachal Pradesh, North-West India. Himalayan Geology 21(1–2): 43–52.

53. Prasad M (2006) Siwalik plant fossils from the Himalayan foot-hills of Himachal Pradesh, India and their significance on palaeoclimate. Phytomorphology 56(1–2): 9–22.

54. Trivedi BS, Chandra R (1971) A palm leaf from the Deccan Intertrappean Series, Mohgaon Kalan (M. P.), India. Current Science 40(19): 526–527.

55. Mahabale TS, Rao SV (1968) Fossil palm remains from Bommuru, Andhra Pradesh. Current Science 37(6): 158–159.

56. Lakhanpal RN, Sah SCD, Sharma KK, Guleria JS (1983) Occurrence of *Livistona* in the Hemis conglomerate Horizon of Ladakh: In: Sharma K K, Thakur V C, Geology of Indus Suture Zone of Ladakh, Wadia Institute of Himalayan Geology, Dehradun. pp. 179–185.

57. Srivastava G, Mehrotra RC, Bauer H (2012) Palm leaves from the Late Oligocene sediments of Makum Coalfield, Assam, India. Journal of Earth System Sciences 121(3): 747–754.

58. Kumaran KPN (1994) *Malpophyllum*, a new name for the fossil genus *Palmophyllum* Conwentz, 1886. Review of Palaeobotany and Palynology 81 (2–4): 337–338.

59. Mehrotra RC, Tiwari RP, Mazumder BI (2003) *Nypa* megafossils from the Tertiary sediments of Northeast India. Geobios 36: 83–92.
60. Singh RR, Patnayak R (2012) A fossil palm leaf impression from ~11.2 Ma old, Siwalik deposits of Kangra Valley, Himachal Pradesh. Journal of the Geological Society of India 79: 85–88.
61. Lakhanpal RN, Guleria JS (1982) Plant remains from Miocene of Kachchh, western India. Palaeobotanist 30(3): 279–296.
62. Paul SK, Ram-Awatar, Mehrotra RC, Sharma A, Phartiyal B, et al. (2007) A new palm leaf from the Hemis Formation of Ladakh, Jammu and Kashmir, India. Current Science 92: 727–729.
63. Lakhanpal RN (1964) A new record of angiospermic leaf impressions from the Garo Hills, Assam. Current Science 33(9): 276.
64. Singh MC, Kushwaha RAS, Srivastava G, Mehrotra RC (2012) New plant remains from the Laisong Formation of Manipur. Journal of the Geological Society of India 79: 287–294.
65. Bonde SD (2008) Indian fossil monocotyledons: Current status, recent developments and future directions. Palaeobotanist 57: 141–164.
66. Sahni B (1964) Revision of Indian fossil plants. Part III, Birbal Sahni Institute of Palaeobotany, Lucknow.
67. Mathur AK, Mishra VP, Mehra S (1996) Systematic study of plant fossils from Dagshai, Kasauli and Dharamsala formations of Himachal Pradesh. Palaeontologia Indica (New Series), Geological Survey of India 1. 1–68.
68. Bonde SD (1986) *Sabalophyllum livistonoides* gen. et sp. nov. a petrified palm leaf segment from Deccan Intertrappean beds at Nawargaon, District Wardha, Maharashtra, India. Biovigyanam 12: 113–118.
69. Lakhanpal RN, Prakash G, Thussu JL, Guleria JS (1984) A fan palm from the Liyan Formation of Ladakh (Jammu and Kashmir). Palaeobotanist 31(3): 201–207.
70. Kapoor R, Singh RY (1987) A note on the geology and distribution of some significant fossils in the lower Tertiary sediments exposed along Kalka- Kasauli road section. Bulletin Indian Geological Association 20(1): 17–23.
71. Barman G, Duara BK (1970) *Zalaccites jaintiensis* gen. et sp. nov. from the plateau (Cherra) sandstones of the Jaintia Hills, United Khasi and Jaintia Hills District, Assam, India. Science and Culture 36(1): 63–64.
72. Daghlian CP (1976) Coryphoid palms from the lower and middle Eocene of southeastern North America. Palaeontographica 166B: 44–82.
73. Zhou WJ, Liu XY, Xu QQ, Huang KY, Jin JH (2013) New coryphoid fossil palm leaves (Arecaceae: Coryphoideae) from the Eocene Changchang Basin of Hainan Island, South China. Science China: Earth Sciences 56: 1493–1501.
74. Huzioka K, Takahashi E (1970) The Eocene flora of the Ube Coal-field, southwest Honshu, Japan. Journal of the Mining College, Akita University Series A 4(5): 1–88.
75. Marmi J, Gomez B, Closas CM, Breva SV (2010) A reconstruction of the fossil palm *Sabalites longirhachis* (Unger) J. Kvaček et Herman from the Maastrichtian of Pyrenees. Review of Palaeobotany and Palynology 163: 73–83.
76. Walther G-R, Gritti ES, Berger S, Hickler T, Tang Z, et al. (2007) Palms tracking climate change. Global Ecology and Biogeography 16: 801–809.
77. Jones DL (1995) Palms throughout the world. Chatswood: Reed Books.
78. Gibbons M (2003) A pocket guide to palms. London: PRC Publishing Ltd.
79. Tomlinson PB (1990) The structural biology of palms. Oxford: Clarendon Press.
80. Woodward FI (1988) Climate and plant distribution. Cambridge: Cambridge University Press.
81. Tomlinson PB (2006) The uniqueness of palms. Botanical Journal of Linnean Society 151: 5–14.
82. Bjorholm S, Svenning JC, Skov F, Balslev H (2005) Environmental and spatial controls of palm (Arecaceae) species richness across the Americas. Global Ecology and Biogeography 14: 423–429.
83. Punyasena SW, Eshel G, McElwain JC (2008) The influence of climate on the spatial patterning of neotropical plant families. Journal of Biogeography 35: 117–130.
84. Jacobs BF (2004) Palaeobotanical studies from tropical Africa: relevance to the evolution of forest, woodland and savannah biomes. Philosophical Transactions of the Royal Society, London B359: 1573–1583.
85. Svenning JC, Borchsenius F, Bjorholm S, Balslev H (2008) High tropical net diversification drives the New World latitudinal gradient in palm (Arecaceae) species richness. Journal of Biogeography 35: 394–406.
86. Bjorholm S, Svenning JC, Baker WJ, Skov F, Balslev H (2006) Historical legacies in the geographical diversity patterns of New World palm (Arecaceae) subfamilies. Botanical Journal of the Linnean Society 151: 113–125.
87. Asmussen CBJ, Dransfield J, Deickmann V, Barfod AS, Pintaud J-C, et al. (2006) A new subfamily classification of the palm family (Arecaceae): evidence from plastid DNA phylogeny. Botanical Journal of Linnean Society 151: 15–38.
88. Manchester SR, Lehman TM, Wheeler EA (2010) Fossil palms (Arecaceae, Coryphoideae) associated with juvenile herbivorous dinosaurs in the upper Cretaceous Aguja Formation, Big Bend National Park, Texas. International Journal of Plant Sciences 171(6): 679–689.
89. Herngreen GFW, Chlonova AF (1981) Cretaceous microfloral provinces. Pollen et Spores 23: 441–555.
90. Herngreen GFW, Kedves M, Rovinina LV, Smirnova SB (1996) Cretaceous palynofloral provinces: a review. In: Jansonius J, Mcgregor D C, Palynology: principles and applications. Dallas: American Association of Stratigraphic Palynologists Foundation. pp. 1157–1188.
91. El-Soughier MI, Mehrotra RC, Zhi-Yan Z, Gong-Le S (2011) *Nypa* fruits and seeds from the Maastrichtian–Danian sediments of Bir Abu Minqar, south western desert, Egypt. Palaeoworld 20: 75–83.
92. Gregor HJ, Hagn H (1982) Fossil fructifications from the Cretaceous–Palaeocene Boundary of SW-Egypt (Danian, Bir Abu Munqar). Tertiary Research 4: 121–147.
93. EL- Saadawi W, Youssef SG, Kamal-El-Din MM (2004) Fossil palm woods of Egypt: II Seven Tertiary *Palmoxylon* species new to the country. Review of Palaeobotany and Palynology 129: 199–211.
94. Ezcurra MD, Agnolin FL (2012) A new global palaeobiogeographical model for the Late Mesozoic and early Tertiary. Systematic Biology 61(4): 553–566.
95. Survey of India (1950) Toposheet No. 64A.

16

New Insights into Phosphorus Mobilisation from Sulphur-Rich Sediments: Time-Dependent Effects of Salinisation

Josepha M. H. van Diggelen[1,2]*, **Leon P. M. Lamers**[2], **Gijs van Dijk**[1,2], **Maarten J. Schaafsma**[1¤], **Jan G. M. Roelofs**[2], **Alfons J. P. Smolders**[1,2]

1 B-WARE Research Centre, Radboud University Nijmegen, Mercator 3, Nijmegen, The Netherlands, **2** Institute for Water and Wetland Research, Department of Aquatic Ecology and Environmental Biology, Radboud University Nijmegen, Nijmegen, The Netherlands

Abstract

Internal phosphorus (P) mobilisation from aquatic sediments is an important process adding to eutrophication problems in wetlands. Salinisation, a fast growing global problem, is thought to affect P behaviour. Although several studies have addressed the effects of salinisation, interactions between salinity changes and nutrient cycling in freshwater systems are not fully understood. To tackle eutrophication, a clear understanding of the interacting effects of sediment characteristics and surface water quality is vital. In the present study, P release from two eutrophic sediments, both characterized by high pore water P and very low pore water iron (Fe^{2+}) concentrations, was studied in a long-term aquarium experiment, using three salinity levels. Sediment P release was expected to be mainly driven by diffusion, due to the eutrophic conditions and low iron availability. Unexpectedly, this only seemed to be the driving mechanism in the short term (0–10 weeks). In the long term (>80 weeks), P mobilisation was absent in most treatments. This can most likely be explained by the oxidation of the sediment-water interface where Fe^{2+} immobilises P, even though it is commonly assumed that free Fe^{2+} concentrations need to be higher for this. Therefore, a controlling mechanism is suggested in which the partial oxidation of iron-sulphides in the sediment plays a key role, releasing extra Fe^{2+} at the sediment-water interface. Although salinisation was shown to lower short-term P mobilisation as a result of increased calcium concentrations, it may increase long-term P mobilisation by the interactions between sulphate reduction and oxygen availability. Our study showed time-dependent responses of sediment P mobilisation in relation to salinity, suggesting that sulphur plays an important role in the release of P from FeS_x-rich sediments, its biogeochemical effect depending on the availability of Fe^{2+} and O_2.

Editor: Todd Miller, University of Wisconsin Milwaukee, United States of America

Funding: This study was part of the National Research Programme "Wormer- en Jisperwater," funded by the Dutch Ministry of Agriculture, Nature and Food Quality (LNV), within the framework of "Nota Ruimte." The water management authority "Hoogheemraadschap Hollands Noorderkwartier" facilitated this programme. The authors did not require further external funding source for this study. The funders had no role in study design, data collection and analysis, decision to publish, or preparation of the manuscript.

Competing Interests: The authors have declared that no competing interests exist.

* Email: J.vanDiggelen@b-ware.eu

¤ Current address: Royal Haskoning DHV, Nijmegen, The Netherlands

Introduction

The eutrophication of surface waters is an urgent problem worldwide [1]. Increased P concentrations have led to a strong decline of the biodiversity in freshwater wetlands, due to the resulting dominance of highly competitive macrophytes, and of algae and cyanobacteria, monopolising light [1–3]. Salinisation of freshwater systems has received increasing attention, especially in relation to climate change and sea level rise [4]. With increasing salinity, higher P concentrations are often found in surface waters (e.g. [5–8]), which may affect P cycling in freshwater systems. Therefore, salinisation is expected to enhance eutrophication in coastal, freshwater wetlands, leading to water quality deterioration and loss of biodiversity.

Internal mobilisation of P from eutrophic aquatic sediments is an important process adding to eutrophication problems in wetlands [9–12]. The classic theoretical framework suggests that sufficiently high oxygen (O_2) concentrations in the surface water can prevent P release from the sediment [13–14]. According to this, the oxidation of dissolved iron (Fe^{2+}) in the sediment will result in the formation of iron oxides and hydroxides ($Fe(OH)_x$) at the sediment surface, effectively binding P and thereby preventing its release to the surface water. Under anaerobic conditions, these ferric compounds will be mobilised by Fe-reducing bacteria, and part of the P is released to the surface water.

Besides anaerobic conditions, increased sulphate (SO_4^{2-}) reduction rates are also known to be able to increase P mobilisation by decoupling Fe - P interactions at the sediment-water interface [12,15–19]. Sulphide (S^{2-}) binds efficiently to dissolved Fe^{2+} in sediment pore water, and most Fe^{2+} can become bound as iron sulphides (FeS_x) in the sediment, strongly decreasing Fe^{2+} sediment pore water concentrations [18,19]. Geurts *et al.* [19] found that, in aerobic surface waters, P mobilisation from sediments with low pore water Fe:P ratios (<1 mol mol^{-1}) was a linear function of sediment pore water P concentrations. As a result, one would expect a release of P irrespective of the O_2 concentration in the surface water of SO_4^{2-} enriched wetlands [10,12,18,20]. In addition, dissolved P concentrations might further increase due to the enhanced anaerobic breakdown of organic matter linked to SO_4^{2-} reduction and concomitant mineralisation of P [2,12,19].

Salinisation of freshwater systems can enhance SO_4^{2-} reduction rates due to a higher SO_4^{2-} availability [21], which may strongly affect P mobilisation as described above. Moreover, increasing Cl$^-$ and SO_4^{2-} concentrations might enhance P release from sediments by competition for anion binding sites [10,22]. At the same time, an increase in salinity also leads to increased Ca^{2+} concentrations [21], which may result in the immobilisation of P by co-precipitation with Ca^{2+} and calcium carbonate ($CaCO_3$) [4,9,23]. Salinity changes affect a suite of biogeochemical processes in freshwater systems, where the net effect on P mobilisation is the combined result of these processes. Moreover, a time-dependent shift in dominance of each process on P release can be expected [18]. Most studies regarding P release focus on relative short-term effects ranging from one day to 90 days [6,19,20,24], while long-term experiments are mostly lacking. In this paper we explore the time-dependent release of P from eutrophic sediments under different salinities, which is highly relevant regarding the worldwide interest in salinisation effects on freshwater wetland functioning.

To test time-dependent interactions between salinisation and P mobilisation, a controlled aquarium experiment was set up that lasted two years. Two FeS$_x$-rich sediments from a coastal freshwater wetland were subjected to three naturally occurring water types characterised by different salinities. Pore waters of the peat sediments were typically rich in P and S, and very poor in Fe, and the low total Fe:S ratios in the sediment suggested that most Fe was bound to reduced S [2]. In such sediments, a very high release of P from the sediment to the surface water can be expected, predominantly depending on pore water P concentrations [10,12,18–20]. By monitoring biogeochemical changes in porewater and surface water under controlled conditions, we try to reveal how salinity affects short-term and long-term P release, in these type of sediments common for coastal wetlands.

Materials and Methods

2.1 Sampling area

In this study, peat sediments were used from the coastal lowland fen area Wormer- and Jisperveld (52° 30′ 42.7644″; 4° 52′ 27.3756″) in the Netherlands. Due to historic intrusion of brackish water, peat rich in minerals such as S, Ca and Fe has accumulated in this area. After more than 50 years of desalinisation resulting from altered hydrological conditions, it gradually became a freshwater system. The peatland comprises ca. 500 ha of open water and ca. 1660 ha of peat meadows, predominantly used for agricultural purposes and partly for nature conservation. Drainage is a standard procedure in this area, leading to peat decomposition and land subsidence. As a result, risks of flooding events and salinisation are increasing in this freshwater peatland.

2.2 Experimental design

On 18 March 2008, two types of submerged peat sediment were collected from a ditch at a depth of 0–20 cm (ca. 25 L in total), using a sediment multi sampler (Eijkelkamp Agrisearch Equipment). Although both sediments were relatively rich in organic S and P, they differed in P availability (sediment characteristics are given in Table 1). To minimise O_2 intrusion, the sediments were

stored anaerobically at 4°C in large, closed containers. The next day, 12 glass cylinders (diameter 15 cm, height 60 cm) were filled with 15 cm of sediment A and another 12 cylinders with 15 cm of sediment B. Next 40 cm of water was carefully poured on top of the sediments, avoiding re-suspension of sediment particles. Artificially composed surface water, based on site conditions (control treatment, Table 2), was used for all sediments during an acclimatisation period of 4 weeks. The experiment was carried out in the dark at a constant and environmentally relevant temperature of 15°C. To allow oxygen diffusion to the surface water, an open cylinder system was used.

After this acclimatisation period, three different surface water types were applied as salinity treatments: rainwater (low salinity; 100 µmol Cl L^{-1}), brackish water (high salinity; 85 mmol Cl L^{-1}) and freshwater (control; 7 mmol Cl L^{-1}). All treatments were artificially composed, based on field measurements (Table 2). Control water simulated water quality in the current conditions that exist in the wetland, brackish water composition was based on the historic conditions reported by Reigersman in 1946 [25]. Rainwater quality equalled the chemical composition of atmospheric deposition as measured in the Netherlands [26]. No P was added in order to be able to estimate the release of P from the sediment. For each treatment and sediment type, 4 replicates were used (24 cylinders in total).

Treatment solutions were stored in polyethylene containers (10 L), from which they were pumped into the cylinders using Masterflex L/S multichannel pumps (model 7535-08). The treatments were started by replacing the control water with the appropriate treatment water during 4 weeks, to ensure that all treatment solutions were added properly. Directly after treatment addition (week 10), stagnant conditions were created in order to measure short-term effects of P and S release from the sediment. Short-term mobilisation rates were calculated from the linear increase of the surface water P and S concentrations (0–10 weeks of the stagnant period). After a stagnant period of 26 weeks, pumps were running with a hydraulic retention time of 25 weeks for the treatment solutions during 48 weeks in order to maintain the appropriate treatment conditions. To measure long-term effects of salinity changes on the release of P and S from the sediment, pumps were stopped again (week 81) to create another stagnant period for 32 weeks. Long-term mobilisation rates were again calculated from the linear increase of the S and P concentrations in the surface water during this stagnant period.

Intact peat cores from the same location as the main experiment were collected separately, to test the effects of aerobic versus anaerobic conditions of the surface water on P release. Water and sediment oxygen (O_2) profiles were measured, using a fixed fiber optical oxygen microsensor (optode) in combination with a Microx TX3 transmitter (PreSens Precision Sensing GmbH). The peat sediment cores were monitored during 18 weeks of either aerobic conditions similar to those of the main experiment, or anaerobic conditions by gently supplying N_2 to the surface water. During both aerobic and anaerobic conditions, P mobilisation rates were calculated from the linear increase in surface water P concentrations.

2.3 Chemical analyses

To monitor water quality, samples of surface water and pore water were collected every 2 months and analysed during the experiment. Pore water was collected anaerobically, using 30 mL vacuum bottles connected to Rhizon SMS-10 cm samplers that were fixed in the upper 10 cm of the sediment (Eijkelkamp Agrisearch Equipment). Disturbance of the sediment and water was minimised by the low frequency of sampling and small sample

Table 1. Characteristics of the two sediments used.

| Sediment | Organic content | Bulk Density | Total amounts bound to sediment | | | | | | | |
| | | | Total - P | Org - P | Inorg - P | Total - Al | Total - Ca | Total - S | Total - Fe | Fe:S ratio |
	%	kg DW L⁻¹ FW	mmol L⁻¹ FW	mmol L⁻¹ FW	mmol L⁻¹ FW	mmol L⁻¹ FW	mmol L⁻¹ FW	mmol L⁻¹ FW	mmol L⁻¹ FW	mol mol⁻¹
A Mean	59.7	0.17	2.9ᵃ	1.8	1.2ᵃ	44.5	54.9	99.3	33.1	0.33ᵃ
SEM	4.8	0.03	0.4	0.3	0.1	12.6	5.4	8.1	3.9	0.024
B Mean	59.6	0.13	4.7ᵇ	2.4	2.3ᵇ	43.4	44.0	83.6	38.3	0.46ᵇ
SEM	7.4	0.02	0.3	0.1	0.1	4.4	0.8	4.7	2.3	0.001

Significant differences between the sediment types are indicated by different letters.

sizes (max. 25 mL). Sulphide concentrations were determined directly after the collection by fixing 10.5 mL pore water with 10.5 mL Sulphide Anti Oxidant Buffer (SAOB), and using an Orion sulphide-electrode and a Consort Ion meter (type C830) [27]. The pH and alkalinity of all samples were measured within 24 hours after sampling, using a combined pH electrode (Radiometer) in combination with a TIM840 pH meter and a Titration Manager Titralab Autoburette. Dissolved total inorganic carbon (TIC) was measured within 24 hours after sampling by injecting 0.2 mL pore water or surface water in a closed chamber containing 0.2 M H_3PO_4 solution, converting all dissolved TIC into CO_2. A continues gas flow (N_2) directly transports the CO_2 to an ABB Advance optima Infrared Gas Analyzer (IRGA) to measure total inorganic C concentrations. A calibration curve was made by injecting different volumes (0.1–1.0 mL) of 1.25 mM HCO_3^- solution. Prior to storage at 4°C until elemental analysis, 0.1 mL HNO_3^- (65%) was added to 10 mL of each sample to prevent metal precipitation. Concentrations of dissolved Ca, Fe, P, S, and Al in these stored samples were measured using an Inductively Coupled Plasma Spectrophotometer (ICP IRIS Intrepid II XDL; Thermo Electron Corporation). Due to the anaerobic sampling of pore water, measured Fe predominantly consisted of dissolved Fe^{2+} rather than far less mobile Fe^{3+}. The remaining samples were stored at -20°C in order to determine the following ion concentrations colourimetrically on Auto Analyzer 3 systems (Bran and Luebbe): NO_3^- [28], NH_4^+ [29], ortho-PO_4^{3-} [30] and Cl^- [31]. Na^+ and K^+ were determined with a Technicon Flame Photometer IV Control (Technicon Corporation).

For both sediments gravimetric water contents were determined by drying for 48 h at 70°C. Organic matter contents were estimated by loss on ignition for 4 h at 550°C. A homogenized portion of 200 mg dry sediment was digested in 5 mL HNO_3 (65%) and 2 mL H_2O_2 (30%), using an Ethos 1 Advanced microwave digestion system (Milestone Inc.). Digestates were diluted and analysed by ICP as described above. In order to distinguish between the organic and inorganic P fraction, a P-fractionation procedure was carried out adapted after Golterman [32].

2.4 Statistical analyses

For statistical analysis, SPSS Statistics for Windows (Version 21.0. IBM Corp. Armonk, NY; 2012) was used. To test for differences among treatments in sediment analyses (single measurements) or differences in calculated mobilisation rates, the General Linear Model (GLM) univariate procedure combined with Tukey's-b post-hoc test was used.

To test for significant differences among treatments in repeated measurements, a GLM mixed model procedure was used. When significant differences between the two sediments were found, using a 2-way GLM mixed model with treatment as fixed factor, sediment as random factor and time as repeated measures, both sediments were analysed separately. In this separate model for sediments, time was used as repeated measures and treatment as fixed factor, with AR(1) heterogeneous as the covariance type. A Bonferroni post-hoc test was used to test for differences between treatments.

2.5 Ethics statement

This study was part of the National Research Programme 'Wormer- en Jisperwater', funded by the Dutch Ministry of Agriculture, Nature and Food Quality (LNV), within the framework of 'Nota Ruimte'. The water management authority 'Hoogheemraadschap Hollands Noorderkwartier' facilitated this

Table 2. Chemical composition of the surface water used for the different treatments (low, normal or high salinity).

Element	Low salinity	Normal salinity	High salinity
	Rain water	Fresh water	Brackish water
	μmol L^{-1}	μmol L^{-1}	μmol L^{-1}
Na$^+$	100	7000	85000
Cl$^-$	100	7000	85000
SO$_4^{2-}$	5	1500	5500
K$^+$	30	500	1000
Ca^{2+}	10	2000	2500
Mg^{2+}	10	1250	3750
HCO$_3^-$	0	4000	4000
NO$_3^-$	50	50	50
NH$_4^+$	50	50	50

programme and the nature management authority 'Natuurmonumenten' gave permission to take samples in their reserve.

Results

3.1 Pore water chemistry

As expected, pore water chemistry was strongly affected by changes in surface water salinity (Fig. 1). Under brackish conditions, Na$^+$ showed a highly significant ($p<0.005$) gradual increase in the pore water over time. For the low salinity and control treatment, no significant changes in pore water Na$^+$ concentrations occurred in sediment A, while Na$^+$ concentrations showed a significant decrease ($p<0.05$) over time at a low salinity in sediment B. An interaction between treatment and sediment type was found for both Na$^+$ and S concentrations, which means that the treatments had a significant, but different, effect on the two sediments. Pore water S concentrations also showed a highly significant ($p<0.005$) gradual increase at the high salinity treatment in both sediments (Fig. 1). At a low salinity, S concentrations remained at a steady level while the control treatment showed a small, but not significant, increase. Moreover, no clear differences in sulphide concentrations were found between sediments or treatments (average values ranged between 0–50 μmol L^{-1} for sediment A, and between 0–500 μmol L^{-1} for sediment B; data not shown).

As a result of a higher salinity, Ca^{2+} was mobilised in the sediment, as shown by significantly ($p<0.005$) increased pore water Ca^{2+} concentrations (Fig. 1). This increase in the pore water, well above the added concentration of 2500 μmol L^{-1} to the surface water, started directly after the onset of the high salinity treatment and Ca^{2+} concentrations remained at a steady high level during the course of the experiment.

Dissolved Fe^{2+} concentrations were low and showed a gradual decrease for all treatments over time in both sediment types (Fig. 2). A significantly higher ($p<0.005$) Fe^{2+} concentration was found for the low salinity treatment at sediment A when compared to the control and higher salinity treatment. In contrast, no differences in pore water Fe^{2+} concentrations between treatments were found for sediment B. Pore water HCO$_3^-$ concentrations were significantly higher ($p<0.05$) under brackish conditions at sediment B, while no differences were found between treatments at sediment A (data not shown).

Pore water P concentrations showed a gradual decrease at sediment A for all treatments. Moreover, a significantly ($p<0.05$) stronger decrease of P in pore water was found for the control treatment, compared to the high and low salinity treatment at sediment A. In strong contrast, P concentrations showed a gradual increase in the pore water of sediment B for all salinity treatments (Fig. 2), with the significantly ($p<0.05$) lowest P concentrations in the high salinity treatment.

3.2 Surface water chemistry

A higher salinity led to gradually increased Na$^+$ and S concentrations in the surface water, and showed significant ($p<0.005$) differences among all treatments, which eventually equalled the concentrations added (S: Fig. 3; Na$^+$: data not shown). For the low salinity treatment, however, S concentrations in the surface water reached much higher concentrations than the concentrations of the treatment water, which suggests S mobilisation from the sediment. These S mobilisation rates were calculated (Table 3) for both a short term, showing significantly ($p<0.005$) negative rates at a high salinity (high S consumption) for sediment A, and for a long term, still showing significantly ($p<0.005$) negative S mobilisation rates at a high salinity in both sediments. In the surface water, Ca^{2+} concentrations also increased and differed significantly ($p<0.005$) among all treatments for both sediments (data not shown). However, both the low and high salinity treatment led to much higher concentrations than the added concentrations.

In the low salinity and control treatment, P concentrations in the surface water increased directly after onset of the treatments (after 10 weeks; t = 0). For the high salinity treatments, P concentrations of the surface water showed a strong and significant (P<0.05) decrease immediately after the onset of the treatments (after 10 weeks; t = 0). After this temporary decrease, P concentrations started to increase gradually. As a result, significantly lower P concentrations ($p<0.05$) were found for the high salinity treatment compared to the low salinity treatment in both sediments at a short term (after 20 weeks; t = 10), and a trend was found when compared to the control treatment ($p<0.1$) at sediment A. When P mobilisation rates were calculated for the short term, however, no differences among salinity treatments were found (Table 3).

More than 80 weeks after the start of the experiment, P concentrations in the surface water above sediment A were

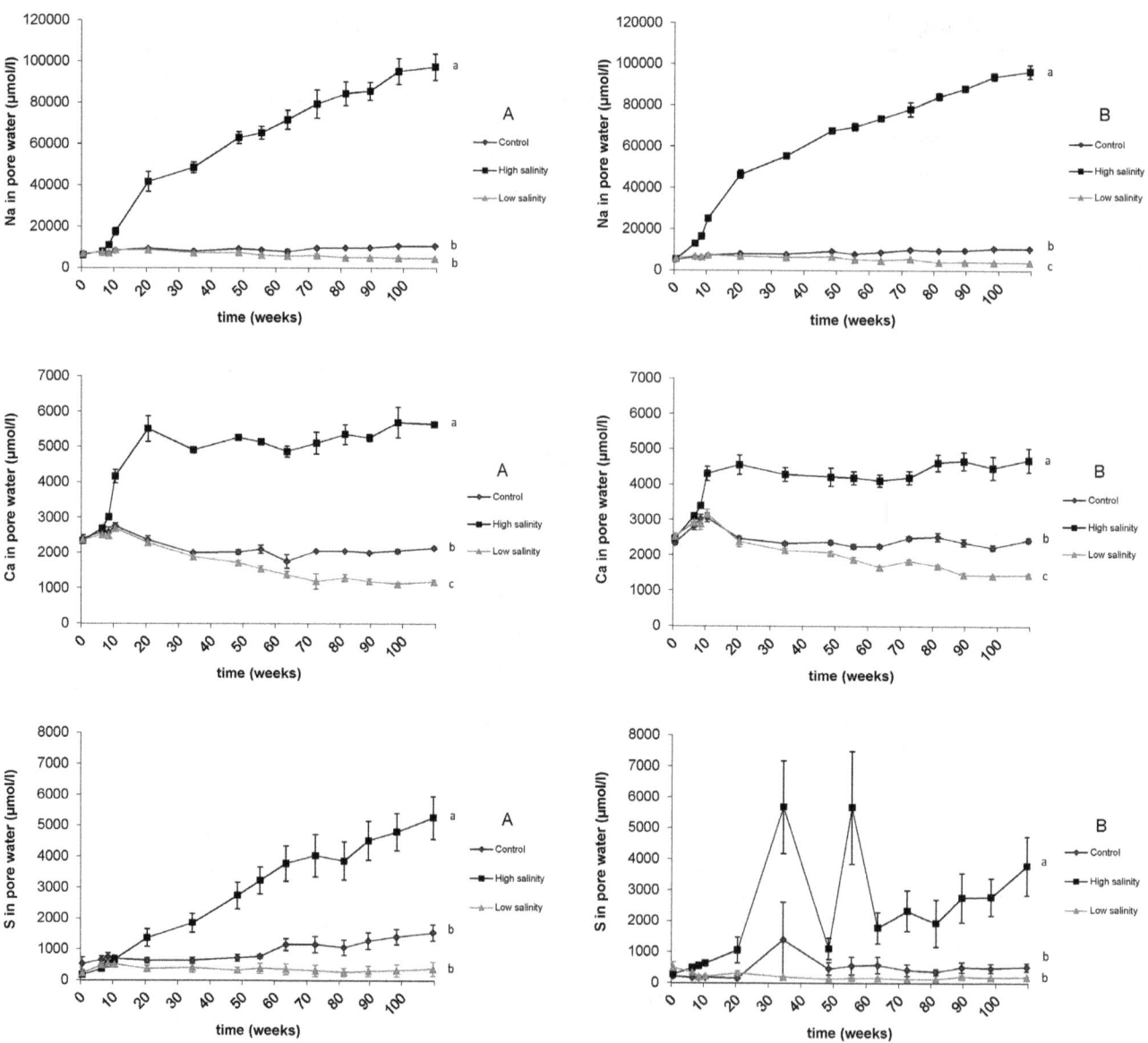

Figure 1. Sodium (Na$^+$), calcium (Ca^{2+}) and sulphur (S) pore water concentrations (μmol L^{-1}) in both sediments (A: left, B: right). Significant differences between treatments are indicated with different letters.

significantly higher (p<0.01) at a high salinity (Fig. 3), which was totally opposite to the short-term effect. Calculated P mobilisation rates were also significantly higher (p<0.05) with a high salinity compared to a low salinity at sediment A. For the control and low salinity treatment, P concentrations in surface water remained low, or even showed a decrease in the long term. At sediment B, however, no change of P in the surface water was found for any of the salinity treatments. The long-term P mobilisation rates with a high salinity were similar to the short-term rates at sediment A, while no long-term P mobilisation was observed for the low salinity and control treatments.

3.3 Aerobic versus anaerobic surface water

The O$_2$ concentration profile (Fig. 4) shows that under aerobic conditions, O$_2$ is still available in the sediment to an average depth of 7 mm (sediment A) and 3 mm (sediment B). The cores of both sediment A and B showed a significant (p<0.001) higher

mobilisation rate of P during anaerobic conditions (Fig. 5). At sediment A, P mobilisation was on average 3 times higher during anaerobic conditions compared to aerobic conditions, while this was almost 4 times higher at sediment B. These aerobic mobilisation rates were well within range of the short-term mobilisation rates found in the main experiment (control treatment; Table 3).

Discussion

4.1 Short-term effects (0–10 weeks)

4.1.1 P mobilisation. In the short term, no differences in net mobilisation rates of P were found among the different treatments. During this first stagnant period, moderate P mobilisation rates of 7–103 μmol m^{-2} d^{-1} were found that fitted within the range of Geurts et al. [19], who found mobilisation rates of 10–150 μmol m^{-2} d^{-1} for sediments of which pore water Fe:P and total sediment Fe:S ratios were <1. Diffusion was most likely the main

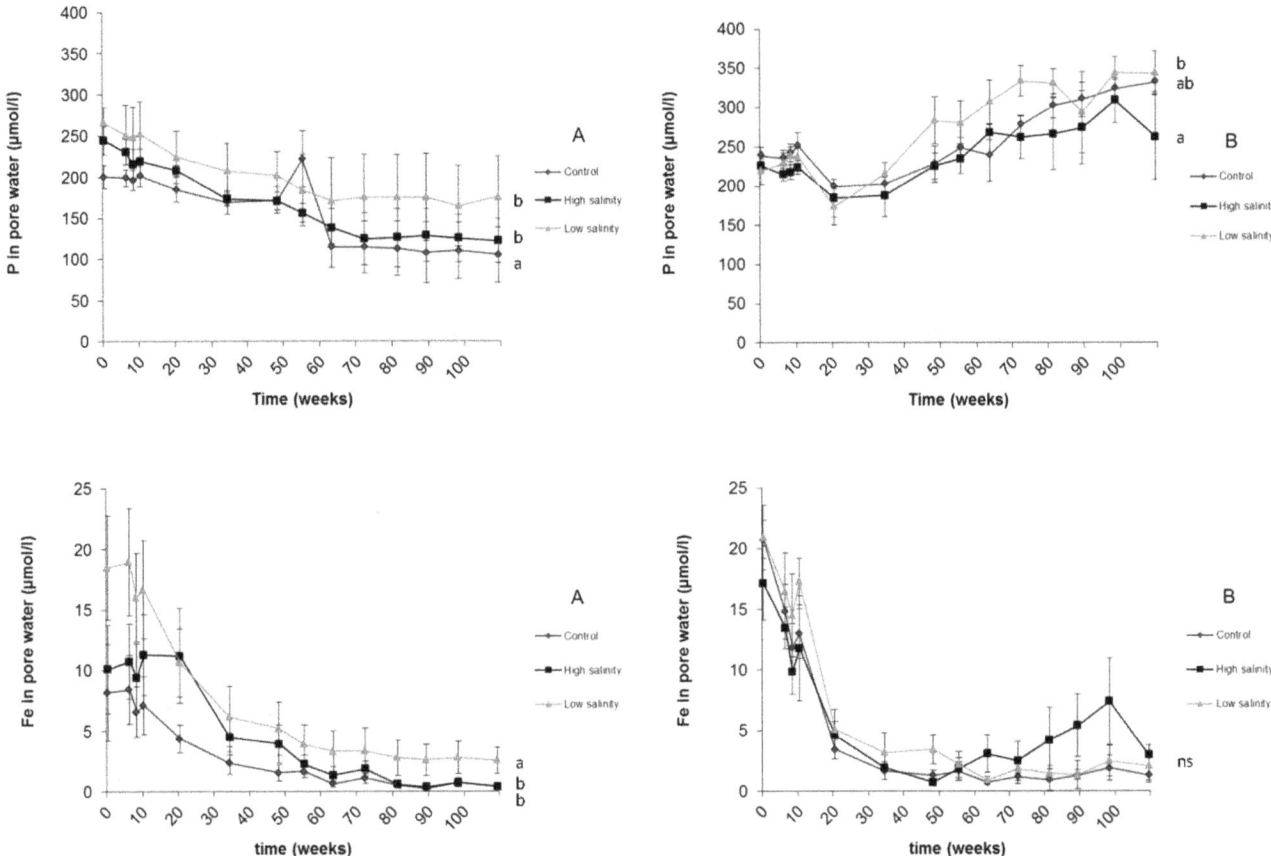

Figure 2. Phosphorus (P) and iron (Fe^{2+}) concentrations (µmol L^{-1}) in pore water of both sediments (A: left, B: right). Significant differences between treatments are indicated with different letters.

mechanism driving P release [19,33], since the sediments used in this experiment were not subjected to bioturbation or resuspension [7], nor to a changed pH or temperature [24]. Moreover, both sediments were characterised by total Fe:S ratios below 0.5 (Table 1), which indicates that most Fe was bound to reduced S [20]. Indeed, dissolved pore water Fe^{2+} concentrations were low in this study (and ranged between 0–20 µmol L^{-1} for both sediments), and showed an even further decrease over time, resulting in very low pore water Fe:P ratios (<0.1) during the entire experimental period.

4.1.2 Salinity effects. Although increased salinity may lead to increased desorption of P from anion exchange sites [10], or by increased S^{2-} production and enhanced mineralisation rates [2,12], we did not find higher pore water P concentrations in the high salinity treatment. Instead, during the addition of the salinity treatments (between week 6 and 10), P concentrations in the surface water showed a short, strong drop for both sediments. This immediate drop of P observed upon a change of the surface water chemistry strongly points at a chemical, rather than a microbiological, explanation. It can most likely be explained by the co-precipitation of P with Ca^{2+} or CaCO$_3$ at the sediment-water interface [4,9], as Ca^{2+} concentrations directly and strongly increased in both surface and pore water upon the high salinity treatment (0–10 weeks). Accordingly, Suzumura et al. [7] found a fast chemical P (im)mobilisation response within minutes, due to adsorption-desorption processes after a changed salinity. Van Dijk et al. [34] found a similar immobilisation of P with increased salinity, explained by co-precipitation with Ca^{2+} in the sediment.

Degassing of carbon dioxide (CO$_2$) and possibly also the presence of microbial mats [35] may well have contributed to the precipitation of CaCO$_3$ at the sediment surface, as HCO$_3^-$ concentrations were up to three times higher in pore water than in the surface water. After the initial drop of P, concentrations started to gradually increase, which shows that the short-term overall net P mobilisation to the surface water was higher than its immobilisation due to co-precipitation with Ca^{2-}.

4.2 Long-term effects (1.5–2 years)

4.2.1 P mobilisation. In contrast to the short-term results, and rather unexpectedly for eutrophic sediments, P mobilisation to the surface water was absent in 5 out of 6 treatments in the longer term (after 80 weeks). This is remarkable, as a strong net diffusive P release in both sediments was expected given the very low pore water Fe^{2+} concentrations and the still very high pore water P concentrations [19]. Although a gradual decrease of P in the pore water of sediment A was observed, concentrations still remained sufficiently high for diffusive P release (>100 µmol L^{-1}) [19,20]. Sediment B even showed a gradual increase of pore water P concentrations during the experiment, without any increase of the P mobilisation to the surface water. Such results can only be explained by assuming that processes preventing net P release at the sediment-water interface become active in the long term, at least under the conditions that were created during our experiment. Possible explanations for this phenomenon are: (1) precipitation of P with Fe^{3+} or Fe(OH)$_x$ by the oxidation of the sediment surface [13,14], (2) storage of P by the microbial

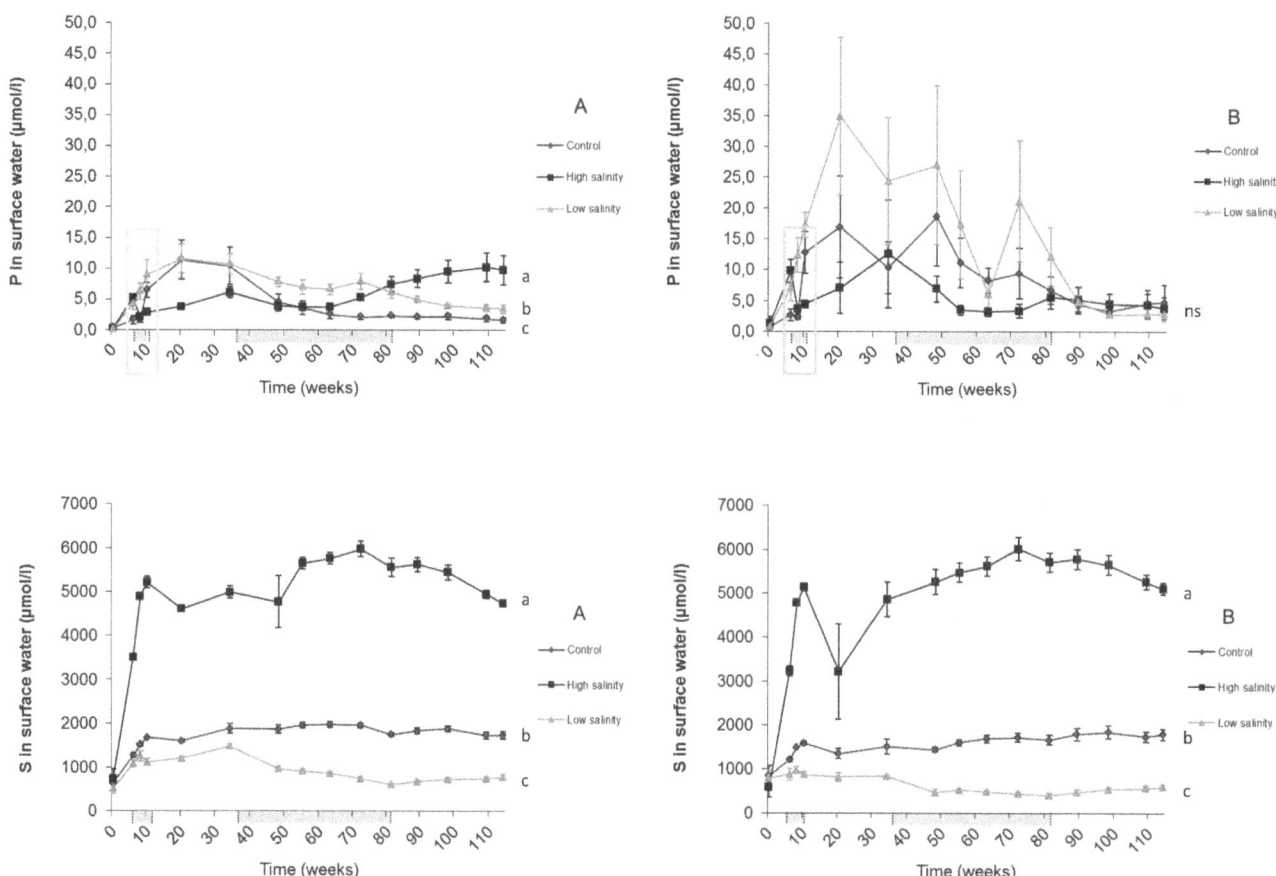

Figure 3. Phosphorus (P) and sulphur (S) concentrations (μmol L⁻¹) in the surface water above both sediments (A: left, B: right). Significant differences between treatments are indicated with different letters. Grey shadings under the x-axis indicate periods with through-flow (see Materials and Methods).

Table 3. P and S mobilisation rates ($\mu mol \, m^{-2} \, day^{-1}$) during stagnant conditions in the short term (0–10 weeks) and in the long term (80–110 weeks).

			P mobilisation ($\mu mol \, m^{-2} \, day^{-1}$)		S mobilisation ($\mu mol \, m^{-2} \, day^{-1}$)	
			short term	long term	short term	long term
A	High salinity	**Mean**	**6.8**	**4.1** [a]	**−1916.4** [a]	**−1447.5** [a]
		SEM	1.6	2.4	86.7	243.0
	Control	**Mean**	**37.7**	**−1.1** [ab]	**140.9** [b]	**−87.0** [b]
		SEM	14.3	0.6	216.0	86.4
	Low salinity	**Mean**	**19.8**	**−4.0** [b]	**−6.5** [b]	**244.2** [b]
		SEM	12.9	2.3	425.0	69.2
B	High salinity	**Mean**	**16.3**	**−2.9**	**−8536.5**	**−1121.7** [a]
		SEM	24.2	1.0	5474.0	141.3
	Control	**Mean**	**53.3**	**−2.2**	**−840.4**	**123.6** [b]
		SEM	36.6	2.5	517.8	91.1
	Low salinity	**Mean**	**102.6**	**−13.1**	**−597.4**	**294.8** [b]
		SEM	74.3	6.5	236.3	83.3

Significant differences between treatments are indicated by different letters.

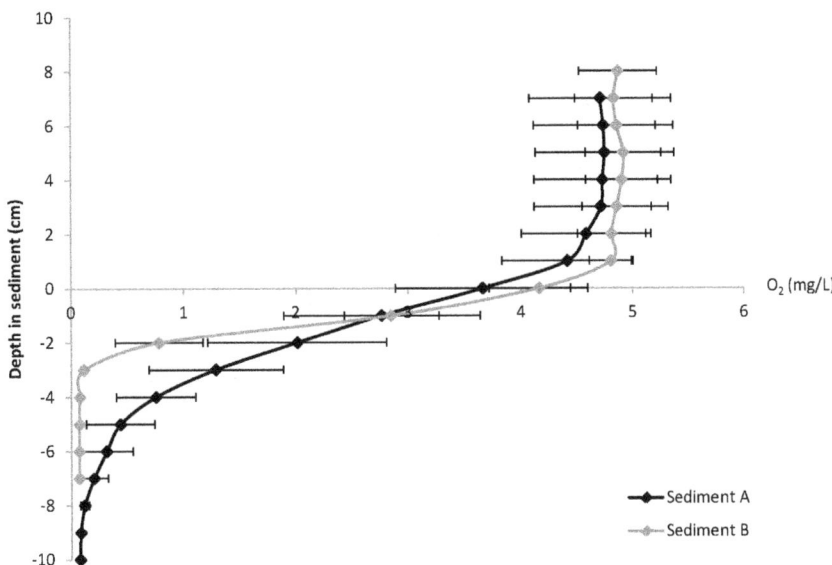

Figure 4. Oxygen (O$_2$) concentration (mg L^{-1}) profile per mm of both sediments (A and B), at the sediment-water interface (indicated by vertical dotted line) during aerobic and anaerobic conditions.

community at the sediment surface during aerobic conditions [36,37], (3) precipitation of P with calcium-minerals [9,23], although the latter would mainly be expected in the high salinity treatment.

An explanation for the lack of P release in the long term might be the uptake of P by microbial mats growing on top of the sediment [35–37]. These mats can develop over time and might also benefit from stable sediment conditions that developed in the experimental set-up. However, our experiment was carried out in the dark, excluding photosynthetically active organisms, and no visible signs of such mats were observed. Nevertheless, the potential role of microbial sequestration of P on the long term cannot be ruled out.

Most likely Fe redox cycling played a dominant role in the absence of P mobilisation, as was also indicated by the strongly increased P release under anaerobic conditions compared to aerobic conditions (Fig. 5). It has been demonstrated that diffusive

P release should be prevented under aerobic conditions if pore water Fe:P ratios are relatively high (at least >1) [19,38,39]. In our sediments, however, pore water Fe:P ratios were very unfavourable. Nevertheless, oxidation processes might be able to mobilise Fe^{2+} from FeS$_x$ at a spatial micro-scale in the sediment surface at relatively low O$_2$ levels [40], catalysed by S oxidising microbes [41]. Our O$_2$ profiles showed that O$_2$ was available in the surface water and in the top millimetres of the sediment. The observed high S mobilisation rates in the low salinity treatment, where no S was added, indeed showed that SO$_4$$^{2-}$ is being mobilised from the sediment by the oxidation of FeS$_x$. Simultaneously, Fe^{2+} thus becomes available to be oxidised [40], and is able to sequester dissolved P. So the intrusion of O$_2$ in reduced sediments may mobilise S bound Fe at a millimetre spatial scale, providing dissolved Fe^{2+} for the formation of ferric Fe(OH)$_x$ at the sediment surface (Fig. 6). This mechanism may very well explain the

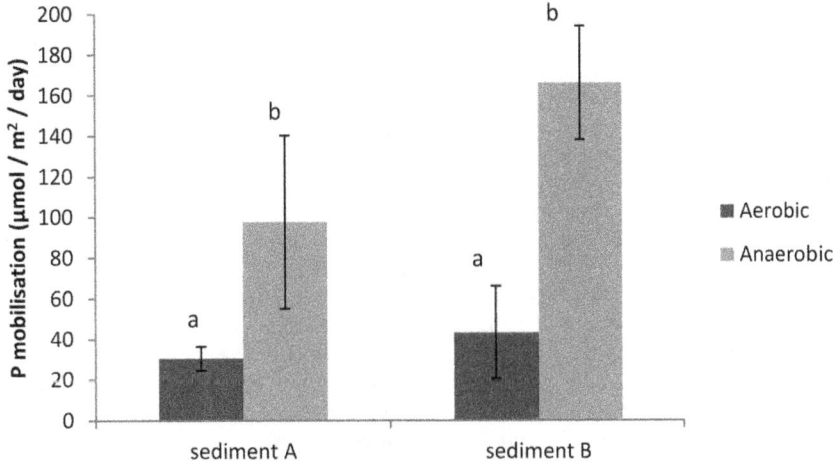

Figure 5. P mobilisation rates (μmol m^{-2} day^{-1}) during aerobic and anaerobic conditions for both sediment cores (A and B). Significant differences between treatments are indicated with different letters.

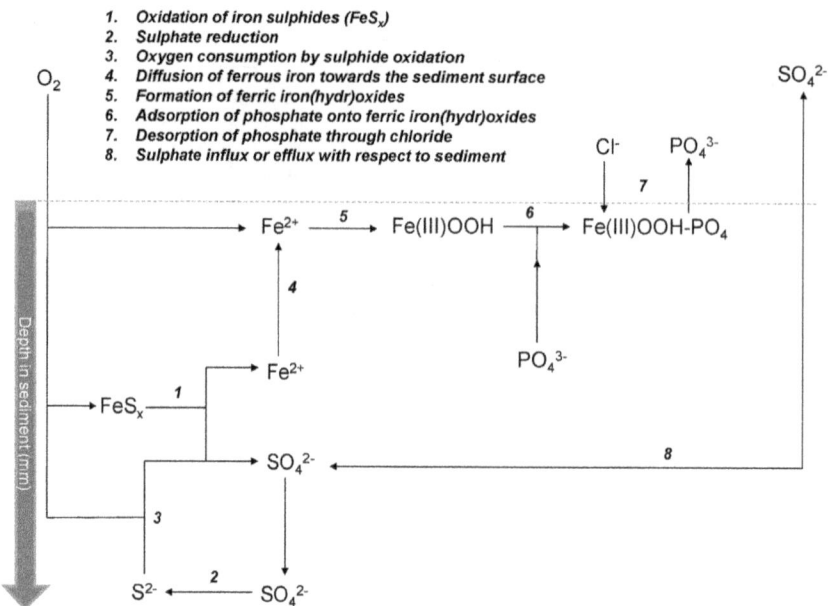

Figure 6. Schematic overview of the proposed mechanism, showing key processes in the upper millimetres of the S-rich, peat sediments involved in P mobilisation. Salinisation leads to an increased SO_4^{2-} influx, affecting Fe diffusion to the sediment surface, enabling increased P mobilisation in the longer term.

unexpected lack of P release from the sediments in the long term under aerobic conditions.

Our experimental set-up, without sediment disturbance and with relatively low biochemical O_2 demand (BOD) due to the absence of fresh organic matter input, will certainly have contributed to the long-term outcome of the experiment. Nevertheless, it seems plausible that it took a relatively long time before the sediment surface became sufficiently oxidised, or before the microbial population was sufficiently developed, to completely prevent P mobilisation in the experiment. These results in the longer term may represent field situations with stable non-bioturbated FeS_x-rich sediments or sediments with stagnant, hypolimnetic water. During anaerobic conditions, P mobilisation was strongly enhanced (Fig. 5), which clearly highlights the importance of O_2 availability to prevent P release. Field experiments are, therefore, necessary to validate our experimental results and suggested mechanism for the lack of P release from S-rich aquatic sediments.

4.2.2 Salinity effects.   For the high salinity treatment, one of the sediments showed an increase of the surface water P concentration also in the long term (80 weeks). In saline or estuarine systems, P is often found to be easily released from soil particles [5,7], and dissolved P concentrations are usually higher with increasing salinity [6,8,39]. At a high salinity, SO_4^{2-} concentrations increased in both surface water and pore water and a considerable part may be reduced deeper in the sediment, since it was not released to the surface water (Fig. 6). Produced S^{2-} will react with O_2 and interfere with the oxidation of FeS_x, or again immobilise Fe^{2+}. As expected, the net mobilisation of Fe^{2+} will be less, leading to insufficient formation of ferric $Fe(OH)_x$ to prevent the release of P to the surface water [18,19]. This decoupling of the Fe and P cycle [10] at a micro-scale diminishes the P-binding capacity at the water-sediment interface. In sediment A, O_2 penetrated deeper into the sediment, suggesting that less O_2 was consumed, less FeS_x was oxidised, and less Fe^{2+} was mobilised. This may partly explain the long-term release of P

from sediment A in the high salinity treatment. Desorption of P from ferric $Fe(OH)_x$ due to the high Cl concentrations [10,22] might have increased this effect.

4.3 Implications for water management

Although the mobilisation of P from the S-rich and relatively Fe-poor sediments (typical for coastal wetlands) was mainly driven by diffusion, the build-up of a stable oxidised sediment surface may have prevented the release of P under the experimental conditions. We hypothesise that the oxidation of FeS_x in the sediment surface delivers the Fe^{2+} necessary for the precipitation of P at the sediment-water interface (Fig. 6). Disturbance of the sediment-water interface due to wind, ebullition of gases from the sediment, and bioturbation can, however, prevent this build-up of a protective Fe-rich sediment surface and potentially increase the release of P [9,33]. Although such processes might also mix the sediment surface with O_2 and have an opposite effect. Moreover, our results indicate that an increased salinity may lead to a long-term P release, probably by interfering with the Fe^{2+} mobilisation due to increased SO_4^{2-} reduction rates in the anaerobic sediment. They also point out that sediments may react differently upon increased salinity. Therefore, O_2 and BOD, but also the actual concentration of SO_4^{2-} play a key role in the mobilisation of P from FeS_x-rich sediments. This might have important implications for water management and nature management of eutrophic peatlands in relation to salinisation.

More research, especially field measurements, is necessary to further confirm the experimental results we found for these FeS_x-rich sediments. Our experiment was carried out at 15°C and without the continuous input of reactive organic material. Warmer conditions, e.g. during warm episodes in summer will lead to increased mineralisation rates, and also to higher O_2 consumption rates and lower solubility of O_2. Especially when there is a high input of reactive organic matter, this will lead to strongly decreased O_2 concentrations in the surface water, which may prevent adequate oxidation of the sediment surface. Under such conditions this biogeochemical mechanism is expected to fail, leading to strong P mobilisation from the sediment as was shown in this study and also found by Smolders *et al.* [12]. As a result, floating-leaved species, or floating beds of algae or cyanobacteria may develop, which will further decrease the O_2 concentrations in the surface water and enhance sediment P mobilisation. This explains why FeS_x-rich sediments that show very high dissolved P concentrations and low dissolved Fe^{2+} concentrations tend to show a high P release mainly in summer, which has important implications for water management.

Conclusions

- Low pore water Fe:P ratios indicated a decoupling of the Fe and P cycle. Although these FeS_x-rich sediments were expected to release significant amounts of P by diffusion, this only seemed to be the case in the short term under aerobic conditions.

- Increased salinity led to co-precipitation of P with Ca^{2+} in the short term, lowering actual P concentrations. However, short-term P mobilisation rates were found to be similar for all treatments, regardless of salinity.

- Our experimental results suggest that the classic theoretical framework of oxidative conditions in the surface water that prevent P release from the sediment, may also hold in sediments showing unfavourable total Fe:S ratios but high FeS_x concentrations. In our FeS_x-rich, eutrophic sediments, typical for coastal wetlands, O_2 availability still seemed to be

the most important determinant of sediment P release, at least under stable sediment conditions.

- We suggest a controlling mechanism in which the partial oxidation of FeS_x mobilises sufficient Fe^{2+} at micro-scale for the precipitation of P at the sediment-water interface.

- Next to O_2, SO_4^{2-} plays a key role in P mobilisation, as high concentrations may counteract the oxidising effect by immobilising Fe^{2+}. In the longer term, an increased salinity may, as a result, led to P mobilisation despite oxidation of the sediment surface.

Acknowledgments

We would like to thank Jeroen Graafland and Rick Kuiperij for their practical assistance in the field and chemical analyses, Jelle Eygensteyn, Paul van der Ven and Sebastian Krosse for their help with chemical analyses, and Leon van den Berg for his help with statistical analyses. We are grateful to the water management authority 'Hoogheemraadschap Hollands Noorderkwartier' for the facilitation of this programme and to the nature management authority 'Natuurmonumenten' for giving their kind permission to take samples in their reserve.

Author Contributions

Conceived and designed the experiments: JMHVD AJPS LPML. Performed the experiments: JMHVD MJS. Analyzed the data: JMHVD AJPS LPML MJS. Contributed reagents/materials/analysis tools: JMHVD AJPS LPML JGMR. Wrote the paper: JMHVD AJPS LPML JGMR GVD.

References

1. Smith VH (2003) Eutrophication of Freshwater and Coastal Marine Ecosystems. A Global Problem. Environ Science & Pollution Research 10 (2): 126–139.

2. Lamers LPM, Falla S-J, Samborska EM, van Dulken IAR, van Hengstum G, et al. (2002) Factors Controlling the Extent of Eutrophication and Toxicity in Sulfate-Polluted Freshwater Wetlands. Limnology and Oceanography 47 (2): 585–593.

3. Geurts JJM, Sarneel JM, Willers BJC, Roelofs JGM, Verhoeven JTA, et al. (2009) Interacting effects of sulphate pollution, sulphide toxicity and eutrophication on vegetation development in fens: A mesocosm experiment. Environmental Pollution 157 (7): 2072–2081.

4. Nielsen DL, Brock MA, Rees GN, Baldwin DS (2003) Effects of increasing salinity on freshwater ecosystems in Australia. Australian Journal of Botany 51: 655–665.

5. Carpenter PD, Smith JD (1984) Effect of pH, iron and humic acid on the estuarine behaviour of phosphate. Environmental Technology Letters 6: 65–72.

6. Gunnars A, Blomqvist S (1997) Phosphate exchange across the sediment-water interface when shifting from anoxic to oxic conditions – an experimental comparison of freshwater and brackish-marine systems. Biogeochemistry 37: 203–226.

7. Suzumura M, Udea S, Sumi E (2000) Control of phosphate concentration through adsorption and desorption processes in groundwater and seawater mixing at sandy beaches in Tokyo Bay, Japan. Journal of Oceanography 56: 667–673.

8. Jordan TE, Cornwell JC, Boynton WR, Anderson JT (2008) Changes in phosphorus biogeochemistry along an estuarine salinity gradient: The iron conveyer belt. Limnology and Oceanography 53 (1): 172–184.

9. Boström B, Andersen JM, Fleisher S, Jansson M (1988) Exchange of phosphorus across the sediment-water interface. Hydrobiologia 170: 229–244.

10. Caraco NF, Cole JJ, Likens GE (1989) Evidence for sulphate-controlled phosphorus release from sediments of aquatic systems. Nature 341: 316–317.

11. Lamers LPM, Tomassen HBM, Roelofs JGM (1998) Sulfate-Induced Eutrophication and Phytotoxicity in Freshwater Wetlands. Environmental Science and Technology 32: 199–205.

12. Smolders AJP, Lamers LPM, Lucassen ECHET, van der Velde G, Roelofs JGM (2006) Internal eutrophication: How it works and what to do about it – a review. Chemistry and Ecology 22 (2): 93–111.

13. Einsele W (1936) Über die Beziehungen des Eisenkreislaufs zum Phosphatkreislauf im eutrophen See. Archiv für Hydrobiologie 29: 664–686.

14. Mortimer CH (1941, 1942) The exchange of dissolved substances between mud and water in lakes. J. Ecology 29: 280–329, Journal of Ecology 30: 147–201.

15. Roelofs JGM (1991) Inlet of alkaline river water into peaty lowlands: effects on water quality and Stratiotes aloides L. stands. Aquatic Botany 39: 267–293.

16. Caraco NF, Cole JJ, Likens GE (1993) Sulfate control of phosphorus availability in lakes. Hydrobiologia 253: 275–280.

17. Smolders AJP, Roelofs JGM (1995) Internal eutrophication, iron limitation and sulphide accumulation due to the inlet of river Rhine water in peaty shallow waters in The Netherlands. Archiv für Hydrobiologie 133: 349–365.

18. Hupfer M, Lewandowski J (2008) Oxygen Controls the Phosphorus Release from Lake Sediments –a Long-Lasting Paradigm in Limnology. International Review of Hydrobiology 93: 414–432.

19. Geurts JJM, Smolders AJP, Banach AM, van de Graaf JPM, Roelofs JGM, et al. (2010) The interaction between decomposition, N and P mineralization and their mobilisation to the surface water in fens. Water Research 44: 3487–3495.

20. Smolders AJP, Lamers LPM, Moonen M, Zwaga K, Roelofs JGM (2001) Controlling phosphate release from phosphate-enriched sediments by adding various iron compounds. Biogeochemistry 54: 219–228.

21. Wetzel RG (2001) Limnology: Lake and River Ecosystems. Academic Press 3, An Imprint of Elsevier, USA: 1006 pag.

22. Beltman B, Rouwenhorst TG, Van Kerkhoven MB, Van Der Krift T, Verhoeven JTA (2000) Internal eutrophication in peat soils through competition between chloride and sulphate with phosphate for binding sites. Biogeochemistry 50: 183–194.

23. House WA (1999) The physio-chemical conditions for the precipitation of phosphate with calcium. Environmental Technology 20: 727–733.

24. Wu Y, Wen Y, Zhou J, Wu Y (2014) Phosphorus release from lake sediments: Effects of pH, temperature and dissolved oxygen. KSCE Journal of Civil Engineering 18 (1): 323–329.

25. Reigersman CJA (1946) Ontzilting van Noord-Holland. (Desalinisation of North-Holland.) Rapport van de Commissie inzake het zoutgehalte der boezemen polderwateren van Noord-Holland, ingesteld bij Besluit van den Minister van Waterstaat van 24 april 1939. Rijksuitgeverij, 's-Gravenhage: 191 pp.

26. Boxman AW, Peters RCJH, Roelofs JGM (2008) Long term changes in atmospheric N and S throughfall deposition and effects on soil solution chemistry in a Scots pine forest in the Netherlands. Environmental Pollution 156: 1252–1259.

27. Van Gemerden H (1984) The sulphide affinity of phototrophic bacteria in relation to the location of elemental sulphur. Archiv für Mikrobiologie 139. 289–294.

28. Kamphake LJ, Hannah SA, Cohen JM (1967) Automated analysis for nitrate by hydrazine reduction. Water Research 1: 205–206.

29. Grasshoff K, Johannsen H (1972) A new sensitive and direct method for the automatic determination of ammonia in sea water. Journal du Conseil Permanent International pour l'Exploration de la Mer 34: 516–521.

30. Henriksen A (1965) An automated method for determining low-level concentrations of phosphate in fresh and saline waters. Analyst 90: 29–34.

31. O'Brien JE (1962) Automation in sanitary chemistry part 4: automatic analysis of chloride in sewage. Wastes Engineering 33: 670–682.

32. Golterman HL (1996) Fractionation of sediment phosphate with chelating compounds: a simplification, and comparison with other methods. Hydrobiologia 335: 87–95.

33. Boström B, Pettersson K (1982) Different patterns of phosphorus release from lake sediments in laboratory experiments. Hydrobiologia 92: 415–429.

34. Van Dijk G, Loeb R, Smolders AJP, Westendorp PJ (2013) Verbrakking in voormalig brak laag Nederland, bedreiging of kans? (Salinisation in former brackish Dutch lowlands, threat or opportunity?) H_2O 46 (3): 1–5.

35. Dupraz C, Reid RP, Braissant O, Decho AW, Normanc RS, et al. (2009) Processes of carbonate precipitation in modern microbial mats. Earth-Science Reviews 96: 141–162.

36. Deinema MH, Habets LHA, Scholten J, Turkstra E, Webers HAAM (1980) The accumulation of polyphosphate in Acinetobacter spp. FEMS Microbiology Letters 9: 275–279.

37. Hupfer M, Uhlmann D (1991) Microbially mediated phosphorus exchange across the mud-water interface. Verhandlungen des Internationalen Verein Limnologie 24: 2999–3003.

38. Gunnars A, Blomqvist S, Johansson P, Andersson C (2002) Formation of Fe(III) oxyhydroxide colloids in freshwater and brackish seawater, with incorporation of phosphate and calcium. Geochimica et Cosmochimica Acta 66 (5): 745–758.

39. Blomqvist S, Gunnars A, Elmgren R (2004) Why the limiting nutrient differs between temperate coastal seas and freshwater lakes: A matter of salt. Limnology and Oceanography 49 (6): 2236–2241.

40. Roden EE (2012) Microbial iron-redox cycling in subsurface environments. Biochemical Society Transactions 40: 1249–1256.

41. Imhoff A, Schneider A, Podgorsek L (1995) Correlation of viable cell counts, metabolic activity of sulphur-oxidizing bacteria and chemical parameters of marine sediments. Helgoländer Meeresunters 49: 223–236.

17

Relationships between Bacterial Community Composition, Functional Trait Composition and Functioning Are Context Dependent – but What Is the Context?

Ina Severin*[¤a], Eva S. Lindström, Örjan Östman[¤b]

Department of Ecology and Genetics/Limnology, Uppsala University, Uppsala, Sweden

Abstract

Bacterial communities are immensely diverse and drive many fundamental ecosystem processes. However, the role of bacterial community composition (BCC) for functioning is still unclear. Here we evaluate the relative importance of BCC (from 454-sequencing), functional traits (from Biolog Ecoplates) and environmental conditions for per cell biomass production (BPC; ^{3}H-leucine incorporation) in six data sets of natural freshwater bacterial communities. BCC explained significant variation of BPC in all six data sets and most variation in four. BCC measures based on 16S rRNA (active bacteria) did not consistently explain more variation in BPC than measures based on the 16S rRNA-gene (total community), and adding phylogenetic information did not, in general, increase the explanatory power of BCC. In contrast to our hypothesis, the importance of BCC for BPC was not related to the anticipated dispersal rates in and out of communities. Functional traits, most notably the ability to use cyclic and aromatic compounds, as well as local environmental conditions, i.e. stoichiometric relationships of nutrients, explained some variation in all six data sets. In general there were weak associations between variation in BCC and variation in the functional traits contributing to productivity. This indicates that additional traits may be important for productivity as well. By comparing several data sets obtained in a similar way we conclude that no single measure of BCC was obviously better than another in explaining BPC. We identified some key functional traits for productivity, but although there was a coupling between BCC, functional traits and productivity, the strength of the coupling seems context dependent. However, the exact context is still unresolved.

Editor: Dionysios A. Antonopoulos, Argonne National Laboratory, United States of America

Funding: This study was supported by a grant from the Carl Tryggers Foundation to ESL and ÖÖ, from a grant from the Olsson-Borgh foundation to IS, and by separate grants from the Swedish Research Council to ESL (project number 2009-5172) and ÖÖ (2007-5932). The funders had no role in study design, data collection and analysis, decision to publish, or preparation of the manuscript.

Competing Interests: The authors have declared that no competing interests exist.

* Email: inaseverin@gmx.de

¤a Current address: Marine Biological Section, Department of Biology, University of Copenhagen, Helsingør, Denmark
¤b Current address: Department of Aquatic Resource, Swedish University of Agricultural Science, Uppsala, Sweden

Introduction

Lakes and reservoirs are central compartments in the global carbon cycle [1], through the processing of organic matter by microorganisms, making resources available for higher trophic levels of the grazer chain [2] or contributing to the outgassing of carbon dioxide to the atmosphere [3,4]. It is well known that local environmental conditions such as temperature and availability of nutrients are related to bacterial production [5]; the question is, however, if also bacterial community composition (BCC) plays a role. Results obtained so far are rather inconclusive, both in field [6,7] and experimental studies [8–12].

Disparate results may rise from methodological differences, for instance due to the fact that in natural bacterial communities a substantial proportion of the cells may be inactive or dormant [13], and BCC measures including such cells may obscure BCC-function relationships. Further, results could be affected by the choice of method to classify operational taxonomic units (OTUs), i.e. whether sequence similarities alone or also phylogenetic distances between taxa [14,15] are taken into account. Using a phylogenetic diversity measure may result in a tighter observed coupling between BCC and functioning if closer related taxa are functionally more similar. However, carbon processing traits tend to be dispersed in the 16S rRNA phylogeny [16], i.e. they are shared in phylogenetically shallow clusters all across the bacterial realm. Therefore, any BCC measure may have a limited explanatory power for productivity compared to measurements of key functional traits. For instance, in a previous study [12], we showed that the community productivity of heterotrophic bacteria is dependent on the community's ability to use certain carbon substrates in habitats where these substrates are abundant. Thus, functional trait composition may have a better explanatory power for functioning than BCC.

In addition to the methodological issues, the strength of BCC-functioning relationships can differ due to ecological reasons, depending for instance on the proportion of generalists and specialists in a community, since a great proportion of generalists should result in weaker BCC-functioning relationships [17]. The degree of generalism may in turn depend on community assembly mechanism and environmental heterogeneity. Dispersal may, for example, favor resource generalists [18], and BCC and functioning should therefore be uncoupled if the dispersal rate is high. In contrast, large differences in environmental conditions and low (but not limiting) dispersal among communities could favor resource specialist taxa via species sorting processes and, thus, stronger BCC-functioning relationships are expected [7,19].

For a better understanding of the importance of BCC for community functioning we need a conceptual framework and systematic studies including both methodological awareness and ecological considerations. Here we used six different data sets of aquatic bacteria, from 7–15 communities each, to explore the relative importance of the local environment, carbon processing trait composition and BCC for the productivity of bacterial communities depending on dispersal rates and environmental heterogeneity among communities. Two of the data sets originate from lake sediment communities, two from the epilimnion communities of lakes and two from stream communities. The ecosystem function of interest in this study is bulk bacterial community production (leucine incorporation) and the different functional traits under consideration are the communities' ability to use different carbon substrates (assessed from Biolog-plates). Bacterial taxonomic community composition (BCCt) and phylogenetic community composition (BCCp) were determined by high-throughput DNA sequencing, hypothesizing that BCCp would explain more variation in productivity than BCCt. Further, our BCC measures were based on the 16S rRNA (rBCC) as well as the 16S rRNA gene (dBCC). Assuming that rBCC would better reflect the active part of the community [20] we hypothesize that rBCC will be more closely linked to productivity than measures based on the total community (dBCC). Finally, we hypothesize that BCC (incorporating all aspects of BCC) would explain less variation in productivity with increasing dispersal rates and/or lower environmental heterogeneity among communities in a data set because these conditions are assumed to favor habitat generalism. In contrast, functional trait composition would not be affected by the degree of generalism and should therefore explain relatively more of the variation in productivity with increasing dispersal rates and/or higher environmental heterogeneity.

Material and Methods

All abbreviations are given as an overview in Table S1.

Ethics Statement

No permits were required to sample any of the water bodies in this study. The authors also confirm that the sampling did not affect endangered or protected species.

Sampling

Three freshwater systems in two geographic regions in Sweden were sampled in summer and autumn 2010; pelagic lake water (epilimnion), lake sediments (upper 1 cm) and stream waters. One of the lake systems is situated in the province of Jämtland (approximately 63°N and 13°E) where all the 14 sampled lakes are connected to the river Indalsälven either by an inlet and/or an outlet. Samples were obtained in June from all pelagic lake waters (Jw) and 7 sediments (Js) (due to harsh weather conditions the sediments in the remaining seven lakes could not be sampled). The other lake-system is situated in Uppland (approximately 60°N and 17°E), where water (Uw) and sediments (Us) were sampled in 15 hydrologically unconnected lakes in June. Stream samples were obtained from 15 sites in River Fibyån, Uppland, in July (S I) and September (S II). We assumed that 1) dispersal of bacterial cells among communities was lowest in sediments since it requires both sediment resuspension within lakes and dispersal among lakes; 2) dispersal rates among stream water communities was highest due to shorter water retention time in streams in relation to lakes, but that dispersal was greater in S II compared to S I since water levels indicated a higher water flow in September; 3) among the lakes, dispersal was higher among the Jämtland lakes since they are all part of the same river system while the Uppland lakes were not. Based on these assumptions the data sets were ranked from 1–6 where 6 denotes the highest dispersal rate (S II) and 1 the lowest (Us).

Environmental data

Non-purgeable total organic carbon (hereafter termed TC) in water samples was determined by measuring organic carbon after acidification with HCl (TOC-5000, Shimadzu, Kyoto, Japan). Total nitrogen (TN) in water samples was measured spectrophotometrically (Hitachi U-2000, Hitachi, Ltd., Tokyo, Japan) as nitrate after oxidation at high temperature. Total phosphorus (TP) in water and sediment samples was also measured spectrophotometrically after oxidative hydrolysis of organically bound phosphorus. Total carbon (TC) and total nitrogen (TN) in sediment samples was determined in freeze-dried and ash-free sediments by combustion with oxygen (elemental combustion system, Costech Analytical Technologies, Inc., Valencia, CA, USA).

The Jämtland lakes are generally more oligotrophic with low levels of total organic carbon compared to all Uppland sites (Table 1). However, the environmental variability, measured as coefficient of variation (CV) among sites, was not consistently higher in any data set but differed between environmental variables (Table 1).

Bacterial abundance

Cell abundances were determined flow-cytometrically [21] (CyFlow space, Partec GmbH, Münster, Germany) for water samples and microscopically (Nikon Eclipse E600 fluorescence microscope, Nikon Corporation, Tokyo, Japan) for sediment samples. Following the protocol by [21] water samples were fixed with filtered formaldehyde (3.7% final concentration) and stored at 4°C for a maximum of two days. The cells were stained with SYTO 13 (Invitrogen, Life Technologies Ltd, Paisley, UK). The sediment samples were diluted 10× with filtered lake water (0.2 μm filters, Supor-200 Membrane Disc Filters, 47 mm; Pall Corporation, East Hills, NY, USA) and fixed with filtered formaldehyde (3.7% final concentration). This sediment slurry was then diluted 500× with an 50/50 mix of tap water and deionized water and sonicated at 100 W for 1 min on ice. After settlement of the particles, the cells in the supernatant were stained with DAPI (4′,6-Diamidino-2-Phenylindole, Dihydrochloride, Invitrogen, Life Technologies Ltd, Paisley, UK) for 15 min and filtered onto 0.2 μm black polycarbonate filters (Sorbent AB, Västra Frölunda, Sweden). At least 10 fields with at least 200 cells in total were counted for each filter.

Table 1. Dispersal, environmental conditions and heterogeneity (measured as coefficient of variation, CV) in the data sets.

Data set	Dispersal category	TC	TN	TP	TN/TC	TP/TC	CV TC	CV TN	CV TP
Jw	1	4	0.1	15	0	4.5	43.5	60.5	60.6
Uw	2	21.7	1.1	46.6	0.03	4.46	32.2	29.6	73.3
Js	3	4.3	0.4	0.1	0.05	2.31	54.8	45.3	38.9
Us	4	15	1.1	0.1	0.1	0.02	44.6	39.1	23.8
S I	5	29.4	1.9	93.9	0.08	0.01	44.5	53.2	78.3
S II	6	31.8	2	103.9	0.07	4.08	50.1	33.3	106.2

TC: total organic carbon [mg l^{-1} for water and % dry weight for sediments], TN: total nitrogen [mg l^{-1} for water and % dry weight for sediments], TP: total phosphorus [µg l^{-1} for water and % dry weight for sediments].

Bacterial production

The incorporation of ^{3}H-labelled leucine into bacterial protein in water samples was determined using the modified method from [22]. In short, for each sample two parallels and a blank (immediate addition of a final concentration of 5% TCA) were incubated in a final concentration of 100 nM ^{3}H-leucine for one hour in the dark at close to ambient temperatures (the same temperature for all samples within a data set). The incubation was stopped by adding a final concentration of 5% TCA to the water samples. After washing with 5% TCA and 80% Ethanol, 0.5 ml of the scintillation cocktail (Optiphase Hisafe 2, PerkinElmer, Inc., Waltham, MA, USA) were added and the samples were kept for at least 24 hours before measurement of the incorporated ^{3}H-leucine (Packard Tri-Carb 2100TR Liquid Scintillation Analyzer, GMI, Inc, Ramsey, MN, USA). For sediment samples, homogenized sediment was diluted 1000× with sterile-filtered (0.2 µm-filter) lake water from the same location. This sediment slurry was incubated with a final concentration of 10 µM ^{3}H-leucine for one hour. Samples and blanks were then treated as described for the water samples.

Average per cell productivity (BPC) was inferred as bacterial production (BP)/bacterial abundance (BA). This was done in order to obtain a measure of bulk community productivity independent of abundance.

Functional trait composition

Functional trait composition of communities was assessed as the capacity of the communities to use different carbon sources. 150 µl of either water or a 1000× dilution of the sediment were incubated on Biolog EcoPlates (Biolog, Inc., Hayward, CA, USA). These 96-well plates contain 3 sets of 31 carbon substrates and a water blank. The use of these substrates was followed by absorbance measurement of the colorless tetrazolium dye which is reduced to a violet formazan during oxidation of the substrates by bacterial metabolism. Changes in color development were measured using a microplate reader (TECAN ULTRA 384, Tecan Group Ltd., Männedorf, Switzerland) at 595 nm. Immediately after inoculation, the zero time-point was measured and measurements were repeated daily. The color development was followed until the maximum color development was reached (no further increase in absorbance). The overall color development of each plate was expressed as average well color development (AWCD, [23]) and the absorbance profiles corresponding to the time at which the AWCD was 1.5 AWCD were used. In the analysis substrates were grouped according to their molecular structure into polymers (tween 40 and 80, α-cyclodextrin and glycogen), aromatic compounds (2-hydroxy benzoic acid, 4-hydroxy benzoic acid, L-phenylalanine, phenyletylamine), non-aromatic amino acids (L-arginine, L-asparagine, L-serine, L-threonine, glycyl L-glutamic acid), cyclic compounds other than aromatics (D-cellobiose, α-D-lactose, β-methyl-D-glucoside, D-xylose, N-acetyl-D-glucosamine, glucose-1-phosphate, D-galactonic acid γ-lactone, D-galaturonic acid), and simpler compounds (the rest). The average AWCD-normalized absorbance scores of the substrates in each group were calculated for the PLS analyses (see below).

Nucleic acid extraction

100 ml of the lake and stream water were filtered onto a 0.2 µm filter (Supor-200 Membrane Disc Filters, 47 mm; Pall Corporation, East Hills, NY, USA). The filters were stored in liquid nitrogen in the field and later at −80°C until further processing. Approximately 0.5 ml of the undiluted sediment was frozen directly. Nucleic acids (DNA and RNA) were extracted using the Easy-DNA kit from Invitrogen (Life Technologies Ltd, Paisley,

UK) according to protocol #3. For the extraction of nucleic acids form the filters, glass beads (0.1 mm zirconia/silica, BioSpec Products, Inc., Bartlesville, OK, USA) were added at the beginning of the extraction procedure. The tubes were then shaken in a vortex for 15 min to break filters and cells. No such step was included for the sediment samples. Extracts were quality-checked on a 1% agarose gel. After completion of the protocol, half of the nucleic acid extract was subjected to DNase treatment (DNase I, Invitrogen, Life Technologies Ltd, Paisley, UK) and reverse transcription (RevertAid HMinus First Strand cDNA Synthesis kit, Fermentas Sweden AB, Helsingborg, Sweden) according to the manufacturers' instructions using random hexamer primers. The resulting cDNA as well as the DNA extracts were stored at $-80°C$ until further processing.

PCR amplification and Template Preparation

The bacterial hypervariable regions V3 and V4 of the 16S rRNA (cDNA) as well as its gene (DNA) were PCR amplified using forward primer 341 5'- CCTACGGGNGGCWGCAG-3' and individually bar-coded reverse primers 805 5'- GAC-TACHVGGGTATCTAATCC-3' [24]. Each 20 μL PCR reaction contained 0.4 U Phusion high-fidelity DNA polymerase (Finnzymes, Espoo, Finland), 1× Phusion HF reaction buffer (Finnzymes), 200 μM of each dNTP (Life Technologies Ltd, Paisley, UK), 250 nM of each primer (Eurofins MWG, Ebersberg, Germany), 0.4 mg mL^{-1} BSA (New England Biolabs, Ipswich, UK) and 5–10 ng of extracted nucleic acid. Thermocycling was conducted with an initial denaturation step at 98°C for 30 sec, followed by 25 cycles of denaturation at 98°C for 10 sec, annealing at 50°C for 30 sec and extension at 72°C for 30 sec, and finalised with a 7-min extension step at 72°C. Three to four technical replicates were run per sample, pooled after PCR amplification and quality-checked on a 1% agarose gel. Purification was carried out using the AMPure XP purification kit (Beckman Coulter Inc., Brea, CA, USA). Nucleic acid yields were then checked on a fluorescence microplate reader (Ultra 384; Tecan Group Ltd., Männedorf, Switzerland) applying the Quant-iT PicoGreen dsDNA quantification kit (Invitrogen, Life Technologies Ltd, Paisley, UK). Finally, PCR amplicons were combined equimolarly, i.e. in equal proportions, to obtain a similar number of 454 pyrosequencing reads per sample.

454 Pyrosequencing

The final, pooled amplicon was 454 pyrosequenced with a 454 GS FLX system (454 Life Sciences) at the Norwegian High-Throughput Sequencing Centre, University of Oslo (NSC; Oslo, Norway; http://www.sequencing.uio.no), using Titanium chemistry. Sequences were, prior to analyses, quality-checked and truncated to 400 bases. Each data set was individually processed with AmpliconNoise to reduce the number of PCR and 454 sequencing artifacts and chimeras [25]. 454 pyrosequencing reads have been deposited in the National Center for Biotechnology Information Sequence Read Archive (NCBI-SRA) under accession number SRP016145. For further analysis, singletons as well as sequences belonging to *Archaea* and chloroplasts were removed from the data set. Operational taxonomic units (OTUs) were defined using complete linkage clustering at a level of 99% sequence identity. Computations were performed on resources provided by SNIC through the Uppsala Multidisciplinary Center for Advanced Computational Science (UPPMAX). Taxonomic affiliation (phylum-level) of OTUs was determined by aligning representative sequences to the Greengenes imputed core reference alignment [26] (http://greengenes.lbl.gov) using PyNAST [27] in Qiime [28].

After processing an average of 1450 (median = 774) and 450 (median = 347) reads per sample of the total community (DNA) and active bacterial community (cDNA), respectively, were obtained. Because of the risk of large sampling errors with few numbers of reads per sample we omitted samples with <200 reads (1 sample in S I and 4 samples in S II for the total commuinity (DNA), and 1 sample in Js, 4 samples in Us and 2 each in the stream samples for the active community (cDNA)). The range in number of reads between cDNA samples was 200–2445 and for DNA 200–4719.

Phylogenetic tree

Due to the limited length of the sequenced region (approx. 400 bp) and the large amount of different taxa (over 30 000) we could only construct a robust phylogenetic tree for a subset of the total community. Otherwise the phylogeny would be a random tree without sufficient node support. Therefore we constructed phylogenetic trees for the most abundant and, hence, likely functionally most important taxa. For the 142 taxa of the cDNA data set and the 153 taxa of the DNA data set that had an average relative abundance of 0.1% across all samples or 0.3% in a single data set (18% and 25% of all reads in total and active community, respectively) we constructed 10 000 single phylogenetic trees and the consensus tree from the 16S RNA sequence using a generalized time reversible (GTR) evolutionary model with gamma-distributed rate variation across variable sites in mrBayes 3.2 [29]. The branch length prior was set to a uniform clock. The standard deviations of splits after 10 000 000 generations was < 0.005 indicating most nodes were well supported. A tree was sampled every 1000th time step and the last 5 000 trees from the two runs were saved and used for calculating the consensus trees. From the consensus trees we caluculated phylogentic similarities of communities using the Phylogenetic Community Dissimilarity [30] in the picante-package for R. It calculates the pairwise phylogenetic distance among nonshared taxa between two communities, i.e this measure is based only on the occurrence of different taxa and their phylogenetic distance, not the relative abundance of taxa.

Bacterial community composition

We calculated four different estimates of bacterial community composition for each of the six data sets, i.e. all BCC measures were only calculated for communities within a data set and do not account for between data set compositional differences (which was much larger than within data set differences). For each DNA and cDNA data set we calculated one measure of taxonomic composition without accounting for phylogenetic distance between taxa (dBCCt and rBCCt for total and active community, respectively), and one measure of phylogenetic community composition accounting for phylogenetic distances (dBCCp and rBCCp for total and active community respectively). To extract one measure of rBCCt and dBCCt for each community we used the site scores from the first axis of a Principal Coordinate Analysis (PCoA) of the Morisita-Horn distance matrix (see 'statistical analyses' below) for each data set. In this case the first axis explained 12–42% of the variation in rBCCt and 33–70% in dBCCt, being highest in the Js data set, and lowest in the S II data set (Table S2). Similarly, using the phylogentic distance matrix (see above) sites scores were obtained from the first axis of PCoA both of the active (rBCCp) and total bacterial communities (dBCCp). Communities with similar PCoA site scores have thus closely related non-shared taxa whereas non-shared taxa are more distantly related between communities with different site scores. The first axis explained similar amounts of variation in community

Table 2. Average Bray-Curtis (BC) and Morisita-Horn (MH) dissimilarities.

Data set	BC Biolog	MH rBCCt	MH dBCCt	MH r/dBCCt
Us	0.32 (0.13)	0.82 (0.11)	0.71 (0.06)	0.78 (0.12)
Js	0.22 (0.03)	0.86 (0.17)	0.70 (0.13)	0.84 (0.14)
Uw	0.22 (0.06)	0.71 (0.11)	0.60 (0.12)	0.52 (0.15)
Jw	0.20 (0.08)	0.63 (0.11)	0.45 (0.09)	0.56 (0.07)
S I	0.13 (0.02)	0.71 (0.09)	0.71 (0.09)	0.62 (0.07)
S II	0.13 (0.03)	0.63 (0.07)	0.72 (0.05)	0.70 (0.09)
Average	0.20 (0.06)	0.72 (0.09)	0.66 (0.10)	0.68 (0.12)

Values are calculated between the sampling sites within each data set for carbon substrate use (Biolog, functional trait composition) and BCC, respectively. Standard deviations (SD) are given in parenthesis. MH r/dBCCt is the average of the within sample Morisita-Horn dissimilarity between rBCCt and dBCCt.

composition in all data sets, 19–40% in rBCCp and 22–38% in dBCCp (Table S2), again being highest in the Js data set in both cases.

Statistical analyses

Average beta-diversity of bacterial communities (OTUs of 99% 16S rRNA similarity) within each data set (rBCCt and dBCCt) was calculated as Morisita-Horn dissimilarities. Morisita-Horn dissimilarities close to zero indicate very similar communnities, whereas values close to 1 indicate completely different communities.

Bray-Curtis dissimilarites of carbon substrate use (AWCD scores of all 31 substrates) were obtained as an estimate of between site variation in trait composition. Bray-Curtis dissimilarities close to zero indicate communities that have a very similar relative use of different carbon substrates, whereas values close to 1 indicate a completely different use of carbon substrates. Non-metric multi-dimensional scaling (nMDS) was performed using PAST version

2.17 [31] to show differences in carbon substrate use and bacterial community composition (BCCt) among data sets.

To study associations between community BPC and local environmental factors, functional trait composition of communities and the different aspects of BCC within data sets, we did partial least square regressions (PLS) with SIMCA 12.0 (Simca 12.0.1., Umetrics AB, Umeå, Sweden). BPC was the dependent variable and TC, TP, TN, TN/TC, TP/TC and average use of the different substrate groups and PCoA sites scores of the first axes of BCC (all BCC measures) were used as explanatory variables (Table 2). PLS has the advantages that it can handle co-variation among variables and is not sensitive to the number of explanatory variables relative sample size as explanatory variables are transformed into one or several latent variables that explain the maximum variance of the dependent variable [32]. From the PLS we extracted Variable Importance for the Projection (VIP-scores) that describe the relative importance of a variable for the

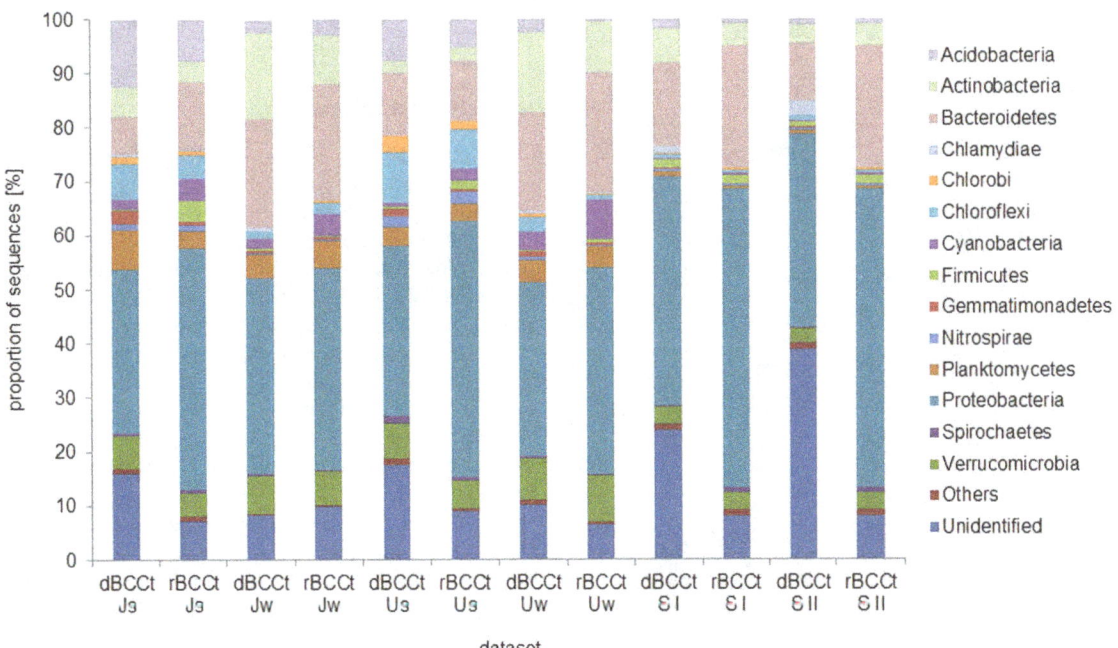

Figure 1. Bar diagrams representing the relative proportion of 16S sequences (rRNA and gene) belonging to the most abundant phyla. 'Others' contain OTUs with a relative abundance below 0.5% in the entire data set. 'Unidentified' denotes OTUs whose taxonomic affiliation is unknown.

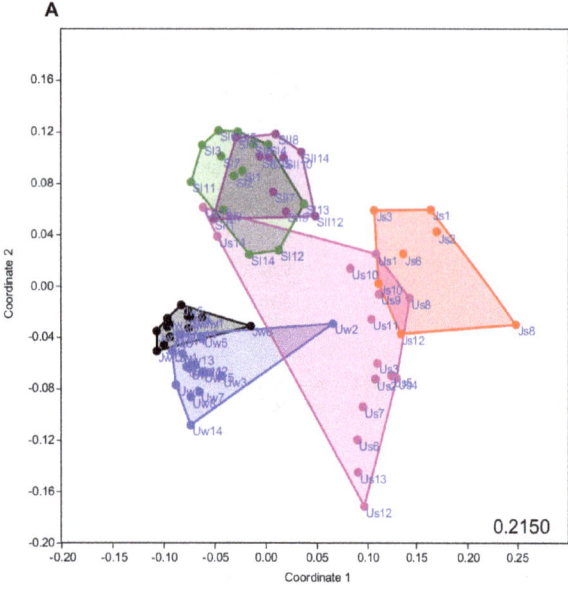

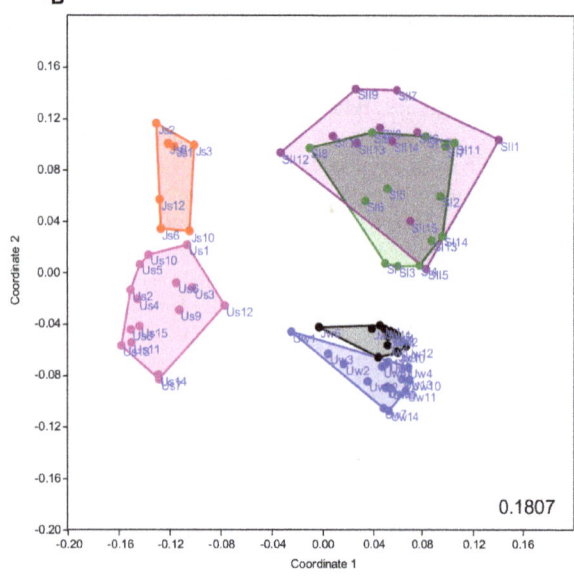

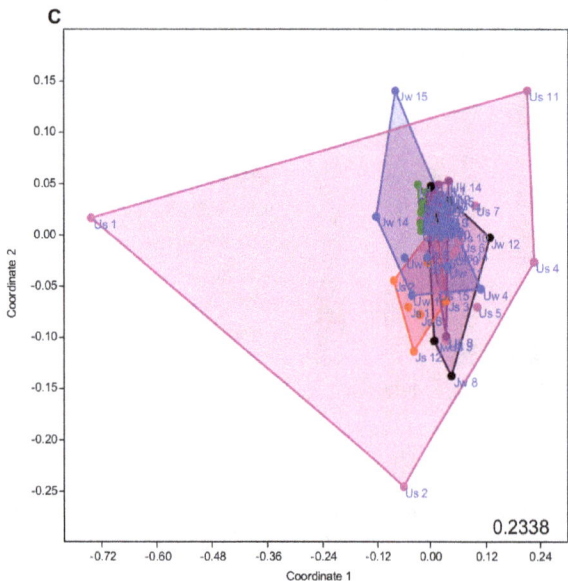

Figure 2. Results from a non-metric multi-dimensional scaling (nMDS) analysis. Depicted are the differences in bacterial community composition between all stations for rBCCt (A) and dBCCt (B), based on Morisita-Horn dissimilarities. The difference in carbon use is based on Bray-Curtis dissimilarities of the Biolog data (C). Stress values are given in the lower right corner.

correlation between the latent explanatory variables and the dependent variable. VIP-scores>1 indicate important variables, and the higher the value the more important is a variable for the correlation between the latent and dependent variable. We used the scaled and centered coefficients between the explanatory variables and the dependent variable to show the direction of the relationship (positive or negative). To infer covariation between different explanatory variables and BPC we plotted the loadings from the PLS. Explanatory variables with high loadings (positively or negatively) explain most of the variation in the latent variables, and explanatory variables close together show high covariation. In the loading plot, explanatory variables close to BPC are positively correlated with BPC whereas explanatory variables distant from BPC are negatively correlated with BPC. However, to get actual values of correlations between explanatory variables we also calculated Pearson correlation coefficients, r. BPC and environmental data was log-transformed prior to the analysis for the data to better fit a normal distribution. We did paired t-tests to compare the explanatory power of active (rBCC) and total (dBCC), respectively phylogenetic based (BCCp) or taxonomic based (BCCt) measures of BCC on BPC.

To study whether dispersal rates and environmental heterogeneity between communities in a data set may contribute to the relative importance of BCC, functional traits and the environment on productivity we did Pearson correlation between the VIP-scores of explanatory variables in a data set and dispersal level or degree of environmental heterogeneity (CV of environmental variables) in each data set. For functional trait composition and environmental variables VIP values could be used directly. For rBCC (active part) and dBCC (total community) we used the highest VIP-value of BCCt and BCCp, respectively.

Results and Discussion

In this field study we investigated the potential importance of bacterial community composition (BCC), functional trait composition (Biolog substrate use), and local environment for functioning (bacterial production per cell, BPC) of freshwater bacterial communities. We hypothesized that these relationships may differ depending on environmental heterogeneity and rates of dispersal among communities. Further, we tested the idea that depending on how BCC was determined, the strength of the relationship to functioning would differ. To enable such an evaluation we obtained six different data sets, each consisting of 7–15 freshwater bacterial communities (water and sediments, lakes and streams), in an identical manner, and statistically evaluated the steering factors for functioning.

The bacterial communities were analyzed by 454 sequencing and their compositions were found to be rather typical for freshwaters [33], e.g., dominated by Proteobacteria and Bacteroidetes (Fig. 1) but there were clear differences in bacterial community composition based both on the 16S rRNA and the 16SrRNA gene (Fig. 2A and B).

The average MH dissimilarity among sites was similar for those two BCC measures, i.e. on average 0.72 among all sites and ranging between 0.63 and 0.86 for rBCCt (Table 2, Fig. 2A) and 0.66 on average and ranging between 0.45 and 0.72 for dBCCt

Table 3. VIP values from PLS analysis between BPC and the explanatory variables for the data sets.

Model fit			Data set			
	Us	Js	Uw	Jw	S I	S II
R2X	23	23	28	56	26	25
R2Y	46	79	54	94	27	51
Explanatory variable						
rBCCt	**1.11 (−)**	0.46 (+)	**1.88 (+)**	**1.76 (−)**	**1.50 (+)**	**1.38 (−)**
rBCCp	**1.02 (−)**	0.36 (+)	0.31 (+)	0.75 (−)	0.11 (−)	0.60 (+)
dBCCt	0.28 (+)	0.56 (+)	0.19 (−)	0.51 (+)	**1.73 (−)**	0.25 (−)
dBCCp	0.95 (−)	**2.25 (−)**	0.45 (−)	0.73 (−)	**2.06 (+)**	**1.13 (−)**
amino acids	0.47 (+)	**1.28 (+)**	0.56 (−)	0.86 (−)	**1.18 (−)**	0.38 (−)
aromatic	0.43 (−)	**1.21 (−)**	**1.11 (+)**	**1.10 (+)**	**1.24 (+)**	0.94 (−)
simple	**1.13 (−)**	**1.31 (−)**	0.4 (−)	**1.06 (−)**	0.26 (−)	0.28 (+)
polymer	0.33 (+)	0.15 (+)	0.5 (+)	0.63 (−)	0.09 (−)	0.31 (−)
cyclic	**1.92 (+)**	**1.53 (+)**	0.33 (−)	0.62 (+)	0.06 (−)	**1.21 (+)**
TN	1.10 (−)	0.43 (−)	**1.22 (+)**	0.87 (+)	0.22 (−)	**1.51 (−)**
TN/TC	0.89 (+)	0.09 (+)	0.67 (+)	**1.25 (+)**	0.80 (+)	0.53 (+)
TC	**1.28 (−)**	0.04 (−)	0.1 (+)	0.64 (−)	0.41 (−)	**1.56 (−)**
TP	0.61 (−)	0.24 (+)	**1.83 (+)**	**1.22 (+)**	0.30 (+)	**1.54 (−)**
TP/TC	**1.13 (+)**	0.92 (+)	**1.65 (+)**	**1.20 (+)**	0.73 (+)	0.58 (−)

VIP>1, identifying variables most relevant for explaining BPC, are shown in bold. The direction of the association between BPC and the explanatory variables is deduced from scaled and centered coefficients (CoeffCS) and given in parenthesis. R2X is the proportion of variation in the explanatory data set explained by the latent factor(s), and R2Y is the proportion of variation in BPC explained by the latent factor(s) from the explanatory data set.

(Table 2, Fig 2B). In contrast, the dissimilarity of functional trait composition (Bray Curtis distance of Biolog substrate use) was much lower, the average being only 0.20 (Table 2). The functional trait composition did also not differ among habitat types in an obvious way (Fig. 2C). This may indicate a redundancy of taxa for processing carbon substrates, i.e. that different taxa can perform similarly on the same carbon substrate.

Assuming that rBCCt should reflect the active community and dBCCt the total community [20] the proportion of active taxa appeared to have been low since the average MH dissimilarity between rBCCt and dBCCt of a given community was 0.68 (Table 2). Especially in sediments, many bacterial cells could have been inactive since MH-distances were on average 0.84 and 0.78 for Jämtland and Uppland respectively, while the lake water communities may have shown a greater proportion of active cells, however still showing MH-distances above 0.5 (Table 2). Therefore it may be expected that rBCCt and dBCCt would be differently related to our functional measure BPC. Our hypothesis was that the active community (rBCC) would be better related to functioning than the total community (dBCC). In our analysis we also included two BCC measures based on phylogenetic distances (rBCCp and dBCCp), hypothesizing that a phylogenetic distance measure would show a stronger coupling between BCC and BCP than the rigid OTU definition used to calculate MH dissimilarities. The potential influential factors on BCP were investigated using PLS which is a statistical method suitable for handling many variables with co-variation among them [32].

The latent variable(s) generated from BCC, carbon substrate use and nutrient levels in the PLS explained most variation in BPC in the Jw data set (94%; R2Y in Table 3) and least variation in S I (27%), and more than 50% in all other data sets. Thus, the latent

variables explained a major part of variation in BPC in most data sets. In all data sets the PLS identified BCC to explain an important part of the variation in BPC, since one or several measures of BCC had VIP>1, but other factors contributed as well (Table 3). In four of the data sets (both lake water data sets, Js and S I) a measure of BCC had the highest VIP value, i.e. explained most of the variation in BCP (Table 3, Fig. S1). In the remaining two data sets (Us and S II) functional trait composition of communities and the local environmental conditions, respectively, were better explanatory variables. Since the most important factor thus varied among data sets this result highlights the importance of not relying on a single set of data to draw general conclusions regarding drivers of bacterial functioning in nature. In line with these results are several previous studies showing variable relationships between BCC and functioning [6,7,8,9,34,35].

A methodological consideration arising from these data is that if we had used only one BCC measure instead of four to infer the general importance of BCC for functioning we would have perceived it to be smaller, since "best" BCC measure varied among data sets. However, contrary to our expectations we did not find rBCC (active community) to be more closely coupled to functioning than dBCC (total community) because rBCC did not have higher VIP-values than dBCC (paired t-test: $t_{10} = 0.3$, p = 0.8). Further, phylogenetic measures of beta-diversity (BCCp) did not explain more variation than those based on the distance of OTUs (BCCt) (paired t-test: $t_{(10)} = 0.1$, P = 0.9). In fact, the BCC measure taking into account the phylogenetic relatedness of the active compartment of the community (rBCCp) performed "worst" with only one VIP-value>1 (Uppland sediment VIP = 1.02). Methodological reasons for these unexpected results may be that rRNA based methods are poor measures of actual

Table 4. Pearson correlation coefficients between BCC measures (in parenthesis) with VIP-values>1 from the PLS.

Data set	Amino acid	Aromatic	Simple	Cyclic	TC	TN	TP	TN/TC	TP/TC
Us (rBCCt)			-0.31	-0.22	-0.32				
Js (dBCCp)	-0.33	0.46	0.75	-0.69					
Uw (rBCCt)		0.30				0.38	0.63		0.74
Jw (rBCCt)		-0.45				-0.60		-0.67	-0.62
S I (dBCCp)	-0.16	0.18							
S II (rBCCt)				-0.05	0.20	0.40	0.33		

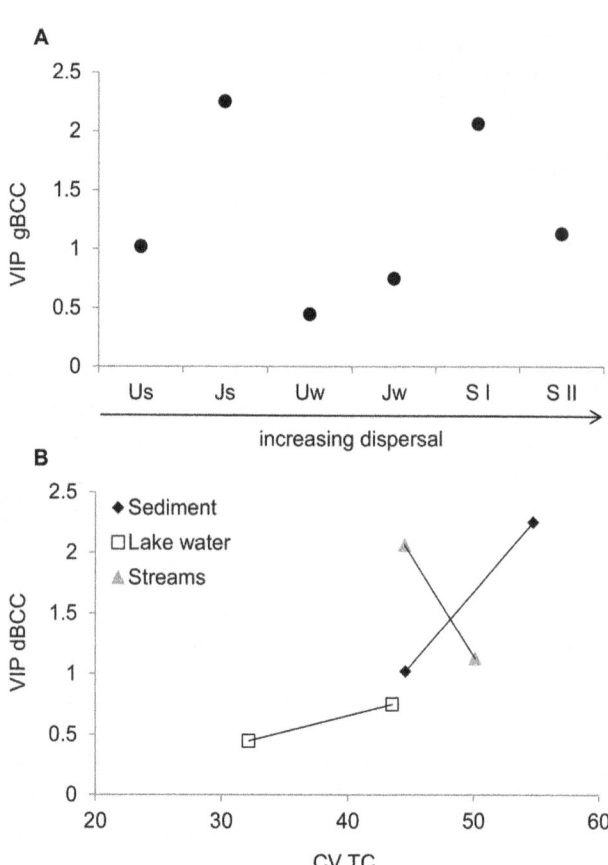

Figure 3. VIP-values of dBCC measures on BPC in relation to dispersal rate (A) and environmental heterogeneity (B). Environmental heterogeneity was measured as coefficient of variation (CV) of total carbon (TC).

activity [36], for instance overestimating the proportion of active cells [37]. Thus, it is unclear what exactly our rBCC measure represents. Moreover, there are limitations in using phylogeny-based estimates of BCC for communities for which the evolutionary history of its constituents is unknown [38]. A problem when assessing phylogenetic diversity of bacterial communities can be that the large number of taxa requires a huge number of variable sites in the gene or genome to get accurate phylogenetic trees to obtain a well-represented estimate of phylogenetic similarity between communities. In our case we could only estimate phylogenetic similarity from less than 25% of total numbers of read (including more taxa would have generated uninformative phylogenetic trees) so our estimates may differ from 'true' phylogenetic similarities between communities. Moreover, the phylogenetic distance estimate we used (PCD) does not account for differences in relative abundance which may be important for community functioning. Thus, we can conclude that methods development is a necessity, both when it comes to accurately define the active proportion of the community, as well as to how phylogenetic diversity should be determined.

At least one of the functional trait groups (i.e. the ability to use certain organic compounds in Biolog plates) showed VIP>1 in each data set (Table 3, Fig. S1). The strongest positive association to any of the functional trait groups was either to aromatic compounds or cyclic compounds (at least three cases with VIP>1 each). This is in line with a previous experimental study where the communities' potential to use aromatic compounds was a key functional trait contributing to variation in bacterial functioning [12]. In contrast, polymers and simple substrates never showed a strong positive association to BPC. Functional traits with high VIP were relatively weakly correlated with BCC measures in both the Uppland and both stream data sets (r<0.31), but showed considerably stronger correlations in the two Jämtland data sets (r>0.45, Table 4). Weak correlations between functional traits important for BPC (high VIP) and BCC in addition to a large variation in BCC among sites compared to the variation in functional traits strongly imply a functional redundancy of the different bacterial communities. However, in most cases BCC explained more variation in productivity than trait composition. This indicates that there are other taxa specific traits than those measured in Biolog plates that also contribute to productivity, e.g., the use of other resources or a very specific resource. Considering that the aquatic dissolved organic carbon pool is complex and poorly characterized [39] it is not an easy task to pinpoint which organic carbon processing traits should be of interest for a mechanistic understanding of bacterial growth, but should be a subject of future research.

BPC in the second stream sampling was best explained by a negative association with TC, TP, and TN (Table 3, Fig. S1). These changes in the environment were not evidently strongly associated with changes in BCC (Table 4). Also in both the lake water data sets BPC was positively correlated with nutrient levels or ratios between different nutrients and rBCC was well correlated with these environmental changes (Table 4). The influence of local environmental variables on bacterial production was strongly reduced by using BPC, and not gross community production as a response variable in the PLS (results not shown). However, local environmental conditions still showed associations with productivity in all data sets except two (Js and S I). Since a limitation of BPC by C or P alone or the combination of C and P or N is often found in Swedish lakes [40,41], it is not surprising that it was the inorganic nutrient availability and its relation to carbon that explained most variation in productivity. This would indicate that the stoichiometry of nutrients is more important for BPC than nutrient levels per se. In the lake water data sets there was a relatively high covariation between environmental variables and BCC variables contributing to variation in BPC, indicating that the environment partly drives changes in BCC that affect BPC. The negative relation between productivity and nutrient levels in the second stream sampling was surprising and could be a case of negative density dependent interactions, e.g., other nutrients (not considered here) become constraining or there are antagonistic interactions between cells.

Since the strength of the association between BCC and BPC differed between data sets, it seems to be context dependent. A previous modeling study linking diversity and functioning suggested this strength to depend on environmental heterogeneity and species specific traits in relation to the environment [42]. We hypothesized that the dispersal rate among communities would determine the degree of species sorting and in turn the strength of the coupling between BCC and functioning. However, the explanatory power of BCC (rBCC or dBCC) for BPC (VIP-scores) showed no consistent association with the anticipated dispersal rates among the datasets (Fig. 3A). There was a stronger coupling between BCC and BPC in data sets with high environmental heterogeneity in organic carbon but it was not significant (r = 0.79; P = 0.06; Fig. 3B), maybe partly due to too low sample size. Thus, we could not show a link between different community assembly processes and strength of BCC-functioning relationships. One explanation to this result could be that the

range in relative importance of different community assembly processes was not very large in our study. For instance none of the communities were truly isolated and, thus, probably did not experience dispersal limitation. It can also be questioned if dispersal rates were high enough to cause mass effects [10]. Thus, the communities investigated here may all have been formed by local species sorting processes to a similar extent. Future studies aiming to explore the effect of community assembly for the connection between community composition and functioning may, thus, aim for greater gradients in dispersal as well as environmental heterogeneity, also including more disturbed environments than we did, thereby including communities being more likely assembled in different ways. For instance isolated sites such as ground water pockets and environmental gradients of greater range and intensity, such as low pH and high salinity could be included.

In summary, BCC was associated with per cell productivity in all six data sets, but which measure of BCC was best associated differed between data sets. We could not find any consistent difference in BCC-functioning relationship using phylogenetic or non-phylogenetic measures of BCC, nor using BCC from active bacteria or the total bacterial community, highlighting the problem of choosing the best method for BCC measurements in such highly diverse communities. Neither could we find any clear pattern in the relative importance of BCC for functioning between data sets. Thus, the coupling between BCC and functioning in aquatic bacteria seemed context-dependent, but here we could not dissect what the context was. We therefore expect that future research in the area will require methodological considerations on how to measure the active compartment of communities as well as which beta-diversity measures to use. The importance of functional carbon processing traits for productivity was to some extent supported by the compounds tested here, suggesting an importance of the capacity to use especially aromatic and cyclic compounds. However, the results also suggest that important traits still remain to be identified.

Supporting Information

Figure S1   Loading (w*c) bi-plots (of the first and second latent factor) between explanatory variables and per cell bacterial productivity (BPC, filled square) from the PLS for each data set. Explanatory (X) variables are split into functional traits of the community (open circle), environment conditions (open triangle) BCC (cross). Us = Uppland sediment, Js = Jämtland sediment, Uw = Uppland lake waters, Jw = Jämtland lake waters, S I = First stream sampling, and S II = Second stream sampling. For Us, Js and Uw the loadings at the second axis are only shown for illustration purposes as only one latent variable was calculated.

Table S1   Overview of abbreviations used throughout the article.

Table S2   Eigenvalues and variation explained by the first Principal Coordinate Axis (PcoA) of the different BCC for each data set.

Data S1   Raw data.

Acknowledgments

We thank Jürg Brendan Logue, Joel Segersten and Jan Johansson for help with field sampling and Daniel Lundin for assistance with sequence data processing. Mercè Berga gave valuable comments on the manuscript.

Author Contributions

Conceived and designed the experiments: IS ESL ÖÖ. Performed the experiments: IS ESL ÖÖ. Analyzed the data: IS ESL ÖÖ. Contributed reagents/materials/analysis tools: ESL ÖÖ. Wrote the paper: IS ESL ÖÖ.

References

1. Tranvik LJ, Downing JA, Cotner JB, Loiselle SA, Striegl RG, et al. (2009) Lakes and reservoirs as regulators of carbon cycling and climate, Limnol Oceanogr 54: 2298–2314.

2. Azam F, Fenchel T, Field JG, Gray JS, Meyerreil LA, et al. (1983) The ecological role of water-column microbes in the sea. Mar Ecol Progr Ser 10: 257–263.

3. Jonsson A, Karlsson J, Jansson M (2003) Sources of carbon dioxide supersaturation in Clearwater and humic lakes in northern Sweden. Ecosystems 6: 224–235.

4. Karlsson J, Jansson M, Jonsson A (2007) Respiration of allochthonous organic carbon in unproductive forest lakes determined by the Keeling plot method. Limnol Oceanogr 52: 603–608.

5. Comte J, Fauteux L, del Giorgio PA (2013) Links between metabolic plasticity and functional redundancy in freshwater bacterioplankton communities. Front Microbiol 4: DOI: 10.3389/fmicb.2013.00112

6. Comte J, del Giorgio PA (2009) Links between resources, C metabolism and the major components of bacterioplankton community structure a range of freshwater ecosystems. Environ Micobiol 11: 1704–1716.

7. Lindström ES, Feng XM, Granéli W Kritzberg ES (2010) The interplay between bacterial community composition and the environment determining the function of inland water bacteria. Limnol Oceanogr 55: 2052–2060.

8. Langenheder S, Lindström ES, Tranvik LJ (2005) Weak coupling between community composition and functioning of aquatic bacteria. Limnol Pceanogr 50: 957–967.

9. Leflaive J, Danger M, Lacroix G, Lyautey E, Oumarou C, et al. (2008) Nutrient effects on the genetic and functional diversity of aquatic bacterial communities. FEMS Microbio Ecol 66: 379–390.

10. Lindström ES, Östman Ö (2011) The importance of dispersal for bacterial community composition and functioning. PLoS ONE 10: e25883. 10.1371/journal.pone.0025883

11. Peter H, Ylla I, Gudasz C, Romani AM, Sabater S, et al. (2011) Multifunctionality and diversity in bacterial fiofilms. PLoS ONE 6: e23225. doi: 10.1371/journal.pone.0023225

12. Severin I, Östman Ö, Lindström ES (2013) Variable effects of dispersal on productivity of bacterial communities due to changes in functional trait composition. PLoS ONE 8: e80825 10.1371/journal.pone.0080825

13. del Giorgio PA, Gasol JM (2008) Marine Ecology of the oceans: Physiological structure and single-cell activity in marine bacterioplankton. John Wiley & Sons Inc., Hobo-ken, NJ. p243–298.

14. Srivastava DS, Cadotte MW, MacDonald AM, Marushia RG, Mirotchnick N (2012) Phylogenetic diversity and the functioning of ecosystems. Ecol Lett 15: 637–648.

15. Tan J, Pu Z, Ryberg WA, Jiang L (2012) Species phylogenetic relatedness, priority effects, and ecosystem functioning. Ecology 93: 1164–1172.

16. Martiny AC, Treseder K, Pusch G (2013) Phylogenetic conservatism of functional traits in microorganisms. ISME J 7: 830–838

17. Gravel D, Bell T, Barbera C, Bouvier T, Pommier T, et al. (2011) Experimental niche evolution alters the strength of the diversity-productivity relationship. Nature 469: 89–92.

18. Venail PA, MacLean RC, Bouvier T, Brockhurst MA, Hochberg ME, et al. (2008) Diversity and productivity peak at intermediate dispersal rate in evolving metacommunities. Nature 452: 210–215.

19. Östman Ö, Drakare S, Kritzberg ES, Langenheder S, Logue JB, et al. (2010) Regional invariance among microbial communities. Ecol Lett 13: 118–127.

20. Jones SE, Lennon JT (2010) Dormancy contributes to the maintenance of microbial diversity. Proc Natl Acad Sci USA 107: 5881–5886.

21. del Giorgio P, Bird DF, Prairie YT, Planas D (1996) Flow cytometric determination of bacterial abundance in lake plankton with the green nucleic acid stain SYTO 13. Limnol Oceanogr 41: 783–789.

22. Smith DC, Azam F (1992) A simple, economical method for measuring bacterial protein synthesis in seawater using tritiated leucine. Mar Microb Food Webs 6: 107–114.

23. Garland JL, Mills AL (1991) Classification and characterization of heterotrophic microbial communities on the basis of patterns of community-level sole-carbon-source utilization. Appl Environ Microbiol 57: 2351–2359.

24. Herlemann DPR, Labrenz M, Juergens K, Bertilsson S, Waniek JJ, et al. (2011) Transitions in bacterial communities along the 2000 km salinity gradient of the Baltic Sea. ISME J 10: 1571–1579.

25. Quince C, Lanzen A, Davenport RJ, Turnbaugh PJ (2011) Removing noise from pyrosequenced amplicons. BMC Bioinformatics 12: doi: 10.1186/1471-2105-12-38

26. DeSantis TZ, Hugenholtz P, Larsen N, Rojas M, Keller K, et al. (2006) Greengenes, a chimera-checked 16S rRNA gene database and workbench compatible with ARB. Appl Environ Microbiol 72: 5069–5072.

27. Caporaso JG, Bittinger K, Bushman FD, DeSantis TZ, Andersen GL, et al. (2010) PyNAST: a flexible tool for aligning sequences to a template alignment. Bioinformatics 26: 266–267.

28. Caporaso JG, Kuczynski J, Stombaugh J, Bittinger K, Bushman FD (2010) QIIME allows analysis of high-throughput community sequencing data. Nature Methods doi: 10.1038/nmeth.f.303.

29. Ronquist F, Teslenko M, van der Mark P, Ayres D, Darling A, et al. (2012) MrBayes 3.2: Efficient Bayesian phylogenetic inference and model choice across a large model space. Syst Biol 61: 539–542

30. Ives AR, Helmus MR (2010) Phylogenetic metrics of community similarity. Am Nat 176: 128–142.

31. Hammer O, Harper DAT, Ryan PD (2001) PAST: Paleontological statistics software package for education and data analysis. Palaeontol Electron 4, art. 4: 9pp. (unpaginated)

32. Carrascal LM, Galván I, Gordo O (2009) Partial least squares regression as an alternative to current regression methods used in ecology. Oikos 118: 681–690

33. Newton RJ, Jones SE, Eiler A, McMahon KD, Bertilsson S (2011) A guide to the natural history of freshwater lake bacteria. Microbiol Mol Biol Rev 75: 14–49.

34. Reed HE, Martiny JBH (2013) Microbial composition affects the functioning of estuarine sediments. ISME J 7: 868–879.

35. Sjöstedt J, Pontarp M, Tinta T, Alfredsson H, Turk V, et al. (2013) Reduced diversity and changed bacterioplankton community composition do not affect the utilization of dissolved organic matter in the Adriatic Sea. Aquat Microb Ecol 71: 15–24.

36. Blazewicz SJ, Barnard RL, Daly RA, Firestone MK (2013) Evaluating rRNA as an indicator of microbial activity in environmental communities: limitations and uses. ISME J 7: 2061–2068.

37. Franklin RB, Luria C, Ozaki LS, Bukaveckas PA (2013) Community composition and activity of estuarine bacterioplankton assessed using differential staining and metagenomic analysis of 16S rDNA and rRNA. Aquat Microb Ecol 69: 247–261.

38. Gravel D, Bell T, Barbera C, Combe M, Pommier T, et al. (2012) Phylogenetic constraints on ecosystem functioning. Nature communications 3: doi: 10.1038/ncomms2123

39. Thurman EM (1985) Developments in Biogeochemistry: Organic geochemistry of natural waters. Martinus Nijhoff/Dr. W. Junk Publishers (Kluwer Academic Publishers Group, Dordrecht). p497

40. Bell RT, Vrede K, Stensdotter-Blomberg U, Blomqvist P (1993) Stimulation of the microbial food web in an oligotrophic, slightly acidified lake. Limnol Oceanogr 38: 1532–1538.

41. Jansson M, Bergström A-K, Lymer D, Vrede K, Karlsson J (2006) Bacterioplankton growth and nutrient use efficiencies under variable organic carbon and inorganic phosphorus ratios. Microb Ecol 52: 358–364.

42. Cardinale BJ, Nelson K, Palmer MA (2000) Linking species diversity to the functioning of ecosystems: in the importance of environmental context. Oikos 91: 175–183.

18

Vegetation Controls on Weathering Intensity during the Last Deglacial Transition in Southeast Africa

Sarah J. Ivory[1]*, **Michael M. McGlue**[2], **Geoffrey S. Ellis**[3], **Anne-Marie Lézine**[4], **Andrew S. Cohen**[5], **Annie Vincens**[6]

1 Brown University, Providence, Rhode Island, United States of America, **2** University of Kentucky, Lexington, Kentucky, United States of America, **3** U.S. Geological Survey, Denver, Colorado, United States of America, **4** LOCEAN, CNRS, Paris, France, **5** University of Arizona, Tucson, Arizona, United States of America, **6** CEREGE, CNRS, Aix-en-Provence, France

Abstract

Tropical climate is rapidly changing, but the effects of these changes on the geosphere are unknown, despite a likelihood of climatically-induced changes on weathering and erosion. The lack of long, continuous paleo-records prevents an examination of terrestrial responses to climate change with sufficient detail to answer questions about how systems behaved in the past and may alter in the future. We use high-resolution records of pollen, clay mineralogy, and particle size from a drill core from Lake Malawi, southeast Africa, to examine atmosphere-biosphere-geosphere interactions during the last deglaciation (~18–9 ka), a period of dramatic temperature and hydrologic changes. The results demonstrate that climatic controls on Lake Malawi vegetation are critically important to weathering processes and erosion patterns during the deglaciation. At 18 ka, afromontane forests dominated but were progressively replaced by tropical seasonal forest, as summer rainfall increased. Despite indication of decreased rainfall, drought-intolerant forest persisted through the Younger Dryas (YD) resulting from a shorter dry season. Following the YD, an intensified summer monsoon and increased rainfall seasonality were coeval with forest decline and expansion of drought-tolerant miombo woodland. Clay minerals closely track the vegetation record, with high ratios of kaolinite to smectite (K/S) indicating heavy leaching when forest predominates, despite variable rainfall. In the early Holocene, when rainfall and temperature increased (effective moisture remained low), open woodlands expansion resulted in decreased K/S, suggesting a reduction in chemical weathering intensity. Terrigenous sediment mass accumulation rates also increased, suggesting critical linkages among open vegetation and erosion during intervals of enhanced summer rainfall. This study shows a strong, direct influence of vegetation composition on weathering intensity in the tropics. As climate change will likely impact this interplay between the biosphere and geosphere, tropical landscape change could lead to deleterious effects on soil and water quality in regions with little infrastructure for mitigation.

Editor: Olivier Boucher, Université Pierre et Marie Curie, France

Funding: Lake Malawi Drilling Project-Earth System History Program (NSF-EAR-0602404) funded field operations, logistics, and some laboratory analysis. NSF Graduate Research Fellowship (2009078688) provided student salary and tuition and some travel support for laboratory analysis. The funders had no role in study design, data collection and analysis, decision to publish, or preparation of the manuscript.

Competing Interests: The authors have declared that no competing interests exist.

* Email: sarah_ivory@brown.edu

Introduction

Tropical climate is rapidly changing resulting in altered atmospheric and oceanic circulation, as well as changing variability of important climatic modes like the El Niño–South Oscillation [1]. Much work is being done within the framework of the IPCC to better understand climate sensitivity and climate-induced changes to the biosphere; however, integrating the effects on and feedbacks from the geosphere has been largely unexplored [2]. Alterations to the geosphere, an important component of the Earth's critical zone, could have costly, hazardous implications for water and soil quality, as well as a host of ecosystems services provided by inland waters (e.g., serving as fisheries, rookeries, and zones of stormwater retention).

Weathering and climate have hypothetically been linked via feedback loops that regulate the Earth system [3]. However, the low resolution of most marine sedimentary records and the strong influence of long time-scale (10^6–10^7 yrs) processes such as orogeny do not provide a scalable framework for evaluating future changes in the critical zone. Additionally, although climate has long been thought to play a strong, direct role, a number of datasets indicate that the role of climate for weathering and erosion may in fact be indirect, mediated by vegetation and soil storage of organic acids [4–7].

Some studies have generated geochemical indicator records of weathering or produced denudation rates for tropical watersheds based on mass balance models; however, such studies are unable to determine how these rates vary over long time-scales [8–11]. Due to the strong interrelation of vegetation and climate, it has been particularly difficult to separate the individual influences of each mechanism. However, it is expected that vegetation plays an important role. Other studies that have examined the relationship between land use and sediment yield in tropical catchments have demonstrated that heavily forested areas generate less sediment than grasslands or areas modified for agriculture or ranching (e.g., [12]). Furthermore, human land-use change could have strong but unclear effects on weathering and the carbon cycle via alteration of vegetation with no climatic change [13]. To answer the question of how climate change influences weathering and erosion in the tropics, well-dated, high-resolution paleo-records from the critical zone are needed. However, records of tropical lowland vegetation and climate are exceedingly rare, despite the great need to better understand these regions and their growing populations, as articulated by the IPCC [1]. Existing Quaternary paleoenvironmental records from tropical Africa attest to the fact that both vegetation and climate have varied greatly over the last 20 ka [14]. However, it is as yet uncertain what the relative importance of their effects on the geosphere may have been.

Our study seeks to address this important knowledge gap using a scientific drill core from Lake Malawi, southeast Africa (Figure 1). The lake is situated at the current southern extent of the Intertropical Convergence Zone (ITCZ) and has been found to be climatically and ecologically sensitive, making it a remarkable natural laboratory to investigate the relationships and feedbacks among vegetation, climate, weathering, and erosion during the last deglaciation (18–9 ka). This period is of interest because of high-amplitude changes in both climate and vegetation that are both progressive and abrupt. This record thus has the potential to assess the influence of climate change on weathering in the tropics.

Modern Setting

Lake Malawi is the southernmost rift lake in the western branch of the East African Rift System (Figure 1). The physical geography of Lake Malawi and its watershed is controlled by Cenozoic extensional tectonics and volcanism [15]. The lake occupies a series of half-graben basins that are linked *en echelon*, such that steeply dipping border faults with opposing polarity are connected by accommodation zones [16]. Subsidence and deformation patterns within individual half-graben basins control lake bathymetry, with maximum depths (~700 m) achieved adjacent to border faults [17].

The drill core used in this study was collected from the northern basin of the lake (Figure 1). Onshore, distinct topography and drainage patterns follow major tectonic features. A border fault margin to the east of the lake is characterized by high-elevation escarpments which form the Livingstone Mountains. The topography of this faulted margin precludes all but short, steep river systems from forming [18]. In contrast, flexural margins and accommodation zones exhibit comparatively lower altitudes and relief, and large river systems are common in these settings [19]. North of the lake, a well-developed axial delta system building along low-gradient plains is sourced by the Songwe-Kiwira River, which drains the Rungwe highlands. Because of the sloping topography and presence of the delta, Johnson and McCave [20] considered the Songwe-Kiriwa River to be the only significant sediment source to the northern basin. The Songwe-Kiriwa delta has a relatively gentle subaqueous slope (1:70) with a broad shelf, reaching 2–3 km offshore [19]. The drill core site is located ~40 km from the mouth of the river. Although the nearest shoreline is 20 km from the drill site, a sub-lacustrine trough to the east probably prevents significant clastic input from the eastern part of the watershed [20]. Offshore, the lake deepens adjacent to the border fault to ~500 m, and sedimentary processes are dominated by debris and turbidity flows [21].

The smaller watersheds that drain into the lake consist primarily of Neogene alkaline volcanic bedrock, with abundant olivine and alkali basalts, phonolites, trachytes, and nephelinites [22]. Permo-Triassic and Cretaceous sedimentary rocks also crop out north and west of the lake (Figure 1). Soils in the woodlands and forests of the lowlands dominantly consist of pellic vertisols and mollic andosols [23]. The higher-gradient watersheds along the northeastern margin of the basin draining the Livingstone Mountains consist primarily of the Neoproterozoic Mozambique belt, with abundant biotite-hornblende-pyroxene gneiss, charnockites, and minor schists and quartzites [22]. Soils in this area are typically thin and weakly developed, consisting of lithosols, chromic cambisols and dystric regosols [23].

Climate within the watershed is primarily controlled by the yearly passage of the ITCZ, imparting a highly seasonal rainfall regime in which rainfall occurs from November to March. A single, long dry season lasts from April to October when little to no rainfall occurs and prevailing winds are strong southeasterlies [24]. Mean annual precipitation varies from ~800 mm/yr in the lowlands to over 2400 mm/yr in the Rungwe highlands [25]. In addition to a marked N-S gradient of precipitation, the rift escarpments bordering the lake create a pronounced local orographic effect. This results in substantial variability in local rainfall over relatively short spatial scales throughout the watershed.

Vegetation within the watershed is controlled by rainfall and dry season length, although temperature plays a role in the subalpine and alpine zones of the highlands [26–28]. Four principal biome types are observed within the Malawi watershed. In the lowlands (<1500 masl), Zambezian miombo woodlands, a low-diversity deciduous tropical woodland, dominate the landscape. These woodlands primarily comprise *Uapaca*, *Brachystegia*, *Isoberlinia*, *Julbernardia*, and species of Combretaceae and are relatively open, with 30–70% canopy cover. Tropical seasonal forests, closed-canopy semi-deciduous forests— with trees such as *Myrica*, *Macaranga*, Ulmaceae, and Moraceae—are not common in the lowlands; however, they are typically found in areas with locally moist conditions and as riparian corridors along streams and rivers. In the highlands (>1500 masl), afromontane vegetation is found. Today, this region consists of discontinuous patches of afromontane forest separated by high-elevation grasslands. Within the forests, composition is controlled by elevation and rainfall with lower montane forests having moister forest taxa such as *Olea capensis* (1500–2000 mm/yr; 0–3 dry months) from 1500 to 2500 m, whereas *Podocarpus*, *Juniperus*, Ericaceae are common above 2000 m in drier sites (800–1700 mm/yr; ~4 dry months) [27]. The most open biome occurs in the southernmost part of the watershed, where rainfall is the lowest (<800 mm/yr) and contains wooded grasslands with Zambezian affinities.

Methods

Core MAL05-2A was collected from the northern basin of Lake Malawi (10°1.1'S, 34°11.2'E; 359 m water depth) during the Lake Malawi Drilling Project (LMDP) in 2005 (permits issued by Malawi Geological Survey) [29]. Coring site, stratigraphy, and age model details can be found in Figures 1 and 2 as well as in Brown et al. [11] and Scholz et al. [29]. We studied a total of 40 samples, which were taken every ~5 cm from a 3-meter core section (~6–9 meters below lake floor [mblf]). The only gap occurred at ~8.1 mblf, where not enough sediment was available in the core archive to maintain the routine sample interval. The age model of the upper 22 m of the core is based on 24 calibrated accelerator mass spectrometry ^{14}C dates fit with a second-order polynomial

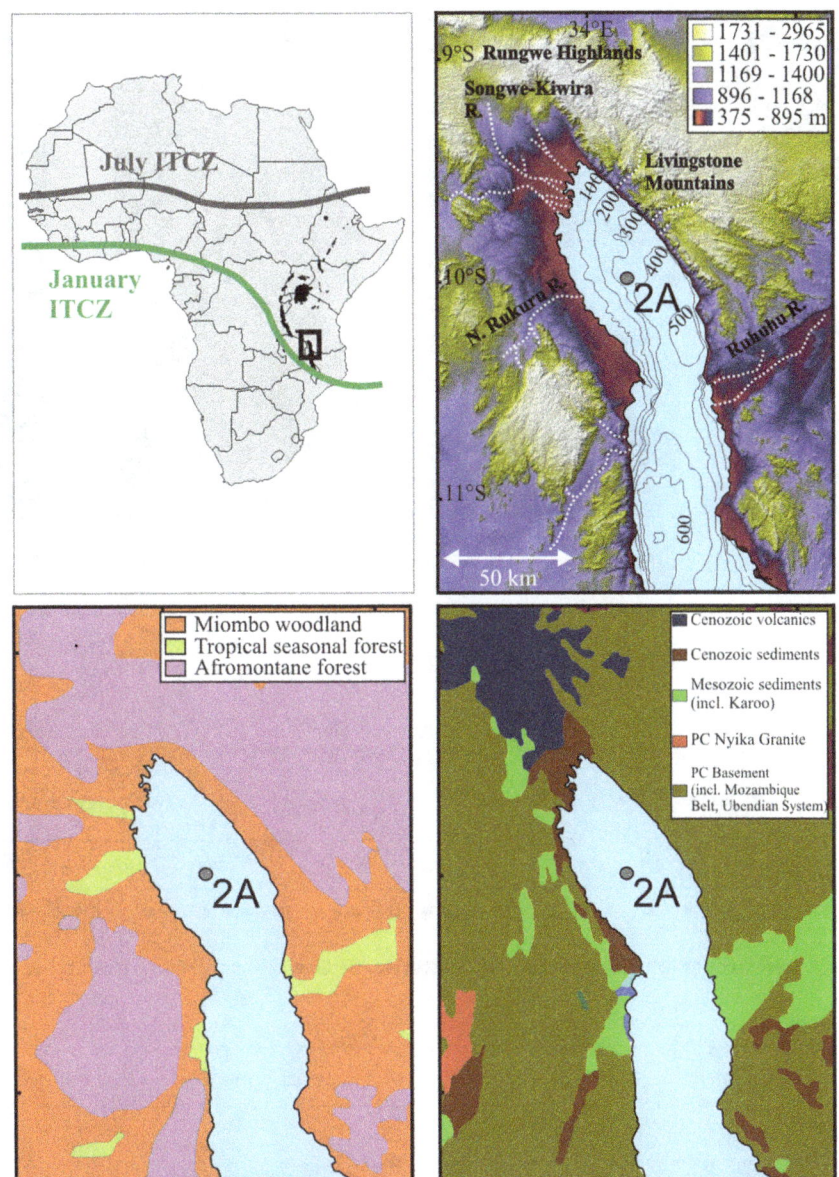

Figure 1. Geography, vegetation, and geology of the study site. (A) showing the July and January positions of the Intertropical Convergence Zone (ITCZ); rectangle represents inset area for watershed maps (modified from Nicholson, 1996). (B) map of the northern basin of Lake Malawi showing topography and bathymetry (isobaths [in meters] modified from Scholz et al., 1989), (C) modern potential vegetation distribution (modified from White, 1983), and (D) bedrock geology (modified from Schlüter, 2006). Scale for panels C and D is the same as for panel B. 2A identifies the location of drill core MAL05-2A.

(calibrated with the "Fairbanks 0107" calibration curve [30] so as to make ages comparable to those of previous studies). The average temporal resolution within the studied section is 208 years, but is ~100 years from 14–11 ka.

Terrigenous mass accumulation rates (TMAR) are a metric of the amount of sediment entering the lake from the watershed and provide insights on the erosion of Lake Malawi's northern watershed. Calculation of TMAR relied on several datasets. Linear sedimentation rates (cm/yr) were calculated using the radiocarbon age model presented in earlier LMDP publications [11]. Dry bulk density was calculated for MAL05-2A sediment horizons using the formulas presented in Dadey et al. [31]. We calculated water content for this purpose and used the gamma-ray density curves from GEOTEK multi-sensor logging of the core

[32]. Sediment mass accumulation rates (g/cm^2/yr) were calculated by multiplying the linear sedimentation rate and dry bulk density. Weight percent terrigenous sediment was determined from X-ray patterns analyzed using the RockJock quantitative mineralogy computer program [33].

Terrigenous particle size data (sand: >62.5 μm; silt: 3.9–62.5 μm; clay: <3.9 μm), interpreted in conjunction with core lithostratigraphy, provide an indicator of landscape sediment flushing, sub-lacustrine hydrodynamic energy, and depositional processes [e.g., [34]]. Particle size analysis was conducted on the terrigenous fraction of the core MAL05-2A samples using a Malvern laser-diffraction particle size analyzer coupled to a Hydro 2000S dispersion bench (see Methods S1).

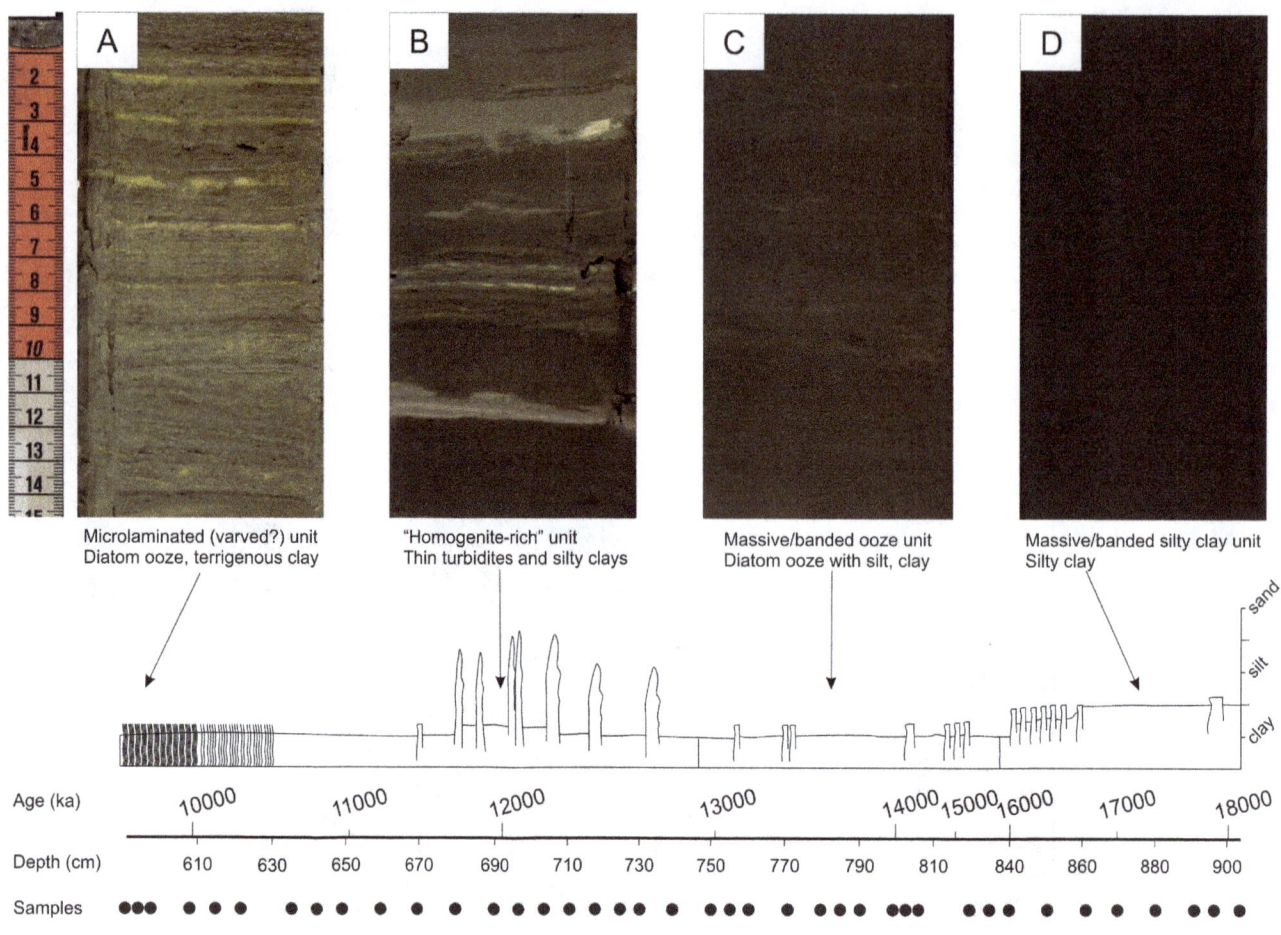

Figure 2. Core MAL05-2A lithostratigraphy within the study interval. Sediment composition and structures vary significantly over the deglacial interval.

Detrital clay minerals in lake sediments provide evidence for the alteration of watershed parent lithologies by physical (disintegration) and chemical (compositional alteration) weathering processes. Quantitative mineralogy was determined using powder X-ray diffraction (XRD) and the computer program RockJock v.11, which has been successfully used for the analysis of fine-grained Cenozoic sediments at a number of locales globally (e.g., [35–37]). For more information on the preparation and processing of these samples, see Methods S1.

In our interpretation of clay mineralogy, we focus on the ratio of kaolinite to smectite (K/S) as a proxy for chemical weathering intensity, which has been used in other similar studies in Africa [38–39]. Kaolinite is produced by the process of leaching in tropical regions marked by high rainfall [40]. In contrast, smectite is typical of tropical semiarid regions and points to increased rainfall seasonality, as it frequently forms during the dry season from the concentration of chemical elements transported to downstream areas by runoff [41–43]. We interpret elevated percentages of micas (illite plus chlorite) to reflect soil formation influenced by physical weathering processes [44].

Pollen samples were taken at the same depths as those for particle size and clay mineralogy and were processed following the standard methods of Faegri and Iverson [45]. More details about sample preparation and pollen identification can be found in Ivory et al. [46]. Vegetation groupings presented in this study are based

on biomes of White [27], as well as prior pollen studies within the watershed by Debusk [25] and Vincens et al. [47].

Results

Vegetation

Detailed descriptions of the ecological dynamics surrounding vegetation change in the Malawi watershed during the last deglaciation appear in Ivory et al. [46]. In this study, we present pollen taxa grouped into biomes (afromontane forest, tropical seasonal forest, Zambezian miombo woodland) in order to present the composition and physiognomy of vegetation on the landscape over time. Following the Last Glacial Maximum (LGM) at 18.1–16.4 ka, afromontane forest taxa percentages were relatively stable at around 20% (Figure 3). Their subsequent decline began in a stepwise manner until 14 ka, at which time, the high-elevation arboreal taxa stabilized at percentages of 7 to 10%. Coeval with the decline of montane taxa, percentages of both lowland arboreal vegetation types increased. Beginning at 14 ka, both tropical seasonal forest and miombo woodland increased progressively in abundance until ~11.8 ka from percentages of <5% up to ~13%. Similarly, a progressive increase in grasses is observed in the record from around 30% to over 45% following the decline of afromontane forest; however, grasses once again returned to lower values around 30% relatively abruptly at 13 ka before the increase of the lowland arboreal vegetation types. The lowland arboreal vegetation continued to increase in a similar manner to a maximum value of 13.5% until 11.8 ka when trends in the tropical seasonal forest and miombo woodland diverged. Forest taxa began a slow decline until a minimum of 3.5 to 7% at 10 ka. At this same time, grasses began to increase at the expense of lowland trees with abundances of nearly 60% by 11 ka. Miombo woodland continued to increase until the end of the record at a maximum value of around 15%.

Mineralogy and Particle Size

Qualitative analysis of X-ray diffraction patterns generated from oriented clay mounts indicated the presence of smectite, kaolinite, illite, and minor chlorite in MAL05-2A (for more information, see the Methods S1). These results are in accord with the findings of Kalindekafe et al. [48] and Branchu et al. [49] for the modern clay mineralogy of northern Lake Malawi. Using RockJock, we determined that smectite was the most abundant clay in the deglacial sequence at all times, ranging from 9 to 29%. Kaolinite ranges from 6 to 13%, and illite ranges from 4 to 15%. At the end of the LGM, kaolinite:smectite (K/S) within the lake sediments is relatively high, with values varying from 0.55 to 0.70 from 18.1 to 15 ka. The maximum value (0.79) of K/S within the lake is reached at 14.5 ka. After this time, sediment K/S begins a two-step decline with values of 0.55 to 0.65 from 14.5 to 13 ka, followed by highly variable values from 0.45 to 0.70 from 13 to 11.8 ka. It is only at the onset of the early Holocene at 11.8 ka that stable minimum values are reached (0.40 to 0.50). In addition, illite plus chlorite (maximum = 18%) increased sharply for the first time in the record at 11.8 ka and remained at stable values for the rest of the record.

Over the entire record (~18.1 to 9.5 ka), there is a decreasing trend in the ratio of silt to clay (silt/clay) which is marked by a two-step transition towards finer material. In the interval from ~18.1 to 14.6 ka, we observe high relative silt/clay values, with a mean of 1.69. Here, particles in the silt-size class (~4.0 to 62.5 μm) dominated the terrigenous fraction. The form of the silt/clay curve is blocky during this early interval, with prolonged periods of invariant conditions on the order of 1000–2000 years in duration. Values of silt/clay from ~18.1 to 16.9 ka range from ~1.84 to 1.95, whereas values from ~16.9 to 14.6 ka range from 1.48 to 1.59. A transition in the record after ~14.6 ka is marked by the saw-tooth form of the silt/clay curve until ~11.9 ka. During this more variable interval, values of silt/clay range from 0.87 to 1.72 with a mean of 1.21. The highest silt/clay values in this interval correlate with high relative values of sand (up to ~3.0%); this was the only instance very fine sand was detected by our analysis. The second important transition in the record took place after ~11.9 ka, where silt/clay ranges from 0.68 to 1.0, with a mean of ~0.82. Over this interval, particles in the clay-size fraction (<4.0 μm) generally exceeded 50% of the terrigenous fraction.

Overall, the record of TMAR is marked by a major transition from low to relatively high rates after 14.3 ka. From ~18.1 to 14.3 ka, the average was ~0.004 gm/cm^2/yr, and TMAR never exceeded ~0.006 gm/cm^2/yr. The transition after ~14.3 ka was initially marked by an abrupt two-fold increase in TMAR. At a finer scale, the TMAR record shows a stepwise change from ~14.3 to 9.5 ka. The initial step, from ~14.3 to 12.8 ka, was characterized by TMAR values ranging from ~0.007 to 0.012 gm/cm^2/yr. This period is followed by brief decline to lower values from ~12.8 to 11.9 ka, when the average TMAR was ~0.006 gm/cm^2/yr. The second TMAR increase was after ~11.9 ka, with a mean value of 0.012 gm/cm^2/yr until ~10.1 ka.

Interpretation

Dense forest and chemical weathering

At the end of the LGM, paleoclimate records from throughout Africa and at Lake Malawi suggest cooler temperatures and aridity [14,50–51]. Although it might be expected that these environmental conditions would slow reaction rates and mineral transformation during the early deglacial period, this does not seem to be the case in northern Malawi. Throughout this phase, K/S is relatively high in our core record (Figure 3). Higher proportions of kaolinite, a clay mineral leached of mobile cations typically associated with soils of the humid tropics, suggests that chemical weathering was relatively intense. The increased delivery of heavily weathered clays may appear counterintuitive given the comparatively cool temperatures and aridity during this time as recorded in the same core based on TEX-86 and leaf wax δ^{13}C, respectively [50–51]; however, the clay mineralogy record is similar to changes in vegetation during the early deglacial period.

At the end of the LGM, many studies have suggested that East African afromontane forest communities expanded to lower altitudes than at present because of cooler temperatures and low atmospheric CO$_2$ [52–54]. This was also the case within the Lake Malawi watershed, where afromontane taxa, dominated by *Juniperus*, Ericaceae, and *Podocarpus*, were found in high percentages at the earliest part of our record beginning around 18.1 ka and lasting until ~15.5 ka (Figure 3) [46]. Although it is unlikely that the afromontane forest reached the lakeshore, Ivory et al. [46] suggest that these normally high-altitude communities expanded down to ~900 masl within several kilometers of the lakeshore. Dense forest communities must have been prevalent within the watershed, given the presence of taxa typical of the lower montane moist forest (*Olea* spp., *Myrica*). Furthermore, grass percentages during this interval are very low, the lowest of the record (Figure 3). In modern lake-floor sediment samples, grass percentages within Lake Malawi contribute ~55% of the total pollen [25], supporting the idea of a much denser canopy cover and larger contribution of the arboreal taxa throughout the watershed and into the lowlands. This may suggest that the presence of dense vegetation on the landscape as a result of lowering of the afromontane belt resulted in higher concentrations of organic acids and increased soil moisture, counteracting any slowdown in weathering due to climate.

The first marked change in both the vegetation and sedimentary records began at 15.5 ka, a period which is coeval with the end of Heinrich Stadial 1 (H1) [55]. Within Africa and at Lake Malawi, this period was characterized by progressive warming [50–51]. During this phase from 15.5 to 14.3 ka, K/S reached the highest values of the record (Figure 3). This suggests that, as in the previous phases of afromontane forest dominance, chemical weathering within the northern Malawi watershed remained relatively intense. Although some wetting occurs in East Africa during this post-H1 period, organic geochemical analysis from leaf wax δ^{13}C implies that Malawi remained relatively arid with respect to modern at this time, despite indications of warming [50–51]. Once again, although chemical weathering reactions would be expected to slow during an arid period, intense weathering continued in the northern Lake Malawi watershed.

Following H1, all of the principal afromontane arboreal taxa began to decline [46]. This decline has been attributed to increasing temperatures which forced afromontane forest to retreat to higher elevations [25]. However, the transition was stepwise. Although the principal afromontane forest taxa that were dominant just following the LGM (Ericaceae, *Juniperus*, and *Podocarpus*) began to decline, from 15.5 to 14.3 ka, the dense

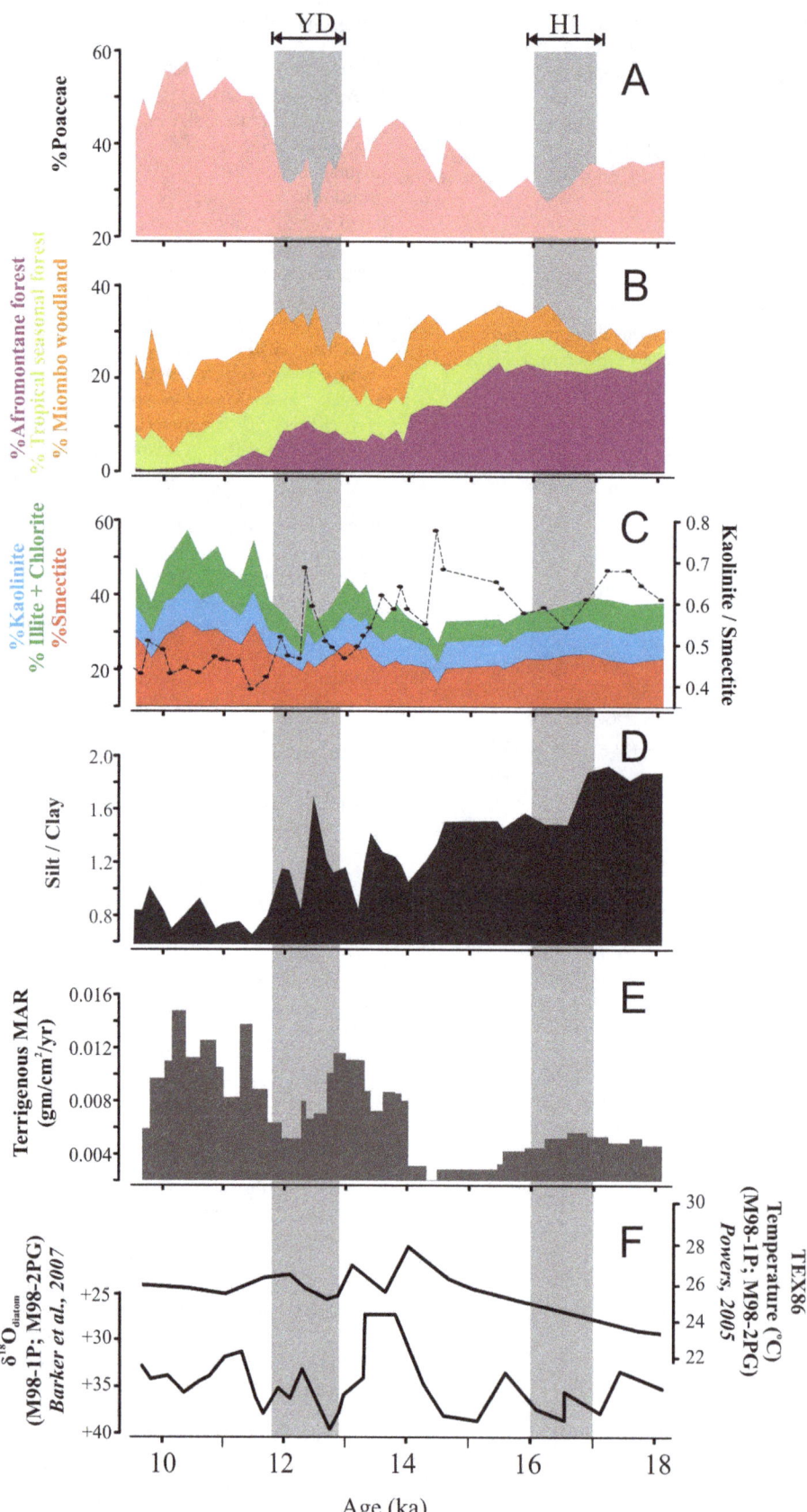

Figure 3. Vegetation and weathering indicators from drill core MAL05-2A.YD, Younger Dryas. H1, Heinrich event 1. (A) pollen percentages of Poaceae and (B) arboreal vegetation types (afromontane forest, tropical seasonal forest, miombo woodland), (C) XRD clay mineral percentages of smectite, kaolinite, and illite+chlorite, ratio of kaolinite to smectite (K/S), (D) ratio of silt to clay (silt/clay), and (E) terrigenous mass

accumulation rate, and (F) $\delta^{18}O$ from diatoms and TEX-86 temperature from piston cores M98-1P and M98-2PG from the northern basin of Lake Malawi from Barker et al. (2007) and Powers (2005), respectively.

lower montane moist forest taxa, such as *Myrica* and *Olea*, known pioneer taxa, continued to expand [46]. The presence of dense forest in the lowlands is the best explanation for the continued high values of K/S at this time.

Throughout the early deglacial period, minimum values of TMAR suggest that erosion was low (Figure 3). Johnson and McCave [20] suggest that TMAR in the northern basin of Lake Malawi is indicative of moisture, and therefore low values would suggest a reduced riverine transport. However, our dataset makes clear that a more likely mechanism for abating the flushing of weathered parent material from the landscape is the presence of dense lowland forests. During the LGM, a slight lake-level regression at Malawi of ~40 to 100 m would have decreased the distance from the mouth of the Songwe-Kiriwa River and other important drainages to the core site [56]. A basinward progradation of the delta during a lake-level lowstand might be expected to show an increase in the amount of terrigenous material deposited; however, we observe the opposite. Thus, despite noted cooler temperature and aridity relative to modern, intense weathering and low erosion are observed over this period by K/S and TMAR. This regime strongly suggests that dense moist forest in the watershed had a strong control on clay transformation and delivery of terrigenous siliciclastic material to the basin.

Transition to the Holocene (14.3 to 11.8 ka)

At 14.3 ka, a transitional period began that was coeval with the onset of the Bolling-Allerod (BA) [57]. In contrast with the predominantly silty detritus in our record lower in the core, a broad range of grain sizes produces highly variable values of silt/clay (1.8 to 0.6) during this time (Figure 3). Over this interval, lake levels were highly variable. Evidence from diatom assemblages implies high-frequency changes on the order of 50 to 100 m [56]. Sand is present in the samples from this interval, but only in significant percentages in one sample. However, there is clear evidence of turbidity flows here as early as ~13.2 ka (Figure 2). The presence of time-equivalent turbidites has also been described in other cores in northern Lake Malawi ("homogenites" of Barry et al. [58]). Thus, the variable silt content is expected to be caused by a change in sub-lacustrine depositional processes and energy. Thin, dominantly fine-grained turbidites may have been produced by wave remobilization of distal deltaic sediments, which move downslope at low velocity under the force of gravity as benthic nepheloid plumes [20].

At this time, the moist montane forest taxa became less abundant in the lowlands, and the afromontane forest as a biome reached very low percentages values, suggesting trees may have retreated to higher elevations by this time (Figure 3). Within the lowlands, however, arboreal taxa of the tropical seasonal forest began to increase in abundance. Miombo woodland abundances began increasing as well; however, the response time is much slower than the other arboreal taxa and values remained relatively low. Although the lowland closed-canopy tropical seasonal forest began expanding at this time, the percentages remained at <10%, values which are not significantly larger than those observed today, suggesting that dense forests were restricted largely to riparian corridors [25]. Ivory et al. [46] attribute this rise of lowland dense forest to the higher temperatures and reinforced monsoon recorded at the time in the basin. The lack of other arboreal vegetation on the landscape away from river banks in the lowlands indicates a marked change of physiognomy from the previous

phase following the LGM to more open vegetation. This openness in the lowlands is reflected in the high percentages of grass pollen during this phase (Figure 3).

Within the basin, the increase in grass is tracked by rising TMAR values in core MAL05-2A (Figure 3). The increase in terrigenous material as the landscape opens supports the argument that TMAR represents erosion from the hinterland into the basin when vegetation is open and forests are scarce. This suggests that the relatively open character of vegetation caused by the retreat of montane forests prevented water storage on the landscape and promoted flushing of clastic material into the basin.

The presence of turbidites makes the clay record in this interval more complex. From 14.3 to 12.7 ka, an overall declining trend in K/S is observed, followed by an abrupt increase at 12.7 to 12.3 ka. Dense tropical seasonal forest and miombo woodland continued to expand in the lowlands at Lake Malawi between 13 and 11.7 ka, and tropical seasonal forest reached its maximum extent at the end of this interval (Figure 3). The beginning of this zone is coeval with the inception of the Younger Dryas (YD) in the Northern Hemisphere [59]. Although many other sites in East Africa record a retreat or slowdown of deglacial forest expansion during this time due to increased aridity, the dense lowland forest became more prevalent at Lake Malawi [60-64].

This indication of dense forest in conjunction with higher K/S values could possibly suggest a return to more intense chemical weathering beginning around 12.7 ka. However, independent particle size evidence, which shows elevated silt/clay and minor sand in the core during the period from 12.7 to 12.3 ka, suggests instead that the elevated K/S may be related to minor reworking by turbidites. Therefore, we interpret the multiple sedimentary indicators from this interval to be influenced by mass wasting. This may be the result of flushing from enhanced precipitation promoting sediment transport (and hyperpycnal sub-lacustrine flows) rather than chemical weathering.

Early Holocene open woodlands

Although records from farther north in tropical and subtropical Africa show an abrupt resumption of the African monsoon and significant wetting during the early Holocene, including at Lake Rukwa (~200 km north of Lake Malawi), the situation at Lake Malawi is slightly different [14,65]. Organic geochemical studies of $\delta^{13}C$ agree that the watershed was wetter than during the late Pleistocene; however, a ~100 m lake-level regression occurred at this time, suggesting a change in hydrology [66-67]. Pollen analysis on a core from Lake Masoko, a small maar lake within the Malawi watershed, suggests that the expansion of open miombo woodlands indicates high summer rainfall, but heightened rainfall seasonality [47]. This conclusion is in agreement with Zr:Ti and biomarker studies from Lake Malawi, which demonstrate wind direction and air mass changes at this time consistent with major reorganization of tropical circulation and a more northerly ITCZ than during the late Pleistocene [11,68].

At this time, total abundance of clays was very high, resulting in a net increase in all dominant clay minerals; however, K/S reached a minimum (Figure 3). Additionally, for the first time in this record, illite plus chlorite reached higher abundances than kaolinite (Figure 3). The dominance of detrital clays retaining mobile cations suggests a very strong decrease in the intensity of chemical weathering beginning at 11.8 ka. Furthermore, the importance of illite, a clay mineral that is only present in low abundances in the modern lake today, is suggestive of increasing physical weathering and reworking during the early Holocene despite increased wetness [48]. In addition, TMAR during this interval reaches maximum values. The agreement of these two independent datasets seems to indicate peak erosion from 11.8 to 10 ka.

This transition in clay mineralogy at 11.8 ka is coeval with the opening of the lowland vegetation (Figure 3). At 11.8 ka, near the onset of the Holocene, an abrupt transition occurred from denser lowland forests to more open miombo woodland in the watershed (Figure 3). Additionally, for the first time in the record, afromontane forest tree pollen is nearly absent. Given the high abundance of woodland and grasses, it is likely that this early Holocene transition represented that largest change in vegetation physiognomy in Malawi during the studied interval.

This increase in erosion was likely caused by the more intense summer rains of the African monsoon and the open character vegetation, conditioning the landscape for the flushing of physically weathered material into the basin. Cecil and Edgar [69] noted a similar relationship between greater coarse siliciclastic sediment transport and strongly seasonal rainfall.

Discussion

Traditional models for the tropics suggest strong climatic controls such as precipitation and temperature on chemical and physical weathering and erosion [44,70]. However, our record from Lake Malawi suggests a more indirect role of climate. During the early part of the last deglaciation, climate within the watershed varied dramatically in mean state and variability, including cooler temperatures and relative aridity from 18.1 to 14.5 ka. This was followed by successive wetting and warming from 14.5 to 12.5 ka, a return to slightly drier conditions with much reduced rainfall seasonality during the YD, and finally an increase in rainfall seasonality at the time of the resumption of the African monsoon at the early Holocene. However, despite the changes in climate, both highland and lowland forests were common in the watershed throughout this time until 11.8 ka. Although changes in rainfall and temperature lead to compositional vegetation changes during this period as high-altitude forest was replaced successively by lowland forest, the physiognomy of the landscape, with dominance of arboreal taxa, remained until the early Holocene. Similarly, although chemical weathering should be more intense during the moister periods, instead we see relative stability in the clay minerals with higher values of kaolinite throughout this period regardless of climatic regime. During the early Holocene, the physiognomic change to more open woodland is coeval with higher mean annual precipitation because of heightened seasonality following the resumption of the African monsoon. It might be expected that this high temperature and high rainfall should have resulted in intense chemical weathering, but the available data do not support this idea. At this time, kaolinite decreases, and high abundance of illite suggests not only less-intense chemical weathering, but higher physical weathering and transport of less-weathered micas into the basin.

The strong, direct control of vegetation on mineral transformation and erosion is not surprising for several reasons as many processes mediated by plants influence chemical and physical breakdown of bedrock. First, although heat and moisture are necessary for driving chemical breakdown of aluminosilicate rocks, organic acids, particularly low-weight organic acids such as those produced by plants and mycorrhizae associated with plant roots, are essential for biological mediation of weathering in soils [71–72]. These acids not only provide H+ for the transformation of primary materials, but also act as ligands that form strong complexes with trivalent cations such as Al^{3+} and Fe^{3+} [6]. This function of organic acids has the potential to dramatically intensify weathering rates by increasing the solubility of minerals containing these elements. Second, root networks, particularly in dense, old-growth forests, are critical for the mechanical breakdown of bedrock [73]. Furthermore, the removal of trees results in an increased response of erosion rates and closer control of abiotic erosion mechanisms as seen in our record.

Perhaps the most important biotic control on weathering is the role of forests and large forest trees in regulating the hydrologic cycle locally within a watershed and at the scale of a soil [74]. In dense forests, canopies intercept rainfall. During intense monsoon rains, this function acts to reduce runoff and increase storage on the landscape, increasing the residence time of moisture available for chemical reactions in soils. Additionally, evapotranspiration within a forest often creates a microclimate that results in higher local precipitation over dark, dense trees as well as more stored water on the landscape and cooler temperatures from evaporation [75]. More importantly however, evapotranspiration controlled by forest trees limits the drainage depth within the weathering profile of a soil, reducing physical weathering processes and allowing time for chemical alteration [76]. Vegetation transitions, as we see in the Malawi record, however, are key as long-lived dense forests, such as those found in the Congo Basin or in the Amazon exhibit very little chemical weathering [77–78]. Early stages of succession, such as the period of intense chemical weathering observed during the deglaciation following the retreat of the afromontane forest as lowland forest colonized the lake shore, are in fact typically the period of most intense chemical weathering as large trees colonize previously unweathered substrate [79–80].

TMAR values show a positive correlation with grass abundances throughout our record. Although TMAR has been interpreted as an indication of moisture availability, we instead suggest that delivery of terrigenous material to the basin is strongly regulated by vegetation on the landscape. This means that more open, grassy woodlands are less effective at trapping material on the landscape leading to greater erosion and deposition of terrigenous material than when forests are dominant in the lowlands [12,81]. This first order control of erosion on the landscape is in agreement with cosmogenic nuclide studies in the Alps by Vanacker et al. [82] who show exponential increases in denudation rates due to modern land use change. Although this suggests rapid landscape conversion once forest is gone, this study, as well as that of Vanacker et al. [82], suggests that once forest returns, stabilization is possible as denudation rates return to background levels. Reducing the residence time of water on the landscape likewise influences chemical weathering intensity, and we note that the inverse correlation of K/S with TMAR and grass percentages, which appears to support this hypothesis.

In our analysis of core records from Lake Malawi, we have observed coarser sediments (high silt/clay or % sand) around the inception of lake-level highstands, which also show high ratios of trees to herb pollen. Several factors may influence this particle size trend. For example, storage release (i.e., flushing) of coarser sediment from low-to-moderate gradient flexural and axial margins concomitant with transgression likely helps to explain some of the variability we have observed. Additionally, increased fluvial discharge implied by the presence of afromontane forest is consistent with what is known about the efficacy of coarse siliciclastic transport [34]. Another potential influence relates to the shortening of fluvial transport networks associated with the higher base level (i.e., less reduction of siliciclastic particle size by transport abrasion and downstream fining), but because fining upward patterns are characteristic of the highstand strata we have examined, the absolute influence of this mechanism is not well known. By contrast, during lowstand intervals such as the early Holocene, when dry season length was long, fluvial networks lengthen and appear to generate finer siliciclastic detritus. Physical weathering of parent rocks and soils, captured in the clay mineral record, may serve as a feedback on reducing particle size during these intervals. Indeed, the lack of forest cover to intercept rainfall during these intervals may also serve to help flush fine sediment from hillslopes and floodplains into river channels. Variability in depositional processes can complicate the interpretation of silt/clay, which is why evidence of mass wasting is critical to identify in the lithostratigraphy. For example, the transition from lowstand to highstand in rift lakes occurs rapidly (10^2 to 10^5 yrs) [83], and as such flexural margins can become prone to gravity flows as pore pressures change and slopes destabilize. Therefore, our study employed particle size and stratigraphic data in support of pollen, clay minerals, and TMAR to provide the most rigorous assessment of weathering and erosion patterns for the deglacial period yet developed for this region of East Africa.

Conclusions

Vegetation composition and structure at Lake Malawi and elsewhere does not unequivocally track simple precipitation amount [28,46–47]. Our record attests to a strong negative relationship between increased rainfall seasonality and vegetation density that results in a specific depositional signature within the lake. When dense forests occupy the watershed, chemical weathering is intense and erosion is low. Leaching within soils leads to generation of highly altered clay minerals like kaolinite. However, during times of open vegetation, chemical weathering is less intense, but erosion is increased. Smectite dominates during these periods, but flashy precipitation on the open landscape leads to flushing of siliciclastics into the lake and thus high relative TMAR.

Finally, these results have important implications for better understanding how weathering and erosion may change in the future. Although little work has been done to quantify potential future alterations in weathering, most models do not take

ecosystem change related to climatic change into account, leading to significant potential biases. Only in the last few years has vegetation been added to projections of soil weathering, thus leading to the suggestion of high sensitivity of weathering to future climate states [84–86]. Although paleostudies hint at the role of biological mediation in weathering over long time scales, no other studies have used paleo-records to quantify the effects of vegetation on weathering processes in southeast Africa [7]. This study points to the importance of vegetation for mediating weathering in the past; however, more and longer records are needed in order to better quantify this effect over a larger range of climatic and vegetation variability in watersheds throughout the tropics.

Acknowledgments

Disclaimer: Any use of trade, product, or firm names is for descriptive purposes only and does not imply endorsement by the U.S. Government.

Thanks to Jean-Pierre Cazet, Guillaume Buchet, Owen Davis, and John Logan for help with sample processing and pollen identification, William Benzel and Adam Boehlke for assistance with x-ray diffraction, and the National Lacustrine Core Repository (LacCore) at the University of Minnesota for sub-sampling and core curation. We would also like to thank three reviewers for helpful comments that improved the manuscript.

Author Contributions

Conceived and designed the experiments: SJI MMM. Performed the experiments: SJI MMM. Analyzed the data: SJI MMM. Contributed reagents/materials/analysis tools: SJI MMM GE AC AML AV. Wrote the paper: SJI MMM.

References

1. IPCC (2013) Climate Change 2013: The Physical Science Basis. Contribution of Working Group I to the Fifth Assessment Report of the Intergovernmental Panel on Climate Change [Stocker TF, Qin D, Plattner GK, Tignor M, Allen SK, et al. (eds.)]. Cambridge University Press, Cambridge, United Kingdom and New York, NY, USA, 1535 pp.

2. Brantley SL, Megonigal JP, Scatena FN, Balogh-Brunstad Z, Barnes RT, et al. (2011) Twelve testable hypotheses on the geobiology of weathering. Geobiology 9: 140–165.

3. Walker JC, Hays PB, Kasting JF (1981) A negative feedback mechanism for the long-term stabilization of Earth's surface temperature. J Geophys Res: Oceans 86: 9776–9782.

4. Langbein WB, Schumm SA (1958) Yield of sediment in relation to mean annual precipitation. Eos 39: 1076–1084.

5. Egli M, Mirabella A, Sartori G (2008) The role of climate and vegetation in weathering and clay mineral formation in late Quaternary soils of the Swiss and Italian Alps. Geomorphology 102: 307–324.

6. Goddéris Y, Roelandt C, Schott J, Pierret M.C, François LM (2009) Towards an integrated model of weathering, climate, and biospheric processes. Rev Mineral Geochem 70: 411–434.

7. Dosseto A, Hesse PP, Maher K, Fryirs K, Turner S (2010) Climatic and vegetation control on sediment dynamics during the last glacial cycle. Geology 38: 395–398.

8. Einsele G, Hinderer M (1998) Quantifying denudation and sediment–accumulation systems (open and closed lakes): basic concepts and first results. Palaeogeogr Palaeoclimatol Palaeoecol 140: 7–21.

9. Sémah AM, Sémah F, Moudrikah R, Fröhlich F, Djubiantono T (2004) A late Pleistocene and Holocene sedimentary record in Central Java and its

palaeoclimatic significance. Modern Quaternary Research in SE Asia, Balkema 19: 63–88.

10. Felton AA, Russell JM, Cohen AS, Baker ME, Chesley JT, et al. (2007) Paleolimnological evidence for the onset and termination of glacial aridity from Lake Tanganyika, Tropical East Africa. Palaeogeogr Palaeoclimatol Palaeoecol 252: 405–423.

11. Brown ET, Johnson TC, Scholz CA, Cohen AS, King JW (2007) Abrupt change in tropical African climate linked to the bipolar seesaw over the past 55,000 years. Geophys Res Lett 34.

12. Dunne T (1979) Sediment yield and land use in tropical catchments. J Hydrol 42: 281–300.

13. Dixon RK, Solomon AM, Brown S, Houghton RA, Trexier MC, et al. (1994) Carbon Pools and Flux of Global Forest Ecosystems. Science 263: 185–190.

14. Gasse F (2000) Hydrological changes in the African tropics since the Last Glacial Maximum. Quat Sci Rev 19: 189–211.

15. Ebinger CJ (1989) Tectonic development of the western branch of the East African rift system. Geol Soc Am Bull 101: 885–903.

16. Rosendahl BR (1987) Architecture of continental rifts with special reference to East Africa. Annu Rev Earth Planet Sci 15: 445–503.

17. Scholz CA, Rosendahl BR, Versfelt JW, Kaczmarick J, Woods LD (1989) Seismic Atlas of Lake Malawi (Nyasa), East Africa. Project PROBE Geophysical Atlas Series, Vol. 2. Duke University, Durham, NC.

18. Soreghan MJ, Scholz CA, Wells JT (1999) Coarse-grained, deep-water sedimentation along a border fault margin of Lake Malawi, Africa: seismic stratigraphic analyses. J Sediment Res 69: 832–846.

19. Scholz CA (1995) Deltas of the Lake Malawi rift, East Africa: seismic expression and exploration implications. AAPG Bull 79: 1679–1697.

20. Johnson T, McCave I (2008) Transport mechanism and paleoclimatic significance of terrigenous silt deposited in varved sediments of an African rift lake, Limnol Oceanogr 5: 1622–1632.

21. Scott DL, Ng'ang'a P, Johnson TC, Rosendahl BR (1991) High-resolution acoustic character of Lake Malawi (Nyasa), East Africa, and its relationship to sedimentary processes. Special Publication of the International Association of Sedimentologists 13: 129–145.

22. Schlüter T (2006) Geological Atlas of Africa. Springer, Berlin. 272 pp.

23. FAO (1988) Soil Map of the World, Revised legend, World Resources Report, 60, FAO, Rome, Italy.

24. Malawi Government (1983) The National Atlas of Malawi. National Atlas Committee and Department of Surveys of Malawi. Natl. Atlas Comm. Dep. Surv. Malawi.

25. DeBusk GH (1994) Transport and stratigraphy of pollen in Lake Malawi, Africa. PhD Thesis, Duke University.

26. Polhill RM (1966) Flora of Tropical East Africa. In: Hubbard, C.E., Milne-Redhead, E. (Eds.), Crown Agents for Oversea Governments and Administrations, London.

27. White F (1983) Vegetation of Africa—a descriptive memoir to accompany the Unesco/ AETFAT/UNSO vegetation map of Africa. Natural Resources Research Report XX. UNESCO, Paris, France.

28. Hély C, Bremond L, Alleaume S, Smith B, Sykes M, Guiot J (2006) Sensitivity of African biomes to changes in the precipitation regime. Global Ecol Biogeogr 15: 258–270.

29. Scholz CA, Johnson TC, Cohen AS, King J, Peck J, et al. (2007) East African megadroughts between 135 and 75 thousand years ago and bearing on early-modern human origins. PNAS 104: 16416–16421.

30. Fairbanks R, Mortlock R, Chiu T, Cao L, Kaplan A, et al. (2005) Radiocarbon calibration curve spanning 0 to 50,000 years BP based on paired 230Th/234U/238U and 14C dates on pristine corals. Quat Sci Rev 25: 1781–1796.

31. Dadey KA, Janecek T, Klaus A (1992) Dry-bulk density: its use and determination. In Proceedings of the Ocean Drilling Program, Scientific Results (Vol. 126, pp. 551–554) College Station, TX: Ocean Drilling Program.

32. Lyons WB, Carey AE, Hicks DM, Nezat CA (2005) Chemical weathering in high-sediment yielding watersheds, New Zealand. J Geophys Res: Earth Surface (2003–2012) 110.

33. Eberl DD (2003) User's guide to RockJock – A program for determining quantitative mineralogy from powder X-ray diffraction data: U.S. Geological Survey Open-File Report 2003–78, 47 p.

34. Heins Keiro (2007), in Sedimentary Provenance and Petrogenesis: Perspectives from petrography and geochemistry, in Arribas, J, Critelli, S. and Johnson, M., eds., GSA Special paper 420, p. 345–379

35. Andrews JT (2008) The role of the Iceland Ice Sheet in the North Atlantic during the late Quaternary: a review and evidence from Denmark Strait. J Quat Sci 23: 3–20.

36. Refsnider KA (2010) Dramatic increase in late Cenozoic alpine erosion rates recorded by cave sediment in the southern Rocky Mountains. Earth Planet Sci Lett 297: 505–511.

37. Dühnforth M, Anderson RS, Ward D J, Blum A (2012) Unsteady late Pleistocene incision of streams bounding the Colorado Front Range from measurements of meteoric and in situ 10Be. J Geophys Res: Earth Surface (2003–2012) 117.

38. Pastouret L, Chamley H, Delibrias G, Duplessy JC, Thiede J (1978) Late Quaternary climatic changes in western tropical Africa deduced from deep-sea sedimentation off Niger Delta. Oceanol Acta 1: 217–232.

39. Lézine AM, Duplessy JC, Cazet JP (2005) West African monsoon variability during the last deglaciation and the Holocene: Evidence from fresh water algae, pollen and isotope data from core KW31, Gulf of Guinea. Palaeogeogr Palaeoclimatol Palaeoecol 219: 225–237.

40. Birkeland PW, (1984) Soils and Geomorphology. Oxford University Press, Oxford. 310 pp.

41. Chamley H (1989) Clay sedimentology (Vol. 623) New York: Springer-Verlag. 623 pp.

42. Weaver CE (1989) Clays, muds, and shales. Elsevier. 818 pp.

43. Alizai A, Hillier S, Clift PD, Giosan L, Hurst A, et al. (2012) Clay mineral variations in Holocene terrestrial sediments from the Indus Basin. Quat Res 77: 368–381.

44. Thiry M (2000) Palaeoclimatic interpretation of clay minerals in marine deposits; an outlook from the continental origin. Earth Sci Rev 49: 201–221

45. Faegri K, Iversen J (1989) Textbook of Pollen Analysis, Wiley, Chichester, UK. 328 pp.

46. Ivory S, Lézine AM, Vincens A, Cohen A (2012) Effect of aridity and rainfall seasonality on vegetation in the southern tropics of East Africa during the Pleistocene/Holocene transition, Quat Res 77: 77–76.

47. Vincens A, Garcin Y, Buchet G (2007) Influence of rainfall seasonality on African lowland vegetation during the late Quaternary: pollen evidence from Lake Masoko, Tanzania. J Biogeogr 34: 1274–1288.

48. Kalindekafe LN, Dolozi MB, Yuretich R (1996) Distribution and origin of clay minerals in the sediments of Lake Malawi, in Johnson, T.C., and Odada, E., eds., The Limnology, Climatology and Paleoclimatology of the East African Lakes: Amsterdam, Gordon and Breach, p. 443–460.

49. Branchu P, Bergonzini L, Delvaux D, De Batist M, Golubev V, et al. (2005) Tectonic, climatic and hydrothermal control on sedimentation and water

50. Powers LA, Johnson TC, Werne JP, Castañeda I, Hopmans E, et al. (2005) Large temperature variability in the southern African tropics since the Last Glacial Maximum. Geophys Res Lett 32: L08706.

51. Castañeda IS, Werne JP, Johnson TC (2007) Wet and arid phases in the southeast African tropics since the Last Glacial Maximum. Geology 35: 823–826.

52. Street-Perrott FA, Huang Y, Perrott RA, Eglinton G, Barker P, et al. (1997) Impact of lower atmospheric carbon dioxide on tropical mountain ecosystems. Science 278: 1422–1426.

53. Wu H, Guiot J, Brewer S, Guo Z (2007) Climatic changes in Eurasia and Africa at the last glacial maximum and mid-Holocene: reconstruction from pollen data using inverse vegetation modelling. Clim Dyn 29: 211–229.

54. Woltering M, Johnson TC, Werne JP, Schouten S, Sinninghe Damsté JS (2011) Late Pleistocene temperature history of southeast Africa: A TEX86 temperature record from Lake Malawi. Palaeogeogr Palaeoclimatol Palaeoecol 303: 93–102.

55. Hemming SR (2004) Heinrich events: Massive late Pleistocene detritus layers of the North Atlantic and their global climate imprint. Rev Geophys 42.

56. Stone JR, Westover KS, Cohen AS (2011) Late Pleistocene paleohydrography and diatom paleoecology of the central basin of Lake Malawi, Africa. Palaeogeogr Palaeoclimatol Palaeoecol 303: 51–70.

57. Wohlfarth B (1996) The chronology of the last termination: A review of radiocarbon date, high resolution terrestrial stratigraphies. Quat Sci Rev 15: 267–284.

58. Barry SL, Filippi ML, Talbot MR, Johnson TC (2002) Sedimentology and geochronology of late Pleistocene and Holocene sediments from northern Lake Malawi. In The East African Great Lakes: Limnology, Palaeolimnology and Biodiversity, Springer Netherlands, 369–391.

59. Alley RB, Meese DA, Shuman CA, Gow AJ, Taylor KC, et al. (1993) Abrupt increase in Greenland snow accumulation at the end of the Younger Dryas event. Nature 362: 527–527.

60. Vincens A (1991) Late quaternary vegetation history of the South-Tanganyika basin. Climatic implications in south central Africa. Palaeogeogr Palaeoclimatol Palaeoecol 86: 207–226.

61. Vincens A (1993) Nouvelle sequence pollinique du Lac Tanganyika: 30,000 ans d'histoire botanique et climatique du Bassin Nord. Rev Palaeobot Palyno 78: 381–394.

62. Bonnefille R, Riollet G, Buchet G, Icole M, Lafont R, et al. (1995) Glacial-interglacial record from intertropical Africa, high resolution pollen and carbon data at Rusaka, Burundi. Quat Sci Rev 14: 917–936.

63. Beuning KR, Talbot MR, Kelts K (1997) A revised 30,000-year paleoclimatic and paleohydrologic history of Lake Albert, East Africa. Palaeogeography, Palaeoclimatology, Palaeoecology 136: 259–279.

64. Ryner M, Gasse F, Rumes B, Verschuren D (2007) Climatic and hydrological instability in semi-arid equatorial East Africa during the late glacial to Holocene transition: a multi-proxy reconstruction of aquatic ecosystem response in northern Tanzania. Palaeogeogr Palaeoclimatol Palaeoecol 248: 440–458.

65. Cohen AS, Van Bocxlaer B, Todd JA, McGlue M, Michel E, et al. (2013) Quaternary ostracodes and molluscs from the Rukwa Basin (Tanzania) and their evolutionary and paleobiogeographic implications. Palaeogeogr Palaeoclimatol Palaeoecol 392: 79–97.

66. Johnson TC, Brown ET, McManus J, Barry S, Barker P, et al. (2002) A high-resolution paleoclimate record spanning the past 25,000 years in southern East Africa. Science 296: 113–132.

67. Barker PA, Leng MJ, Gasse F, Huang Y (2007) Century-to-millennial scale climatic variability in Lake Malawi revealed by isotope records. Earth Planet Sci Lett 261: 93–103.

68. Konecky BL, Russell JM, Johnson TC, Brown ET, Berke MA, et al. (2011) Atmospheric circulation patterns during late Pleistocene climate changes at Lake Malawi, Africa. Earth Planet Sci Lett 312: 318–326.

69. Cecil CB, Edgar NT (2003) Climate Controls on Stratigraphy: Society for Sedimentary Geology Special Publication 77, 275 p.

70. Burnett AP, Soreghan MJ, Scholz CA, Brown ET (2011) Tropical East African climate change and its relation to global climate: a record from Lake Tanganyika, Tropical East Africa, over the past 90+ kyr. Palaeogeogr Palaeoclimatol Palaeoecol 303: 155–167.

71. Landeweert R, Hoffland E, Finlay RD, Kuyper TW, van Breemen N (2001) Linking plants to rocks: ectomycorrhizal fungi mobilize nutrients from minerals. Trends Ecol Evol 16: 248–254.

72. Bonneville S, Smits MM, Brown A, Harrington J, Leake JR, et al. (2009) Plant-driven fungal weathering: early stages of mineral alteration at the nanometer scale. Geology 37: 615–618.

73. Roering JJ, Marshall J, Booth AM, Mort M, Jin Q (2010) Evidence for biotic controls on topography and soil production. Earth Planet Sci Lett 298: 183–190.

74. Rodríguez-Iturbe I, Porporato A (2004) Ecohydrology of water-controlled ecosystems. In Soil Moisture and Plant Dynamics.

75. Spracklen DV, Arnold SR, Taylor CM (2012) Observations of increased tropical rainfall preceded by air passage over forests. Nature 489: 282–285.

76. Roelandt C, Goddéris Y, Bonnet MP, Sondag F (2010) Coupled modeling of biospheric and chemical weathering processes at the continental scale. Global Biogeochem Cy 24: GB2004.

77. Gaillardet J, Dupré B, Louvat P, Allegre CJ (1999) Global silicate weathering and CO2 consumption rates deduced from the chemistry of large rivers. Chem Geo159: 3–30.

78. Moquet JS, Crave A, Viers J, Seyler P (2011) Chemical weathering and atmospheric/soil CO2 uptake in the Andean and foreland Amazon basins. Chem Geo 287: 1–26.

79. Moulton KL, West J, Berner RA (2000) Solute flux and mineral mass balance approaches to the quantification of plant effects on silicate weathering. Am J Sci 300: 539–570.

80. Goddéris Y, Donnadieu Y, Tombozafy M, Dessert C (2008) Shield effect on continental weathering: implication for climatic evolution of the Earth at the geological timescale. Geoderma 145: 439–448.

81. Hay WW (1998) Detrital sediment fluxes from continents to oceans. Chem Geo 145: 287–323.

82. Vanacker V, von Blanckenburg F, Govers G, Molina A, Poesen J, et al. (2007) Restoring dense vegetation can slow mountain erosion to near natural benchmark levels. Geology 35: 303–306.

83. McGlue MM, Lezzar KL, Cohen AS, Russell JM, Tiercelin JJ, et al. (2008) Seismic records of late Pleistocene aridity in Lake Tanganyika, tropical East Africa. J Paleolimn 40: 635–653.

84. Beaulieu E, Goddéris Y, Labat D, Roelandt C, Oliva P, et al. (2010) Impact of atmospheric CO2 levels on continental silicate weathering. Geochem Geophys Geosyst 11.

85. Beaulieu E, Goddéris Y, Donnadieu Y, Labat D, Roelandt C (2012) High sensitivity of the continental-weathering carbon dioxide sink to future climate change. Nature Climate Change 2: 346–349.

86. Goddéris Y, Brantley S, François LM, Schott J, Pollard D, et al. (2013) Rates of consumption of atmospheric CO 2 through the weathering of loess during the next 100 yr of climate change. Biogeosciences 10: 135–148.

19

Tiny Is Mighty: Seagrass Beds Have a Large Role in the Export of Organic Material in the Tropical Coastal Zone

Lucy G. Gillis[1]*, Alan D. Ziegler[2], Dick van Oevelen[3], Cecile Cathalot[4], Peter M. J. Herman[1], Jan W. Wolters[5], Tjeerd J. Bouma[1]

1 Spatial Ecology Department, Royal Netherlands Institute for Sea Research (NIOZ), Yerseke, Zealand, The Netherlands, 2 Geography Department, National University of Singapore (NUS), Singapore, Singapore, 3 Ecosystems Studies Department, Royal Netherlands Institute for Sea Research (NIOZ), Yerseke, Zealand, The Netherlands, 4 Laboratoire Environnement Profond (LEP), French Research Institute for Exploitation of the Sea, Polouzane, Brittany, France, 5 Department of Biology, University of Antwerp, Antwerp, Flanders, Belgium

Abstract

Ecosystems in the tropical coastal zone exchange particulate organic matter (POM) with adjacent systems, but differences in this function among ecosystems remain poorly quantified. Seagrass beds are often a relatively small section of this coastal zone, but have a potentially much larger ecological influence than suggested by their surface area. Using stable isotopes as tracers of oceanic, terrestrial, mangrove and seagrass sources, we investigated the origin of particulate organic matter in nine mangrove bays around the island of Phuket (Thailand). We used a linear mixing model based on bulk organic carbon, total nitrogen and $\delta^{13}C$ and $\delta^{15}N$ and found that oceanic sources dominated suspended particulate organic matter samples along the mangrove-seagrass-ocean gradient. Sediment trap samples showed contributions from four sources oceanic, mangrove forest/terrestrial and seagrass beds where oceanic had the strongest contribution and seagrass beds the smallest. Based on ecosystem area, however, the contribution of suspended particulate organic matter derived from seagrass beds was disproportionally high, relative to the entire area occupied by mangrove forests, the catchment area (terrestrial) and seagrass beds. The contribution from mangrove forests was approximately equal to their surface area, whereas terrestrial contributions to suspended organic matter under contributed compared to their relative catchment area. Interestingly, mangrove forest contribution at 0 m on the transects showed a positive relationship with the exposed frontal width of the mangrove, indicating that mangrove forest exposure to hydrodynamic energy may be a controlling factor in mangrove outwelling. However we found no relationship between seagrass bed contribution and any physical factors, which we measured. Our results indicate that although seagrass beds occupy a relatively small area of the coastal zone, their role in the export of organic matter is disproportional and should be considered in coastal management especially with respect to their importance as a nutrient source for other ecosystems and organisms.

Editor: David William Pond, Scottish Association for Marine Science, United Kingdom

Funding: This research was funded by Ecoshape, Building with Nature (SDWI/Eco-Shape/Building with Nature grant R-303-001-020-414). The funders had no role in study design, data collection and analysis, decision to publish, or preparation of the manuscript.

Competing Interests: The authors have declared that no competing interests exist.

* Email: lucygwen.gillis@nioz.nl

Introduction

The tropical coastal zone is composed of ecosystems emerging from the open ocean, extending over coral reefs and seagrass beds towards the tidal zone, which may include mangrove forests. This zone forms a well-structured and gradual interface between the land and the sea that contains some of the most productive and biogeochemically active ecosystems in the world [1]. Part of this productivity is maintained by particulate and dissolved organic matter (POM and DOM) inputs from both terrestrial and oceanic sources [2].

The importance of external inputs to the coastal zone and the exchanges between coastal ecosystems depends on various factors. For example, the quantity of terrestrial organic matter (OM) input to the coastal zone has been shown to depend on the size of the catchment area and the land-use within [3]. Tidally-dominated estuaries have large exchanges of water with the ocean, and therefore, receive ocean-derived DOM and POM [4–6]. The recognized mangrove outwelling hypothesis [7–8] postulates that export of mangrove-derive OM supports adjacent ecosystems and food webs.

Another potentially important OM source and sink within the tropical coastal zone are seagrass beds. Seagrass plants trap suspended POM originating from both external sources and from leaf shedding inside the seagrass bed [9–10]. Trapping results directly from the physical structure of the bed and from settling induced by changes in the hydrodynamic conditions - both are related in part to plant density and leaf characteristics [11–12]. Once POM is deposited within seagrass beds, the matter is 'protected' from re-suspension, and consequently, most POM degrades within the bed, releasing dissolved nutrients that may in turn be taken up by the seagrass [13–15] or released to the water column [9,16]. Seagrass beds however also export particulate nutrients via leaf shedding enhanced during strong hydrodynamic events and via marine herbivore consumption [17].

Due to the different ecosystems in the seascape, the high production of ecosystems and the potential for import and export from a variety of sources, it is challenging to constrain the contribution of the different resources to the exchange of or OM in the tropical coastal zone. One classical approach is to use a chemical tracer mixing model based on a combination of carbon to nitrogen ratios (C:N) with carbon and nitrogen stable isotope ratios (δ^{13}C and δ^{15}N, respectively) [18–19]. Each of these tracers has a specific signature for each source (oceanic, seagrass, mangrove and terrestrial), which can be used to identify the contribution of the different sources in a mixture [18–19].

Investigations of source contributions in Southern Thailand and found that seagrasses contributed between 36–42% of OM to sediment trap samples [4]. Other studies in Gazi Bay, Kenya discovered that mangrove and seagrass matter dominated suspended POM in the water column [17,20]. These studies indicate that contributions from individual sources likely depend on the local conditions. In this paper, we build upon the foundations of previous work to investigate the origin of POM in trap and suspended sediment samples in the coastal zone in Phang-nga Bay in southern Thailand. We studied nine sites with contrasting physical attributes of bays (width and length), of ecosystems (aerial extent and width) and of catchment areas (aerial extent and land use). This allowed us to gain a greater understanding of export of POM from different ecosystems at the tropical seascape scale. Specifically, we compared the amount of OM originating from seagrass beds versus mangrove, oceanic, and terrestrial sources across a number of different sub-habitats in the region, comparing the contributions to their respective surface areas.

Methods

Ethics statement

The work was conducted in collaboration with Rajabhat University (Phuket) on public lands. No permit was required for sampling. No live samples were collected.

Study Area and Physical attributes

Within the Andaman Sea, along the southwest coast of Thailand is the province of Phuket, which borders Phang-nga bay (Figure 1). Our sampling sites are located in Phang-nga Bay, which is 68 km long (head to mouth) and has a surface area of 3000 km^2. Mean tidal range is 1.8 m (Figure 1). The southern part of the bay is open to the Andaman Sea and the northwest area is open to the sea via the Pak Pra Inlet, which separates Phuket Island from the mainland (Figure 1). Estuarine salinity conditions dominate in the north, whilst marine conditions dominate in the south [21]. Circulation in the bay changes depending on the dry (May to October) or wet season (November to April), which alters the wind direction from northeast in the dry season to southwest in the wet season. Mean annual rainfall is about 2300 mm and mean temperature is 28°C.

Land-use around Phang-nga Bay changed from natural forest to initially tin mining (1600–1800) [21]. Other land-uses have gained greater importance such as rubber and palm oil plantations as well as shrimp (located on land) and fish farms (situated in coastal zones) [21]. In recent decades, rapid urbanisation related to tourism has occurred in Phuket.

Samples were collected from all the fully accessible mangrove sites located in the west of Phang-nga Bay, on the islands of Phuket, Yao Yai and Yao Noi. A total of 9 sites were sampled (Figure 1) during March 2011. All sites were outside the reach of direct river discharge and experienced a strong tidal exchange (tidal range 0.5–2.5 m) with the ocean. Therefore, all sites are classified as tidal mangroves [22].

Sample collection and data processing

At each site we established transects beginning at the edge of the mangrove, extending towards the ocean through an adjacent intertidal seagrass patch. Samples were taken, when logistically possible, at distances of 0 (seaward edge of mangrove forest), 50, 100, 200, 300, 400, 500, 1000, 1500 and 2000 m along the transect, perpendicular to the seaward edge of the mangrove forest towards the open sea. Two types of samples were collected: trap samples of suspended matter above the bed (0.05 m) and suspended particulate matter in the water column. At least half of the sampling points (for trap and water samples) were located in the seagrass beds; the other half were taken at points landward or seaward of the bed, depending on the physical constraints of the different sites. Sediment traps at each point were secured to steel rods and anchored to the substrate at a height of 0.05 m above the sediment. They were emptied once after approximately 24 hours. On the day of trap installation, 2-liter surface water samples were taken at each point above the traps at the water surface during high tide.

Four POM sources were considered in this study: terrestrial vegetation, mangrove leaves, seagrass leaves and oceanic plankton. For the terrestrial vegetation, three replicates of 3–4 leaves from rice paddies, rubber trees (*Hevea sp.*) and native vegetation (*Delonix sp.* and *Terminalia sp.*) were collected on the island of Koh Yai. These were considered to be representative for terrestrial vegetation of the province. This matter represents the source value for the terrestrial OM used in the mixing model. Mangrove and seagrass leaves were collected during the trap deployment. Plankton and suspended particulate matter (SPM) samples were collected at a point believed to have the majority of oceanic influence (7°52.573′N and 98°35.635′E) and taken as proxy for oceanic matter. Plankton samples were collected with weighted nets of mesh sizes 400 µm and SPM with a five 1-litre water sample. The mean of the C and N content and δ^{13}C and δ^{15}N value of each POM source was taken as the respective end-member to determine the contribution of each source to the OM along the transect.

All solid samples (sediment and leaves) were placed in separate sample bags, stored in a cooling box and transported immediately to the field laboratory where they were dried at 60°C for 48 hours. Water samples for SPM and plankton were filtered through pre-combusted glass fiber filters (GF/F, 0.7 µm pore diameter). The GF/F filters were dried at 60°C for 48 hours. Both solid material (sediment and leaves) and GF/F filters were packed in airtight containers and transported to the laboratory of Royal Netherlands Institute for Sea Research (NIOZ) for elemental and isotope analysis.

Elemental and isotope analysis

Sediment and leaf material samples were ground for homogenisation. The trap, plankton and SPM samples were acidified to remove carbonates [23]; the leaf samples were not acidified. All samples were analysed for total organic carbon (TOC), total nitrogen (TN), and the isotopes δ^{13}C and δ^{15}N by means of elemental analysis isotope ratio mass spectrometry (EA-IRMS) using a Thermo Finnigan Flash 1112.

Stable isotope ratios are expressed as δ values (‰) relative to conventional standards (VPDB limestone for C and atmospheric N$_2$ for N):

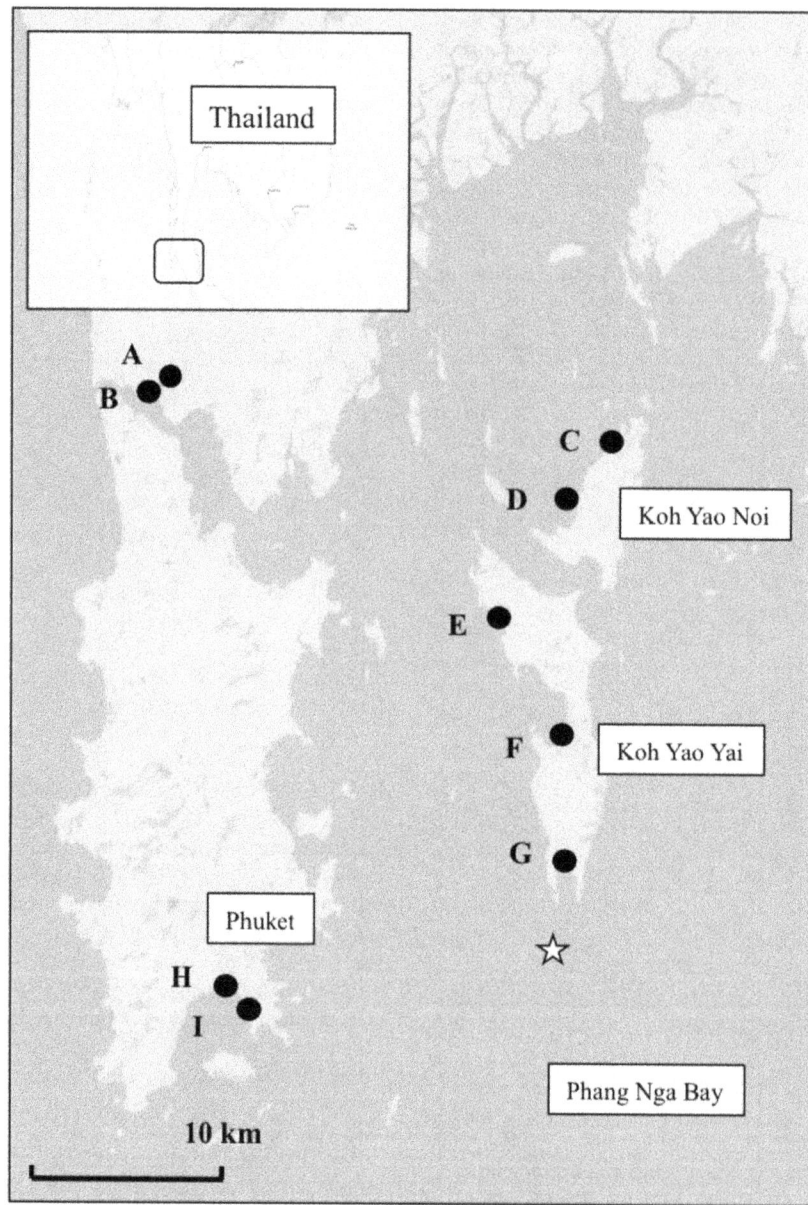

Figure 1. Map showing location of the nine sampling sites (black circles) in Phang Nga Bay, southern Thailand. The white star represents the point oceanic samples where collected.

$$\delta X = \left(\left(X_{sample} - X_{standard} \right) / X_{standard} \right) * 1000(‰) \qquad (1)$$

where δX is either $\delta^{13}C$ or $\delta^{15}N$; X_{sample} is the $^{13}C/^{12}C$ or $^{15}N/^{14}N$ ratio of the sample; $X_{standard}$ is the $^{13}C/^{12}C$ or $^{15}N/^{14}N$ ratio of the standards (0.01118 for C and 0.00368 for N).

Mixing model

We considered the following potential sources (Table 1): terrestrial vegetation (*Delonix sp.*, *Terminalia sp.* and *Hevea sp.*), mangrove leaves (*Rhizophora sp.*), seagrass leaves (*Enhalus sp.*, *Halodule sp.* and *Halophile sp.*) and oceanic production (plankton and SPM). Mean values were taken as end member for each

source (Table 1). Only SPM and trap samples were analysed with the mixing model (as discussed below).

A linear mixing model was used to determine the contributions of the different sources to the SPM and trap samples [24–25]. The mixing model uses the means of $\delta^{13}C$, $\delta^{15}N$ and the C:N ratios (corrected for differences in %C and %N between sources) of individual trap or SPM samples to determine the contribution (f) of each of the four sources (Table 1) to the mixture. The variation between replicates is small justifying the use of the means. The mixing model has the following four equations that are solved simultaneously to recover a unique solution of the four source contributions (f_{oce}, f_{ter}, f_{man}, f_{sea}):

$$f_{oce} + f_{ter} + f_{man} + f_{sea} = 1 \qquad (2)$$

Table 1. Mean and standard error values of isotopes carbon isotope ($\delta^{13}C$), nitrogen isotope ($\delta^{15}N$), bulk organic carbon (C) and total nitogen (N) for terrestrial trees, mangroves trees, seagrass plants and oceanic sources.

Sources	Species	$\delta^{13}C$		$\delta^{15}N$		C		N		n
Group		Mean	SE	Mean	SE	Mean	SE	Mean	SE	
		($^o/_{oo}$)		($^o/_{oo}$)						
Terrestrial	Havea sp.	−31.8	0.5	0.4	0.001	46.8	0.2	4.2	0.04	3
	Delomix sp.	−30.1	0.5	0.6	0.001	46.9	0.3	2.6	0.2	3
	Terminalia sp.	−31.2	0.5	1.9	0.001	44.8	0.2	3.6	0.006	3
Overall terrestrial	-	**−31.0**	**0.4**	**1.0**	**0.0004**	**46.1**	**0.2**	**4.01**	**0.08**	**3**
Overall Mangrove trees	Rhizophora sp.	**−26.9**	**0.6**	**3.6**	**0.2**	**37.5**	**0.6**	**1.2**	**0.07**	**45**
Seagrass plants	Halodule sp.	−11.4	0.7	2.9	0.15	21.4	2.6	1.4	0.2	7
	Halophile sp.	−11.6	0.5	2.8	0.9	21.4	2.1	1.3	0.2	19
	Enhalus sp.	−10.1	0.3	5.2	0.3	25.6	1.6	1.9	0.1	17
Overall seagrass plants	-	**−11.0**	**0.6**	**3.6**	**0.2**	**21.7**	**1.4**	**1.5**	**0.1**	**3**
Oceanic	SPM	−24.1	0.13	18.3	3	1.2	0.07	0.2	0.01	3
	Plankton	−24	0.04	7.9	0.03	4.6	0.5	0.8	0.1	3
Overall oceanic	-	**−24.1**	**0.06**	**13.1**	**3.5**	**2.9**	**0.8**	**0.5**	**0.1**	**6**

The values for each source (highlighted) are the means of the different contributions (species for terresrtial, mangrove and seagrasses & SPM/plankton for oceanic). These values are defined as the end members (organic matter sources), which are used in the mixing model to determine the different mixture of fractions for organic matter. The organic matter was collected in the core, sediment trap and suspended sediment matter samples. n, is the number of samples and SE is standard error.

$$\delta^{13}C_{mix} = \delta^{13}C_{oce} \cdot f_{oce} + \delta^{13}C_{ter} \cdot f_{ter} + \delta^{13}C_{man} \cdot f_{man} + \delta^{13}C_{sea} \cdot f_{sea} \tag{3}$$

$$\delta^{15}N_{mix} = \delta^{15}N_{oce} \cdot f_{oce} + \delta^{15}N_{ter} \cdot f_{ter} + \delta^{15}N_{man} \cdot f_{man} + \delta^{15}N_{sea} \cdot f_{sea} \tag{4}$$

$$0 = (\%C_{oce} - CN_{mix} \cdot \%N_{oca}) \cdot f_{oce} + (\%C_{ter} - CN_{mix} \cdot \%N_{ter}) \cdot f_{ter} + (\%C_{man} - CN_{mix} \cdot \%N_{man}) \cdot f_{man} + (\%C_{sea} - CN_{mix} \cdot \%N_{sea}) \cdot f_{sea} \tag{5}$$

Where oce = oceanic, ter = terrestrial, man = mangrove trees and sea = seagrass plants. The model was implemented and solved in R [26], using the *lsei* function available in the LIM package [27]. Using the contribution for each source as returned by the mixing model, we calculated if each source was under-contributing or over-contributing compared to its relative surface:

$$F_x / A_{fx} \tag{6}$$

This index was not calculated for POM derived from oceanic sources, because no representative surface area for the 'ocean' could be defined. The contribution (F_x; %) per source in each sample was then divided by the percent of total surface area (A_{fx}) occupied by the particular system (catchment area-terrestrial, mangrove forest, seagrass bed). This gives a dimensionless number, for which 1 implies that the ecosystem contributes proportionally to its relative surface area; values below or above 1 indicates that the proportional contribution is lower or higher, respectively.

Statistical analysis

From the results of equation 6 we used step-wise regression analysis to determine if the contribution of each resource was related to relevant physical aspects of the nine sites. For this analysis we only used the mixing model results at 0 and 100 m along the transect, as data for these distances were available across all nine sites. Contributions from terrestrial, mangrove and seagrass to SPM and trap samples were correlated with physical attributes of each site. The mangrove contribution was tested against area of mangrove forest (m^2), area of bay (m^2), urbanisation in the catchment area (%) and width of mangrove forest (m) (Table 2). Terrestrial contribution was analysed against catchment area (m^2), area of bay (m^2) and urbanisation in the catchment area (%) (Table 2). Finally, seagrass was tested with area of the seagrass bed (m^2), area of bay (m^2) and urbanisation in the catchment area (%) (Table 2).

The physical characteristics (bay area, bay width/length), surface areas of the mangrove forests and seagrass beds were determined with ground truthing during the sampling campaign. Land use in the catchment area, width of the bay, and length of the bay were determined by analysing Quickbird, WorldView-1 and WorldView-2 satellite imagery. All statistical analysis was completed in R [26], where probabilities (p) <0.05 were considered significant.

Results

All samples (both trap and SPM) along the gradient showed different contributions of the different sources of OM (Figure 2). At sites A and B, the signatures for the majority of samples were between those of terrestrial, mangrove and oceanic sources (Figure 2). The trap sample signatures from sites C and I were spread from seagrass to the oceanic end members (Figure 2). At sites D and F, the trap and SPM samples indicated a pattern from terrestrial and mangrove sources to the mid-point between all the sources (Figure 2). The majority of site E trap samples similarly

Table 2. Physical description of the sites and the physical attributes of the marine ecosystems for each of the nine sites (A to I) and their associated bay.

Physical attributes	Site								
	A	B	C	D	E	F	G	H	I
Urbanized area (km²)	1.4	1.4	0.9	0.4	0.1	0.1	0.6	0.7	5.8
Mangrove forest area (km²)	3.3	3.3	10	0.6	7.5	3.2	2	9	4.7
Catchment area (km²)	29.6	29.6	3	0.2	0.6	0.7	1.5	2.1	0.4
Seagrass bed area (km²)	1	0.5	0.1	0.3	0.8	0.04	0.8	0.9	0.8
Width of bay (km)	0.97	0.97	NA	1.77	1.44	3.76	1.22	2	2
Length of bay (km)	1.47	1.47	NA	1.11	0.72	3.31	2.94	1.06	1.06
Width of mangrove forest (km)	3.4	3.4	0.06	0.71	1.18	0.89	1.84	0.96	0.31

NA means that these sites are situated in an open coast, all other sites are bays.

plotted in the centre of the sources; and sites C, E and I SPM samples followed this pattern too (Figure 2). Terrestrial sources dominated trap samples from sites G and H, but in the case of SPM samples they where spaced between terrestrial and mangrove end numbers and the oceanic end number (Figure 2).

The mixing model determined the fraction of each source end-member (mangrove and seagrass plants, terrestrial and oceanic) in the mean of 2–3 individual sediment traps and SPM samples for the three sites (Table 3). The majority of SPM samples (for sites B, C, D, F, G, H and I) showed a spatial pattern for oceanic sources, where the oceanic source contribution increased from 0 m to the end of the transect (Table 3). This pattern (for oceanic sources) was similarly repeated in six of the trap samples (sites A, B, C, D, F and I) (Table 3). The bulk of mangrove plant contributions for SPM samples showed a spatial pattern where the source influence decreased with distance from 0 m to the end of the transect (B, C, D, E, F, G, I). This same pattern was seen in the trap samples (A, B, C, D, E, H and I) (Table 3). Only approximately half of terrestrial plant contributions for SPM samples showed a gradient through the transect (C, D, H and I) (Table 3). However, over half of trap samples (A, B, C, D, G and I) did show a spatial pattern, where terrestrial plant contribution was higher nearer the landward side of the transect (Table 3). The seagrass contribution was highest in the seagrass beds for SPM samples at sites A, B, F and for trap samples at sites E and F (Figures 3, 4 and Table 3). The remainder of the trap and SPM samples showed no spatial pattern in the seagrass plant contribution (Figures 3, 4 and Table 3). For SPM samples, variation of contributions across the transect was very small (0.8–6%); and this was repeated for trap samples (0.5–12%) (Table 3).

Step-wise regression indicated only two significant correlations. A negative correlation between terrestrial trap contributions (y) and catchment area (x) at 100 m (y = −0.78x+27.2, R^2 = 0.5, p = 0.04) and a positive correlation were shown for mangrove contribution to trap samples at 0 m contributions with width of mangrove forest (y = 12.8x+2.8, R^2 = 0.6, p = 0.03).

Discussion

This study was designed to clarify the origin and exchange of different OM sources within tropical coastal waters. Despite the high local primary production in mangrove and seagrass stands, we found that oceanic sources dominated trap and SPM samples along the entire transect. The mixing model results however showed that mangrove forests and terrestrial sources did contrib-

ute, especially to the trap samples. Interestingly we found a correlation between the width of the mangrove forest and their contribution, which provides new insights into physical processes (such as exposure to hydrodynamics) associated with mangrove outwelling.

However when taking into account the relative size of the ecosystem (mangrove forest, catchment area and seagrass beds) we found that seagrass beds, although occupying the smallest area of all the potential sources, contributed significantly to the OM in stations along land- to seaward transects. This provides further evidence of their importance as a nutrient (dissolved and particulate) source in the tropical coastal seascape. Before moving to the implications of these findings, we first discuss the sampling and modeling approach that we undertook.

Source contributions to organic matter

Mixing models are often used to estimate contributions to a mixture [24–25]. Some samples could not be solved with the model, which indicates that they violate some of the implicit assumptions of the linear mixing model formulation, such as imprecise measurement of end-members values, the existence of unidentified sources, or degradation processes that have altered the isotope or C:N ratio of the OM. For our study site, we are confident that we included the dominant POM sources in our design and the end-member values were consistent and estimated with a low uncertainty (Table 1). We therefore believe that the few samples that fell outside the mixing polygons (Figure 2) had already undergone substantial biogeochemical modification that lead to a decrease of δ^{15}N values and/or N content through mineralization and denitrification [4,28]. We think that the trap samples that fell out of the end-member polygon contained predominantly re-suspended bottom sediment (not unlikely given the close proximity to the sediment). Re-suspension of bottom sediment is a potential source of matter caught in the trap that had been subjected to biogeochemical modification.

There was a strong oceanic signal in the trap samples for the nine sites, suggesting that oceanic-derived OM settled out (Figure 2). The SPM and trap samples indicated a substantial oceanic input, which is in agreement with previously studies on coastal wetlands [29–30]. The OM of the trap samples at sites A, B, G and H showed a strong contribution from mangrove sources. These sites have large channels coming from the inner mangrove area, which may explain the increased contribution of mangrove OM. Sites G, H and I trap samples within the seagrass bed show an strong inclination towards the seagrass plant end-member and

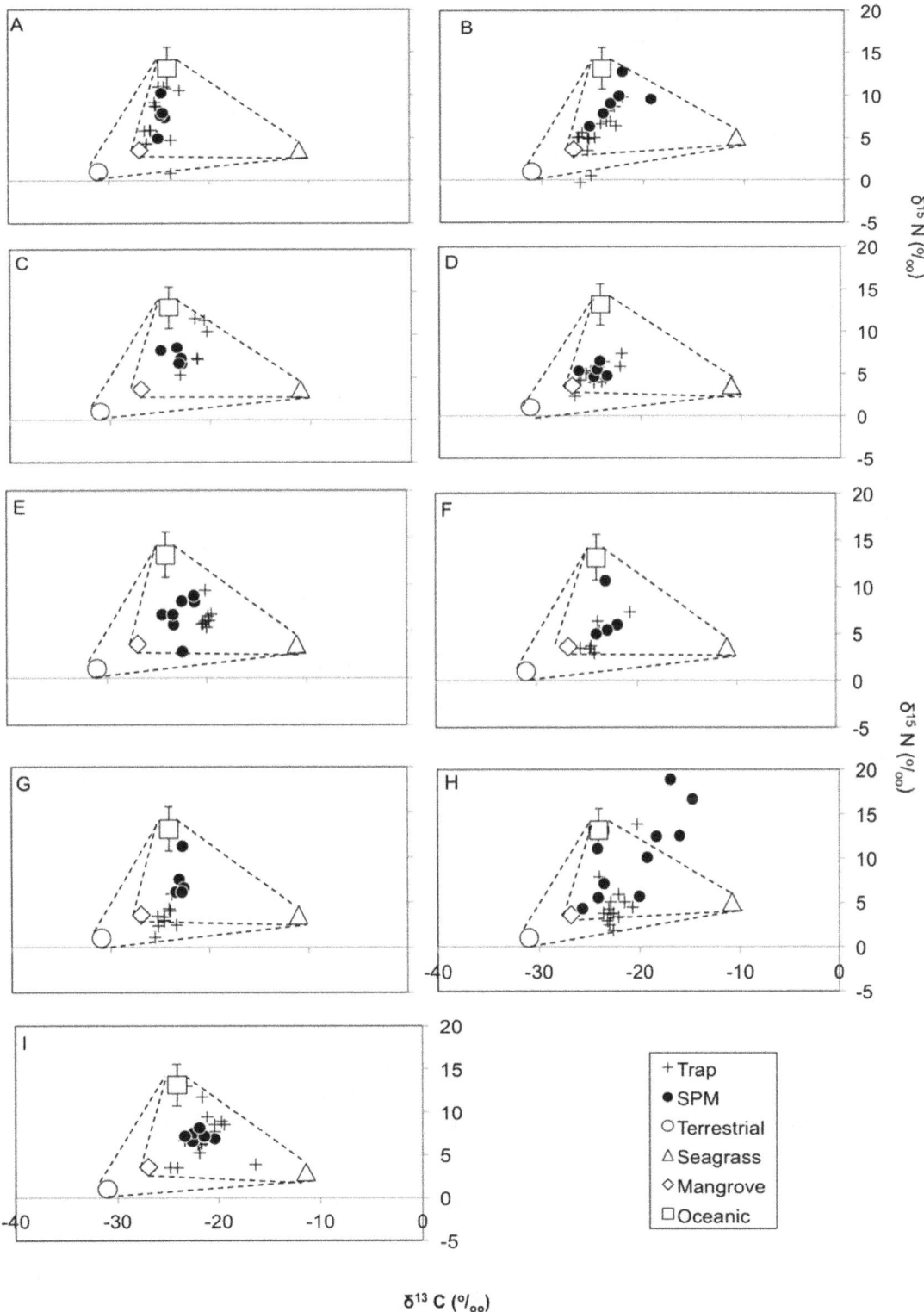

Figure 2. Carbon (δ^{13}C $^{o}/_{oo}$) and nitrogen (δ^{15}N $^{o}/_{oo}$) isotopic composition for organic matter from trap and suspended particulate matter (SPM) samples. The end members shown are: white circles represent the terrestrial value; white squares represent oceanic sources; white triangles represent seagrass plants; and white diamonds are mangrove trees. The trap and SPM samples are represented by crosses and black circles, respectively. Values for the end members are means (+/− SE). For the end members of the oceanic, terrestrial, seagrass plants and mangrove plants sources, n = 6, 9, 44 and 45, respectively. Note that this is a 2-dimensional representation of isotope values only. Table 3 shows results using all three components, i.e. carbon, nitrogen isotopes and C:N ratios.

Table 3. End member contributions of each source (terrestrial, mangrove, seagrass and oceanic) to each sediment sample (SPM & trap), as determined from the mixing model.

Site	Sample	Distance (m)	0	50	100	200	300	400	500	1000	1500
		Source	Contribution (%)								
A	SPM	Mangrove	17	3	1		1^	1			
		Oceanic	66	86	90		83^	85			
		Terrestrial	12	7	5		12^	4			
		Seagrass	5	4	4		4^	10			
	Trap	Mangrove	33	6	1		0.9^	2.8^			
		Oceanic	60	90	97		98^	95^			
		Terrestrial	5	3	1		0.5^	0.8^			
		Seagrass	2	1	1		0.6^	0.9^			
B	SPM	Mangrove	2	2	8	18	11^	5^	ns	ns	ns
		Oceanic	93	94	86	73	84^	88^	ns	ns	ns
		Terrestrial	1	1	3	4	3^	4^	ns	ns	ns
		Seagrass	3	3	3	5	2^	3^	ns	ns	ns
	Trap	Mangrove	65	ns	21	6	6^	1^	14	ns	ns
		Oceanic	22	ns	65	86	91^	95^	80	ns	ns
		Terrestrial	9	ns	9	4	1^	2^	2	ns	ns
		Seagrass	4	ns	5	4	2^	1^	4	ns	ns
C	SPM	Mangrove	13	10	3	11	12^				
		Oceanic	65	78	77	67	78^				
		Terrestrial	10	5	13	11	6^				
		Seagrass	12	7	7	11	4^				
	Trap	Mangrove	5	0.01	3	0.01	1^				
		Oceanic	64	78	60	98	80^				
		Terrestrial	27	20	33	1	17^				
		Seagrass	4	2	4	0.9	2^				
D	SPM	Mangrove	1	0.3^	0.2^	0.1^	0.6^				
		Oceanic	75	85^	79^	75^	83^				
		Terrestrial	23	14^	20^	24^	16^				
		Seagrass	1	0.7^	0.8^	0.9^	0.4^				
	Trap	Mangrove	11	6^	5^	2^	0.2^				
		Oceanic	52	57^	59^	67^	80^				
		Terrestrial	36	35^	33^	29^	19^				
		Seagrass	1	2^	3^	2^	0.8^				
E	SPM	Mangrove	7	1	1	0.2^	ns	ns			
		Oceanic	79	69	34	76^	ns	ns			
		Terrestrial	11	28	63	23^	ns	ns			
		Seagrass	3	1	2	0.8^	ns	ns			
	Trap	Mangrove	3	3	1	1^	2^	3^			
		Oceanic	77	65	68	72^	82^	70^			
		Terrestrial	17	29	29	25^	13^	25^			
		Seagrass	3	3	2	2^	3^	2^			
F	SPM	Mangrove	1	0.5	1	0.3^	0.8^				
		Oceanic	74	72	81	75^	72^				
		Terrestrial	24	27	18	24^	26^				
		Seagrass	1	1	1	1^	2^				
	Trap	Mangrove	7	3	0	7^					
		Oceanic	43	65	80	30^					
		Terrestrial	47	30	18	59^					
		Seagrass	3	2	2	4^					

Table 3. Cont.

Site	Sample	Distance (m) Source	0 Contribution (%)	50	100	200	300	400	500	1000	1500
G	SPM	Mangrove	1.0	ns	1.6	ns	ns	1.3^			1^
		Oceanic	74	ns	80	ns	ns	94^			71^
		Terrestrial	17	ns	12	ns	ns	2^			19^
		Seagrass	8	ns	6	ns	ns	3^			9^
	Trap	Mangrove	26	30	24	32	38^				
		Oceanic	62	34	50	41	40^				
		Terrestrial	5	17	10	9	6^				
		Seagrass	7	19	16	18	16^				
H	SPM	Mangrove	1	6	9	11	7	15	22^	15^	
		Oceanic	68	58	54	56	75	71	67^	74^	
		Terrestrial	23	25	24	20	7	4	3^	1^	
		Seagrass	8	11	13	13	11	10	8^	10^	
	Trap	Mangrove	20	ns	20	ns	ns	ns	16^	8^	7
		Oceanic	48	ns	23	ns	ns	ns	31^	58^	40
		Terrestrial	17	ns	34	ns	ns	ns	29^	19^	34
		Seagrass	15	ns	23	ns	ns	ns	24^	15^	19
I	SPM	Mangrove	6	4	5	1	6	ns	4^	1^	
		Oceanic	69	77	71	72	72	ns	76^	82^	
		Terrestrial	14	11	11	14	10	ns	13^	10^	
		Seagrass	11	8	13	13	13	ns	7^	7^	
	Trap	Mangrove	22	ns	9	10	3	4^	2^	3^	
		Oceanic	59	ns	68	62	71	71^	84^	77^	
		Terrestrial	6	ns	10	12	15	8^	5^	8^	
		Seagrass	13	ns	13	16	11	17^	9^	12^	

Distance (m) represents the distance along the transect from land (0 m) to the ocean for each site. SPM and trap are sample type, SPM is suspended particulate matter taken from the water column. (ns) means that the model did not find a solution. (^) represents sediment trap and SPM samples located in seagrass beds. Blank spaces indicate that no samples were collected.

this, indicates that the seagrass canopy traps seagrass OM [10,31–32].

Landscape patterns of organic matter fluxes

Terrestrial-derived organic material showed an under contribution relative to the size of the catchment area. Different land uses may donate particulate and dissolved nutrients to downstream mangrove forests, but it is plausible that mangrove forests trap much of the nutrients that are flowing into them [33–34]. The negative correlation between catchment area and terrestrial OM is a data artefact. This has occurred due to sites A and B having catchment areas that are an order of magnitude bigger than the others, conversely sites A and B display a minor terrestrial contribution thus these two sites skew the correlation.

Mangrove derived organic matter contributed in proportion to their surface area. The latter finding may be surprising because mangrove forests are known to have a high productivity [7–8]. However, mangrove forests are also known to retain much of their POM and associated nutrients within the forests due to retention mechanisms as supported by other studies [33–35]. Our results indicate that the magnitude of outwelling of OM at the mangrove boundary (0 m) is correlated with the sea-facing width of the mangrove forest and therefore the extent that the forest is exposed to waves. Waves have been shown to decrease the trapping capacity of mangrove roots and therefore increase the transport of

POM through mangrove roots [34]. The shape of the mangrove forest may have a strong influence on retention of POM. This suggests that mangrove forests that are distributed thinly along a coastline or large delta would export more POM because of their extensive frontal width, which is exposed to greater levels of hydrodynamic energy. In contrast, mangrove forests that extend inland or within protected bays may retain much of their POM and other sources of nutrients (dissolved and particulate) flowing into them. Our study sites are a mixture of these two categories category (Figure 1).

If one compares the potential amount of influx of OM sources in relation to their areal extent, seagrass is "relatively" a much more important source than mangrove forests or terrestrial sources (Figures 3 and 4). If seagrass beds are an important source of OM they may donate this material to other ecosystems and organisms. For example, stony corals have been found to assimilate particulate seagrass as a food source [36] and seagrass leaves are known to be an important food source for herbivore organisms [40]. No relationships were found between seagrass plant contribution and physical factors: hence there may be another controlling factor that has not been measured in the present study. Local hydrodynamics are a potential one. Indeed, as turbulence and wave energy can be a structuring factor in seagrass habitats. For instance, waves with an overlaying current have been found to

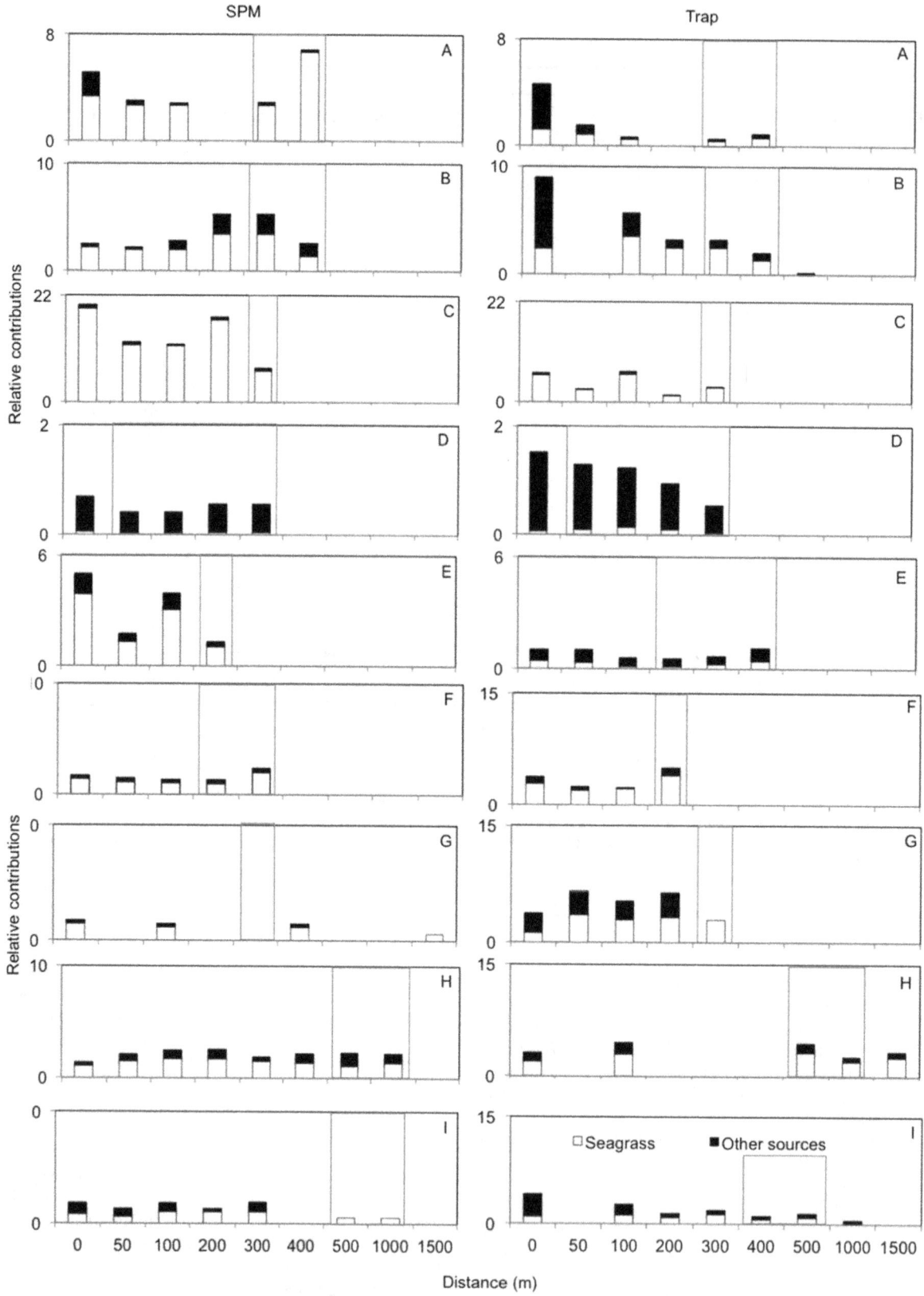

Figure 3. A dimensionless index indicating the relative contribution of seagrass plants and other sources (mangrove trees + terrestrial) for suspended particulate matter (SPM) and sediment trap samples (trap). The values were determined by dividing the mixing model contribution result (Table3) with the per cent of total surface area occupied by a particular ecosystem at each site. Distance (m) represents the distance along the length of the transect from the land (0 m) to the ocean for each site. SPM and trap are sample type, SPM is suspended particulate matter taken from the water column. Whilst trap represents sediment samples captured on the sediment floor over a 48-hour period. Each chart represents a site location; see letter at the top left right side of chart (Figure1). White areas represent seagrass beds, black areas symbolize other sources (mangrove trees + terrestrial). Green boxes around columns indicate when samples were taken in a seagrass bed. This index was not calculated for POM derived from oceanic sources because a representative surface area for the 'ocean' could not be defined.

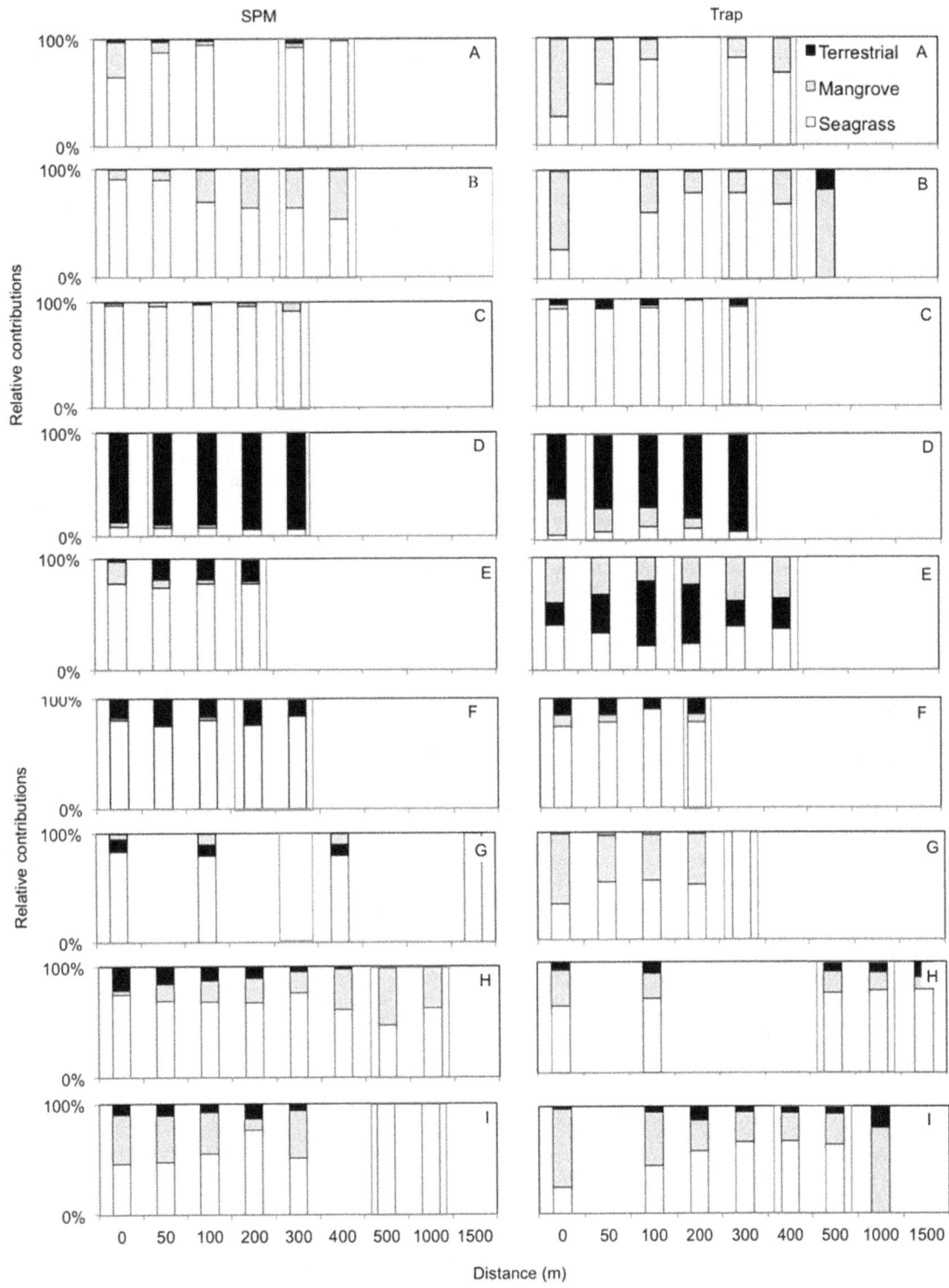

Figure 4. Dimensionless index indicating the relative contribution of terrestrial, mangrove and seagrass sources for suspended particulate matter (SPM) and sediment trap samples (trap), plotted in a stacked column chart. Distance (m) represents the distance along the length of the transect from the land (0 m) to the ocean for each site. SPM and trap are sample type, SPM is suspended particulate matter taken from the water column. Whilst trap represents sediment samples captured on the sediment floor over a 48 hour period. Each chart represents a site location, see letter at top left right side of chart (Figure 1). White areas represent seagrass beds, grey and black areas symbolise mangrove forests and terrestrial sources respectively. Green boxes around columns indicate when samples were taken in a seagrass bed. This index was not calculated for POM derived from oceanic sources, because no representative surface area for the 'ocean' could be defined.

increase litter movement through seagrass beds [34]. Herbivores digesting the seagrass plants and transferring nutrients via faeces across the tropical coastal seascape [37–39] may be another undetected factor.

Approximately one quarter of the samples showed a higher seagrass contribution within the bed but otherwise, the seagrass contribution seemed unrelated to the distance from or within the seagrass bed (Figures 3 and 4). This indicates that seagrasses have a spatially extended influence beyond the physical boundaries of the bed. Although the mechanisms are still unknown of what controls seagrass contribution, our results clearly show that seagrasses are a important contributor of organic matter in the tropical seascape of Phangnga bay relative to their spatial coverage.

Future Perspective and Management Implications

The high oceanic contribution (Table 3) shows how important oceanic derived nutrients are to ecosystems and organisms in these locations. We reason that this may be because of the high tidal flushing in all of our sites allowing for a large exchange of water. This could have implications for pollutants from industries such as fish farms being easily flushed into seagrass beds or mangrove forests from this tidal exchange.

Terrestrial input had the smallest contribution to the OM (Figure 3 and Figure 4). The differences in urbanizations across the 9 sites showed no significant correlation with terrestrial derived OM. Additionally, mangrove forests were not found to be an important input for OM in the coastal zone, compared to oceanic or seagrass organic matter. Recent research has found that mangrove forests may be an effective POM trap and therefore a strong nutrient sink [34]. Mangrove forests are effective filters, allowing them to retain most nutrients fluxes from terrestrial sources. For this reason the integrity and a critical size of the forest should always be ensured. Although many mangroves have been reduced in aerial extent in the Phuket region of Thailand, our results indicate that, though impoverished, mangroves may still maintain important ecosystem services of retaining mangrove and terrestrial derived nutrients in this region. However, continued deforestation or non-sustainable use of the forests may jeopardize this function.

An important consideration is that changes in mangrove forest aerial extent or biomass could reduce buffering and therefore increase the flux of terrestrial-derived nutrients to the coastal waters. This in turn could cause nutrient enrichment or increased turbidity for seagrass beds, causing physiological problems for the seagrass plants that could reduce the area and biomass of this ecosystem [42]. We have shown that seagrass beds are a comparatively large source of POM for adjacent ecosystems. However, seagrass beds can become unstable, for example following eutrophication [41] or hurricane damage [15]. This instability could have implications for the productivity of the seagrass beds and thus, their role as a nutrient source for other ecosystems and organisms could be affected. These alterations to the integrity of the system could have consequences on their ecosystem function as a nutrient sink or buffer for excess nutrients for sensitive adjacent ecosystems such as coral reefs. Therefore, a key role in management of these areas is to ensure the health and physical/physiological structure of seagrass beds. Both factors (health and physical/physiological structure of beds) can be related to quantities of POM being trapped and outwelled [41,43].

This study provides further evidence that the existence of connective particulate fluxes, which occur between ecosystems in the tropical coastal seascape, do exist. These connections could have implications for strengthening management especially from an ecosystem-based perspective.

Author Contributions

Conceived and designed the experiments: LG Gillis: principle author, data collection and analyses. AD Ziegler: daily supervision of paper, D Van Oevelen: data analyses and feedback on paper, C. Cathalot: data collection, analyses and feedback on paper, JW Wolters: data collection and analyses, PMJ Herman: feedback on paper and idea, TJ Bouma: perceived idea, daily supervision. Performed the experiments: LGG CC JWW. Analyzed the data: LGG DVO CC JWW. Contributed reagents/materials/analysis tools: LGG DVO CC. Wrote the paper: LGG ADZ DVO CC PMJ TJB. Principal author: LGG. Data collection: LGG CC JWW. Daily supervision of paper: ADZ TJB. Feedback on paper: DVO CC. Feedback on paper and idea: PMJH. Perceived idea: TJB.

References

1. Barbier EB, Hacker SD, Kennedy C, Koch EW, Stier AC, et al. (2011) The value of estuarine and coastal ecosystem services. Ecol Monog 81(2): 169–193.
2. Gattuso JP, Frankignoulle M, Wollast R (1998) Carbon and carbonate metabolism in coastal aquatic ecosystems. Annu Rev Ecol Syst 29: 405–434.
3. Krusche AV, Martinelli LA, Victoria RL, Bernardes M, de Camargo PB, et al. (2002) Composition of particulate and dissolved organic matter in a disturbed watershed of souteast Brazil (Piracicaba River basin). Wat Res 36: 2743–2752.
4. Kuramoto T, Minagawa M (2001) Stable carbon and nitrogen isotopic characterization of organic matter in a mangrove eocystem on the southwestern coast of Thailand. 2001. J Oceanogr 57: 421–431.
5. Thimdee W, Deein G, Sangrungruang C, Nishioka J, Matsunaga K (2003) Sources and fate of organic matter in Khung Krabaen Bay (Thailand) as traced by δ13C and C/N atomic ratios. Wetlands 23(4): 739–738.
6. Meksumpun S, Meksumpun C, Hoshika A, Mishima Y, Tanimoto T (2005) Stable carbon and nitrogen isotope ratios of sediment in the gulf of Thailand: Evidence for understanding of marine environment. Cont Shelf Res 25: 1905–1915.
7. Odum EP (1968) A research challenge: evaluating the productivity of coastal and estuarine water. In Proceedings of the Second Sea grant Conference, University of Rhode Island, Kingston, USA: 63–64.
8. Lee SY (1995) Mangrove outwelling- A Review. Hydrobiologia 295(1–3): 203–212.
9. Terrados J, Duarte CM (2000) Experimental evidence of reduced particle resuspension within a seagrass (Posidonia oceanica L.) meadow. J Exp Mar Biol Ecol 243: 45–53.
10. Vonk JA, Middelburg JJ, Stapel J, Bouma TJ (2008) Dissolved organic nitrogen uptake by seagrasses. Limnol Oceanogr 53: 542–548

11. Fonseca MS, Cahalan JA (1992) A preliminary evaluation of wave attentuation for four species of seagrass. Est Coast Shelf Sci 35: 565–576.
12. Hendriks IE, Sintes T, Bouma TJ, Duarte CM (2008) Experimental assessment and modeling evaluation of the effects of the seagrass Posidonia oceanica on flow and particle trapping. Mar Ecol Prog Ser 356: 163–173.
13. Hemminga MA, Marba N, Stapel J (1999) Leaf nutrient resorption, leaf lifespan and the retention of nutrients in seagrass systems. Aquat Bot 65: 141–158.
14. Koch EW, Verduin JJ (2001) Measurements of physical parameters in seagrass habitats. In Short FT, Short CA, Coles RG (Eds.). Global Seagrass Research Methods. Amsterdam: Elsevier Science pp. 325–344.
15. Infantes E, Terrados J, Orfila A, Canellas B, Alvarez-Ellacuria A (2009) Wave energy and the upper depth limit distribution of Posidonia oceanica. Bot Mar 52: 419–427.
16. Wilkie L, O'Hare MT, Davidson I (2012) Particle trapping and retention by Zostera noltii: A flume and field study. Aquat Bot 102: 15–22.
17. Hemminga MA, Slim FJ, Kazungu J, Ganssen GM, Nieuwenhuize J, et al. (1994) Carbon outwelling from a mangrove forest with adjacent seagrass beds and coral reefs (Gazi Bay, Kenya). Mar Ecol Prog Ser 106(3): 291–301.
18. McConnachie JL, Petticrew EL (2006) Tracing organic matter sources in riverine suspended sediment: Implications for fine sediment transfers. Geomorphology 79: 13–26.
19. Schindler Wildhaber Y, Liechti R, Alwell C (2012) Organic matter dynamics and stable isotope signature as tracers of the sources of suspended sediment. Biogeosciences 9: doi: 10.5194/bg-9-1985-2012.
20. Bouillon S, Dehairs F, Velimirov B, Abril G, Vieira Borges A (2007) Dynamics of organic and inorganic carbon across contiguous mangrove and seagrass

systems (Gazi Bay, Kenya). J Geophys Res G02018, doi:10.1029/2006JG000325.

21. Limpsaichol P, Khotiattiwong S, Bussarwait N, Sojisuporn P (1998) Environmental factors influencing the health and productivity of Phang-nga Bay. In Nickerson D. J. (Ed.) 1992: Community-based fisheries management in the Phang-nga bay, Thailand. Proceedings of the National Workshop on Community-based Management. Phuket, Thailand (BOBP Report No 78): 85–120.

22. Woodroffe C (1992) Mangrove Sediments and Geomorphology. In: Robertson AI and DM Alongi (eds.) Trophical Mangrove Ecosystems. American Geophysical Union Washington, USA: 7–41.

23. Nieuwenhuize J, Maas EMY, Middelburg JJ (1994) Rapid analysis of organic carbon and nitrogen in particulate materials. Mar Chem 45: 217–244.

24. Phillips DL, Gregg JW (2001) Uncertainty in source partioning using stablis isotopes. Oecologia 127: 171–179.

25. Phillips DL, Koch PL (2002) Incorating concentration dependence in stable isotope mixing modes. Oecologia 120: 114–125.

26. R Core Team (2012) R: A language and environment for statistical computing. R Foundation for Statistical Computing, Vienna, Austria. ISBN 3-900051-07-0, Available: http://www.R-project.org/. Accessed 2014 Oct 1.

27. Van Oevelen D, Van den Meersche K, Meysman F, Soetaert K, Middelburg JJ, et al. (2010) Quantitative reconstruction of food webs using linear inverse models. Ecosystems 13: 32–45.

28. Cline JD, Kaplan IR (1975) Isotopic fractionation of dissolved nitrate during denitrification in the eastern tropical North Pacific Ocean. Mar Chem 3: 271–299.

29. Dittmar T, Lara RJ (2001) Driving forces behind nutrient and organic matter dynamics in a mangrove tidal creek in North Brazil. Est Coast Shelf Sci 52: 249–259.

30. Hunsinger GB, Mitra S, Findlay SEG, Fischer DT (2010) Wetland-driven shifts in suspended particulate organic matter compoistion of the Hudson River estuary, New York. Limnol Oceanogr 55(4): 1653–1667.

31. Evrard V, Kiswara W, Bouma TJ, Middelburg JJ (2005) Nutrient dynamics of seagrass ecosystems: N-15 evidence for the importance of particulate organic matter and root systems. Mar Ecol Prog Ser 295: 49–55.

32. Van Engeland T, Bouma TJ, Morries EP, Brun FG, Peralta G, et al. (2011) Potential uptake of dissolved organic matter by seagrasses and macroalage. Mar Ecol Prog Ser 427: 71–81.

33. Alongi DM (1990) Effect of mangrove detrital outwelling on nutrient regeneration and oxygen fluxes in coastal sediments of the central-greater-barrier-reef lagoon. Est Coast Shelf Sci 31(5): 581–598.

34. Gillis LG, Bouma TJ, Kiswara W, Ziegler AD, Herman PMJ (2014) Leaf transport in mimic mangrove forests and seagrass beds. Mar Ecol Prog Ser 498: 95–102.

35. Bouillon S, Borges AV, Castaneda-Moya E, Diele K, Dittmar T, et al. (2008) Mangrove production and carbon sinks: A revision of global budget estimates. Global Biogeochem cycles 22, doi: 10.1029/2007GB003052.

36. Lai S, Gillis LG, Mueller C, Bouma TB, Guest JR, et al. (2013) First experimental evidence of corals feeding on seagrass matter. Coral Reefs 32 (4): 1061–1064.

37. Krumme U (2009) Dial and Tidal Movements by Fish and Decapods Linking Tropical Coastal Ecosystems. In: Nagelkerken I (Ed.). Ecological Connectivity among Tropical Coastal Ecosystems, Springer Netherlands: 271–324.

38. Forward RB, Tankersley RA (2001) Selective tidal-stream transport of marine animals. Oceanogr Mar Biol 39: 305–353.

39. Kimirei IA, Nagelkerken I, Mgaya YD, Huijbers CM (2013) The Mangrove Nursery Paradigm Revisited: Otolith Stable Isotopes Support Nursery-to-Reef Movements by Indo-Pacific Fishes. PLOS ONE 8 (6): e66320.

40. Heck KL, Valentine JF (2006) Plant-herbivore interactions in seagrass meadows. J Exp Mar Biol 330: 420–436.

41. Burkholder JM, Tomasko DA, Touchette BW (2007) Seagrasses and eutrophication. J Exp Mar Biol Ecol 350(1–2): 46–72.

42. Todd PA, Ong XY, Chou LM (2010) Impacts of pollution on marine life in Southeast Asia. Biodiversity Conserv 19: 1063–1082.

43. Perez M, Invers O, Ruiz JM, Frederiksen MS, Holmer M (2007) Physiological responses of the seagrass Posidonia oceanica to elevated organic matter content in sediments: An experimental assessment. J Exp Mar Biol Ecol 344: 149–160.

20

Biologically Induced Deposition of Fine Suspended Particles by Filter-Feeding Bivalves in Land-Based Industrial Marine Aquaculture Wastewater

Yi Zhou*, Shaojun Zhang, Ying Liu, Hongsheng Yang

Key Laboratory of Marine Ecology and Environmental Sciences, Institute of Oceanology, Chinese Academy of Sciences, Qingdao, P. R. China

Abstract

Industrial aquaculture wastewater contains large quantities of suspended particles that can be easily broken down physically. Introduction of macro-bio-filters, such as bivalve filter feeders, may offer the potential for treatment of fine suspended matter in industrial aquaculture wastewater. In this study, we employed two kinds of bivalve filter feeders, the Pacific oyster *Crassostrea gigas* and the blue mussel *Mytilus galloprovincialis*, to deposit suspended solids from marine fish aquaculture wastewater in flow-through systems. Results showed that the biodeposition rate of suspended particles by *C. gigas* (shell height: 8.67 ± 0.99 cm) and *M. galloprovincialis* (shell height: 4.43 ± 0.98 cm) was 77.84 ± 7.77 and 6.37 ± 0.67 mg ind^{-1}·d^{-1}, respectively. The total solid suspension (TSS) deposition rates of oyster and mussel treatments were 3.73 ± 0.27 and 2.76 ± 0.20 times higher than that of the control treatment without bivalves, respectively. The TSS deposition rates of bivalve treatments were significantly higher than the natural sedimentation rate of the control treatment ($P < 0.001$). Furthermore, organic matter and C, N in the sediments of bivalve treatments were significantly lower than those in the sediments of the control ($P < 0.05$). It was suggested that the filter feeders *C. gigas* and *M. galloprovincialis* had considerable potential to filter and accelerate the deposition of suspended particles from industrial aquaculture wastewater, and simultaneously yield value-added biological products.

Editor: Vanesa Magar, Centro de Investigacion Cientifica y Educacion Superior de Ensenada, Mexico

Funding: This research was supported by the Hi-Tech Research and Development Program (863) of China (No. 2006AA100305), the National Natural Science Foundation of China (No. 41176140/30972268), and the Key Projects in the National Science & Technology Pillar Program during the 12th Five-year Plan Period (2011BAD13B06). The funders had no role in study design, data collection and analysis, decision to publish, or preparation of the manuscript.

Competing Interests: The authors have declared that no competing interests exist.

* Email: yizhou@qdio.ac.cn

Introduction

Land-based flow-through (FT) aquaculture production systems are being widely used around the world. The N and P nutrient discharge from FT aquaculture should be regulated to mitigate effluent nutrient contributions to the receiving waters, which may lead to degradation and eutrophication [1]. Land-based intensive marine recirculating aquaculture system (RAS), as the newest form of fish farming production system, has been rapidly developing in the past decade [2]. RAS is typically an indoor system that allows farmers to control environmental conditions throughout the year. Suspended waste solids, including uneaten feed and fish feces, have to be removed as quickly as possible to prevent their accumulation to unsafe levels within the RAS. If left in the system, the suspended solids could generate additional oxygen demand and ammonia nitrogen as a result of bacterial decomposition. Thus, removal of suspended solids is one of the critical processes in RAS [3–5]. Large particles are easily removed from industrial aquaculture wastewater via filter techniques or settling tanks, while small particles are too small to be retained by traditional filter techniques, and they have a settling velocity too small to be trapped in sedimentation basins. In addition to the general mechanical filtration that removes larger suspended solids, methods such as microfiltration or sieve-bend screen have been

applied to remove the suspended solids of medium particle size. Although the removal accuracy of suspended solids has generally ranged from 100 to 200 µm, the removal rate of suspended solids of smaller particle size less than 100 µm has been relatively low. Chen et al. [3] found that, using conventional filter techniques, it is difficult to remove particulate matters with a diameter less than 30 µm, which account for 80–90% of the total mass of particulate matters in the water body of high-density RAS. In particular, the feces in the discharged water from the finfish culture of *Cynoglossus semilaevis* Günther easily breaks down into flocculent suspended solids which are difficult to be precipitated and removed. This not only increases the processing load of the water treatment unit of the RAS, but also reduces the contact area of the biofilm with the dissolved pollutants in wastewater and influences the processing efficiency of the biofilm due to the adhesion of fine suspended solids [3–5].

Filter-feeding bivalves, such as oysters and mussels, have strong water-filtration ability. They are able to filter a large number of fine particulate matters, including phytoplanktons, zooplanktons, microorganisms (e.g. bacteria [6]), and other particulate organic debris [7], and cause sedimentation of the suspended solids in the form of feces and pseudofeces. This process is known as biodeposition [8–9]. Furthermore, the organic components in

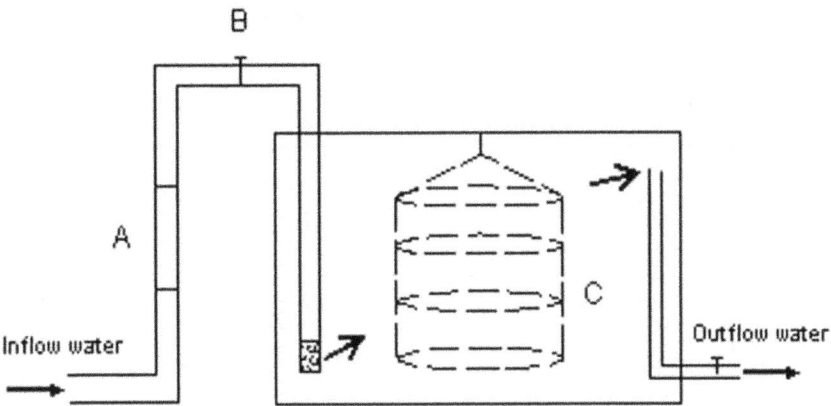

Figure 1. Schematic diagram of individual experimental biotanks. Each biotank measured 75×50×50 cm in size. A, Flowmeter; B, Control valve; C, Lantern net deployed in a biotank for bivalve culture.

the suspended matters can be assimilated and utilized by filter-feeding bivalves. Many studies have shown that mussels, oysters, clams, and other filter-feeding bivalves are very important components of a healthy coastal ecosystem, and act as key species and functional group [10–11]. These bivalves are not only economically vital, but also ecologically important. They significantly enhance the coupling between water layer and benthos through physiological and ecological processes such as filter-feeding and biodeposition, and play an important role in material cycling and energy flow in coastal ecosystems [7,9,12–13]. Recently, the filter-feeding capacity of bivalves has been proposed to have the potential to purify organic effluents released from open-water fish farms and shrimp ponds; in the Integrated Multi-Trophic Aquaculture (IMTA), the filter-feeding bivalves not only remove the suspended particulate matters in the water body, but also transform these wastes into economically valuable products [14]. Nowadays, there is an increasing interest in the IMTA as a sustainable approach to fed aquaculture (e.g. finfish, shrimp). In the IMTA, fed aquaculture is combined with inorganic (nutrient) extractive aquaculture (e.g. seaweed [14–16]), suspended organic extractive aquaculture (e.g. filter-feeding bivalves [17–18]), and deposited organic extractive aquaculture (e.g. sea cucumber [19–20]) to create balanced systems that provide environment bioremediation, mutual benefits to cocultured organisms, and economic diversification by producing other value-added products [14,21].

For the occurrence of large quantities of fine suspended particles in industrial aquaculture wastewater, filter feeders may offer an approach for treatment of those suspended matter. Surprisingly, little work has been done on removal of fine suspended particles from industrial marine aquaculture wastewater using macro-bio-filters. Our previous work showed that suspension-cultured filter-feeding bivalves (*Chlamys farreri*) can greatly increase the deposition of suspended particulate matter in coastal eutrophic

waters [9,13]. In the present study, our hypothesis was that bivalve filter feeders could filter and accelerate the deposition of suspended particles from industrial aquaculture wastewater. We tested the use of two filter-feeding bivalve species (the Pacific oyster *Crassostrea gigas* and the blue mussel *Mytilus galloprovincialis*) with strong adaptability and water-filtration ability to remove small suspended particles from the marine RAS. The treatment potential of the two bivalves to deposit suspended particulate matters in saline wastewater from the RAS was experimentally determined.

Material and Methods

Ethics Statement

The collecting of the bivalves (*Crassostrea gigas* and *Mytilus galloprovincialis*) used in this experiment from an adjacent coastal area (Jiaozhou) was permitted by Benshan Wang, manager of that area. Ethical approval was not required for this study because no endangered animals were involved. However, specimen collection and maintenance were performed in strict accordance with the recommendations of Animal Care Quality Assurance in China.

2.1. Experimental design. The FT experiment included three treatments, i.e. oyster treatment, mussel treatment, and control treatment. Each treatment consisted of five plastic biotanks (Figure 1), with each biotank measuring 75×50×50 cm in size; and a total of 15 plastic biotanks were used. In each bivalve treatment, the experimental bivalves were evenly placed in a four-tray lantern net, while in the control treatment, only an empty lantern net was placed. In each biotank, the bottom of the inflow pipe was sealed, and small holes were perforated at approximately 10 cm on the lower part of the pipe to allow advection of water into the biotank.

The experiment was conducted at a land-based aquaculture site (Haifa Company, Tianjin, norh China) for the finfish *Cynoglossus semilaevis* Günther. The fish (40–50 cm in body length, 550–

Table 1. Biological parameters of bivalves used in the experimental systems.

Treatments	Shell height (cm)	Total wet weight (g)	Total dry tissue weight (g)	Individual dry tissue weight (g·ind⁻¹)
Oyster	8.67±0.99	1038.96±14.01	27.29±1.10	2.10±0.12
Mussel	4.43±0.98	765.61±24.05	22.01±1.73	0.20±0.01

Table 2. Biological filter-removing rates of suspended particles in aquaculture wastewater by experimental bivalves.

Treatments	Sedimentation rates (mg·d^{-1})	Ratios of sedimentation rates(Bivalve/Control)	Biodeposition rates (mg·ind^{-1}·d^{-1})	Percentage of TSS reduced (%)
Oyster	1381.9±101.0[a]	3.73±0.27[a]	77.84±7.77[a]	21.3±3.7[a]
Mussel	1020.7±75.8[b]	2.76±0.20[b]	6.37±0.67[b]	14.1±2.6[b]
Control	370.0±79.3[c]	-	-	-

Sedimentation rates (mg·d^{-1}): the sedimentation rate of suspended solids in the FT system; Ratios of sedimentation rates: ratios of the sedimentation rate of suspended solids in a bivalve treatment to that in the control (without bivalves); Percentage of TSS reduced (%): percentage of TSS reduced in the outflow water in oyster or mussel treatment, compared with control treatment without bivalves. Values are given as means±SD. Values with different superscripted letters in the same column are significantly different from each other ($P<0.05$).

900 g ind^{-1} in wet weight) was intensively cultured in a density of 25 ind m^2 in 38-m^2 culture tanks (0.8 m in depth). During the culture, formula feed with an amount of 1–2% that of wet weight of fish was fed twice a day. The bivalves (*Crassostrea gigas* and *Mytilus galloprovincialis*) used in the experiments were collected on 18 November 2012 from the adjacent coastal area. The experimental wastewater was the discharged water from the intensive culture of *C. semilaevis*, which was filtered through sieve-bend screen. The water temperature was 20.1–24.6°C, dissolved oxygen (DO) was 6.04–7.83 mg L^{-1}, and pH was 7.60–7.81. The experimental wastewater was pumped into an elevated cistern and then piped into the systems; the volume of each experimental biotank was 150 L. The experiment was conducted from 26 November 2012 and lasted for 7 days. The flow rate of the inflow water was 80 L h^{-1}, and the stocking densities in each biotank were 13 individuals with total wet weight 1038.96±14.01 g for oysters and 110 ind with total wet weight 765.61±24.05 g for mussels. The total dry tissue weight of bivalves in each oyster biotank and mussel biotank was 27.29±1.10 g, and 22.01±1.73 g (Table 1), respectively. Before the initiation of the experiment, the bivalves were acclimated in the experimental systems for one week.

2.2. Sampling and determination. At the end of the experiment, the inflow water was stopped and the shellfishes were taken out. The biotank was left to stand for 5 h. The sediments were collected using siphonage, desalted and baked at 60°C, and weighed to obtain the dry weights. The contents of organic matter (OM), organic carbon (OC), organic nitrogen (ON), inorganic phosphorus (IP), organic phosphorus (OP), and total phosphorus (TP) in the sediments were measured. The OM was expressed as the difference in weight before and after ashing at 500°C for 3 h. After the sediments were decarbonated using acid mist of concentrated hydrochloric acid for 5 h, the OC and ON were measured using a CHN elemental analyzer. The IP, OP, and TP in the sediments were measured using the ashing method [22]. The shell heights and wet weights of the oysters and mussels in each cage were measured.

The formula for determining the sedimentation rate (mg·d^{-1}) of suspended solids in the system was as follows:

$$Sedimentation\ rate = W_{O\,(M,\,C)}/t$$

where $W_{O\,(M,\,C)}$ is the dry weight (mg) of total sediments collected from the bottom of the single biotank of oyster (or mussel/control) treatment, and t is the experiment duration (days). The biodeposition rates (BDR; mg·ind^{-1}·d^{-1}) of the suspended solids by the two kinds of bivalves were calculated using the following formula:

$$BDR = \left(W_{M(O)} - W_C\right)/N\cdot t$$

where $W_{M\,(O)}$ is the dry weight (mg) of the total sediments collected from the bottom of the biotank of oyster (or mussel) treatment, W_C is the dry weight (mg) of the total sediments collected from the bottom of the biotank in the control, t is the experiment duration (days), and N is the number of bivalves used in each treatment.

During the experiment, the total suspended solids (TSS; mg L^{-1}) in the inflow and outflow water of the system were determined 3 times every day. For TSS determination, 2 L of water sample was filtered through a pre-weighed mixed fiber membrane with a diameter of 1.0 μm. After filtration, the filter was rinsed with 100 mL of distilled water to remove the salts, dried at 60°C for 48 h, and weighed.

In the present study, results are presented as mean ± SD. The differences between the treatments were tested using one-way analysis of variance (ANOVA). Prior to analysis, data were examined for homogeneity of variances using Levene's tests. The differences were considered significant at a probability level of 0.05. Statistics were performed using the software SPSS 16.0.

Results

The concentration of the total suspended solids (TSS; mg L^{-1}) in inflow water in the experimental system was 3.57±2.37 mg L^{-1}. The TSS sedimentation rates of the bivalve treatments were significantly higher than the natural sedimentation rate of the control treatment ($df = 9$, $F = 331.0$, $P<0.001$; $df = 9$, $F = 166.6$, $P<0.001$). The TSS sedimentation rates of oyster and mussel treatments were 3.73±0.27 and 2.76±0.20 times higher than that of the control treatment, respectively. The biodeposition rates of the oyster *C. gigas* (shell height: 8.67±0.99 cm; Table 1) and the mussel *M. galloprovincialis* (shell height: 4.43±0.98 cm) were 77.84±7.77 and 6.37±0.67 mg·ind^{-1}·d^{-1}, respectively (Table 2). Compared with control treatment, TSS in the outflow water in oyster and mussel treatments was averagely reduced 21.3±3.7% and 14.1±2.6%, respectively.

Table 3 shows the contents of organic matter (OM) and C, N, and P in the sediments of bivalve and control treatments. The OM content in the sediments of oyster treatment was 29.46±0.45%, which was significantly lower than that in the sediments of the control (38.44±3.19%; $P<0.001$). Furthermore, the contents of C, N, and P in the sediments of oyster treatment were 11.72±0.29, 1.70±0.06, and 4.30±0.06%, respectively, while those in the sediments of control treatment were 15.29±1.79, 2.34±0.19, and 5.85±0.43%, respectively. These results showed that the OM and C, N, and P contents were significantly lower in the sediments of

Table 3. Comparisons in chemical compositions of sediments between bivalve (oyster or mussel) treatment and the control (without bivalves) in the FT experiments.

Treatments	OM (%)	OC (%)	ON (%)	TP (%)	OP (%)	C/N	C/OP
Control	38.44a	15.29a	2.34a	5.85a	0.64a	7.65a	58.0c
(±SD)	(3.19)	(1.79)	(0.19)	(0.43)	(0.11)	(0.28)	(6.4)
Oyster	29.46c	11.72c	1.70c	4.30bc	0.40b	8.04a	76.1a
(±SD)	(0.45)	(0.29)	(0.06)	(0.06)	(0.07)	(0.24)	(10.0)
Mussel	32.70b	12.79b	1.94b	4.40b	0.47ab	7.69a	71.9ab
(±SD)	(1.63)	(0.50)	(0.14)	(0.13)	(0.07)	(0.28)	(9.3)

Values are given as means±SD. OM, organic matter; OC, organic carbon; ON, organic nitrogen; TP, total phosphorus; OP, organic phosphorus. Values with different superscripted letters in the same column are significantly different from each other ($P<0.05$).

oyster treatment than in the sediments of the control treatment (Figure 2). Also, there were significant differences in the OM and C, N, and P contents in the sediments of the mussel and control treatments (all $P<0.05$).

Discussion

In recent years, application of filter-feeding bivalves for the removal of suspended particulate matter from the fish and shrimp aquaculture system in coastal waters has been reported [14]. In these systems, the food sources of bivalves include phytoplankton, uneaten feed, and fish feces. However, there have been no reports on the application of the macro-bio-filters for the removal of suspended particles from land-based industrial aquaculture systems, which are difficult to be removed using conventional mechanical filtration. In the present study, we conducted the FT experiment and found that the bivalve filter feeders, *Crassostrea gigas* and *Mytilus galloprovincialis*, had marked potential to deposit suspended solids in discharged wastewater from fish aquaculture.

Under the FT condition, the TSS sedimentation rates of the oyster and mussel treatments were 1381.9 ± 101.1 mg d^{-1}, and 1020.7 ± 75.8 mg d^{-1}, respectively, with the former significantly higher than the latter (Table 2). The difference might be due to the higher total dry tissue biomass (27.29 ± 1.10 g) in each oyster treatment than that in each mussel treatment (22.01 ± 1.73 g; Table 1). Also, dry tissue weight of bivalves (2.10 ± 0.12 g ind^{-1}) in oyster treatments was almost ten times higher than that in oyster treatments (0.20 ± 0.01 g ind^{-1}; Table 1); this might explain the much higher biodeposition rate of oysters (77.84 ± 7.77 mg·ind^{-1}·d^{-1}) than that of mussels (6.37 ± 0.67 mg·ind^{-1}·d^{-1}; Table 2) in this experiment.

Many studies have shown that the filter-feeding bivalves have strong water-filtration abilities and can exert profound impact on the suspended particulate matter in coastal waters. Hatcher et al. [23] determined the deposition rates in the Upper South Cove mussel aquaculture area in Canada and compared the results with those in the neighboring non-culture areas. They found that the amount of deposited materials in the former area was more than two times higher than that in the latter area. In many bivalve aquaculture areas, the biodeposition rate has been noted to be very impressive. For example, in the oyster aquaculture area in Hiroshima bay (Japan), 420,000 oysters could produce 16 metric tons of feces and pseudofeces in a 9-month culture period [24]. Since faeces and pseudofaeces are voided as mucus-bound aggregates, they are larger and more prone to sedimentation than the nonaggregated particles from which they are formed, and deposit at rates up to 40 times that of nonaggregated particles [25–27]. In a mussel farm covering an area of 1500 m^2 in Sweden, the biodeposition of dry matter was estimated to about 10 metric tons, and sediment under the rafts would accumulate to about 10 cm during a farming period of 1.5–2 years, and mussel farming was suggested to be a potential way to improve water quality in a eutrophied system [28–29].

The present study conducted in a land-based industrial aquaculture site showed that under the FT condition, the bivalves, *C. gigas* and *M. galloprovincialis*, both exhibited high suspended solid filtration rate in the effluent from *C. semilaevis* culture. The average biodeposition rate of individual oysters and mussels were 77.84 ± 7.77 and 6.37 ± 0.67 mg·ind^{-1}·d^{-1}, respectively. By contrast, the biodeposition rate of *C. gigas* (shell length: 95–110 mm) in natural aquaculture conditions in the Xuanmen port (north China) was 37.5–83.7 mg·ind^{-1}·d^{-1} [30]. The main factors affecting biodeposition rates of bivalves include body size, water

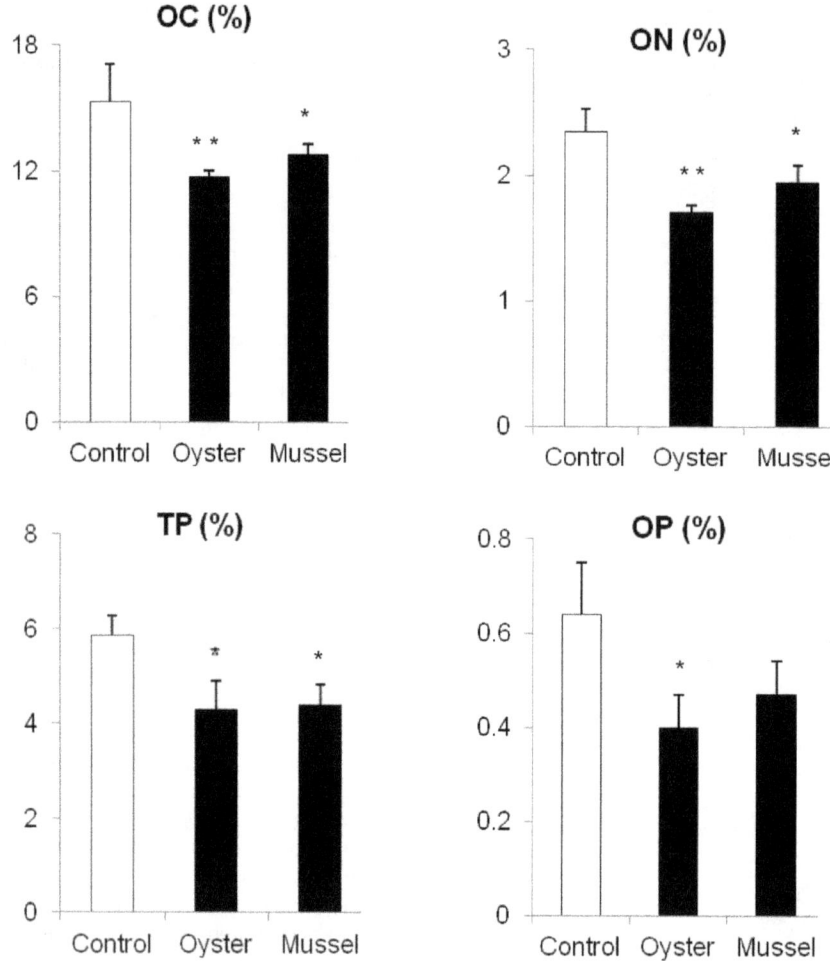

Figure 2. Comparisons in C, N, and P in the deposited sediments between a bivalve (oyster or mussel) treatment and the control (without bivalves). OC, organic carbon; ON, organic nitrogen; TP, total phosphorus; OP, organic phosphorus. *, significant difference between control and bivalve treatment ($P<0.05$); **, $P<0.01$. Values are means ± SD.

temperature, seston concentration, and composition [9,13]. The wastewater used in this study was the effluent from the *C. semilaevis* culture. The main components of the suspended solids were feed wastage and fish feces. In coastal fish-bivalve polyculture, the particulate organic matter from feed wastage and fish feces has been shown to be the potential food source for filter-feeding bivalves [17,31–32]. In the present study, the OM and C, N, and P contents in the sediments of bivalve treatments were significantly lower than those in the sediments of the control (Table 3; Figure 2), indicating that the bivalves could assimilate the organic matter in the suspended solids in aquaculture wastewater.

In summary, considering the fact that large quantities of fish suspended solids in discharged water from intensive finfish culture (*C. semilaevis*) are difficult to remove using conventional mechanical filtration, therefore we tried bivalve filter feeders (oysters and mussels) in this study to perform biological filtration and deposition. The present study demonstrated that both bivalve species had strong biological treatment potential for accelerating the deposition of suspended solids in effluent discharged from fish

culture tanks. The biologically induced deposition of suspended solids in industrial mariculture wastewater by filter-feeding bivalves might achieve the goals of removing the suspended solids as well as transforming the aquaculture wastes into biological resources. Before promoting the use of these bivalves as biofilters, it is necessary to evaluate the possible content of pathogens in them. It is suggested that future work should be focused on developing an effective way to employ bivalve filter feeders to remove fish suspended solids in land-based industrial marine aquaculture wastewater.

Acknowledgments

We would like to thank many people from Haifa Aquaculture Company (Tianjin, norh China) for their help in the flow-through experiment.

Author Contributions

Conceived and designed the experiments: YZ YL HSY. Performed the experiments: SJZ YZ. Analyzed the data: YZ SJZ. Contributed reagents/materials/analysis tools: YL. Wrote the paper: YZ SJZ.

References

1. MacMillan JR, Huddleston T, Woolley M, Fothergill K (2003) Best management practice development to minimize environmental impact from large flow-through trout farms. Aquaculture 226: 91–99.

2. Tal Y, Schreier HJ, Sowers KR, Stubblefield JD, Place AR, et al. (2009) Environmentally sustainable land-based marine aquaculture. Aquaculture 286: 28–35.

3. Chen S, Timmons MB, Aneshansley DJ, Bisogni JH (1993) Suspended solids characteristics from recirculating aquaculture systems and design implications. Aquaculture 112: 143–155.

4. Cripps SJ, Bergheim A (2000) Solids management and removal for intensive land-based aquaculture production systems. Aquacultural Engineering 22: 33–56.

5. Ni Q, Zhang Y (2007) Suspended solids removal technology in recirculating aquaculture systems. Fishery Modernization 34: 7–10 (in Chinese, with English abstract).

6. Stabili L, Acquaviva MI, Cavallo RA (2005) *Mytilus galloprovincialis* filter feeding on the bacterial community in a Mediterranean coastal area (Northern Ionian Sea, Italy). Water Research 39: 469–477.

7. Newell RIE (2004) Ecosystem influences of natural and cultivated populations of suspension-feeding bivalve molluscs: a review. Journal of Shellfish Research 23: 51–61.

8. Haven S, Morales-Alamo R (1966) Aspects of biodeposition by oysters and other invertebrate filter feeders. Limnology & Oceanogrraphy 11: 487–498.

9. Zhou Y, Yang H, Zhang T, Liu S, Zhang S, et al. (2006a) Influence of filtering and biodeposition by the cultured scallop *Chlamys farreri* on benthic-pelagic coupling in a eutrophic bay in China. Marine Ecology Progress Series 317: 127–141.

10. Dame RF (1996). Ecology of marine bivalves: an ecosystem approach. CRC Press, Boca Raton, FL.

11. Manganaro A, Pulicanò G, Reale A, Sanfilippo M, Sarà G (2009) Filtration pressure by bivalves affects the trophic conditions in Mediterranean shallow ecosystems. Chemistry and Ecology 25: 467–478.

12. Prins TC, Smaal AC, Dame RF (1998) A review of the feedbacks between bivalve grazing and ecosystem processes. Aquatic Ecology 31: 349–359.

13. Zhou Y, Yang H, Zhang T, Qin P, Zhang F (2006b) Density-dependent effects on seston dynamics and rates of filtering and biodeposition of the suspension-cultured scallop *Chlamys farreri* in a eutrophic bay (northern China): An experimental study in semi-in situ flow-through systems. Journal of Marine Systems 59: 143–158.

14. Neori A, Chopin T, Troell M, Buschmann AH, Kraemer GP, et al. (2004.) Integrated aquaculture: rationale, evolution and state of the art emphasizing seaweed biofiltration in modern mariculture. Aquaculture 231: 361–391.

15. Zhou Y, Yang H, Hu H, Liu Y, Mao Y, et al. (2006c) Bioremediation potential of the macroalga *Gracilaria lemaneiformis* (Rhodophyta) integrated into fed fish culture in coastal waters of north China. Aquaculture 252: 264–276.

16. Troell M, Joyce A, Chopin T, Neori A, Buschmann AH, et al. (2009) Ecological engineering in aquaculture — Potential for integrated multi-trophic aquaculture (IMTA) in marine offshore systems. Aquaculture 297: 1–9.

17. Handå A, Min H, Wang X, Broch OJ, Reitan KI, et al. (2012) Incorporation of fish feed and growth of blue mussels (*Mytilus edulis*) in close proximity to salmon (*Salmo salar*) aquaculture: implications for integrated multi-trophic aquaculture in Norwegian coastal waters. Aquaculture 356–357: 328–341.

18. Sarà G, Reid G, Rinaldi A, Palmeri V, Troell M, et al. (2012) Growth and reproductive simulation of candidate shellfish species at fish cages in the southern Mediterranean: Dynamic Energy Budget (DEB) modelling for integrated multi-trophic aquaculture. Aquaculture 324–325: 259–266.

19. Zhou Y, Yang H, Liu S, Yuan X, Mao Y, et al. (2006d) Feeding and growth on bivalve biodeposits by the deposit feeder *Stichopus japonicus* Selenka (Echinidermata: Holothuroidea) co-cultured in lantern nets. Aquaculture 256: 510–520.

20. Yu Z, Zhou Y, Yang H, Ma Y, Hu C (2014) Survival, growth, food availability and assimilation efficiency of the sea cucumber *Apostichopus japonicus* bottom-cultured under a fish farm in southern China. Aquaculture 426–427: 238–248.

21. Chopin T, Buschmann AH, Halling G, Troell M, Kautsky N, et al. (2001) Integrating seaweeds into marine aquaculture systems: a key towards sustainability. Journal of Phycology 37: 975–986.

22. Zhou Y, Zhang F, Yang H, Zhang S, Ma X (2003) Comparison of effectiveness of different ashing auxiliaries for determination of phosphorus in natural waters, aquatic organisms and sediments by ignition method. Water Research 37: 3875–3882.

23. Hatcher A, Grant J, Schofield B (1994) Effects of suspended mussel culture (*Mytilus* spp.) on sedimentation, benthic respiration and sediment nutrient dynamics in a coastal bay. Marine Ecology Progress Series 115: 219–235.

24. Arakawa KY, Kusuki Y, Kamigaki M (1971) Studies on biodeposition in oyster beds(I)—economic density for oyster culture. Venus 30: 113–128 (In Japanese, with English abstract).

25. Kautsky N, Evans S (1987) Role of biodeposition by Mytilus edulis in the circulation of matter and nutrients in a Baltic coastal ecosystem. Marine Ecology Progress Series 38: 201–212.

26. Widdows J, Brinsley MB, Salkeld PN, Elliott M (1998) Use of annular flumes to determine the influence of current velocity and biota on material flux at the sediment-water interface. Estuaries 21: 552–559.

27. Giles H, Pilditch CA (2004) Effects of diet on sinking rates and erosion thresholds of mussel Perna canaliculus biodeposits. Marine Ecology Progress Series 282: 205–219.

28. Haamer J (1996) Improving water quality in a eutrophied fjord system with mussel farming. AMBIO 25: 356–362.

29. Petersen J.K., Hasler B, Timmermann K, Nielsen P, Tørring DB, et al. (2014) Mussels as a tool for mitigation of nutrients in the marine environment. Marine Pollution Bulletin. http://dx.doi.org/10.1016/j.marpolbul.2014.03.006

30. Wang J, Jiang Z, Chen R (2005) Assimilation efficiency and biodeposition of mussel *Mytilus crassitesta*. Journal of Fishery Sciences of China 12: 150–155 (in Chinese, with English abstract).

31. Reid GK, Liutkus M, Bennett A, Robinson SMC, MacDonald B, et al. (2010) Absorption efficiency of blue mussels (*Mytilus edulis* and *M. trossulus*) feeding on Atlantic salmon (*Salmo salar*) feed and fecal particulates, implications for integrated multi-trophic aquaculture. Aquaculture 299: 165–169.

32. MacDonald BA, Robinson SMC, Barrington KA (2011) Feeding activity of mussels (*Mytilus edulis*) held in the field at an Integrated Multi-Trophic Aquaculture (IMTA) site (Salmo salar) and exposed to fish food in the laboratory. Aquaculture 314: 244–251.

21

Biodegradation of the Alkaline Cellulose Degradation Products Generated during Radioactive Waste Disposal

Simon P. Rout, Jessica Radford, Andrew P. Laws, Francis Sweeney, Ahmed Elmekawy, Lisa J. Gillie, Paul N. Humphreys*

Department of Chemical and Biological Sciences, School of Applied Sciences, University of Huddersfield, Huddersfield, United Kingdom

Abstract

The anoxic, alkaline hydrolysis of cellulosic materials generates a range of cellulose degradation products (CDP) including α and β forms of isosaccharinic acid (ISA) and is expected to occur in radioactive waste disposal sites receiving intermediate level radioactive wastes. The generation of ISA's is of particular relevance to the disposal of these wastes since they are able to form complexes with radioelements such as Pu enhancing their migration. This study demonstrates that microbial communities present in near-surface anoxic sediments are able to degrade CDP including both forms of ISA via iron reduction, sulphate reduction and methanogenesis, without any prior exposure to these substrates. No significant difference ($n = 6$, $p = 0.118$) in α and β ISA degradation rates were seen under either iron reducing, sulphate reducing or methanogenic conditions, giving an overall mean degradation rate of 4.7×10^{-2} hr^{-1} ($SE \pm 2.9 \times 10^{-3}$). These results suggest that a radioactive waste disposal site is likely to be colonised by organisms able to degrade CDP and associated ISA's during the construction and operational phase of the facility.

Editor: Paul Jaak Janssen, Belgian Nuclear Research Centre SCK•CEN, Belgium

Funding: The research was funded by the University of Huddersfield. The funder had no role in study design, data collection and analysis, decision to publish, or preparation of the manuscript.

Competing Interests: The authors have declared that no competing interests exist.

* Email: p.n.humphreys@hud.ac.uk

Introduction

The current strategy for the management of the UK's radioactive waste is a single Geological Disposal Facility (GDF) providing suitable, safe containment of the national waste inventory. One illustrative concept for the disposal of long lived, Intermediate Level Wastes (ILW) and some Low Level Wastes (LLW) is that of a cementitious backfilled facility which will re-saturate post closure [1]. Although the facility in general will be backfilled with a cementitious grout, not all the waste will be encapsulated with cement allowing lower pH environments to be present within the waste. Such a facility is expected to develop anoxic conditions soon after closure due to the removal of oxygen by the corrosion of the steel waste containers. This will result in an anoxic, alkaline environment which in combination with the host rock, will provide a multi barrier system for the containment of the radionuclide inventory [1].

By 2010, the U.K. held an estimated 2,000 tonnes of cellulosic ILW composed of packaging materials, disposable clothing and surface wipes and a further 100,000 tonnes of cellulosic LLW [2]. The chemical hydrolysis of cellulose under anoxic, alkaline conditions is a well described process [3] in which amorphous cellulose is degraded via the peeling reaction to the α and β forms of isosaccharinic acid (ISA) and a range of organic acids including formic, glycolic and acetic [4]. Of these water soluble products, ISA has received considerable attention on account of its ability to form complexes with radionuclides present in the wastes [5–7].

The construction and operational phases of a GDF provide an opportunity for the microbial contamination and colonisation of the facility by microorganisms from the near-surface environment [8]. The microbial degradation of ISA may have a significant impact on the evolution of GDF and the migration of the radioelements present.

Within the wider environment, ISA is generally absent, although it is present in the black liquor resulting from the Kraft paper pulping process [9]. Studies of industrially contaminated sites suggest that ISA degrading microbial populations may evolve within decades [10–12]. Other studies have shown that ISA is capable of being degraded under aerobic and denitrifying conditions, conditions unlikely to be seen in the near field of a GDF [9,11]. The presence of significant amounts of steel (construction materials and waste packages) within a GDF mean that corrosion processes will promote the generation of reducing conditions and generate ferric iron phases that may support microbial iron reduction [1]. In addition, groundwater ingress is likely to provide sulphates that will be able to support microbial sulphate reduction [13]. Consequently, both iron reduction and sulphate reduction along with methanogenic processes may play an important role in the development of the ambient geochemistry within a GDF [8]. ISA and other organic compounds generated by the chemical hydrolysis of cellulose are potential substrates for these microbial processes.

The aim of this study was to determine the ability of the microbial communities in anoxic near surface sediments to

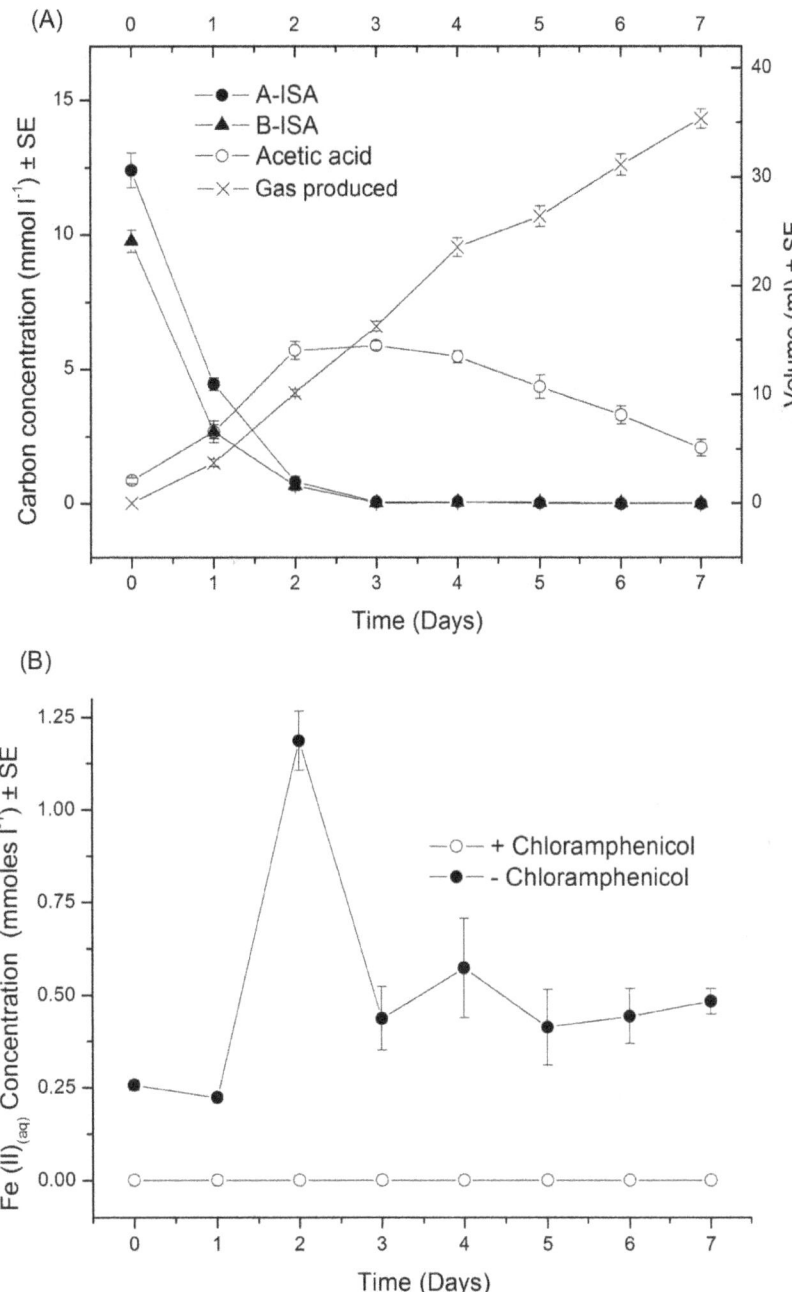

Figure 1. Organic chemical evolution of iron reducing reactors (A) and iron (II) (aq) evolution versus chloramphenicol treated control (B) (n = 6).

degrade CDP and associated ISA under iron reducing, sulphate reducing and methanogenic conditions. As such, this study investigates the microbial processes that may take place at the interface between ungrouted cellulosic wastes and the cementitious backfill and at the interface between the repository and the host rock that is receiving an ISA containing plume.

Experimental Procedures

Sediment samples

Sediment samples were taken from the Leeds/Liverpool canal at the University of Huddersfield (ordnance co-ordinates, SE 14890 16416); further samples were taken from reed beds at the National Coal Mining museum Wakefield (SE 25076 16368). Samples were taken using a weighted sampler and stored under anoxic conditions in sealed plastic containers at room temperature; samples were transferred into microcosms within 14 days of collection. The permissions of the University of Huddersfield and National Coal Mining Museum were acquired prior to sampling.

Production of cellulose degradation products (CDP)

CDP was prepared in a similar fashion to that of Cowper et al [21] and standard laboratory tissue (Pristine paper hygiene, London, UK) was used as a cellulose source for degradation. Laboratory tissue (200 g) was added to 1.8 l of N_2 flushed 0.1 M

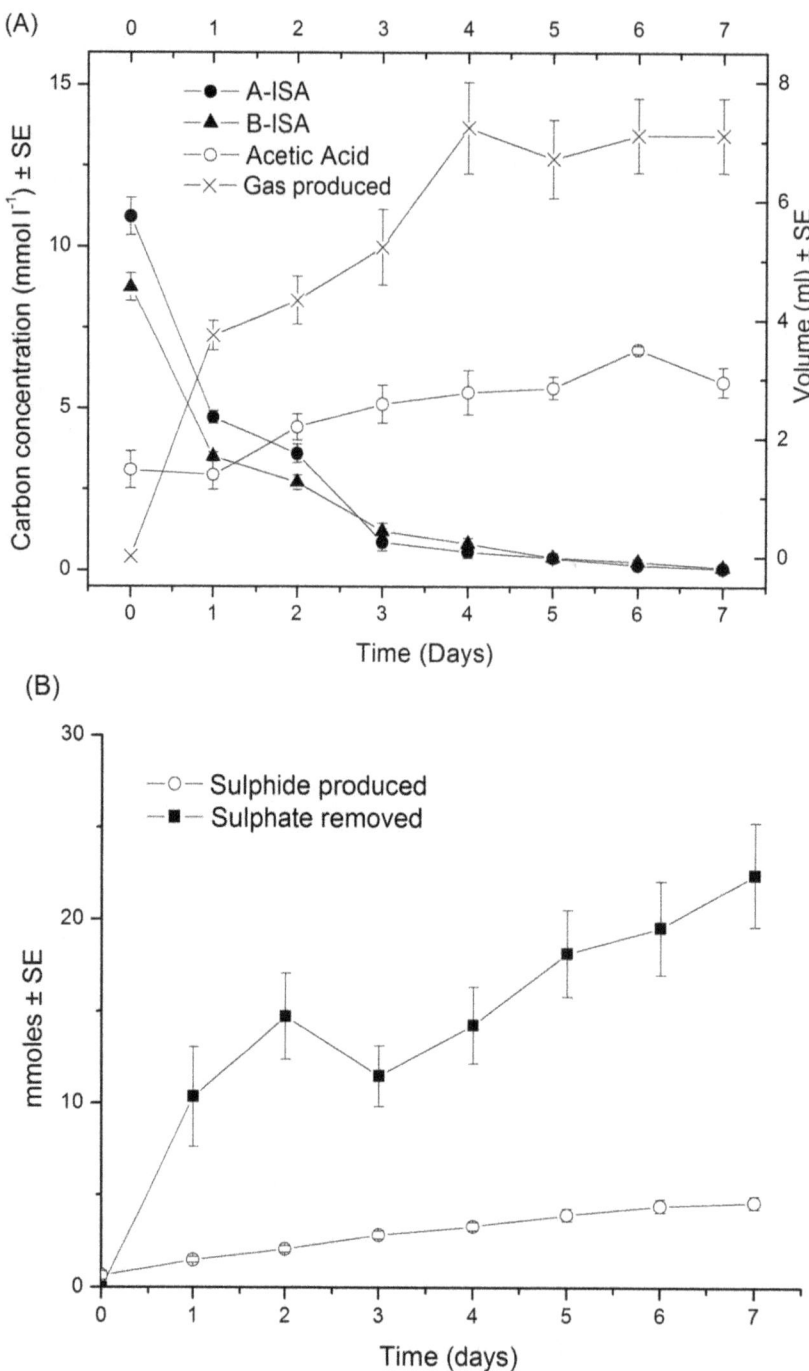

Figure 2. Organic chemical evolution of sulphate reducing reactors (A) and sulphate removal and sulphide production (B) (n = 6).

NaOH and 10 g l^{-1} Ca(OH)$_2$ in a pressure vessel. The pressure vessel was sealed and made anoxic by flushing the headspace with nitrogen for 30 minutes and then placed in an incubator at 80°C for 30 days. After 30 days, the vessel was allowed to cool before the resultant liquor was sterile filtered under a nitrogen atmosphere. Bottles of CDP were covered with foil to exclude light and stored under a nitrogen atmosphere. The composition of the synthesised CDP can be found in supplementary table (Table S1 in File S1).

Microcosm set up

Sediment samples were diluted 2-fold in the anaerobic mineral media specified in BS14853 [22] to a volume of 450 ml in 500 ml reaction vessels fitted with inlet and outlet ports to allow for the addition and removal of samples and a third port fitted with a septum to allow headspace gas sampling. Following an initial feed of 50 ml of CDP, microcosms were batch fed under nitrogen on a weekly cycle by replacing 50 ml of the microcosm contents with 50 ml of fresh CDP followed by a further 20 minutes of nitrogen flushing. This feeding cycle was continued for 24 weeks prior to sampling. Iron and sulphate reducing conditions were established

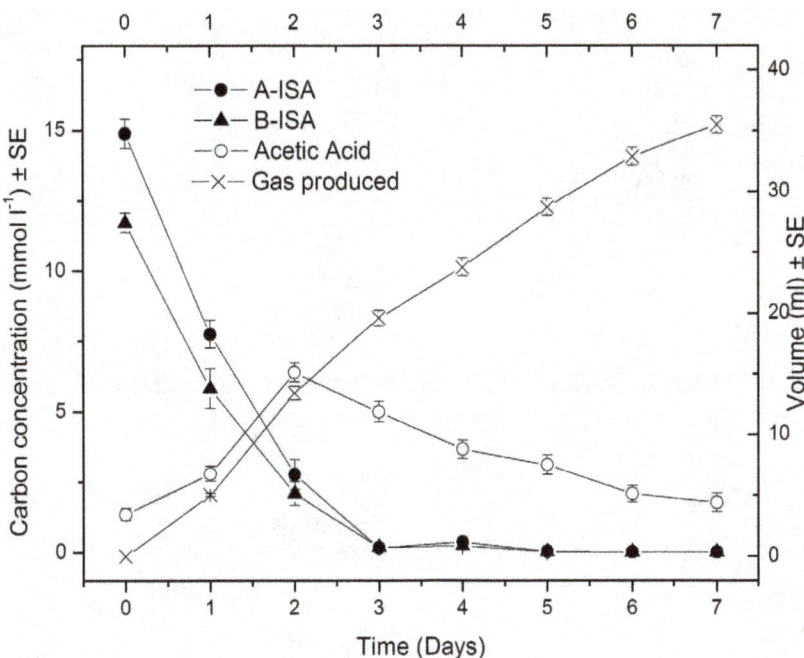

Figure 3. Chemical evolution of methanogenic reactors (n = 6).

through the additions of excess quantities of calcined iron (III) oxide (Fisher Scientific Ltd, UK) and sodium sulphate (Fisher Scientific Ltd, UK) relative to the moles of organic carbon present in the CDP. Methanogenic conditions were established through

the absence of iron (III) or sulphate. The iron (III) oxide employed was identified as haematite via X-ray diffraction analysis (Bruker D2 phaser and diffraction patterns recorded using Cu-K$_\alpha$ radiation ($\lambda = 1.54184$ Å) utilising a LYNXEYE detector) and

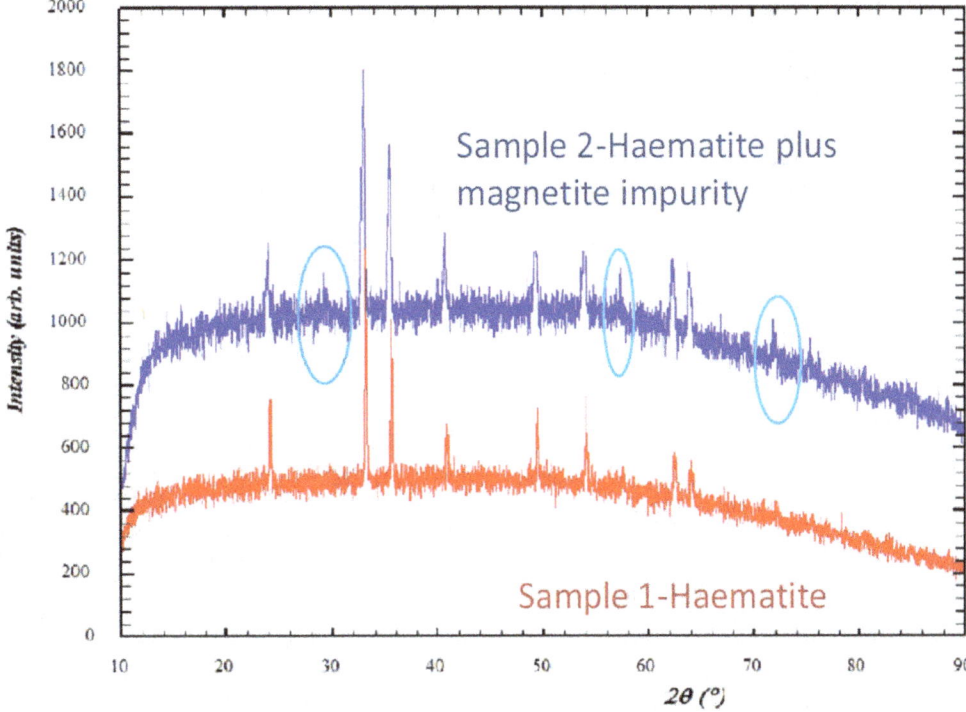

Figure 4. XRD patterns for the iron oxide used in microcosms (Sample 1), identified as haematite through comparison with diffraction database (peaks at 24, 33, 36, 41, 49 54, 62 and 64 2 Theta) and pattern at the end of the sampling period (Sample 2), where haematite contained an impurity, determined as magnetite through comparison with diffraction database (peaks at 30, 58 and 74 2 Theta, circled).

Figure 5. Scanning electron micrographs of virgin calcined iron (III) oxide (A, C) and iron (III) oxide following incubation under iron reducing conditions (B, D).

comparison with Bragg peaks obtained from the Powder Diffraction File database (Figure S1 in File S1). Previous authors have noted that sorption reactions do not occur between ISA and other CDP components and haematite [21], supporting the selection of this source of iron (III) for these microcosm experiments. Surface areas and pore sizes were determined by nitrogen adsorption at 77K using an ASAP2020 (Micromeritics Instrument Corp).

In addition, three control microcosms amended with 50 µg ml^{-1} chloramphenicol were set up containing the same proportions of mineral media, sediment and CDP and sampled on a daily basis.

Chemical analysis

In brief, 5 mL samples were taken on a daily basis over 3 consecutive feed cycles under a nitrogen atmosphere. Samples were centrifuged at $9000\times g$ for 10 minutes, the resulting supernatant was then sterile filtered using a 0.45 µm filter and kept at 4°C prior to use, subsequent analysis of samples was carried out under ambient conditions. In addition, 0.9 ml of sample was acidified with 0.1 ml of phosphoric acid and frozen at $-20°C$ for volatile fatty acid analysis. The presence and concentration of volatile fatty acids was determined using a gas chromatograph (HP GC6890, Hewlett Packard, UK) fitted with a HP-FFAP column (Agilent Technologies) and a flame ionization detector under the following conditions: an initial temperature of 95°C for 2 minutes, followed by an increase to 140°C at a ramp rate of 10°C min^{-1} with no hold, followed by a second ramp to 200°C at a ramp rate of 40°C min^{-1} with a hold of 10 minutes, falling to a post run temperature of 50°C. Total organic carbon

analysis was carried out using a Shimadzu TOC 5050A. ISA concentrations in both the alpha and beta conformations were measured using high performance anion exchange chromatography and pulsed amperometric detection on a Dionex 3000 Ion chromatography system (Dionex, Camberly, UK) employing a Dionex Carbopac PA20 column (3×150 mm, 6.5 µm particle size) and eluting with aqueous sodium hydroxide (0.05 mol l^{-1}) against a range of standards [23]. The volume of gas produced was measured using a Quick Scan 1.8c apparatus (Challenge Technology, Arkansas, US), the headspace gas composition was determined using an Agilent 6850 gas chromatograph with a thermal conductivity detector and GS-Q column operating at a column temperature of 30°C and a detector temperature of 200°C. The soluble iron concentration was measured spectrophotometrically using a ferrozine extraction method described previously [24]. The sulphate concentration was measured via ion chromatography using amperometric detection on a Metrohm 850 Professional IC (Metrohm, Cheshire, UK) employing a Metrohm Metrosep A Supp 5 column (4×150 mm, 5 µm particle size) eluted with sodium carbonate and sodium hydrogen carbonate (3.2 mmol l^{-1}, 1.0 mmol l^{-1} respectively) alongside a range of standards. The sulphide concentration was measured using a micro ION electrode LIS146AGSCM (Lazar Labs, US) calibrated against a range of standards. SEM analysis was carried out using an FEI Quanta FEG 250 equipped with energy-dispersive x-ray spectroscopy.

DNA extraction and direct/nested PCR

A 50 ml sample was removed from each microcosm and centrifuged at $9000\times g$ for 10 minutes at 4°C, 5 ml of supernatant

Table 1. DNA analysis by direct and nested PCR techniques.

Species	Size	Terminal Electron Acceptor					
		Iron		Sulphate		Carbon Dioxide	
		D	N	D	N	D	N
Clostridium I	820	-	-	-	-	-	-
Clostridium III	720	-	+	+	+	+	+
Clostridium IV	580	+	+	+	+	-	-
Clostridium XIV	620	-	+	+	+	+	+
Methanococcales	340	-	+	-	-	-	+
Methanobacteriales	340	+	+	-	-	+	+
Methanomicrobiales	550	+	+	-	-	+	+
Methanosarcinales	350	+	+	-	-	+	+
Methanosaeta	250	+	+	-	-	+	+
SRB group 1	702	-	+	+	+	-	-
SRB group 2	1120	-	-	-	-	-	-
SRB group 3	840	+	+	+	+	-	-
SRB group 4	1150	-	-	+	+	-	-
SRB group 5	860	+	+	+	+	-	-
SRB group 6	620	+	+	+	+	-	-
Geobacter	300	+	+	+	+	N	N
Shewanella	1040	-	-	-	-	N	N

N-not sampled.

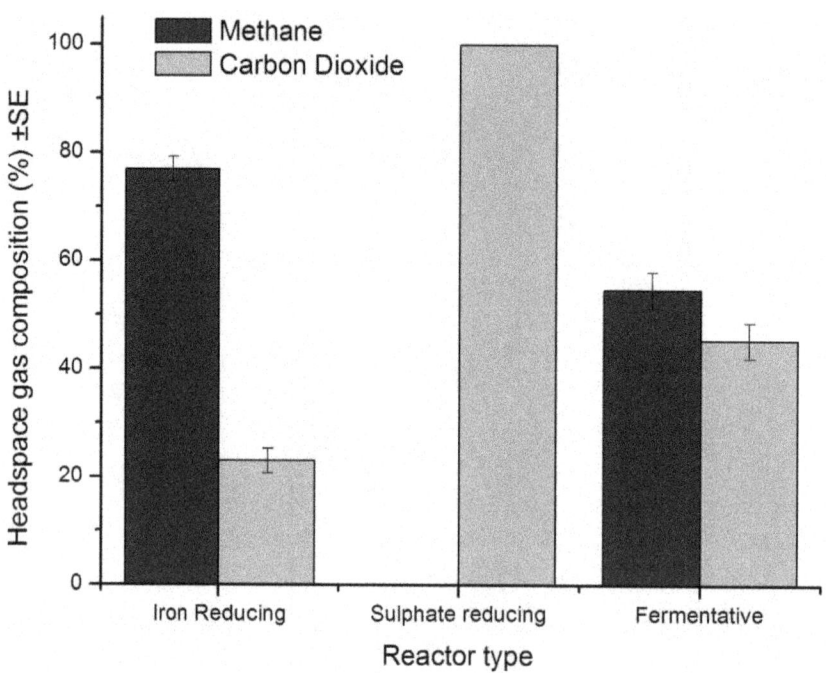

Figure 6. Composition of microcosm headspace gases (n = 6).

was retained and mixed with the pellet to give a concentrated suspension. Microbial DNA was extracted from each sample using a PowerSoil DNA isolation kit (MoBio laboratories, California, US). Extracted DNA was diluted to a concentration of approximately 100 ng/µl and PCR carried out using a range of primers (Table S2 in File S1) in accordance with previously published methods [25]. When PCR product was not observed through direct PCR, the relevant 16S rDNA amplification product was used to perform a nested reaction using the associated primers. In addition, a range of control DNA samples (Table S3 in File S1) were used to validate the results of each PCR step.

Statistical Analysis. All statistical analysis was carried out using IBM SPSS V 20 for Windows, data were checked for their normality of distribution and equality of variance prior to ANOVA.

Results and Discussion

Across all microcosms, under iron reducing, sulphate reducing and methanogenic conditions, a significant proportion of organic carbon removal was associated with α and β ISA metabolism with no apparent difference between α and β ISA consumption profiles (Figures 1A, 2A and 3). Fermentation processes were evident by the generation of acetic acid, which was the most prevalent volatile fatty acid formed, although other longer chain fatty acids including propionic, isobutyric, butyric and isovaleric acids were produced in sub mM concentrations (Figure S2 in File S1).

In microcosms amended with haematite, iron reduction was indicated by the generation of Fe (II)(B) which coincided with the removal of both forms of ISA (Figure 1A). This contrasts with the associated control reactors where no Fe (II) generation or ISA removal was observed. In these iron reducing systems the fermentation of at least a portion of the ISA was illustrated by the initial generation of acetic acid. However, by the end of the incubation period, acetic acid levels had significantly reduced (p< 0.05) indicating its subsequent degradation (Figure 1A, B). The Fe

(II) profiles indicate an initial generation followed by a reduction to a lower resting level. This profile is consistent with the precipitation of Fe (II) containing mineral phases, with the final solution phase concentrations determined by precipitation/dissolution reactions. This profile and the resting Fe (II) concentrations are also consistent with previously published data on haematite driven iron reduction systems [14]. XRD analysis confirmed the generation of Fe (II) mineral phases, in particular magnetite, that were absent from the original haematite (Figure 4). The presence of biogenic magnetite in bulk Fe (III) oxides has also been observed by previous authors employing XRD [15]. The surface area (from 4.4 $m^2 g^{-1}$ to 13.8 $m^2 g^{-1}$) and associated porosity (0.02 cm^3/g to 0.04 cm^3/g) of the haematite increased following iron reduction. This increased porosity was confirmed by SEM (Figure 5 A, B) which in conjunction with energy-dispersive x-ray spectroscopy (Figure S3 in File S1) confirmed the formation of calcite on the haematite surface (Figure 5 C, D). This suggests that calcite formation is occurring due to biogenic CO_2 reacting with calcium present in the CDP. In contrast, both magnetite and calcite were absent from the sediment remaining in the control microcosms.

The formation of methane indicated that methanogens were also active alongside fermentative and iron reducing communities, suggesting that the crystalline nature of the Fe (III) source facilitates the presence of methanogenesis by limiting the rate of iron reduction. Haematite is known to support a lower rate of iron reduction than more amorphous Fe (III) phases or complexed Fe (III) [16]. Stoichiometric calculations [17] indicated that methanogenesis and accumulated acetic acid accounted for only 18% of the degraded ISA, confirming the role of iron reduction as the primary metabolic process within the system.

Of the Clostridia clusters investigated, direct and nested PCR approaches indicated that cluster IV was more abundant than clusters III and XIV, with cluster I being undetectable (Table 1). Iron reduction may be attributed to a mixture of *Geobacter* sp and organisms from sulphate reducing bacteria (SRB) groups 1, 3, 4, and 5. Previous authors have noted the ability of SRBs to

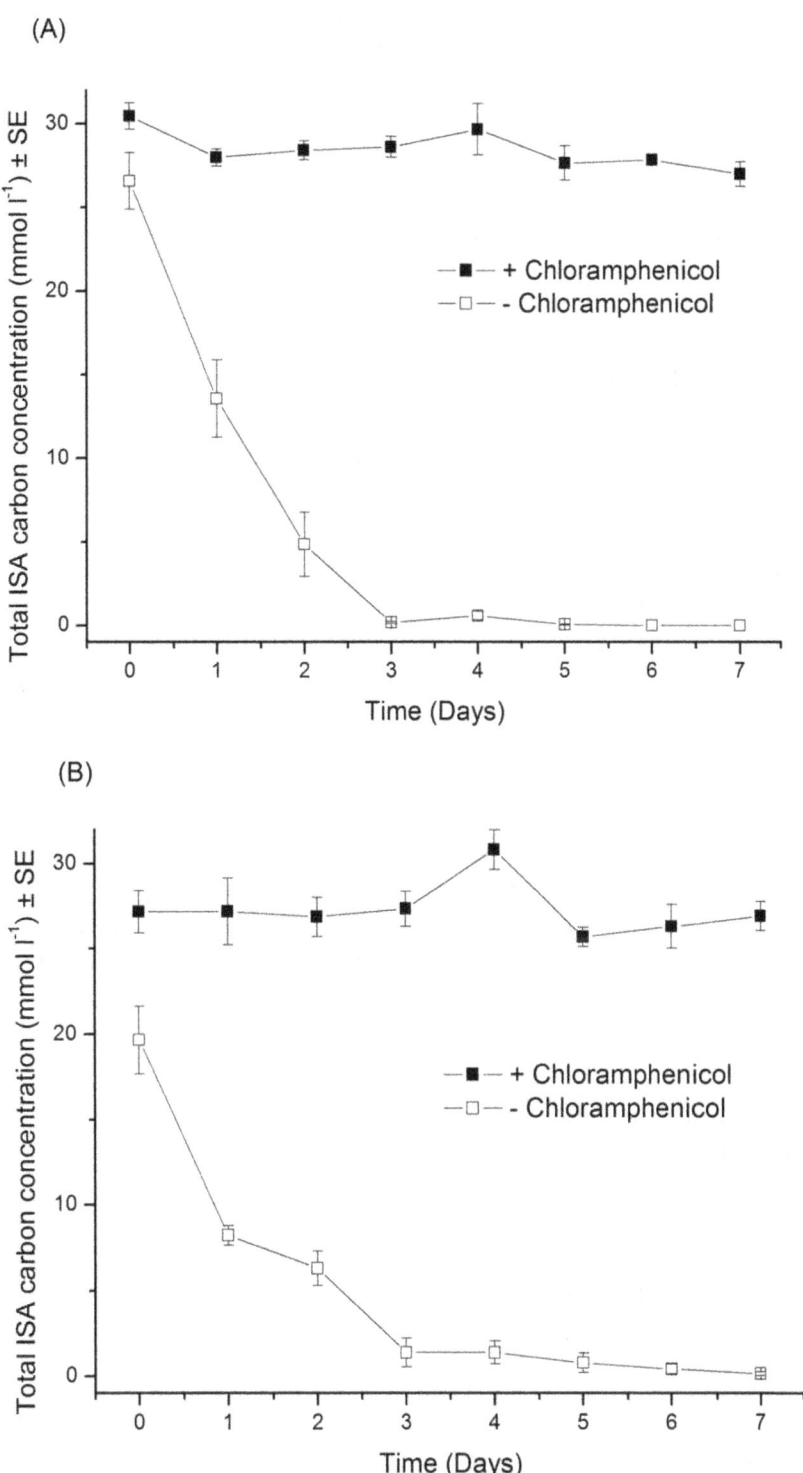

Figure 7. Fate of ISA's in microcosms treated with chloramphenicol when compared with untreated microcosm (A) ISA in presence of canal sediment and (B) reed bed sediment (n = 3).

enzymatically reduce Fe (III) from these groups [18]. Methanogenic bacteria capable of acetoclastic and hydrogenotophic metabolism were also present within this community (Table 1).

Unlike iron reduction, the presence of sulphate allowed SRBs to dominate the terminal electron accepting processes as indicated by the absence of evolved methane within the headspace of these

microcosms (Figure 6). Sulphide was generated in the aqueous phase as sulphate was removed (Figure 2B); no free sulphide was detected in the associated control microcosms. The accumulation of acetic acid up to day 6 suggests that sulphate reduction of acetic acid is occurring at a slower rate than its generation. Through direct PCR the presence of groups 1, 3, 4, 5 and 6 sulphate reducing bacteria as described by [19] were observed alongside groups III, IV and XIV of the Clostridia.

In methanogenic microcosms (Figure 3) the removal of both forms of ISA was associated with the production and removal of acetic acid and the generation of methane which comprised $54.7\% \pm 3.3$ of the gas generated. Direct and nested PCR confirmed the presence of Clostridia groups III and XIV and all five methanogen groups investigated.

Degradation rates for ISA under anoxic conditions are not available in the literature, consequently first order degradation rates were calculated from the iron reducing, sulphate reducing and methanogenic α and β ISA removal data. No significant difference (ANOVA, $n = 6$, $p = 0.118$) was found between the degradation rates of either α and β ISA under iron reduction, sulphate reduction or methanogenic conditions, giving an overall ISA degradation rate of 4.7×10^{-2} hr^{-1} ($n = 36$, $SE \pm 2.9 \times 10^{-3}$). These data support a two stage degradation model for ISA with fermentation to short chain volatile fatty acids being the dominant, rate limiting step across all three consortia, followed by the iron reduction, sulphate reduction and methanogenesis of the fermentation end products.

ISA is known to be subject to sorption and precipitation reactions [7,20], consequently a set of control microcosms treated with 50 μg ml^{-1} chloramphenicol were sampled over the same period and analysed for ISA content. In this instance, ISA was not removed over the seven day sample period (Figure 7), suggesting that the removal previously seen was microbially mediated rather than through sorption or precipitation processes. Other organic carbon sources were present within the CDP feed stock (<30% of total carbon) including the xylo-isosaccharinic acid and the α and β metasaccharinic acids. These minor components are degraded in all three systems (data not shown), however the CDP did contain recalcitrant components that remained un-degraded throughout the incubation period.

Conclusions

Although the α and β forms of ISA are not naturally observed in the wider environment, bacteria found in anoxic sediments are capable of degrading these compounds by utilising a range of terminal electron acceptors at circa neutral pH. Under iron reducing, sulphate reducing and methanogenic conditions the degradation of ISA followed the pathway seen in anoxic environments driven by the degradation of polymeric organic materials; i.e. the fermentation of polymer monomers followed by the degradation of fermentation end products by terminal electron accepting processes. In this case, however, hydrolysis is a chemical rather than a microbial process. The persistence of bacteria commonly associated with the anaerobic degradation of cellulose (the Clostridia) in these batch fed microcosms suggests that they may play an important role in the metabolism of ISA into common fermentation end products allowing electron and carbon flow within these systems. In summary, these findings indicate that the ability to degrade ISA is common in near-surface microbial communities and consequently such communities represent a potential source of ISA degrading consortia for the colonisation of a GDF during the operational and pre-closure period.

The observed rates of ISA degradation suggest that at the interface between neutral and alkaline environments (e.g. within ungrouted wastes) ISA production will be the rate limiting step and that microbial activity will prevent the accumulation and transport of ISA and therefore prevent the enhanced migration of radionuclides. However, the activity of these communities within a GDF will be dependent on either the establishment of low pH environments within ungrouted wastes and/or their ability to adapt to the prevailing alkaline conditions. Consequently, ISA may persist, migrate and complex in regions where the pH inhibits microbial activity.

Supporting Information

File S1 Combined file containing supporting figures and tables. Figure S1: XRD pattern from iron (III) oxide used in this study. Overlaid red lines indicate the allowed positions of the Bragg peaks for hematite, from the Powder Diffraction file database (Joint Committee of Powder Diffraction, JCPDS card number 89–0599. Figure S2: Non acetic volatile fatty acid concentrations in (A) iron reducing reactors, (B) sulphate reducing reactors and (C) methanogenic reactors. Figure S3: EDS output from analysis of calcite deposit in Figure 3, D. Table S1: Composition of cellulose degradation products. Table S2: PCR primers used in this study. Table S3: Organisms used as positive controls for PCR studies.

Acknowledgments

We would like to thank L. Dawson for their assistance with the BET analysis and M. Stirling for their assistance with IC.

Author Contributions

Conceived and designed the experiments: PNH APL SR. Performed the experiments: SR JR AE FS LG. Analyzed the data: PNH APL SR JR AE FS LG. Wrote the paper: PNH APL SR LG.

References

1. NDA (2010) Near-field Evolution Status Report. NDA/RWMD/033, Nuclear Decommissioning Authority (Radioactive Waste Management Directorate), Harwell, Didcot, Oxfordshire, UK.
2. NDA (2011) The 2010 UK Radioactive Waste Inventory. NDA/ST/STY(11)0004 Nuclear Decommissioning Authority (Radioactive Waste Management Directorate), Harwell, Didcot, Oxfordshire, UK
3. Humphreys P, Laws A, Dawson J (2010) A Review of Cellulose Degradation and the Fate of Degradation Products under Repository Conditions. Serco technical report prepared for the Nuclear Decommissioning Authority (Radioactive Waste Management Directorate), Harwell, Didcot, Oxfordshire, UK
4. Knill CJ, Kennedy JF (2003) Degradation of cellulose under alkaline conditions. Carbohydrate Polymers 51: 281–300.
5. Allard S, Ekberg C (2006) Complexing properties of α-isosaccharinate: stability constants, enthalpies and entropies of Th-complexation with uncertainty analysis. Journal of Solution Chemistry 35: 1173–1186.
6. Askarieh MM, Chambers AV, Daniel FBD, FitzGerald PL, Holtom GJ, et al. (2000) The chemical and microbial degradation of cellulose in the near field of a repository for radioactive wastes. Waste Management 20: 93–106.
7. Warwick P, Evans N, Hall T, Vines S (2003) Complexation of Ni(II) by α-isosaccharinic acid and gluconic acid from pH 7 to pH 13. Radiochimica Acta 91: 233–240.
8. Humphreys P, West J, Metcalfe R (2010) Microbial Effects on Repository Performance. Quintessa contractors report prepared for the Nuclear Decommissioning Authority (Radioactive Waste Management Directorate), Harwell, Didcot, Oxfordshire, UK
9. Strand SE, Dykes J, Chiang V (1984) Aerobic microbial degradation of glucoisosaccharinic acid. Applied and Environmental Microbiology 47: 268–271.
10. Bailey MJ (1986) Utilization of glucoisosaccharinic acid by a bacterial isolate unable to metabolize glucose. Applied Microbiology and Biotechnology 24: 493–498.
11. Grant WD, Holtom GJ, O'Kelly N, Malpass J, Rosevear A, et al. (2002) Microbial Degradation of Cellulose-derived Complexants Under Repository Conditions. AEAT/ERRA-0301. AEA Technology and University of Leicester Report for UK Nirex Ltd.

12. Wang S, McCarthy J, Ferguson J (1993) Utilization of glucoisosaccharinic acid and components of Kraft black liquor as energy sources for growth of anaerobic bacteria. Holzforschung 47: 141–148.

13. Nirex (1997) Sellafield Geological and Hydrogeological Investigations: The Hydrochemistry of Sellafield, 1997 Update. Nirex Science Report, S/97/089. UK Nirex Ltd.

14. Lovley DR, Phillips EJ (1988) Novel mode of microbial energy metabolism: organic carbon oxidation coupled to dissimilatory reduction of iron or manganese. Applied and Environmental Microbiology 54: 1472–1480.

15. Chaudhuri SK, Lack JG, Coates JD (2001) Biogenic magnetite formation through anaerobic biooxidation of Fe (II). Applied and Environmental Microbiology 67: 2844–2848.

16. Lovley DR (1991) Dissimilatory Fe(III) and Mn(IV) reduction. Microbiological Reviews 55: 259–287.

17. Rittmann BE, McCarty PL (2001) Environmental Biotechnology, McGraw Hill, New York.

18. Lovley DR, Roden EE, Phillips E, Woodward J (1993) Enzymatic iron and uranium reduction by sulfate-reducing bacteria. Marine Geology 113: 41–53.

19. Daly K, Sharp RJ, McCarthy AJ (2000) Development of oligonucleotide probes and PCR primers for detecting phylogenetic subgroups of sulfate-reducing bacteria. Microbiology 146: 1693–1705.

20. Greenfield B, Hurdus M, Spindler M, Thomason H (1997) The Effects of the Products from the Anaerobic Degradation of Cellulose on the Solubility and Sorption of Radioelements in the Near Field. NSS/R375. AEA Technology Report for UK Nirex Ltd.

21. Cowper M, Marshall T, Swanton S (2011) Sorption Detriments in the Geosphere: the Effect of Cellulose Degradation Products. Phase 1 Experimental Study. Serco technical report prepared for the Nuclear Decommissioning Authority (Radioactive Waste Management Directorate), Harwell, Didcot, Oxfordshire, UK.

22. B.S.I. (2005) BS ISO 14853: 2005: Plastics-Determination of the Ultimate Anaerobic Biodegradation of Plastic Materials in an Aqueous System- Method by Measurement of Biogas Production. British Standards Institute, London, UK.

23. Shaw PB, Robinson GF, Rice CR, Humphreys PN, Laws AP (2012) A robust method for the synthesis and isolation of β-gluco-isosaccharinic acid ((2R,4S)-2,4,5-trihydroxy-2-(hydroxymethyl)pentanoic acid) from cellulose and measurement of its aqueous pKa. Carbohydrate Research 349: 6–11.

24. Lovley D, Phillips E (1987) Rapid assay for microbially reducible ferric iron in aquatic sediments. Applied and Environmental Microbiology 53: 1536–1540.

25. McDonald JE, Lockhart RJ, Cox MJ, Allison HE, McCarthy AJ (2008) Detection of novel Fibrobacter populations in landfill sites and determination of their relative abundance via quantitative PCR. Environmental Microbiology 10: 1310–1319.

22

Seagrass Canopy Photosynthetic Response Is a Function of Canopy Density and Light Environment: A Model for *Amphibolis griffithii*

John D. Hedley[1]*, Kathryn McMahon[2], Peter Fearns[3]

1 Environmental Computer Science Ltd., Tiverton, Devon, United Kingdom, **2** School of Natural Sciences and Centre for Marine Ecosystems Research, Edith Cowan University, Joondalup, Western Australia, **3** Department of Imaging and Applied Physics, Curtin University of Technology, Perth, Western Australia

Abstract

A three-dimensional computer model of canopies of the seagrass *Amphibolis griffithii* was used to investigate the consequences of variations in canopy structure and benthic light environment on leaf-level photosynthetic saturation state. The model was constructed using empirical data of plant morphometrics from a previously conducted shading experiment and validated well to *in-situ* data on light attenuation in canopies of different densities. Using published values of the leaf-level saturating irradiance for photosynthesis, results show that the interaction of canopy density and canopy-scale photosynthetic response is complex and non-linear, due to the combination of self-shading and the non-linearity of photosynthesis versus irradiance (P-I) curves near saturating irradiance. Therefore studies of light limitation in seagrasses should consider variation in canopy structure and density. Based on empirical work, we propose a number of possible measures for canopy scale photosynthetic response that can be plotted to yield isoclines in the space of canopy density and light environment. These plots can be used to interpret the significance of canopy changes induced as a response to decreases in the benthic light environment: in some cases canopy thinning can lead to an equivalent leaf level light environment, in others physiological changes may also be required but these alone may be inadequate for canopy survival. By providing insight to these processes the methods developed here could be a valuable management tool for seagrass conservation during dredging or other coastal developments.

Editor: Kay C. Vopel, Auckland University of Technology, New Zealand

Funding: This research was funded by an Edith Cowan University (ECU) Industry Collaboration Scheme with BMT Oceanica Pty. Ltd., the industry partner, awarded to KM, JH, and PF. ECU Faculty Visiting Fellow Scheme supported JH for travel to ECU. KM was supported by the ECU Collaborative Research Network. Environmental Computer Science Ltd. provided support in the form of salary for author JH. The funders had no role in the study design, data collection and analysis, decision to publish, or preparation of the manuscript. The specific roles of these authors are articulated in the "author contributions" section.

Competing Interests: This work was partly funded by BMT Oceanica. The author J Hedley is an employee of Environmental Computer Science Ltd.

* Email: j.d.hedley@envirocs.com

Introduction

Seagrass meadows are a dominant habitat of most coastal environments and provide important ecosystem services such as primary production, nutrient cycling, sediment stabilization, food and habitat for other organisms and trophic transfers to adjacent habitats [1]. Globally, these ecosystem services have been valued at an approximated US$ 19000 ha^{-1} yr^{-1} [2] but emerging understanding of the carbon storage capability of seagrass meadows implies this may be an underestimate [3]. Despite these recognized values, the area of seagrass is reducing world-wide at an increasing rate. Waycott et al. [4] estimated 29% of the known areal extent has disappeared since seagrass areas were initially recorded in 1879, and the rate of decline has accelerated in the last two decades.

The key anthropogenic pressures impacting seagrass meadows at local scales are urban, industrial and agricultural runoff, infrastructure development and dredging [5]. These pressures impact seagrasses directly via physical removal or indirectly through the introduction of pollutants such as nutrients, or suspended sediments that result in a reduction of light reaching seagrass meadows. Seagrasses are sensitive to light reduction as they are typically adapted to high light environments [1].

Increasing research is being undertaken to improve the management and conservation of seagrass meadows through improved understanding of the risks they face (e.g. [6]), developing bioindicators of the pressures they are exposed to [7] and thresholds of stressors such as light reduction which may differentiate sub-lethal effects from permanent loss of seagrass [8,9]. In general, leaf-level photosynthetic activity in response to irradiance follows a 'photosynthesis versus irradiance curve', which is linear for subsaturating irradiances but becomes non-linear, as progressively increasing irradiance causes saturation of the photosynthetic electron transport chain, and finally attains a plateau phase, which is defined as maximal photosynthesis rate (P_{max}) [10,11]. A key physiological parameter that represents a species response to a given light level is E_k, defined as the intersection between the initial linear slope and P_{max} on a P-I curve. E_k is frequently referred to as the 'saturating irradiance' [12,13] although technically it is slightly below the irradiance at which full photosynthetic saturation occurs, and above the irradiance at which saturation starts to cause deviation from linearity. E_k can be empirically determined and for each species may vary over a restricted range due to physiological acclimatization or factors such as temperature [14].

Various light threshold analyses have been proposed as having predictive capability for seagrass mortality. Dependent on available data, light levels can be assessed with respect to different factors or components of the environment, including the water column light attenuation coefficients [15] or Secchi disk depths [16]; light at the top of the seagrass canopy expressed as percentage of surface irradiance [17,18]; instantaneous or mean daily irradiance [8,19] or the number of hours of irradiance above E_k per day, H_{sat}, [8,20]. These thresholds can also be integrated over time, which is relevant to management when pressures persist over particular durations, e.g., dredging or flood plumes. The percentage of days below a particular mean daily irradiance [8] or the sum of the hours of irradiance below E_k compared to reference conditions [9] are two examples for which thresholds have been proposed to predict the onset of seagrass mortality.

One important component that all of these thresholds do not consider is the interaction of the seagrass canopy itself with the benthic light field, since it is the amount of light reaching individual leaves of a seagrass that governs the plants photosynthetic response [21]. The photosynthetic activity in turn influences how the seagrass meadow responds to the changes in light [12] and overall plant productivity [22]. Canopy structure of seagrass meadows can also vary markedly due to natural variations in light [23] or in response to light perturbations [9]. Due to canopy self-shading, light levels at the top of the canopy may be very different to light levels within the canopy, and will vary throughout the canopy in a manner dependent on the incident benthic light field, canopy structure and bending angle of the leaves, which vary under water motion [24,25,26]. Therefore, a mechanistic explanation of how light levels affect canopy sustainability must include the interaction of the canopy structure with the incident light field.

In this study we developed a 3D model of a complex seagrass canopy (*Amphibolis griffithii*) of varying structure, from low to high leaf area index (LAI), by adapting the model described in [25] and [27]. We modeled the exposure of these virtual canopies to a number of environmentally relevant levels of light reduction to assess the amount of light reaching each leaf surface and how this varies under different canopy densities and positions due to movement associated with water motion. Finally, we assessed the canopy saturation state by relating the light each leaf receives to values of leaf-level E_k for *A. griffithii* found in the literature. The modeling scenarios were based on empirically quantified canopy structures from specific plant morphologies, and were designed to be comparable to a shading experiment that was conducted on *A. griffithii* in 2005 [9].

In summary, the objectives were:

- To develop a 3D canopy model for a seagrass species with a complex canopy, hence demonstrating an advance in technical capability with respect to the simple *Thalassia* morphology model of Hedley and Enríquez [25].

- To understand the consequences on within-canopy light capture and canopy saturation state of 1) canopy position: upright vs. moving under high wave action, 2) canopy structure: low to high LAI (1.27 to 7.65), and 3) light reduction: 0–95% shading

- To identify potential descriptors of canopy light levels which could have use for the management of seagrass beds under light reduction events such as dredging or coastal pollution.

Methods

Canopy structures

The modelling experiment was designed to mirror aspects of a previously published empirical shading manipulation experiment [9,28]. The empirical study utilised an extensive (>6 ha) meadow of *Amphibolis griffithii* in 4.5 m water depth at Jurien Bay, Western Australia (30° 18′ 34″ S, 115° 00′ 26″ E; WGS84 datum). A control plus two-treatment shading experiment was conducted, the first phase of which ran from 10th March to 14th June 2005. Before and during the experiment individual *A. griffithii* plants were sampled and characterised in terms of stem and branch lengths, internodal distances, and number and dimensions of terminal leaves (Fig. 1a).

In the computer model, ten sets of individual plant data from the initial control sampling were replicated as vector mesh structures (Fig. 1f). The model plants were assembled into five canopies of leaf area index (LAI) from 1.27 to 7.65, by varying the choice and number of plants in a 20 cm×20 cm segment of substrate (Table 1, Fig. 2). The leaves and stems of the vector mesh structures were modelled as a point-mass and force system according to methods typically used for modelling cloth in the computer graphics industry [29]. A dynamic numerical integrator modelled the plant structures flexing naturalistically under a simple wave-action force model. Two wave actions, 'high' and 'low', were employed. In the dynamic model the low wave energy treatment plants were allowed to assume a typical upright position with no wave induced movement (Fig. 2). Under high wave energy plants underwent a vigorous cycle of forward and backward motion (Figs. 2f–i). From these dynamic models canopy structure treatments were extracted as instantaneous snapshots for each of the five LAI treatments: 1) a single snapshot for low wave action, 2) 14 snapshots through a cycle of movement for high wave action (Fig. 2). The 14 snapshots for high wave action were individually passed to the optical model (see below) and the results were averaged, thereby assuming the canopies undergo this movement continuously and photosynthetic response is the mean of the responses at any instant in time.

Water column optical model and shading

The canopy structures were input to the optical model for estimating diurnal leaf-incident irradiance. The model framework, previously described in [27] and [25], propagates sky radiance distributions through the canopy to give leaf incident irradiance in 17 wavebands of 20 nm width from 400–740 nm. Spectral irradiance can then be reduced to photosynthetically available radiation (PAR) at leaf level, and related to leaf tissue photosynthetic saturating irradiance, approximated by E_k (Fig. 1). To parameterise the model, hourly clear sky radiance distributions were produced using libRadtran and a directional radiance model [30,31] corresponding to the Jurien Bay site on 27th April; the middle of the post summer 3 month trial phase in Lavery et al. [9] (Fig. 1c).

The sky radiance distributions were input to PlanarRad (http://www.planarrad.com), a plane parallel water column model functionally similar to HydroLight [32,33] to estimate the hourly top of canopy radiance distribution (Fig. 1e). The model provides directional radiance tabulated over a hemisphere of zenith and azimuth angles, but to remove any dependency on sun azimuth and canopy orientation downwelling irradiances were azimuthally averaged to have only a zenith angle dependency (Fig. 1c, e). The water column utilised a library set of spectral inherent optical properties (IOPs, for details see [34]) which when input to the model produced a diffuse attenuation of planar PAR irradiance,

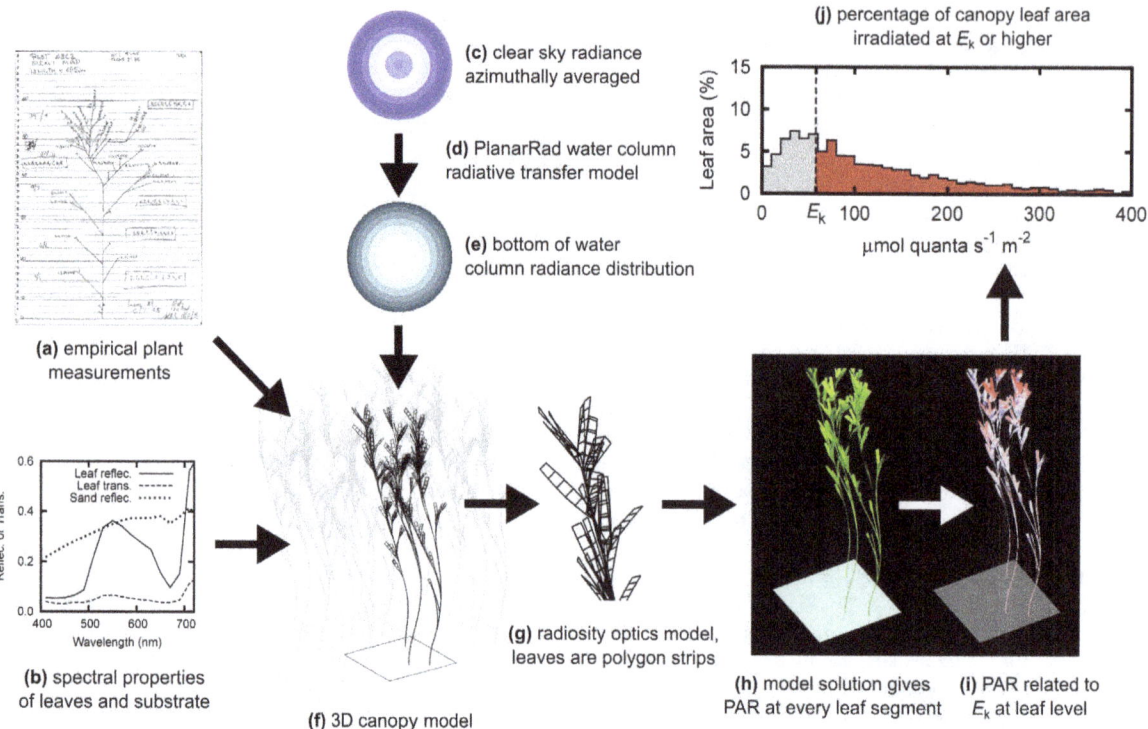

Figure 1. Overview of modelling system. (a, b) empirical data informs construction of 3D canopy model (f), (c, d, e) A plane-parallel model estimates directional radiance incident on the top of the canopy, (f, g, h) a geometric optical model handles radiative transfer to and between leaf segments, (i, j) PAR distribution over leaf area is reduced to the percentage of the canopy irradiated above leaf-level photosynthetic saturation, E_k.

k_d, of approximately 0.2 m^{-1}. In comparison, k_d values measured at the time of the empirical shading experiment ranged from 0.07 to 0.19 over a four month period but were 0.19 in April (Department of Parks and Wildlife, unpublished data). A set of nine modelled shading treatments were implemented by taking the top of canopy radiance distribution and reducing the values by 10%, 20%, etc. up to 95% (Table 1). Hence shading was spectrally neutral as was the shade cloth used in Lavery et al. [9], where the shading treatments were equivalent to 81–87% and 89–95% in our notation. The empirical study therefore represented quite a strong shading effect with respect to the modelled range. The water column optical model was additionally evaluated by comparing modelled top of canopy daily PAR irradiance against *in-situ* measurements from the associated study [28].

Canopy structure optical model

The top of canopy irradiance was propagated through a geometrical optics model [25,27] that accounts for inter-reflection and transmission between leaf segments. The spectral reflectance and transmittance of *A. griffithii* leaves was taken from the paper of Durako [35]. In this study we did not attempt to capture inter or intra-plant variability in leaf absorptance. This can be done [25], but the data collection requirements are onerous. All surfaces were considered Lambertian reflectors and transmitters. The underlying substrate reflectance was set from a library sand spectral reflectance that had a mean value of 0.33. The 20×20 cm modelled canopy segment was repeated periodically horizontally so the modelled canopy was of uniform LAI and has no edge (Fig. 1f).

Empirical measurements of PAR irradiance close to midday at both the canopy top and on the substrate underneath canopies of *A. griffithii* of differing LAIs were available for validation of the

canopy optical model from the study of McMahon and Lavery [28]. Canopy transmission was measured in control and treatment plots of varying but known LAI through measuring the instantaneous photosynthetic photon flux density (PPFD, μmol m^{-2} s^{-1}) at the top and base of the canopy. The light sensor (Odyssey PAR sensor) was calibrated against a standard calibration light source (Quartz Tungsten Halogen Reference Lamp operated at 3150°K from a LI-1800-02 Optical Radiation Calibrator). The low wave energy structure model treatments at the hour closest to midday were used to perform this validation, and a number of additional runs with different LAIs to those in Table 1 were added to further populate the validation data. An additional quality assurance protocol for the canopy optical model is to set the within-canopy water absorptance to zero and then verify energy conservation between the top of canopy incident and exitant irradiances and energy absorbed by all surfaces in the model [25]. This was performed for a subset of the runs in Table 1.

Relation to photosynthetic properties

The model solution provided incident PAR at every point on every leaf at a resolution of approximately 0.5 cm^2. This was then related to the leaf level saturation irradiance for *A. griffithii*, approximated by E_k. Masini and Manning [12] evaluated E_k in *A. griffithii* as ranging from 25 to 55 μmol quanta m^{-2} s^{-1} for temperatures of 13°C to 23°C, of which the upper value is closer to the conditions of the empirical data from the associated study here. While Masini and Manning [12] did not assess physiological variation in E_k in *A. griffithii*, in the same study *Posidonia sinuosa* was shown to have E_k that varied from 50.4 to 39.1 μmol quanta m^{-2} s^{-1} in depths of 4 m and 12–15 m respectively. Therefore to accommodate a realistic variation in E_k in the absence of data, we

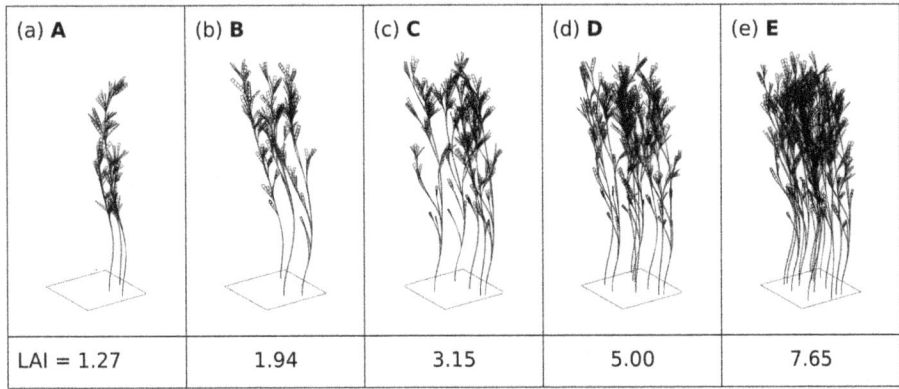

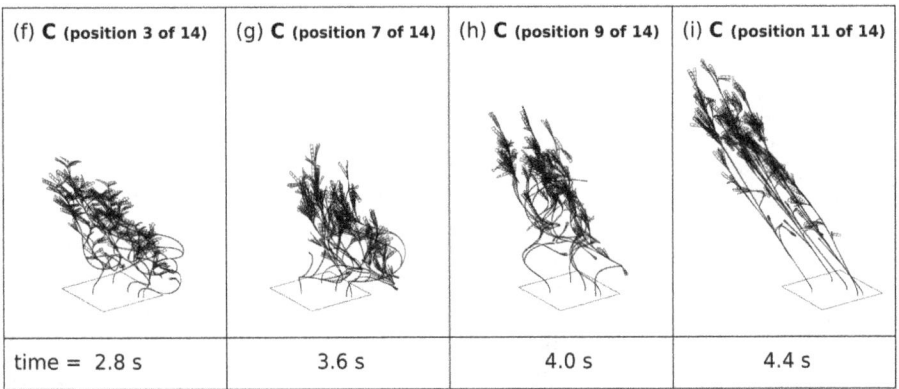

Figure 2. Example canopy structures and positions used in the model treatments. (a)–(e) low wave action canopy position for the five LAI treatments A to E. (f)–(i) subset of time sequence positions under high wave action for canopy C, all 14 positions were used in the optical model. In all cases the canopy structure is notionally repeated in all horizontal directions such that the square substrate section tessellates.

have used the comparable range of 45–55 μmol quanta m^{-2} s^{-1} to put all of our results into the context of potential physiological variation in E_k. To produce plots including the range and mid-point specific values of 45, 50 and 55 were used. Based on the value of E_k of 45 or 55 μmol quanta m^{-2} s^{-1}, the model can report the instantaneous proportion of the leaf area of the canopy that is irradiated at or above the saturation irradiance. To interpret these results, for each canopy structure and shading treatment, a value termed H^A_{sat} was calculated as the time integral of the percentage leaf area above saturation in a 24 hour period, with units % leaf area×hour. This measure is discussed later, but was intended to be analogous to H_{sat}, the daily top of canopy irradiated hours above saturation [9,20] but also factoring in the canopy self-shading.

Table 1. Modelling experiment design.

LAI	Structure	Shading (%)	Hour
A −1.27	low wave energy (1 position)	0	×12
B −1.94	high wave energy (14 positions)	10	
C −3.15		25	
D −5.00		40	
E −7.65		50	
		60	
		70	
		85	
		95	

A fully-factored set of model runs were performed for each of five LAI treatments, 15 canopy structures and nine shading treatments over 12 hourly diurnal intervals, a total of 8100 runs.

Results

Optical model validation

The sky radiance and water column model produced a daily top of canopy PAR dose of 11.0 mol quanta m^{-2}, whereas the comparable *in-situ* measured average daily PAR irradiance over 3 months was 19.0 mol quanta m^{-2} [9]. Since our study was primarily concerned with the relative effect of the shading treatments and LAI this discrepancy is not of great importance, but could be due to: 1) the accuracy of the libRadtran sky radiance model (no validation data available); 2) the accuracy of the Odyssey PAR sensors, which can have issues in long-term stability (Slivkoff, pers. comm.), or; 3) the model water column $k_d(PAR)$, which was at the upper range compared to measurements taken during the empirical study (0.2 vs. 0.070.19). This deviation in $k_d(PAR)$ does provide an almost exact explanation for the discrepancy, but since $k_d(PAR)$ is a wavelength-integrated output of the model parameterised on spectral IOPs for absorption and backscatter it is not trivial to set an arbitrary value of $k_d(PAR)$. In the scope of this study, using the closest IOP set from actual measured data [34] was considered adequate. In reality the daily measured PAR was sometimes above and sometimes below the model value, so all things considered the modelled canopy PAR dose was reasonable and the discrepancy is inconsequential to the subsequent interpretation of the results.

The percentage of the incident top of canopy PAR irradiance transmitted to the substrate, as a function of leaf area index, validated well against empirical data (Fig. 3). The empirical data showed wide variation, but the modelled transmitted irradiances corresponded very closely to the upper bound of the empirical data. This is to be expected, since some of the real canopies contained free standing and epiphytic macroalgae which would have reduced the transmission beyond that described by the *A. griffithii* LAI alone. The upper bound points most likely represent the most monospecific *A. griffithii* canopies and correspond best to the model. An exponential function fit to the model data ($n = 28$) gave r^2 of 0.96, the fit of all 27 empirical data points to that same function gives an r^2 of 0.73. However, if only four outliers are removed (Fig. 3) the empirical data r^2 rises to 0.90. The model and empirical data therefore compare well, especially given the practical difficulty in making accurate within-canopy light measurements.

The performance of the model in terms of energy conservation was demonstrated in the subset of runs for which water absorption was set to zero. For the majority of runs energy losses were less than 2% and for all runs they were less than 3%. In practice, when water absorption is non-zero, energy conservation performance would be better than these figures suggest. The current model implementation requires water absorption to be set to zero for energy accounting, but this in itself removes a damping effect on the multiple scattering and increases energy losses through numerical errors. Therefore the model solutions for leaf incident irradiance can be considered, at worst, slight underestimates by around 2%.

Effect of canopy structure and position on leaf level PAR

As expected, the distribution of leaf level PAR irradiance became increasingly skewed to lower values as LAI increased (Figs. 4a, e, i, m, q). In low LAI canopies the distribution of PAR over leaf area was almost flat: leaf tissue received a wide range of PAR with almost equal probability, and much of it was above saturating irradiance at mid-day under the model conditions of clear sky and moderately clear water. In denser canopies the leaf level light distribution had a long high-end tail: many leaves

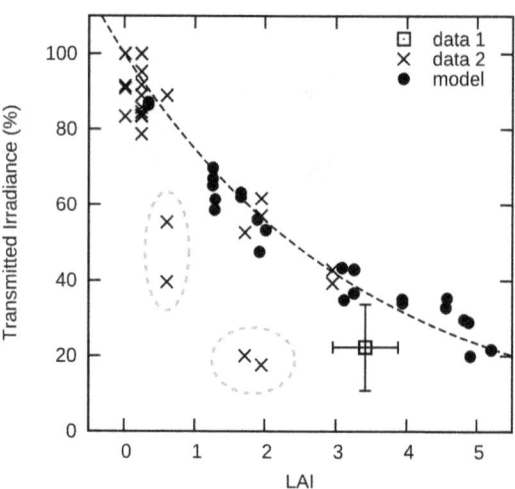

Figure 3. Percentage of downwelling top of canopy PAR irradiance reaching the substrate, as a function of LAI. Results for empirical in-situ measurements (data 1 and 2) and modelled estimates (model) are shown. Curve fit to model data points is $y = 100 * \exp(-0.29 \times LAI)$. $\times$(data 2) - are 26 individual coincident measurements of LAI and irradiance above and below the canopy. $\square$ (data 1) is a single point based on a site mean of 13 LAI determinations from 2.84 to 4.09 and a set of associated but not spatially co-incident light measurements, error bars are one S.D. The four encircled points are the outliers referred to in the text. Transmittance data collection is described in McMahon and Lavery 2014 and corresponding LAI is unpublished data from Lavery et al. 2009. ● are 25 model runs including both those described in detail and some additional runs. In the models runs solar zenith angle was approximately 28°.

received light below E_k, but a few leaves received very high light (Fig. 4i). Overall the pattern was clearly linked to the relative openness or self-shading within the canopies. The range of E_k of 45 to 55 µmol quanta m^{-2} s^{-1} was generally small compared to range of irradiances the leaves experienced, but this was more true for the lower LAI and unshaded treatments (Fig. 4).

The treatments of canopy position of upright or moving under wave action appeared to have little effect on leaf level PAR irradiance (e.g. Fig. 4a vs. 4b). Numerically the canopy movement slightly reduced the daily integrated percentage of saturated leaf area for all but the lowest LAI (Fig. 5). However overall there was not a statistically significant difference at either E_k of 45 or 55 µmol quanta m^{-2} s^{-1} (paired value t-test, $p > 0.05$). Therefore there is no evidence from our data that canopy movement affects time-integrated light capture. However the instantaneous light capture has a high variation under movement. While the standard deviation was 10–20% of the mean (Fig. 5) at some individual time points the saturated leaf area was up to 50% more or less than the mean. As expected, shading scaled the x-axis position of leaf-level irradiance distribution plots by the corresponding factor. That is, halving the top of canopy irradiance halved the leaf level incident irradiance at every leaf (Figs. 4b, d, f, h, etc.).

Diurnally accumulated saturated leaf area

The accumulated percentage of leaf area above saturation over a twelve-hour day showed a complex relationship between both shading and LAI (Fig. 6). While, as expected, increasing either shading or LAI monotonically decreased the accumulated percentage of leaf area above saturation, the shape of the function was non-linear and there was an interaction between shading and LAI (Fig. 6a). The contour lines in Figure 6a make clear the trade-

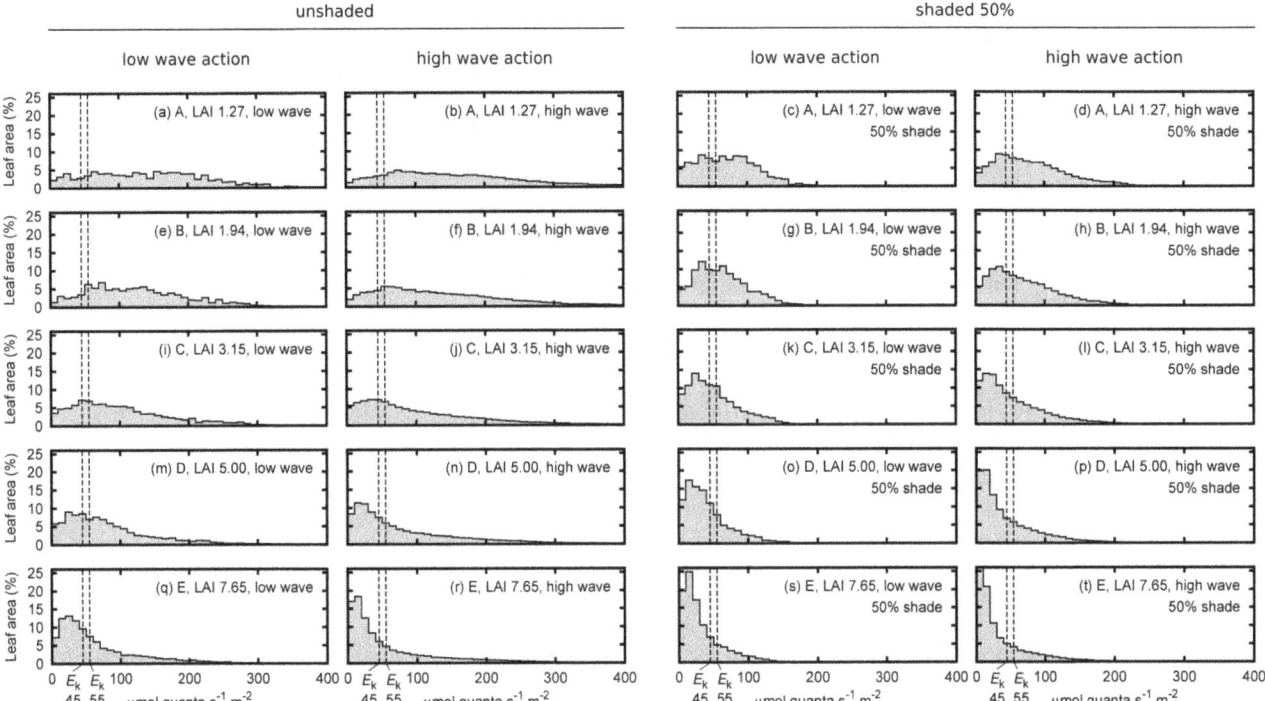

Figure 4. Distribution of leaf saturation state in the canopy in terms of percentage of leaf area at midday. Treatments are upright low wave action canopy positions and the average over the high wave action movement positions (left columns), and the same for 50% shading (right columns), for the five LAI treatments. The estimated photosynthetic saturation irradiances, E_k, of 45 and 55 µmol quanta m^{-2} s^{-1} inferred from Masini and Manning (1997) are shown as vertical dotted lines.

off between leaf area index and light with respect to the saturation state of the canopy. These lines show equal points in the LAI–shading function space, so for example a canopy of LAI 5.5 with no shading was equivalent to LAI 2.0 with 50% shading, with respect to the diurnally accumulated saturation of relative leaf area. The potential acclimation range of E_k from 45 to 55 µmol quanta m^{-2} s^{-1} (assuming water temperature at approx. 23°C) added a degree of freedom to the LAI-shading relationship approximately equivalent to 1 unit of LAI at low shading (e.g. along the x-axis of Fig. 6a), but this increased as shading increased to 60% or more (Fig. 6a). Therefore at low shading modifying E_k

over the suggested range is equivalent to changing LAI by plus or minus one half.

For the high wave action treatment there is a small qualitative difference in the position of the contour lines in low LAI and low shading region as compared to the upright low wave action treatment (Fig. 6b vs. 6a). However a sensitivity analysis of the data tables underlying Figures 6a and 6b showed that these differences are equivalent to an error in the shading percentage of only 6 points or less. In other words, if in a practical application shading were quantified at discrete levels of 0, 5, 10, 15% etc. then the difference between upright and moving canopies would be negligible.

Discussion

Geometric optical modelling of seagrass canopies and validation

In terms of the geometrical optical modelling of seagrass canopies, the results presented here corroborate those of Hedley and Enriquez [25], showing that it is possible to construct a physical three dimensional model of a seagrass canopy and obtain acceptable validation against *in-situ* light measurements. Through-canopy transmission was estimated accurately for pure *Amphibolis* canopies, but the importance of considering epiphytes or other canopy constituents was underlined by the high variability of the empirical data, which in some cases had lower light penetration than the model predicted based on *Amphibolis* LAI alone.

With respect to morphological complexity, *A. griffithii* is on the more complex end of the spectrum in comparison to strap leaf morphologies of *Thalassia* and many other seagrass species [36], to which this modelling framework was previously confined.

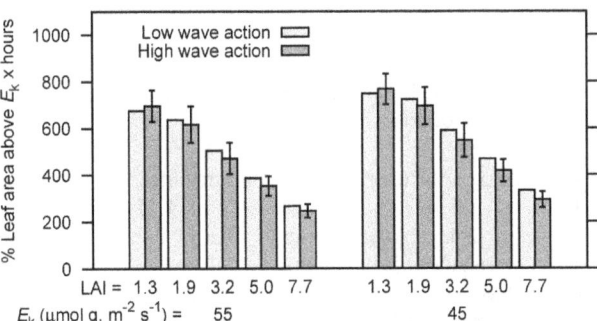

Figure 5. Time accumulated percentage of leaf area irradiated above photosynthetic saturation irradiance for low wave action and high wave action treatments, for E_k of 45 and 55 µmol quanta m^{-2} s^{-1}. Error bars on high wave action treatment are the standard deviation over all 14 movement positions and hence indicate the range in the instantaneous canopy saturation state.

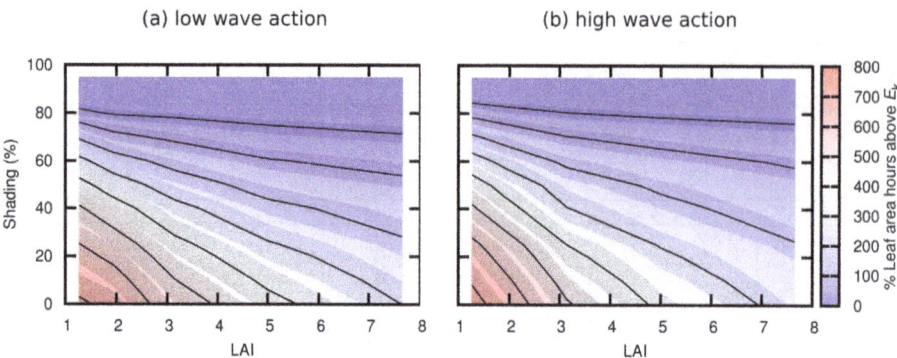

Figure 6. Time accumulated percentage of leaf area irradiated above photosynthetic saturation irradiance. The colour scale shows H^A_{sat}, the "percentage leaf area hours" above E_k, as a function of LAI and shading. (a) is for upright low wave action canopy structures, (b) is for the average over the high wave energy canopy positions. Contour lines are isoclines based on the mid value of E_k equal to 50 μmol quanta m^{-2} s^{-1} while the surrounding greyed region shows the limits for E_k of 45 to 55 μmol quanta m^{-2} s^{-1}. The isoclines are located at H^A_{sat} of 50, 100, 200, and then steps of 100 up to 800% leaf area hours.

Therefore the potential for future models of other seagrass species is good as these two examples capture the range in canopy complexity. Hedley and Enríquez [25] used profiles of light through the canopy to derive a diffuse attenuation profile, k_d, for validation. In this study only light measurements at the top and bottom of the canopy were available. However, this simpler validation may be preferable and adequate. In practice, empirical measurements of light profiles within canopies are difficult to make, and rarely fit well to exponential attenuation with depth. The measurements at the top and bottom of the canopy are the strongest "signal" for within-canopy attenuation and can be used to derive k_d if canopy height is known. So in future empirical work to which such modelling may be subsequently applied, we recommend measuring downwelling irradiance at the top and bottom of canopies, together with canopy height and LAI.

Influence of shading, LAI and position on diurnal leaf photosynthetic saturation

The previous empirical shading study on *A. griffithii* [9] quantified the change in H_{sat} induced by shading, i.e. the total number of hours of top-of-canopy irradiance that was above photosynthetic saturation, as compared to the unshaded treatments. This quantity, summed over time, was demonstrated as a good indicator of changes in canopy biomass and capacity for subsequent recovery. However H_{sat} relates the top of canopy irradiance to leaf-level photosynthetic saturation irradiance and so ignores canopy self-shading and other structural factors, which therefore introduce an additional degree of freedom. Here, we factor in the canopy structure by considering the percentage of the canopy leaf area above saturating irradiance accumulated over time, H^A_{sat}, with units of % leaf area×hour. This descriptor extends H_{sat} by reducing to a single number the interaction of the duration of saturating irradiance and the canopy self-shading. It can be roughly interpreted as the daily ratio of saturated photosynthesis to leaf area at canopy scale, and ranges from 0 to 1200 for plants completely saturated for 12 hours of daylight.

Considering the variation in H^A_{sat} with leaf area index and shading, as expected LAI has a strong effect on diurnal leaf saturation state (Fig. 6.). A change in LAI from 1 to 7 has as much effect as 60% shading (Fig. 6a), so the ambient light field cannot be treated independently from the canopy structure when photosynthetic processes at leaf level are of interest. Furthermore, the relationship between LAI and both shading and leaf saturation

state is a non-linear interaction; Figure 6 represents a curved surface in both axes of shading and LAI. This occurs because while leaf level irradiance is a linear function of canopy level irradiance, the leaf level photosynthetic response is not a linear function of irradiance when the irradiance approaches or exceeds E_k. As leaf level irradiance approaches and exceeds E_k the photosynthetic response levels off at P_{max}. In general, since photosynthesis versus irradiance curves are non-linear in the region of the saturating irradiance any derived measure of leaf level photosynthetic activity will have a complex relationship with LAI unless all leaves are well below saturating irradiance.

Within this study there is no statistically significant evidence to support the statement that canopy movement effects light capture and photosynthetic response. Qualitatively it is interesting that under movement the lowest LAI canopy experienced an increase in daily saturation whereas the higher LAI canopies were systematically lower (Fig. 5). To test the statistical significance of this observation would require substantial further modelling effort and was outside the scope of this study, however at low LAIs sideways movement may serve to enhance light collection by spreading leaves out horizontally and making them insensitive to the directionality of incident light. Under wave action such flattening is intermittent and as we have shown here is not a factor of great photosynthetic significance. Additionally, under these conditions the optical consequences of surface waves and sediment resuspension should also be considered and may be more significant [37]. However, in other systems and species, canopy flattening can be a result of shallow water depth or tidal or estuarine flow [38]. In this case the semi-constant flattening of the canopy may be of optical significance.

Potential for LAI modification as an acclimation response

Figure 6 indicates that modification of LAI is a possible response to maintain the saturation state of the canopy under reduced or enhanced light conditions. This role of morphological plasticity has been demonstrated in a number of experimental studies (e.g. [39]) and hypothesized as a regulatory mechanism in *Thalassia* [13]. From our model data (Fig. 6), if a canopy of LAI 7 is observed to reduce to LAI 4 after a period of 40% reduction in light, this loss of biomass might be interpreted as a trajectory of canopy decline but alternative interpretation is that of an acclimation response to maintain the leaf level photosynthetic state. This interpretation is independent of the mechanism by which it occurs. Leaf mortality might be considered just a by-product of inadequate light to maintain respiration, but if the net effect is a return to an unstressed leaf-level light regime then the distinction between a compensatory morphological adjustment and a decline is at best ambiguous. This argument is of course dependent on the definition of 'decline'; it does not apply if net productivity per unit area rather than biomass is the criteria. Here, by 'decline' we mean an implied trajectory toward canopy eradication.

Under the interpretation of potential acclimation a key question is whether the reduction in LAI remains on the isocline for canopy saturation state (i.e. the contour lines in Fig. 6). A canopy that moves on a trajectory through LAI–shading 'space' such that it stays on a contour line is experiencing the same time integrated percentage of its leaf area at saturating irradiance (see A to E in Fig. 7). That is, it experiences the same daily photosynthetic saturation in relation to its leaf biomass. We might therefore hypothesise that if a canopy can sustainably exist at one point on an H^A_{sat} isocline, canopies can also sustainably exist at other points on that line, all other things being equal. Movement along an isocline can occur purely by modification of the LAI, alternative acclimation responses such as modification of E_k at the leaf-level will enable movement perpendicular to the isoclines, illustrated by the range around the isoclines in Figures 6 and 7 delimited by E_k of 45 to 55 μmol quanta m^{-2} s^{-1}. Ignoring the latter possibility, and focussing on LAI modification alone, the prospects for long term survival of a canopy under a change in light environment can be estimated by following the isocline from its current location in LAI-shading space to the new light environment (y-axis) location. If at this location the LAI is greater than zero (judged by extrapolation), then the canopy could survive by thinning out to this LAI, at which point it will have the same relative photosynthetic saturation of its leaf area. In the following section we use this concept to interpret the results from the previous empirical shading experiment [9].

Canopy trajectories in LAI and shading space

In the post-summer treatment of the shading experiment of Lavery et al. [9] canopies with an LAI of ~4 had reduced to an LAI of ~2 after three months of 84% shading. This change in LAI and shading can be represented as a trajectory on the H^A_{sat} map: Figure 7, point A to B. At 6 months of shading the LAI had reduced to one and by 9 months the canopy was almost eradicated

and did not subsequently recover (Fig. 7, point C). Assuming for the moment that H^A_{sat}, the accumulated percentage area of leaf saturation above E_k, is a measure relevant to canopy sustainability then Figure 7 indicates that such a measure could have predictive power for canopy survival. In the previous section we postulated that canopies can move along the isoclines by modification of LAI alone. The initial reduction of LAI in the empirical data (Fig. 7, point A to B) occurred in the first three months in response to 84% shading. The trajectory cuts across the isoclines because there is a time lag as the canopy cannot become thinner instantaneously. At three months (Fig. 7, point B) the LAI has reduced to 2 but the shading is extreme so at the leaf level the light environment is still very much reduced. There can be two response pathways, either physiological changes may allow the canopy to exist on the new isocline, such as the mobilisation of stored reserves [40] or reductions in the saturating irradiance [23], or if such processes cannot bring about sufficient change then further LAI reduction is required in an attempt to return to an isocline closer to the original. In this case following the isocline to the right and extrapolating to the intercept with the x-axis it is clear that the light environment is equivalent to a canopy with huge LAI of at least 20+ in the original un-shaded situation. For the canopy to survive would require physiological changes that would permit canopies of these high LAIs to exist normally in this environment. Such canopies did not exist, hence physiological changes are insufficient (it is clear from Fig. 7 that variation in E_k is inadequate), hence the LAI continued to decrease and eventually the canopy was eradicated (Fig. 6, Point C).

The previous example is a straightforward case of severe light limitation, but with moderate shading (for example 40%, Fig. 7) the situation is more complicated. The empirical data of Lavery et al. [9] only contained shading at a minimum of 81% so the example of 40% in Figure 7 is hypothetical. If a situation of 40% shading is introduced, assuming the validity of H^A_{sat}, it is clear that the canopy could survive by reducing LAI from 4 to 1 (moving along the isocline from point A to point E, Fig. 7). However, because of the time lag in reducing LAI, an initial trajectory in which LAI partially reduces (A to D, Fig. 7) is realistic and is likely to include physiological responses in tandem. For example E_k could decrease, but at LAI around 3 the range of E_k from 45 to 55 only allows accommodation of up to 20% shading at the most (Fig. 7). The existence of a time lag is supported by studies on different seagrass species that have incorporated less extreme shading treatments over short time-scales; physiological changes such as increases in chlorophyll and reduction in LAI occur after longer durations of reduced light [39]. At point D in Figure 7 the canopy lies on an isocline that represents a canopy of LAI 5.5 in the unshaded environment. If such canopies can sustainably exist, at the same depth, water clarity etc. then the canopy may survive at LAI of 2 to 3 (point D), rather than reducing LAI to 1 (point E). Either way, Figure 7 has predictive power for the canopy response in that if it is anticipated a 40% shading event may occur, e.g., from dredging activities, then it is clear that a canopy that is sustainable at LAI of 4 could reduce to LAI of 1, or, could induce physiological changes to maintain an LAI higher than 1, but in the latter case only if canopies greater than LAI of 4 currently exist in that environment.

The possible trajectories in LAI-shading space of Figure 7 are dependent on the capability and time constants of other physiological acclimation mechanisms. These mechanisms could include adjustments to photosystem kinetics to increase the efficiency (i.e. lower E_k), increases in chlorophyll content and a:b ratio to enhance light capture, or mobilisation of stored carbohydrates for maintenance and growth of the existing leaf

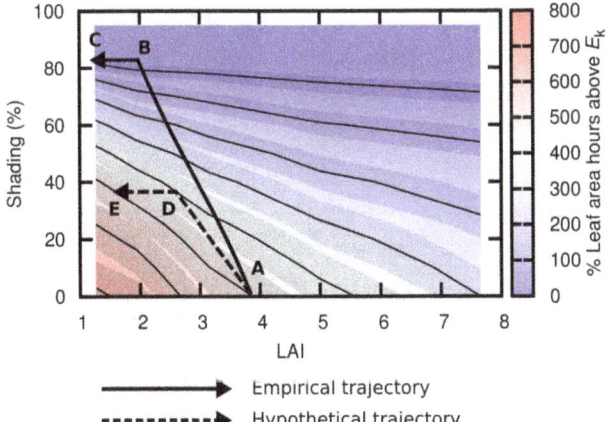

Figure 7. Trajectories of canopies from empirical shading experiment and hypothetical example. Underlying plot is as described in the caption of Figure 6a.

biomass [7]. To our knowledge there is no published data on photo-acclimation in *A. griffithii* under changing light conditions. However, unpublished data by co-author McMahon shows that under high levels of shading there are reductions in the saturating irradiance and other photo-acclimation responses, which maintain electron transport rates at unshaded values, but there is a time-limit over which this photo-acclimation is maintained of around 21 days. Therefore, the model as we have developed here is very relevant for predicting impacts associated with longer term reductions in light of over three weeks or more.

Other measures of canopy scale photosynthetic response to light

In the previous discussion we have assumed that the concept of isoclines of equal light environment' with respect to H^A_{sat} is valid. Alternative measures may be more appropriate but this does not affect the primary concept that canopy self-shading can be equivalent to environmental shading, and that there are two mechanisms of photosynthetic acclimation: physiological and via canopy structure. Any alternative measure of the photosynthetically relevant light environment would likely have a similar form to that of Figures 6 and 7. The interaction of self-shading and the non-linearity of leaf-level photosynthesis must inevitably result in a complex canopy scale response to LAI and the light environment for all canopies that are subjected to irradiances above photosynthetic saturation. Another candidate measure would be the integration of photosynthesis over time, i.e. to propagate the leaf-level light through a photosynthesis versus irradiance (P-I) curve to give an integrated photosynthesis measure equivalent to μmol O_2 evolution. In addition, plots of actual top of canopy PAR light levels may have greater descriptive power than percentage shading (Figures 6 and 7). Lavery et al. [9] observed different canopy changes at similar shading levels and interpreted these as being due to differences in the absolute light levels. In this study we suggested H^A_{sat} as a simple extension of the top-of canopy H_{sat}, since that measure has been demonstrated to have predictive power for canopy sustainability [8,9] and has been used in management contexts, and percentage shading was employed as a mirror of the empirical treatments. Clearly, there are many opportunities for further experiments and modelling to determine the most relevant measure of canopy photosynthetic response, the key point being that that measure needs to include within-canopy light propagation.

Conclusions

Three dimensional canopy modelling of *Amphibolis griffithii* has revealed that the interaction of light levels and canopy density on canopy-scale photosynthetic activity is complex and non-linear, in particular due to the non-linearity of leaf-level photosynthesis at saturating irradiance. The accumulated percentage area of leaf saturation above saturating irradiance, H^A_{sat}, was proposed as a measure relevant to canopy sustainability, based on extension of the equivalent top-of-canopy measure H_{sat} that has previously proved useful. The available empirical data were not sufficient to evaluate the efficacy of H^A_{sat} due to lack of lower shading treatments. Evaluating this measure and other candidates such as integrated leaf-level photosynthesis requires further experimental work. Nevertheless the principle has been demonstrated that plots of equal light environment' (Fig. 6) produced for different seagrass species, water depths, and water column optical properties could have practical management applications for predicting and interpreting canopy changes under light reduction events. Reduction in seagrass density in response to shading must be interpreted in terms of the leaf-level light environment. While physiological responses are also important, existing canopies in the same environment can provide information of the limits of physiological acclimation, and indicate if change in light levels will induce a trajectory to steady state sustainability, or to eradication. An important future step is to understand the time constants in change and recovery trajectories, to determine how long shading events can be tolerated and the required recovery periods. This information will be invaluable to coastal management.

Acknowledgments

J Hedley wishes to acknowledge Susana Enríquez, whose ideas and insight have informed the work described here. K McMahon acknowledges the collaboration with P S Lavery, which provided background for this research, L Twomey and M Westera from BMT Oceanica Pty Ltd for insights contributing to this work and N Dunham for canopy measurements.

Author Contributions

Conceived and designed the experiments: JH KM PF. Performed the experiments: JH KM. Analyzed the data: JH KM PF. Contributed reagents/materials/analysis tools: JH KM. Contributed to the writing of the manuscript: JH KM PF.

References

1. Orth RJ, Carruthers TJB, Dennison WC, Duarte CM, Forqurean JW, et al. (2006) A global crisis for seagrass ecosystems. Bioscience 56: 987–996.

2. Costanza R, d'Arge R, de Groot R, Farber S, Grasso M, et al. (1997) The value of the world's ecosystem services and natural capital. Nature 387: 253–260.

3. Fourqurean JW, Duarte CM, Kennedy N, Marbà N, Holmer M, et al. (2012) Seagrass ecosystems as a globally significant carbon stock. Nature Geoscience 5: 505–509.

4. Waycott M, Duarte CM, Carruthers TJB, Orth RJ, Dennison WC, et al. (2009) Accelerating loss of seagrasses across the globe threatens coastal ecosystems. Proceedings of the National Academy of Sciences 106: 12377–12381.

5. Grech A, Chartrand-Miller K, Erftemeijer P, Fonseca M, McKenzie L, et al. (2012) A comparison of threats, vulnerabilities and management approaches in global seagrass bioregions. Environmental Research Letters 7: 024006.

6. Grech A, Coles R, Marsh H (2011) A broad-scale assessment of the risk to coastal seagrasses from cumulative threats. Marine Policy 35: 560–567.

7. McMahon K, Collier C, Lavery PS (2013) Identifying robust bioindicators of light stress in seagrasses: A meta-analysis. Ecological Indicators 30: 7–15.

8. Collier CJ, Waycott M, McKenzie LJ (2012) Light thresholds derived from seagrass loss in the coastal zone of the northern Great Barrier Reef, Australia. Ecological Indicators 23: 211–219.

9. Lavery PS, McMahon K, Mulligan M, Tennyson A (2009) Interactive effects of timing, intensity and duration of experimental shading on *Amphibolis griffithii*. Marine Ecology Progress Series 394: 21–33.

10. Lambers H, Chapin FS, Pons TL (1998) Plant physiological ecology. New York: Springer. 540 pp.

11. Kirk JTO (1994). Light and photosynthesis in aquatic ecosystems. Cambridge: Cambridge Press. 528 pp.

12. Ralph PJ, Durako MJ, Enríquez S, Collier CJ, Doblin MA (2007) Impact of light limitation on seagrasses. Journal of Experimental Marine Biology and Ecology 350: 176–193.

13. Cayabyab NM, Enriquez S (2007) Leaf photoacclimatory responses of the tropical seagrass *Thalassia testudinum* under mesocosm conditions: a mechanistic scaling-up study. New Phytol 176: 108–123.

14. Masini RJ, Manning CR (1997) The photosynthetic responses to irradiance and temperature of four meadow-forming seagrasses. Aquatic Botany 58: 21–36.

15. Duarte CM, Marba N, Krause-Jensen D, Sanchez-Camacho M (2007) Testing the predictive power of seagrass depth limit models. Estuaries and Coasts 30: 652–656.

16. O'Brien KR, Grinham A, Roelfsema CM, Saunders MI, Dennison WC (2011) Viability criteria for the presence of the seagrass *Zostera muelleri* in Moreton Bay, based on benthic light dose. In: MODSIM 2011: International Congress on Modelling and Simulation, Proceedings. Modelling and Simulation Society of Australia and New Zealand (MODSIM 2011), Perth, Australia, 12–16 December 2011. 4127–4133.

17. Kemp WM, Batiuk R, Bartleson R, Bergstrom P, Carter V, et al. (2004) Habitat requirements for submerged aquatic vegetation in Chesapeake Bay: Water quality, light regime, and physical-chemical factors. Estuaries 27: 363–377.

18. Dennison WC, Orth RJ, Moore KA, Court Stevenson J, Carter V, et al. (1993) Assessing water quality with submersed aquatic vegetation: habitat requirements as barometers of Chesapeake Bay health. BioScience 43: 86–94.

19. Gacia E, Marba N, Cebrian J, Vaquer-Sunyer R, Garcias-Bonet N, et al. (2012) Thresholds of irradiance for seagrass *Posidonia oceanica* meadow metabolism. Mar Ecol Prog Ser 466: 69–79.

20. Dennison WC, Alberte RS (1985) Role of daily light period in the depth distribution of *Zostera marina* (eelgrass). Mar Ecol Prog Ser 25: 51–61.

21. Enríquez S, Merino M, Iglesias-Prieto R (2002) Variations in the photosynthetic performance along the leaves of the tropical seagrass *Thalassia testudinum*. Marine Biology 140: 891–900.

22. Fourqurean JW, Zieman JC (1991) Photosynthesis, respiration and whole plant carbon budget of the seagrass *Thalassia testudinum*. Mar Ecol Prog Ser 69: 161–170.

23. Collier CJ, Lavery PS, Masini RJ, Ralph PJ (2007) Morphological, growth and meadow characteristics of the seagrass *Posidonia sinuosa* along a depth-related gradient of light availability. Mar Ecol Prog Ser 337: 103–115.

24. Carruthers TJB, Walker DI (1997) Light climate and energy flow in the seagrass canopy of *Amphibolis griffithii* (J.M.Black) den Hartog. Oecologia 109: 335–341.

25. Hedley JD, Enríquez S (2010) Optical properties of canopies of the tropical seagrass *Thalassia testudinum* estimated by a three-dimensional radiative transfer model. Limnology and Oceanography 55: 1537–1550.

26. Zimmerman R (2006) Light and photosynthesis in seagrass meadows. In: Larkum AWD, Orth RJ, Duarte C, editors. Seagrasses: Biology, ecology and conservation. Springer. 303–321.

27. Hedley JD (2008) A three-dimensional radiative transfer model for shallow water environments. Optics Express 16: 21887–21902.

28. McMahon K, Lavery PS (2014) Canopy-scale modifications of the seagrass *Amphibolis griffithii* in response to and recovery from light reduction. Journal of Experimental Marine Biology and Ecology 455: 38–44.

29. House DH, Breen DE (2000) Cloth modeling and animation. Massachusetts: A. K. Peters. 344 p.

30. Mayer B, Kylling A (2005) The libRadtran software package for radiative transfer calculations – description and examples of use. Atmospheric Chemistry and Physics 5: 1855–1877.

31. Grant RH, Heisler GM, Gao W (1996) Photosynthetically-active radiation: Sky radiance distributions under clear and overcast conditions. Agricultural and Forest Meteorology 82: 267–292.

32. Mobley CD (1994) Light and Water. San Diego: Academic Press. 608 p.

33. Mobley CD, Sundman L (2000) HydroLight 4.1 user's guide. Sequoia Scientific. Available: http://www.sequoiasci.com/products/Hydrolight.aspx.

34. Hedley JD, Roelfsema CM, Phinn SR, Mumby PJ (2012). Environmental and sensor limitations in optical remote sensing of coral reefs: Implications for monitoring and sensor design. Remote Sensing 4: 271–302.

35. Durako MJ (2007) Leaf optical properties and photosynthetic leaf absorptances in several Australian seagrasses. Aquatic Botany 87: 83–89.

36. Green EP, Short F (2004) World atlas of seagrasses. Berkeley: University of California Press. 320 p.

37. Pedersen TM, Gallegos CL, Nielsen SL (2012) Influence of near-bottom re-suspended sediment on benthic light availability. Estuarine, Coastal and Shelf Science 106: 93–101.

38. Koch E, Gust G (1999) Water flow in tide and wave dominated beds of the seagrass *Thalassia testudinum*. Mar Ecol Prog Ser 184: 63–72.

39. Collier CJ, Waycott M, Ospina AG (2012b) Responses of four Indo-West Pacific seagrass species to shading. Marine Pollution Bulletin 65: 342–354.

40. Brun FG, Vergara JJ, Hernádez I, Pérez-Lloréns JL (2003) Growth, carbon allocation and proteolytic activity in the seagrass *Zostera noltii* shaded by *Ulva* canopies. Funct Plant Biol 30: 551–560.

23

Methane-Derived Carbon in the Benthic Food Web in Stream Impoundments

John Gichimu Mbaka*, Celia Somlai, Denis Köpfer, Andreas Maeck, Andreas Lorke, Ralf B. Schäfer

Institute for Environmental Sciences, University of Koblenz-Landau, Landau, Rhineland-Palatinate State, Germany

Abstract

Methane gas (CH_4) has been identified as an important alternative source of carbon and energy in some freshwater food webs. CH_4 is oxidized by methane oxidizing bacteria (MOB), and subsequently utilized by chironomid larvae, which may exhibit low $\delta^{13}C$ values. This has been shown for chironomid larvae collected from lakes, streams and backwater pools. However, the relationship between CH_4 concentrations and $\delta^{13}C$ values of chironomid larvae for in-stream impoundments is unknown. CH_4 concentrations were measured in eleven in-stream impoundments located in the Queich River catchment area, South-western Germany. Furthermore, the $\delta^{13}C$ values of two subfamilies of chironomid larvae (i.e. Chironomini and Tanypodinae) were determined and correlated with CH_4 concentrations. Chironomini larvae had lower mean $\delta^{13}C$ values (− 29.2 to −25.5 ‰), than Tanypodinae larvae (−26.9 to −25.3 ‰). No significant relationships were established between CH_4 concentrations and $\delta^{13}C$ values of chironomids (p>0.05). Mean $\delta^{13}C$ values of chironomid larvae (mean: −26.8‰, range: − 29.2‰ to −25.3‰) were similar to those of sedimentary organic matter (SOM) (mean: −28.4‰, range: −29.3‰ to −27.1‰) and tree leaf litter (mean: −29.8 ‰, range: −30.5‰ to −29.1‰). We suggest that CH_4 concentration has limited influence on the benthic food web in stream impoundments.

Editor: David William Pond, Scottish Association for Marine Science, United Kingdom

Funding: JGM was funded by the German Academic Exchange service (DAAD) (Grant number: A/12/91652). The funder had no role in study design, data collection and analysis, decision to publish, or preparation of the manuscript.

Competing Interests: The authors have declared that no competing interests exist.

* Email: mbakagichimu@gmail.com

Introduction

Allochthonous and autochthonous plant organic matter are major sources of carbon and energy for freshwater ecosystems [1]. Recent studies have revealed that also methane, which can be produced by microbial degradation of organic matter under anoxic conditions, can significantly contribute to the carbon budget of freshwater ecosystems [2]. Part of this gas is released to the atmosphere, where it contributes to the pool of green house gases [3], or is oxidized by methane oxidizing bacteria (MOB) [4]. The biogenic methane in MOB can contribute to the biomass of chironomid larvae [5]. Chironomids are one of the most dominant invertebrate groups in the soft sediments in freshwater ecosystems and their larvae feed mainly on algae or allochthonous organic material and associated microorganisms [6]. CH_4 derived organic carbon may constitute a crucial source of carbon and energy for chironomids, in comparison to other aquatic invertebrates, because their burrowing habit creates and exposes them to oxyclines at the sediment-water interface, where MOB density and CH_4 oxidation rates are usually high [7–10]. For example, chironomid larvae collected from some lakes were sustained (up to 70%) by CH_4 derived carbon [11].

These quantitative estimates are based on the stable carbon isotope signature ($\delta^{13}C$) of CH_4 which is highly depleted due to carbon isotopic fractionation related to methanogenesis [12]. Additionally, MOB that oxidize CH_4 are usually characterized by further depletion in $\delta^{13}C$ [13]. Therefore, organisms that consume

MOB have lower $\delta^{13}C$ values (typically $<-40‰$; [14]), in comparison to organisms that feed on plant organic matter (-32 to -21 ‰; [15]). Bunn & Boon [16] determined the $\delta^{13}C$ values of invertebrates in backwater pools and found Chironominae larvae to have $\delta^{13}C$ values ($<-35‰$) that were lower than for particulate organic matter (-29 to -25 ‰). Kiyashko et al. [18] and Jones & Grey [19] also reported lower (-64 to -55 ‰) $\delta^{13}C$ values for some chironomid larvae than for particulate organic matter in lakes. The observed differences between $\delta^{13}C$ values of chironomid larvae and potential food resources led to the conclusion that the chironomid larvae might have fed on MOB, which has very low $\delta^{13}C$ values. The carbon isotope composition of consumers (e.g. insects) is determined by their diet and usually potrays an enrichment by about 1 ‰, even though the $\delta^{13}C$ can deviate from -3 ‰ to +3 ‰ [20]. Given that methane is isotopically very distinct, stable carbon isotopes are particularly useful for tracing methane derived carbon [17].

Although most existing studies on the importance of CH_4 derived carbon in freshwater food webs mainly focused on lakes [2], a wide array of anoxic habitats with high potential for CH_4 production also exist in rivers and streams [21–23]. Particularly, impoundments increases the residence time of water, promotes accumulation of organic matter and sediment, and have been identified as hot spots of CH_4 emissions [24]. Maeck et al. [24] measured CH_4 concentrations in riverine and impoundment reaches and found sediment accumulation in dams to be the main source of CH_4 emissions. Guérin et al. [25] reported an increase in CH_4 emissions at the downstream sides of impoundments as a result of release of water enriched with CH_4.

In shallow aquatic systems such as rice paddies and small lakes, CH_4 has been shown to be an important source of energy in the benthic food webs [26–28]. In spite of the high abundance of small in-stream ponds in smaller streams [29], the relationship between CH_4 concentrations and stable carbon isotope ratios ($\delta^{13}C$) of chironomid larvae in such systems has not been examined. Globally, there exist millions of small impoundments (height < 15 m; [30]). Within this study, we assessed (i) CH_4 concentrations in stream and pore-water and (ii) the relationship between CH_4 concentrations and $\delta^{13}C$ values of chironomid larvae in impoundments located in the Queich River catchment area, South-western Germany. We hypothesized that CH_4 would have a significant influence on the $\delta^{13}C$ of chironomid larvae.

Methods

Ethics statement

This study was conducted in the Queich River catchment area (see coordinates below) and was not conducted in an area requiring research permit (e.g. national park) or private land. This study did not involve endangered or protected species.

Study area and sites

The study was conducted in the Queich River catchment area, Rhineland-Palatinate State, South-western Germany. The Queich River (length: 52 km) originates from the Palatinate forest (49°10′6″N 7°50′48″E) and flows (mean discharge: 1.31 m³ s⁻¹; www.geoportal-wasser.rlp.de) through the upper Rhine Valley to its confluence with the Rhine River in Germersheim (49°13′39″N 8°23′4″E). The catchment (area: 271 km²) is primarily covered by sandstone and is between 100 m and 673 m above sea level. The Rhineland-Palatinate region has dry climate conditions in summer.

Typical for most stream networks in central Europe, 67 small in-stream impoundments (www.geoportal-wasser.rlp.de) have been constructed on the main stem and tributaries of the Queich River, South-western Germany, for various purposes such as hydropower generation and flood control. Here, we selected eleven study sites (e.g. Figure S1 in File S1) located from the downstream to the upstream reaches of the Queich River catchment area (Figure 1, Table S1 in File S1). The study sites were located at five different streams. The sampled impoundments were approximately 0.5–2.0 metres deep and had water bypasses that transported water to the downstream reaches.

Water chemistry and physical characteristics

Field measurements and sampling were conducted between 9th and 24th June, 2013. Data collection was done between 9 a.m. and 4 p.m. Electrical conductivity, temperature, dissolved oxygen concentration and pH of stream water were measured *in situ* with a WTW Multi 340i/SET (Wissenschaftlich Werkstätten GmbH, Weilheim, Germany). Average water depth was computed from three measurements taken on a transect across the river channel and current speed was estimated by timing a float over a distance of 5 metres [31]. Water discharge was calculated from velocity, width and depth [31]. Water residence time was calculated as follows:

$$T = \frac{V}{Q} \qquad \text{(i)}$$

where: T is the water residence time, V is the volume of water stored in the impoundment, and Q is the water discharge [32]. Nitrate and phosphate concentrations in stream water were determined in the laboratory using Macherey-Nagel viscolor kits (Macherey-Nagel, Düren, Germany).

CH_4 concentrations

Concentrations of dissolved CH_4 in stream and pore-water were measured at the impoundments. Water samples were collected from each study impoundment using 20 mL serum bottles. The samples for stream water CH_4 analysis were collected by filling water to the sample bottles from the bottom to top, and overflowing the sample bottles several times over. Three bottles were completely filled with water at each sampling site and several drops (250 µL) of mercuric chloride were added to each bottle as preservative [33]. The bottles were capped and sealed and transported to the laboratory. A headspace was prepared by replacing 10% of the bottle (i.e. 2 mL) with nitrogen gas. To generate the headspace, each sample bottle was held upside down and a 20 gauge needle was inserted through the septum. Then 2 mL nitrogen gas was added to each bottle using a syringe, while the replaced water sample escaped through the needle. The samples were manually shaken, for 1 minute, to equilibrate the gas between the headspace and the water [34]. The samples were analyzed using a CH_4 analyser (Los Gatos Research Inc., Mountain view, CA, U.S.A.). A closed loop was created between the gas inlet and outlet of the analyser. A gas tight syringe was then used to inject 0.5 mL gas sample into the closed loop. The CH_4 concentrations were averaged over 30 seconds before and after gas injection. Concentration of the injected gas was computed as:

$$c_{sample} = \frac{\Delta c_{LosGatos}(V_{LosGatos} + V_{Injection})}{V_{Injection}} \qquad \text{(ii)}$$

where: c_{sample} is the mol fraction of the sampled gas in parts per million, $\Delta c_{LosGatos}$ is the change in mole fraction before and after gas sample injection, $V_{LosGatos} = 92.5$ mL and $V_{Injection} = 0.5$ mL.

Sediment samples for pore-water CH_4 analysis were obtained at the impoundments, where fine sediment accumulated. Pore-water CH_4 concentrations were assessed from sediment cores (1 core per site). Cores were taken at location of soft sediment using a piston corer and analyzed for porosity, carbon:nitrogen (C:N) ratio, total organic carbon (TOC) content and pore-water CH_4 concentration. Cut-off syringes (3 mL) were used to extract sediment sub-samples which were immediately placed into crimp capped 20 mL vials containing 3 mL of 2.5% NaCl solution for conservation of the CH_4. Pore-water was sampled in the cores from the homogenized upper (0–10 cm) sediment layer, where chironomid larvae are found [35]. The pore-water CH_4 samples were measured as described for water samples. For C:N ratio, TOC and porosity, three sediment sub-samples were extracted from the cores (0–10 cm) and placed into glass tubes before analysis in the laboratory.

Chironomid sampling and processing

Chironomid samples were collected from the deepest point in each impoundment using an Ekman grab sampler (Hydro-Bios, Kiel, Germany). Sediments were sieved by passing them through

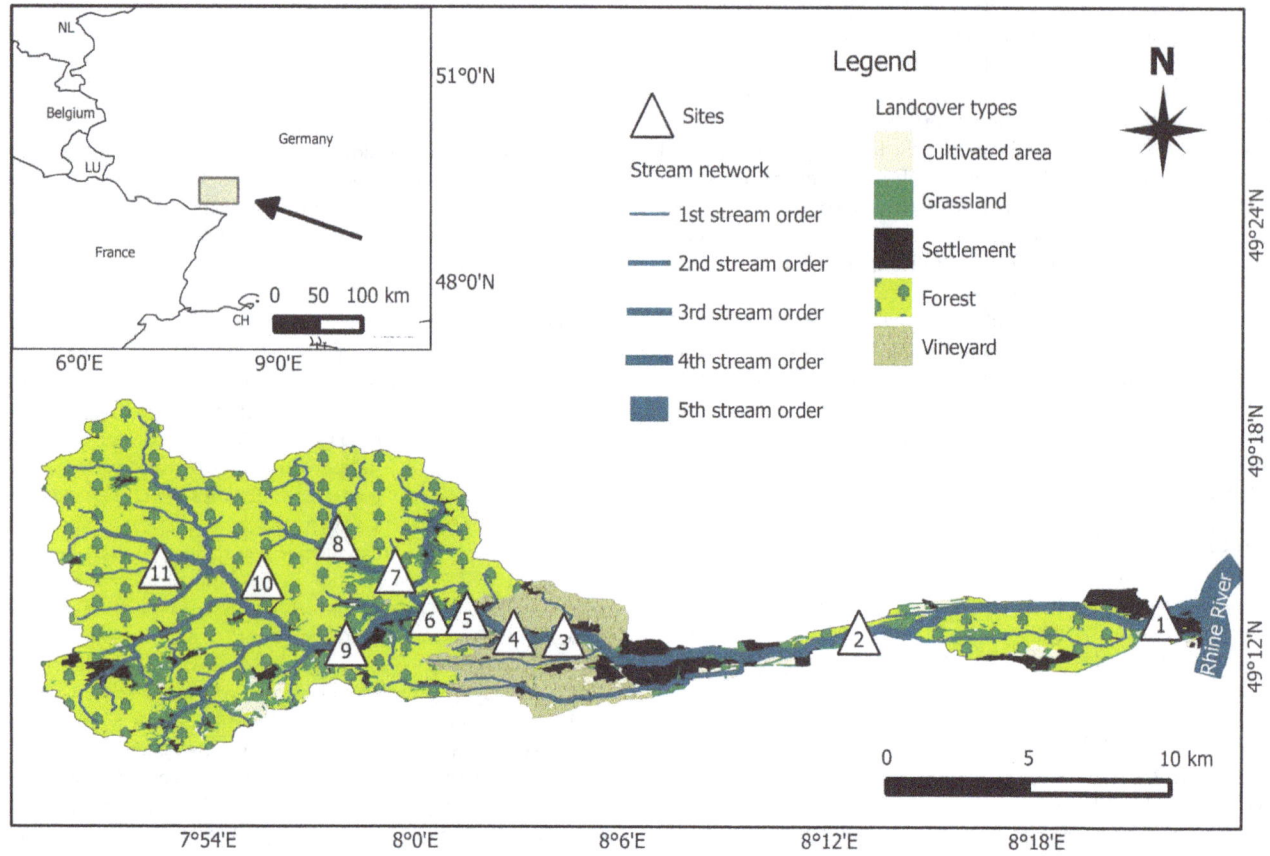

Figure 1. Locations of the study impoundments in the Queich River catchment area.

two metal sieves (mesh size: 500 μm and 2 mm). Materials such as stones and large pieces of organic matter (>5 cm) were removed and chironomid larvae were picked from a sorting tray using forceps and placed into 500 mL sample bottles containing river water. In some sites only few chironomids (<5 chironomids) were found. The chironomid samples were transported to the laboratory and transferred to sample bottles containing clean tap water for 24 hours to allow gut clearance. Faecal materials were periodically removed to prevent ingestion by chironomids [36]. Chironomids were sorted by tribe and subfamily [37–40] and size (instar) [41,42] under a dissecting microscope (magnification: x 40-100). Chironomids of the same tribe, subfamily and size were pooled to obtain sufficient mass (0.5–1 mg) for isotope analysis [43]. Sorting of chironomids by size was done to detect the potential effect of body size on the isotopic signal as demonstrated by Grey et al. [44]. In most cases the limited number of specimens excluded replicate analyses. Therefore, we collected individuals from a site with a high abundance of chironomids to exemplarily determine the $\delta^{13}C$ variability from 9 replicates. Second instar larvae were discarded as they were too small for identification and their mass was insufficient for isotope analysis. Before isotope analysis, chironomids were placed into glass tubes, oven dried at 60°C for 24 hours and subsequently stored in a desiccator.

Stable isotope, TOC and C:N analyses

Three replicate sediment samples, from each site were analysed, for $\delta^{13}C$ of sedimentary organic matter (SOM), TOC and C:N ratios. They were rinsed with a 2.5% HCL solution for four hours to remove carbonates [45], rinsed three times with demineralised

water, oven dried at 60°C for 24 hours and ground using a mortar and pestle. Leaves, for analysis of $\delta^{13}C$ of potential allochthonous food resources, were collected from trees near the impoundments, washed with demineralised water, rinsed, oven dried and ground before analysis.

Sediment samples for C:N ratios and TOC were weighed into tin cups (15–20 g) and analysed using a Vario Microcube elemental analyser (Elementar Analysensysteme, Hanau, Germany). The chironomid, SOM and leaf litter samples for stable isotope analysis were also weighed (approximately 0.5–1.0 mg for chironomids and 5–20 mg for SOM and leaf litter) into tin cups before their combustion in an isotope ratio mass spectrometer (ThermoScientific, Bremen, Germany). Stable isotope ratios were expressed in per mille (‰).

Statistical analysis

Relations between variables were tested using Spearman's rank correlation test [46]. Comparisons of CH_4 concentrations in pore and stream water, and between $\delta^{13}C$ values of chironomids, and leaf litter and SOM were done using paired t-test. A value of $p <$ 0.05 was considered as statistically significant. Homogeneity of variances was examined with Bartlett's test and data were square root transformed to improve normality. Statistical analyses were done using the R statistical package [47] and all data are provided in File S1.

Results

Water chemistry and physical characteristics

Water residence time varied from 0.5 to 6.0 minutes. Nutrient concentrations in stream water ranged from 3.5 to 5.0 mg NO_3 L^{-1}, and from 0.1 to 0.3 mg PO_4 L^{-1}. Water temperature ranged from 10.2 to 18.6°C, whereas electrical conductivity, dissolved oxygen concentrations and water discharge ranged from 61 to 380 μS cm^{-1}, 8.5 to 11.9 mg L^{-1} and 0.2 to 7.8 m^3 s^{-1} (Table S1 in File S1).

Methane gas concentrations

Average values of dissolved CH_4 concentrations in stream water ranged from 0.07 μmol L^{-1} at Site 6 to 0.7 μmol L^{-1} at Site 10 (Table 1). Pore-water CH_4 concentrations ranged from 0.3 μmol L^{-1} at Site 7 to 1657.5 μmol L^{-1} at Site 9, and were statistically significantly (t-value $= 3.1$, $p = 0.005$) higher than the stream water CH_4 concentrations. Some pore-water CH_4 measurements at sites 3 and 6 differed greatly from the other measurements, either due to disturbance during sampling or gas leakage during analysis, and were therefore taken to be unreliable and excluded from further analysis.

$\delta^{13}C$ of SOM and leaf litter, C:N ratios and TOC

$\delta^{13}C$ values of SOM ranged from $-29.3‰$ at site 7 to $-27.1‰$ at site 3 (Table 1), whereas $\delta^{13}C$ values of leaf litter ranged from $-30.5 ‰$ to $-29.1‰$. C:N ratios of sediment and the TOC ranged from 12.0 at site 3 to 23.2 at site 2, and from 0.2% at site 3 to 3.1% at site 2, respectively (Table 1).

$\delta^{13}C$ of chironomids

The chironomids were identified as Chironomini, Tanypodinae and *Chironomus* sp. *Chironomus* sp. were only found at site 10 and were not analyzed because they did not provide an adequate pooled mass for isotope analysis. The lowest $\delta^{13}C$ value, $-29.2 ‰$, was measured in a third instar Chironomini larvae collected from site 8, whereas the highest $\delta^{13}C$ values, $-25.3 ‰$, were measured in third instar Tanypodinae larvae collected from Sites 9 and 11 (Table S2 in File S1). Generally, the highest mean $\delta^{13}C$ values were measured in third ($-25.3 \pm 0.01‰$) and fourth (-26.3 ± 0.14 ‰) instar Tanypodinae, whereas slightly lower mean $\delta^{13}C$ values were measured in fourth (-27.2 ± 0.16 ‰) and third (-26.9 ± 0.25 ‰) instar Chironomini. $\delta^{13}C$ values did not differ significantly between third and fourth instar Chironomini larvae (t-value $= 1.4$, $p = 0.17$). The analysis of 9 replicates of fourth instar Chironomini larvae from site 4 had a standard deviation as low as 0.19 ‰ for $\delta^{13}C$ (Table S2 in File S1). No significant correlations were observed between CH_4 concentrations and $\delta^{13}C$ values of chironomids (Figure 2, Table 2). However, $\delta^{13}C$ values of chironomid larvae were significantly correlated to those of the SOM and to each other within site ($p<0.05$) (Table 2). Mean $\delta^{13}C$ of chironomid larvae ($-26.8 ‰$) was more similar to that of the SOM ($-28.4 ‰$) than leaf litter ($-29.8 ‰$) and there were significant differences ($p<0.05$) between $\delta^{13}C$ values of chironomids, and SOM and leaf litter.

Discussion

Methane gas concentrations

CH_4 concentrations measured in this study are comparable to those measured in other aquatic ecosystems [48–50]. Pore-water CH_4 concentrations showed a highly variable pattern among the impoundments (Table 1). These differences can be attributed to heterogeneity in the distribution of sedimentary organic materials within the impoundments. Sanders et al. [51] found pore-water CH_4 concentrations to be influenced by sediment heterogeneity.

Table 1. Average CH_4 concentrations (μmol L^{-1}) and bulk sediment characteristics (TOC, C:N ratio, porosity and $\delta^{13}C$) of samples from the studied impoundments (in parentheses are standard errors, $\pm$ SE (when applicable).

Name	Code	Pore-water CH_4	Dissolved CH_4	TOC (%)	C:N ratio	Porosity	$\delta^{13}C$ (‰)
Germersheim	1	27.8 (4.6)	0.2 (0.006)	2.8 (0.1)	17.6 (0.8)	0.6 (0.02)	-28.7 (0.01)
Fuchsbach	2	658.1 (244.4)	0.4 (0.004)	3.1 (0.1)	23.2 (0.9)	0.7 (0.02)	-28.2 (0.3)
Godramstein	3	142.5	0.3 (0.04)	0.2 (0.02)	12.0 (1.0)	0.4 (0.02)	-27.1 (0.02)
Siebeldingen	4	1499.6 (784.6)	0.1 (0.06)	2.8 (0.03)	17.1 (0.6)	0.7 (0.003)	-28.8 (0.04)
Albersweiler Pfalz	5	199.6 (8.2)	0.1 (0.004)	0.6 (0.04)	16.1 (1.2)	0.5 (0.02)	-28.9 (0.2)
Rosenfeldt Mill	6	263.0	0.07 (0.001)	1.8 (0.04)	16.4 (0.6)	0.5 (0.03)	-28.5 (0.1)
Eußerbach	7	0.3 (0.01)	0.3 (0.01)	0.3 (0.03)	15.2 (1.4)	0.4 (0.03)	-29.3 (0.2)
Eisbach	8	45.2 (8.7)	0.2 (0.03)	1.6 (0.2)	14.7 (0.6)	0.6 (0.02)	-28.6 (0.1)
Annweiler Am Trifels	9	1657.5 (128.9)	0.1 (0.002)	2.4 (0.2)	19.5 (1.2)	0.6 (0.01)	-28.7 (0.1)
Langenbächel	10	4.6 (2.2)	0.7 (0.07)	0.9 (0.1)	17.9 (1.5)	0.6 (0.02)	-28.9 (0.04)
Modenbach	11	105.9 (18.3)	0.3 (0.01)	0.5 (0.02)	16.1 (0.6)	0.3 (0.01)	-27.3 (0.1)

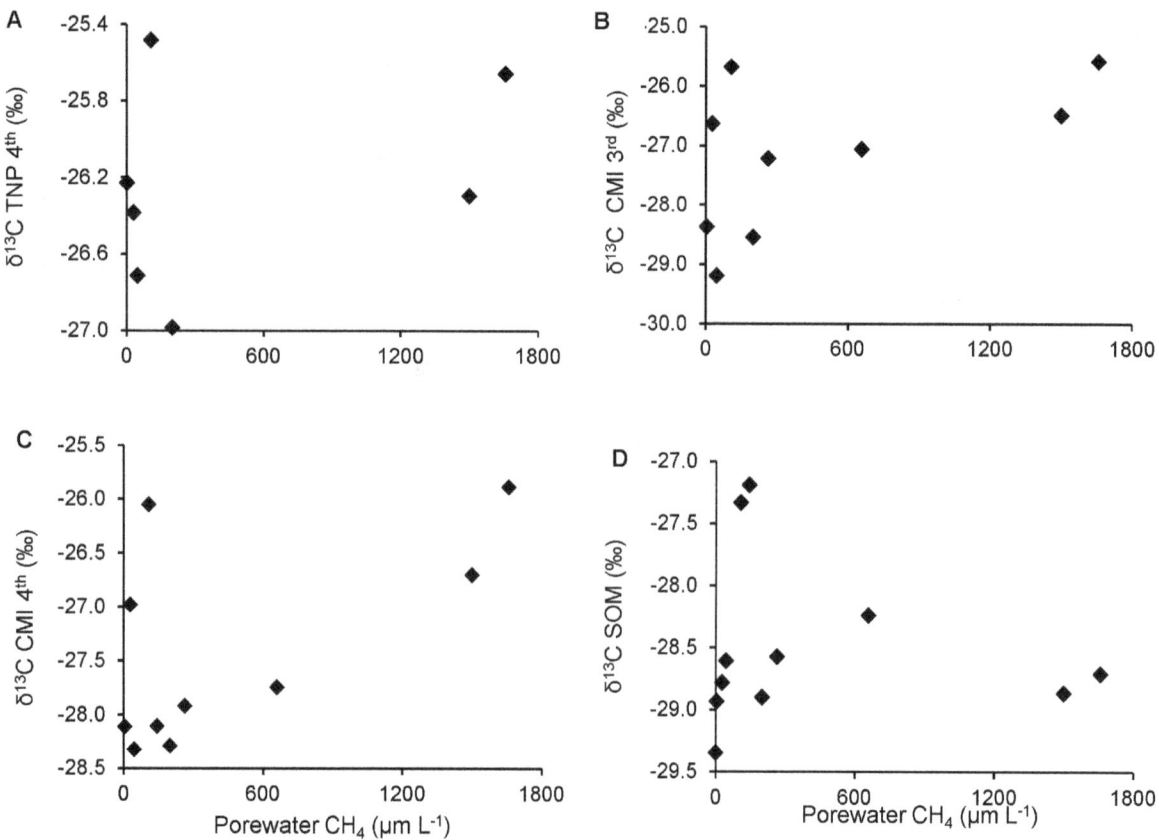

Figure 2. Relationships between pore-water CH₄ concentrations and δ¹³C values. TNP 4th, CMI 3rd, CMI 4th and SOM stands for Tanypodinae fourth generation, Chironomini third and fourth generation, and sedimentary organic matter, respectively.

The authors reported that enhanced retention of sediment by macrophytes (*Ranunculus penicillatus*) increased pore-water CH_4 production in streams. Significantly higher mean CH_4 concentrations were recorded in pore than stream water. This can be explained by the fact that CH_4 is usually produced in sediments, where anoxic conditions are likely to develop, whereas the stream water is rather well oxygenated (Table S1 in File S1) [21].

Relationship between δ¹³C of chironomids and methane gas

Few studies examined CH_4 as a source of carbon and energy for stream invertebrates (e.g. [52]). In these studies, mean $\delta^{13}C$ values of invertebrates supported by CH_4 derived carbon were lower (<-40 ‰) than those of potential photoautotrophic food resources, indicating ingestion and assimilation of MOB, which had oxidized isotopically light CH_4. This was also demonstrated for lake invertebrates [7,43,53], where significant negative relationships between CH_4 concentrations and $\delta^{13}C$ values of some invertebates indicated ingestion and assimilation of MOB.

Mean $\delta^{13}C$ of chironomid larvae ranged from -27.2 (fourth instar Chironomini) to -25.3 ‰ (fourth instar Tanypodinae). The mean $\delta^{13}C$ values of potential food resources ranged from -28.4 (SOM) to -29.8 ‰ (leaf litter) and were significantly correlated to those of the chironomids. The similarity of $\delta^{13}C$ of Chironomini larvae and SOM can be attributed to the fact that Chironomini larvae are either filterers or gathering-collectors, feeding on fine particulate organic matter in aquatic systems [54]. In comparison to Chironomini, Tanypodinae utilize different types of food (e.g. detritus, oligochaetes, diatoms; [55]) and their $\delta^{13}C$ values may be

difficult to interpret when compared with the other chironomids. The differences between the $\delta^{13}C$ values of chironomids, SOM and leaf litter were within the reported range (±3 ‰) of $\delta^{13}C$ values for consumers and their food resources [20].

In the current study, no significant relationships were established between CH_4 concentrations in stream and pore-water and $\delta^{13}C$ values of chironomid larvae. In the UK, Trimmer et al. [52] found river water to have higher mean (0.16 μmol L^{-1}) CH_4 concentrations than pore-water (0.07 μmol L^{-1}). Highly depleted mean $\delta^{13}C$ values (<-42 ‰) of Trichoptera larvae (*Agapetus fuscipes*), relative to those of potential food sources (-38 ‰), were only measured in areas with low pore-water CH_4 concentrations. In comparison to Trimmer et al., we recorded higher mean CH_4 concentrations in stream and pore-water and the mean $\delta^{13}C$ values of chironomid larvae were not low. The type of food consumed by chironomids and presence of MOB influence the $\delta^{13}C$ values of chironomid larvae. The short water residence times and the shallow nature of the studied impoundments could have enhanced water turn over rates, mixing of the entire water column and supply of organic matter or periphyton into the benthic zone. Thus, in the case of chironomid larvae feeding on sedimenting organic matter, we would anticipate their $\delta^{13}C$ to be similar to that of their food resources. Mixing of the water column may also increase oxygen concentration and CH_4 oxidation rates, and reduce CH_4 production and the biomass of MOB available to chironomids [9]. For example, Eller et al. [9] recorded two fold higher MOB density in the anoxic waters (0.1 mg L^{-1} O₂) of a stratified lake than in the well mixed and oxygenated waters (9 mg L^{-1} O₂) of a polymictic lake. Additionally, the contribution of

Table 2. Correlations (*rho*-values) between $\delta^{13}C$ values, C:N ratio of bulk sediment, water residence time (min) and CH_4 concentrations (*$p<0.05$).

	$\delta^{13}C_{SOM}$	Dissolved methane	Pore-water methane	CMI 4th	TNP 4th	Water residence time	CMI 3rd
C:N$_{SOM}$	−0.16	−0.28	0.35	−0.46	−0.15	−0.55	−0.29
$\delta^{13}C_{SOM}$		−0.80*	0.30	0.66*	0.92*	−0.28	0.76*
Dissolved CH_4			−0.78*	−0.11	−0.57	0.22	−0.28
Pore-water CH_4				−0.50	−0.05	0.21	−0.35
CMI 4th					0.87*	−0.41	0.97*
TNP 4th						−0.53	0.95*
Water residence time							−0.55

$\delta^{13}C_{SOM}$, $\delta^{13}C$ of bulk sediment; CMI 4th, fourth instar Chironomini; CMI 3rd, third instar Chironomini; TNP 4th, fourth instar Tanypodinae.

MOB to chironomid larvae biomass was higher in the anoxic than oxygenated waters. Jones et al. [11] found that the contribution of MOB to the chironomid larvae biomass was highest at sites with low dissolved oxygen content ($2–4 \text{ mg L}^{-1}$). Within the above mentioned studies, the isotopic values of chironomids were more similar to those of the SOM, when they were collected from sediments overlaid by well oxygenated waters. Use of MOB was particularly pronounced under hypoxia or post mixing following on from stratification.

$\delta^{13}C$ values of chironomid larvae (mean = ∼−27 ‰; −29 to −25 ‰) were similar to those of SOM (mean = ∼−28 ‰; −29 to −27 ‰) and leaf litter (mean = ∼−29 ‰; −30 to −29 ‰). Additionally, $\delta^{13}C$ values of chironomid larvae were significantly correlated to each other. Given that the $\delta^{13}C$ values varied little between consumers and their food sources [56], the chironomid larvae collected from the sampled impoundments most likely obtained their carbon through ingestion and assimilation of SOM or allochthonous leaf litter. Other studies using stable carbon isotope analysis also demonstrated SOM and allochthonous plant organic matter as significant sources of carbon and energy for freshwater invertebrates [57]. The C:N ratio can be used to determine the source of organic matter in aquatic ecosystems because autochthonous organic matter generally has lower C:N ratios (e.g. algae: 4–10; [58]) than allochthonous organic matter. The measured C:N ratios (12.0–23.2) of sediments indicated elevated proportions of allochthonous organic matter.

Although the $\delta^{13}C$ values of chironomid larvae did not indicate utilization of methane derived carbon, other invertebrates could have used it as a source of energy. For example, Kohzu et al. [59] found coleopterans collected from backwater pools to have lower mean (−40 to −67 ‰) $\delta^{13}C$ values than the other invertebrates (e.g. chironomids; −36 ‰), suggesting increased utilization of methane derived carbon. In summary, this study reveals that methane derived carbon did not contribute substantially to chironomid larval biomass in small impoundments, rather that allochthonous organic matter was the main source of energy. Future studies assessing the role of methane derived carbon in stream impoundments should include MOB community characterization, CH_4 oxidation rates and fluxes, and $\delta^{13}C$ values of other invertebrates.

Supporting Information

File S1 Supporting figure and tables. Figure S1, Example of a weir that impounded the studied rivers. The white arrow shows the direction of water movement. Table S1, Location and environmental characteristics recorded from the studied impoundments. Table S2, $\delta^{13}C$ values (‰) of chironomid larvae from the studied impoundments. For each of the impoundments, between 1 and 9 replicates (*n*) were made from pooled samples.

Acknowledgments

We thank Vivian Rhee, Sebastian Geissler, Zeyad Alshboul and Anne König for invaluable practical assistance in the field. We also thank Markus Kurtz for C:N and TOC analyses and Andreas Hirsch for stable isotope analysis.

Author Contributions

Conceived and designed the experiments: JGM RS AL. Performed the experiments: JGM CS AM DK. Analyzed the data: JGM RS. Contributed reagents/materials/analysis tools: RBS AL. Wrote the paper: JGM RBS AL AM CS.

References

1. Meili M (1992) Sources, concentrations and characteristics of organic matter in softwater lakes and streams of the Swedish forest region. Hydrobiologia 229: 23–41.

2. Jones RI, Grey J (2011) Biogenic methane in freshwater food webs. Freshwater Biol 56: 213–229.

3. Bastviken D, Tranvik LJ, Downing JA, Crill PM, Enrich-Prast A (2011) Freshwater methane emissions offset the continental carbon sink. Science 331: 50.

4. Kajan R, Frenzel P (1999) The effect of chironomid larvae on production, oxidation and fluxes of methane in a flooded rice soil. FEMS Microbiol Ecol 28: 121–129.

5. Kiyashko SI (2001) Contribution of methanotrophs to freshwater macroinvertebrates: evidence from stable isotope ratios. Aquat Microb Ecol 24: 203–207.

6. Berg H (1995) Larval food and feeding behaviour. In: Armitage, P.D., Cranston, P.S. & Pinder, L.C.V. (eds), *The Chironomidae: biology and ecology of non biting midges*. Chapman& Hall, London. pp 136–168.

7. Deines P, Grey J (2006) Site-specific methane production and subsequent midge mediation within Esthwaite Water, UK. Arch Hydrobiol 167: 317–334.

8. Sanseverino AM, Bastviken D, Sundh I, Pickova J, Enrich-Prast A (2012) Methane carbon supports aquatic food webs to the fish level. PLoS ONE 7: e42723–e42723.

9. Eller G, Deines P, Grey J, Richnow H, Krager M (2005) Methane cycling in lake sediments and its influence on chironomid larval δ13C. FEMS Microbiol Ecol 54: 339–350.

10. Yasuno N, Shikano S, Shimada T, Shindo K, Kikuchi E (2013) Comparison of the exploitation of methane-derived carbon by tubicolous and non-tubicolous chironomid larvae in a temperate eutrophic lake. Limnology 14: 239–246.

11. Jones RI, Carter CE, Kelly A, Ward S, Kelly DJ, et al. (2008) Widespread contribution of methane-cycle bacteria to the diets of lake profundal chironomid larvae. Ecology 89: 857–864.

12. Teh YA, Silver WL, Conrad ME, Borglin SE, Carlson CM (2006) Carbon isotope fractionation by methane-oxidizing bacteria in tropical rain forest soils. J Geophys Res-Biogeo 111: G2.

13. Summons RE, Jahnke LL, Roksandic Z (1994) Carbon isotopic fractionation in lipids from methanotrophic bacteria: relevance for interpretation of the geochemical record of biomarkers. Geochim Cosmochim Acta 58: 2853–2863.

14. Zemskaya TI, Sitnikova TY, Kiyashko SI, Kalmychkov GV, Pogodaeva TV, et al. (2012) Faunal communities at sites of gas-and oil-bearing fluids in Lake Baikal. Geo-Marine Lett 32: 437–451.

15. Hershey A, Binkley E, Fortino K, Keyse M, Medvedeff C, et al. (2010) Use of allochthonous and autochthonous carbon sources by Chironomus in arctic lakes. Verh Internat Verein Limnol 30: 1321–1325.

16. Bunn SE, Boon PI (1993) What sources of organic carbon drive food webs in billabongs? A study based on stable isotope analysis. Oecologia 96: 85–94.

17. Zanden MJV, Rasmussen JB (1999) Primary consumer δ13C and δ15N and the trophic position of aquatic consumers. Ecology 80: 1395–1404.

18. Kiyashko S, Narita T, Wada E (2001) Contribution of methanotrophs to freshwater macroinvertebrates: evidence from stable isotope ratios. Aquat Microb Ecol 24: 203–207.

19. Jones RI, Grey J (2004) Stable isotope analysis of chironomid larvae from some Finnish forest lakes indicates dietary contribution from biogenic methane. Boreal Environ Res 9: 17–24.

20. DeNiro MJ, Epstein S (1978) Influence of diet on the distribution of carbon isotopes in animals. Geochim Cosmochim Acta 42: 495–506.

21. Trimmer M, Maanoja S, Hildrew AG, Pretty JL, Grey J (2010) Potential carbon fixation via methane oxidation in well-oxygenated riverbed gravels. Limnol Oceanogr 55: 560–568.

22. Shelley F, Grey J, Trimmer M (2014) Widespread methanotrophic primary production in lowland chalk rivers. Proc R Soc Lond B: Biol Sci 281: 20132854.

23. Crawford JT, Stanley EH, Spawn SA, Finlay JC, Loken LC, et al. (2014) Ebullitive methane emissions from oxygenated wetland streams. Glob Change Biol Doi:10.1111/gcb.12614.

24. Maeck A, DelSontro T, McGinnis DF, Fischer H, Flury S, et al. (2013) Sediment trapping by dams creates methane emission hot spots. Environ Sci Technol 47: 8130–8137.

25. Guérin F, Abril G, Richard S, Burban B, Reynouard C, et al. (2006) Methane and carbon dioxide emissions from tropical reservoirs: significance of downstream rivers. Geophys Res Lett 33.

26. Jones RI, Grey J (2004) Stable isotope analysis of chironomid larvae from some Finnish forest lakes indicates dietary contribution from biogenic methane. Boreal Environmen Res 9: 17–24.

27. Khalil M, Shearer MJ, Rasmussen R, Duan C, Ren L (2008) Production, oxidation, and emissions of methane from rice fields in China. J Geophys Res-Biogeo 113: G00A04.

28. Minami K, Neue HU (1994) Rice paddies as a methane source. Climate Change 27: 13–26.

29. Van Looy K, Tormos T, Souchon Y (2014) Disentangling dam impacts in river networks. Ecol Ind 37: 10–20.

30. WCD (2000) Dams and development. A new framework for decision making, World commision on Dams. Earthscan publications ltd. London Sterling, VA. 399 pages.

31. Gordon ND, McMahon TA, Finlayson BL (1994) Stream hydrology, an introduction for ecologists. Wiley, New York. 526 pages.

32. Rueda F, Moreno-Ostos E, Armengol J (2006) The residence time of river water in reservoirs. Ecol model 191: 260–274.

33. Christian JR, Karl DM (1995) Measuring bacterial ectoenzyme activities in marine waters using mercuric chloride as a preservative and a control. Mar Ecol Prog Ser 123: 217–224.

34. Baird AJ, Stamp I, Heppell CM, Green SM (2010) CH4 flux from peatlands: a new measurement method. Ecohydrol 3: 360–367.

35. Deines P, Grey J, Richnow H-H, Eller G (2007) Linking larval chironomids to methane: seasonal variation of the microbial methane cycle and chironomid δ13C. Aquat Microb Ecol 46: 273–282.

36. Feuchtmayr H, Grey J (2003) Effect of preparation and preservation procedures on carbon and nitrogen stable isotope determinations from zooplankton. Rapid Commun Mass Spectrom 17: 2605–2610.

37. Ashe P, O'Connor JP (2009) A world catalogue of chironomidae (Diptera). Part 1. Buchonomyiinae, chilenomyiinae, Podominae, Aphoratemiinae, Tanypodinae, Usambaromyiinae, Dimesinae, Prodiamesinae and Telmatogetominae. Irish Biogeographical Society & National Museum of Ireland, Dublin, 445 pages.

38. Orendt C, Dettinger-klemm A, Spies M (2012) Chironomidae larvae in brakish waters of Germany and adjacent areas. Federal Environment Agency, Berlin, 214 pages.

39. Orendt C, Spies M (2012) Chironomini (Diptera: Chironomidae). Keys to Central European larvae using mainly macroscopic characters. 2nd revised edition, 64 pages.

40. Orendt C, Spies M (2012) *Chironomus* (Meigen). Key to the larvae of importance to biological water analysis in Germany and adjacent areas. 2nd revised edition, 24 pages.

41. Dillon PM (1985) Chironomid larval size and case presence influence capture success achieved by dragonfly larvae. Freshwat Biol 4: 22–29.

42. Frouz J, Ali A, Lobinske RJ (2002) Suitability of morphological parameters for instar determination of pestiferous midges *Chironomus crassicaudatus* and *Glyptotendipes paripes* (Diptera: Chironomidae) under laboratory conditions. J Am Mosq Control 18: 222–227.

43. van Hardenbroek M, Lotter AF, Bastviken D, Duc NT, Heiri O (2012) Relationship between δ13C of chironomid remains and methane flux in Swedish lakes. Freshw Biol 57: 166–177.

44. Grey J, Kelly A, Jones RI (2004) High intraspecific variability in carbon and nitrogen stable isotope ratios of lake chironomid larvae. Limnol Oceanogr 49: 239–244.

45. Kelly A, Jones RI, Grey J (2004) Stable isotope analysis provides fresh insights into dietary separation between *Chironomus anthracinus* and *C. plumosus*. JNABS 23: 287–296.

46. Hauke J, Kossowski T (2011) Comparison of Values of Pearson's and Spearman's Correlation Coefficients on the Same Sets of Data. Quaestiones Geographicae 30: 87–93.

47. R Core Team (2013) R: A language and environment for statistical computing. R Foundation for Statistical Computing, Vienna, Austria. ISBN 3-900051-07-0, Available: http://www.R-project.org/. Accessed 2014 Jan 6.

48. Bastviken D, Santoro AL, Marotta H, Pinho LQ, Calheiros DF, et al. (2010) Methane emissions from Pantanal, South America, during the low water season: toward more comprehensive sampling. Environ Sci Technol 44: 5450–5455.

49. Deborde J, Anschutz P, Guérin F, Poirier D, Marty D, et al. (2010) Methane sources, sinks and fluxes in a temperate tidal Lagoon: The Arcachon lagoon (SW France). Estuar Coast Shelf S 89: 256–266.

50. Hofmann H, Federwisch L, Peeters F (2010) Wave-induced release of methane: Littoral zones as a source of methane in lakes. Limnol Oceanogr 55: 1990–2000.

51. Sanders I, Heppell C, Cotton J, Wharton G, Hildrew A, et al. (2007) Emission of methane from chalk streams has potential implications for agricultural practices. Freshw Biol 52: 1176–1186.

52. Trimmer M, Hildrew AG, Jackson MC, Pretty JL, Grey J (2009) Evidence for the role of methane-derived carbon in a free-flowing, lowland river food web. Limnol Oceanogr 54: 1541.

53. van Hardenbroek M, Heiri O, Parmentier FJW, Bastviken D, Ilyashuk BP, et al. (2013) Evidence for past variations in methane availability in a Siberian thermokarst lake based on δ13C of chitinous invertebrate remains. Quat Sci Rev 66: 74–84.

54. Armitage PD, Cranston P, Pinder L (1995) The Chironomidae: biology and ecology of non-biting midges. Chapman and Hall, London. 572 pages.

55. Baker A, McLachlan A (1979) Food preferences of Tanypodinae larvae (Diptera: Chironomidae). Hydrobiol 62: 283–288.

56. Peterson BJ, Fry B (1987) Stable Isotopes in Ecosystem Studies. Annu Rev Ecol Evol Syst 18: 293–320.

57. Fuentes N, Güde H, Straile D (2013) Importance of allochthonous matter for profundal macrozoobenthic communities in a deep oligotrophic lake. Int Rev Hydrobiol 98: 1–13.

58. Meyers PA (1994) Preservation of elemental and isotopic source identification of sedimentary organic matter. Chem Geol 114: 289–302.

59. Kohzu A, Kato C, Iwata T, Kishi D, Murakami M, et al. (2004) Stream food web fueled by methane-derived carbon. Aquat Microb Ecol 36: 189–194.

24

Linking Environmental Forcing and Trophic Supply to Benthic Communities in the Vercelli Seamount Area (Tyrrhenian Sea)

Anabella Covazzi Harriague[1]*, Giorgio Bavestrello[1], Marzia Bo[1], Mireno Borghini[2], Michela Castellano[1], Margherita Majorana[1], Francesco Massa[1], Alessandro Montella[1], Paolo Povero[1], Cristina Misic[1]

1 Dipartimento di Scienze della Terra, dell'Ambiente e della Vita - DiSTAV – University of Genoa, Italy, 2 CNR-ISMAR, Institute of Marine Sciences, National Research Council, Section of La Spezia, Pozzuolo di Lerici, Italy

Abstract

Seamounts and their influence on the surrounding environment are currently being extensively debated but, surprisingly, scant information is available for the Mediterranean area. Furthermore, although the deep Tyrrhenian Sea is characterised by a complex bottom morphology and peculiar hydrodynamic features, which would suggest a variable influence on the benthic domain, few studies have been carried out there, especially for soft-bottom macrofaunal assemblages. In order to fill this gap, the structure of the meio-and macrofaunal assemblages of the Vercelli Seamount and the surrounding deep area (northern Tyrrhenian Sea – western Mediterranean) were studied in relation to environmental features. Sediment was collected with a box-corer from the seamount summit and flanks and at two far-field sites in spring 2009, in order to analyse the metazoan communities, the sediment texture and the sedimentary organic matter. At the summit station, the heterogeneity of the habitat, the shallowness of the site and the higher trophic supply (water column phytopigments and macroalgal detritus, for instance) supported a very rich macrofaunal community, with high abundance, biomass and diversity. In fact, its trophic features resembled those observed in coastal environments next to seagrass meadows. At the flank and far-field stations, sediment heterogeneity and depth especially influenced the meiofaunal distribution. From a trophic point of view, the low content of the valuable sedimentary proteins that was found confirmed the general oligotrophy of the Tyrrhenian Sea, and exerted a limiting influence on the abundance and biomass of the assemblages. In this scenario, the rather refractory sedimentary carbohydrates became a food source for metazoans, which increased their abundance and biomass at the stations where the hydrolytic-enzyme-mediated turnover of carbohydrates was faster, highlighting high lability.

Editor: Erik V. Thuesen, The Evergreen State College, United States of America

Funding: This work was undertaken within the PRIN (Progetti di Rilevante Interesse Nazionale) project "Tyrrhenian Seamount Ecosystems: An Integrated Study (TySEc)", financed by the Italian Ministry of Research and Instruction. The funders had no role in study design, data collection and analysis, decision to publish, or preparation of the manuscript.

Competing Interests: The authors have declared that no competing interests exist.

* Email: anabella7@hotmail.com

Introduction

The deep-sea communities of the Mediterranean Sea (namely those living below a 200 m depth) have been investigated rather intensively, but these studies have typically been characterised by a limited spatial or temporal scale of investigation [1,2,3,4,5,6]. Focusing on the deep Tyrrhenian Sea, very few papers have been published on microbial benthic communities [7] and meiofaunal communities [8] and studies on deep, soft-bottom macrofauna are lacking. This is surprising as the Tyrrhenian Sea hosts a number of morphological peculiarities (seep, vent, slope, and abyssal plain habitats, and seamounts etc.) that have led us to suppose that an interesting part of the Mediterranean's biodiversity could be hidden there.

The northwestern Tyrrhenian Sea is characterised by a complex hydrology [9,10], which responds to the bottom morphology. A submerged ridge, called the Vercelli Seamount, with its main axis SW-NE, reaches the photic layer from the bathyal plain [11,12].

Almost permanent frontal zones exist on the main Vercelli Seamount axis, modifying the pelagic-benthic coupling, while cyclonic and anticyclonic gyres move the water masses around it [13].

In this complex scenario, the forcings that usually shape benthic communities (depth, sediment texture, trophic supply) may change their roles suddenly. The presence of a Taylor column [14] and the impinging of water circulation on the seamount's flanks have been suggested as pivotal factors for the development of the benthic communities, for instance where the presence of a rather stable Taylor column isolates the summit and limits foraging of the down-current zones [15,16]. Higher and lower current speeds have been invoked to explain community trophic differences on different flanks of the seamount [11]. In such systems a constant pelagic-benthic coupling, providing the aphotic sediment with food, is hardly possible, strongly influencing the benthic assemblages [14,17,18,19]. All these variables and the variable slope of the flanks, that may exert a certain influence on the sediment texture, influence the characteristics of the benthic community, and have encouraged the hypothesis that seamounts and the surrounding area are peculiar hotspots of marine life [20,21,22].

This study aims to link the metazoan benthic community (meiofauna and macrofauna) to the environmental constraints, highlighting how the parameters considered (morphological as well as trophic) may have a role in community characterisation.

Material and Methods

Study area, sampling sites and sampling strategy

The sampling area lies in the NW Tyrrhenian Sea (NW Mediterranean), near Sardinia (Fig. 1), and it is located within the following coordinates: 40°46′N/10°39′E, 40°47′N/11°34′E, 41°24′N/11°34′E and 41°24′N/10°38′E. It is centred on the Vercelli Seamount, an elongated, chain seamount whose axis is oriented SW-NE, and whose summit (41°06′N/10°54′E) rises from the bathyal plain to a depth of 55 m. The seamount summit covers an area of about 0.36 km^2, characterised by alternating rocky and sandy surfaces of variable depths. About half of the area lies between 100 and 120 m, while only 15% is shallower than 80 m.

This area has been studied previously for its hydrodynamic features, within the framework of research on the Tyrrhenian Sea circulation and water mass fluxes. The previous studies showed the presence of a large cyclonic structure (the Bonifacio gyre) [23] that displayed permanent features, centred NW of the sampling area and crossing it in its W sector [9,13]. Krivosheya [24] noted the presence of an anticyclonic companion of the Bonifacio gyre to the southeast. The Vercelli Seamount is placed within the transition area between the two gyres [9], whose boundaries are frontal zones [10]. In addition, Vetrano et al. [13] noted the presence of another mesoscale structure in the NE section of the sampling area, although less stable than the others. All these structures extend from the surface to considerable depths [13].

Sediment and water samples were collected from the R/V Urania during May 2009. During our field studies endangered or protected species were not involved and the permit for sampling activity was issued by the Italian Ministry of Defence (Military Navy) and the Italian Ministry of Communications. Nine stations were visited (Fig. 1). Station 28 was only sampled for organic matter (OM), and enzymatic and granulometric parameters due to the steepness of the seafloor, which prevented proper closure of the sampling device, and station 0 only for macrofauna due to the risk of damage to the sampling device on the irregular sandy-rocky bottom.

The stations (Table 1) were located on the summit (station 0) and around the seamount at different distances from the summit. Stations 14 and 28 were located on the upper-flank area, NE and NW of the summit respectively, where the flanks were rather steep. The other stations were located at different depths on the seamount's flanks, stations 9 and 16 at medium depths (1100–1200 m) and stations 25 and 32 at lower depths (1700–1800 m), close to the bathyal plain. Station 41 was located in the extreme SE sector of the study area and up-current from the seamount. Station 53 was, instead, placed in the extreme NW sector, influenced by the permanent Bonifacio gyre. Both stations were considered control sites, being about 50 km from the seamount summit, although having very different depths and features.

Undisturbed sediment cores were collected with independent deployments using a box-corer with a 29 cm internal diameter. Immediately after the arrival of the box-corer on board, the overlaying bottom water was gently removed without notably

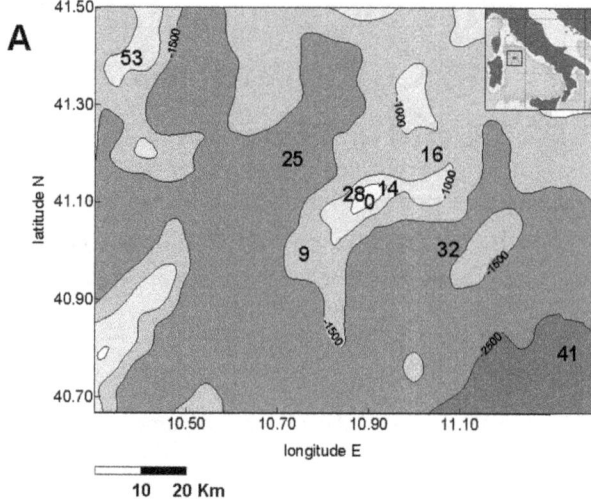

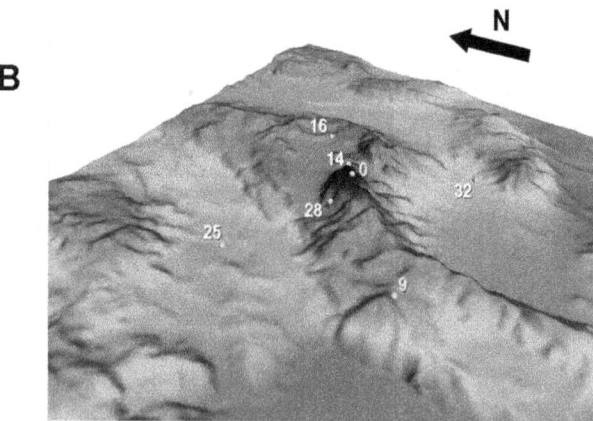

Figure 1. View of the Vercelli Seamount area (Tyrrhenian Sea). The bathymetric map (A) shows that the seamount has a SW-NE oriented axis and the summit (41°06′N/10°54′E) rises from the bathyal plain to a depth of 55 m. All the stations where sediment has been sampled are shown (see Table 1 for further details). The three-dimensional detail of the seamount (map B) is centred on the summit station 0 (at 115 m depth), the upper flank stations 14 and 28 (at 400 and 877 m depth), the medium-flank stations 9 and 16 (at 1232 and 1166 m depth), and the deeper-flank stations 25 and 32 (at 1833 and 1728 m depth).

disturbing the sediment surface. For the granulometric, OM, enzymatic activity and meiofaunal analyses, the sediment in the box-corer was subsampled by pushing PVC cores (5 cm internal diameter) into it, in duplicate for each analysis except meiofaunal, which was collected in triplicate. Each core was then aseptically sliced (at 0–2 and 2–10 cm depths). Each sediment slice was frozen until analysis, except for the enzymatic activity determinations, that were performed immediately. For the macrofaunal analyses, 3 deployments were performed for each station and entirely sorted with a 500 μm mesh net. The limited sampling for meiofauna could have reduced the representativeness of the data. In particular, the rare species could have been underestimated, thus leading only to preliminary considerations on meiofaunal diversity.

In addition to the sediment sampling, we also performed surface-water layer sampling at 25 stations distributed throughout the sampling area. Four depths were sampled with Niskin bottles: surface, oxygen minimum (between 30 and 50 m), fluorescence

Table 1. Location of sampling stations with respect to the seamount morphology and to the general current flow and evaluated variables.

station	latitude N	longitude E	depth (m)	location	bottom inclination (°)	exposition (°N)	current from (a)	sampling parameters
0	41.108	10.907	115	summit	1,70	77	SE	Gr, Ma
9 *	40.993	10.759	1232	medium flank	8,49	128	SW	Gr, OM, E, Me, Ma
14	41.128	10.940	400	upper flank	14,61	20	SE	Gr, OM, E, Me, Ma
16 *	41.200	11.038	1166	medium flank	4,91	261	N-NW	Gr, OM, E, Me, Ma
25	41.188	10.733	1833	deeper flank	0,36	352	SE	Gr, OM, E, Me, Ma
28	41.117	10.867	877	upper flank	17,46	332	SE	Gr, OM, E
32 *	41.003	11.074	1728	deeper flank	2,84	287	SE	Gr, OM, E, Me, Ma
41 *	40.790	11.339	2646	far-field	5,23	302	SE	Gr, OM, E, Me, Ma
53 *	41.394	10.378	887	far-field	5,84	238	W-SW	Gr, OM, E, Me, Ma

Asterisks denote those stations whose bottom inclination doesn't follow the seamount flank general pendency.
Gr: granulometry, OM: organic matter, E: enzymatic activity, Me: meiofauna, Ma: macrofauna.
(a): general current flows as reported in Vetrano et al. [13]

maximum (between 55 and 75 m), depth of extinction of the fluorescence signal (at ca. 120 m). Seawater was filtered through Whatman GF/F filters (in triplicate) for the chlorophyll-a concentration determination.

Analytical procedures

Benthic community. Meiofauna was extracted from the sediment by sieving through 500 μm and 45 μm mesh nets. The fraction retained was resuspended and processed according to the protocol reported by Danovaro [25]. All meiofaunal organisms were counted and classified to phylum or class level under a stereomicroscope. In order to obtain the functional parameter (biomass) of the meiofaunal component, the organisms were weighed after drying at 60°C for 24 h.

The macrofaunal specimens were recognised, when possible, down to species level. All the organisms, divided by species, were weighed to obtain the biomass value expressed as DW (drying at 60°C for 24 h). Molluscs and echinoderms were treated with 30% HCl prior to weighing.

Environmental variables. The grain size analysis was performed following Buchanan and Kain [26]; briefly, sediments were sieved (9 mesh sizes from 3.35 to 0.063 mm) after H_2O_2 treatment and drying (60°C, 48 h) and each fraction was weighed. Sediment particle-size diversity (Sed-H) was calculated from the percent dry weight of 5 size classes (<0.063 mm, 0.063–0.212 mm, 0.212–0.5 mm, 0.5–2 mm and >2 mm) using the Shannon–Wiener diversity index [27].

The chlorophyll-a concentrations in the water column were measured on board following the method of Holm-Hansen et al. [28], using a Perkin Elmer LS50B spectrofluorometer calibrated with chlorophyll-a from spinach (Sigma C5753). The specific standard deviation of the replicates was on average 4%.

The protein content of the sediment was determined following Hartree [29], the carbohydrate content was determined following Dubois et al. [30] and the lipid content was determined following Bligh and Dyer [31] and Marsh and Weinstein [32]. A Jasco V-500 spectrophotometer was calibrated with bovine serum albumin, glucose and tripalmitine solutions, respectively. Labile organic phosphorus was determined following the first step of the sequential extraction (SEDEX) proposed by Ruttenberg [33]. Briefly, sediment samples (3 to 5 g) were shaken for 2 h at 50°C in a 1 M $MgCl_2$ solution in order to detach the loosely absorbed P from the sediment. The supernatant was then treated with an oxidising solution ($K_2S_2O_8$) [34] in order to transform all the P into inorganic phosphates, which were then detected following Hansen and Grasshoff [35] with a SYSTEA Nutrient Probe Analyser. Organic carbon and total nitrogen were determined following Hedges and Stern [36] with a Carlo Erba Mod. 1110 CHN Elemental Analyser after acidification with hydrochloric acid to remove the inorganic carbonate fraction. Cyclohexanone-2,4-dinitrophenyl hydrazone was chosen as standard.

The hydrolytic enzymatic activities (β-glucosidase – BG, alkaline phosphatase – AP and leucine aminopeptidase – LA) were determined following Hoppe [37], using artificial substrates: 4-methylumbelliferyl β-D glucopyranoside and 4-methylumbelliferyl phosphate (excitation at 365 nm and emission at 460 nm) for BG and AP, respectively, and L-leucine 7-amido-4-methylcoumarin hydrochloride (excitation at 380 nm and emission at 440 nm) for LA. The samples and controls (sample sediment boiled as a blank for accidental contamination due to handling and for abiotic cleavage of the artificial substrates) were incubated in duplicate with 0.5 ml of substrate solution for 3 h. Incubations in the dark respected in-situ temperatures. Fluorescence was measured with a Perkin-Elmer 50 L spectrofluorometer previously calibrated with 4-methylumbelliferone and 7-amino-4-methylcoumarin solutions. The LA and BG activities were converted into equivalents of mobilised C assuming that 1 nmol of substrate hydrolysed enzymatically corresponded to 72 ng of mobilised C [38].

The two degradative enzymes were associated to their respective OM component: the LA with proteins, because the enzymatic hydrolysis of polypeptides is the preliminary step to amino-acid mineralisation [39], and the BG with carbohydrates, because the enzyme is involved in cellulose degradation [40]. AP activity has generally been associated with remineralisation of dissolved inorganic P, but it has bi-functional features [41]. We related it to labile P. The OM turnover times were calculated by converting the proteins and carbohydrates into C equivalents (factors of 0.49 and 0.40 for proteins and carbohydrates, respectively, according to the C content of the standard) and then dividing by the LA and BG activities transformed into their equivalents of mobilised C. The labile phosphorus turnover time was calculated following the same procedure, but assuming that 1 nmol of substrate hydrolysed enzymatically by AP corresponded to 31 ng of potentially released phosphate [38].

Statistics

We tested the differences in the same variable between different samplings with the one-way ANOVA test followed by the Newman-Kneuls post-hoc test (ANOVA+NK test) (Statistica software). To test the relationships between the various parameters, a Spearman-rank correlation analysis was performed. We used the PRIMER 6β programme package to perform SIMPER analysis on the metazoan abundance data, separately for meiofauna and macrofauna. The data matrices have been transformed using presence/absence. DISTML (distance-based linear model) routines were performed with the PERMANOVA+ programme package for PRIMER to analyse and model the relationship between the meiofaunal abundance and the environmental variables. Of the original set of environmental parameters, 9 were retained for further analysis. The variables with correlation R^2 values >0.9 (considered redundant) were omitted for the DISTLM procedures. The meiofaunal DISTML was constructed using the step-wise selection procedure and the adjusted R^2 as selection criterion to enable the fitting of the best explanatory environmental variables in the model. Euclidean distance was used as resemblance measure.

A PCA analysis, based on to the trophic quality of the OM, was performed of the 4th-root-transformed data (reported in Table 2) to reveal similarities between stations.

Results

Benthic communities

The meiofaunal communities had their highest densities at the stations situated on the upper and medium flanks of the seamount (Fig. 2A). The total abundance, although it has to be considered as preliminary data, decreased significantly with water depth (R = −0.82, n = 7, p<0.05), a trend confirmed also by the DISTLM analysis (Table 3). The first 2 cm depth was the layer mainly inhabited by the organisms (85–99% of the total abundance) at most of the stations, although in the northwestern area (stations 53 and 25) the vertical distribution was more homogeneous: only 62% and 58%, respectively, preferred the surface sediments (Fig. 2A). Overall 13 taxa were found in the study area, 11 at the seamount stations (Table 4), although the limited sampling procedure could have not highlighted all the rare species in the different sites. The number of taxa ranged between 5 (stations 9

Table 2. Quality indexes of the OM: C/N ratio, protein/carbohydrate ratio, turnover times (days) in the two sediment layers.

station	area	C/N ratio		protein/carbohydrate ratio		carbohydrate turnover		protein turnover		labile P turnover	
		0–2 cm	2–10 cm	0–2 cm	2–10 cm	0–2 cm	2–10 cm	0–2 cm	2–10 cm	0–2 cm	2–10 cm
53	far-field	14.8±0.8	13.5±1.6	0.23	0.15	379	823	2.6	1.6	0.09	0.19
25	deeper flank	20.7±2.8	10.6±0.1	0.34	0.24	218	840	1.9	3.2	0.04	0.25
9	medium flank	19.7±2.7	15.2±0.4	0.11	0.30	549	653	1.5	3.9	0.02	0.10
28	upper flank	12.4±1.1	15.0±2.9	0.16	0.32	382	768	0.8	2.1	nd	0.09
14	upper flank	11.7±0.9	10.5±1.8	0.11	0.10	179	997	3.7	3.0	0.01	0.06
16	medium flank	11.8±0.7	11.0±1.0	0.14	0.20	223	264	1.2	1.6	0.03	0.07
32	deeper flank	13.6±0.8	13.6±0.2	0.20	0.24	826	1332	1.8	2.4	0.08	0.18
41	far-field	12.0±0.8	12.6±0.3	0.43	0.27	427	854	7.6	8.2	0.09	0.11

and 25) and 8 (stations 14 and 16, Fig. 2A). The assemblage abundances were dominated by nematodes and copepods with nauplii (61–76% for nematodes and 5–11% for copepods). The community structure was similar in the different parts of the seamount and far field (SIMPER: dissimilarities from 33.6% to 23.8%). The biomass patterns were similar to the abundance ones (Fig. 2B), except at station 53, which had a higher biomass in the deeper sediments due to the high number of polychaetes found. The higher biomass contributions were given by nematodes (26–71%) and polychaetes (0–69%), while copepods represented only 4–9%.

Macrofauna was only found at four stations (53, 9, 16 and 0, Fig. 3A), characterised by depths shallower than 1500 m. It seems that the presence of macrofauna was related to depth, but actually station 14 (390 m depth) showed no macrofaunal organisms, indicating that features other than depth must be involved. Macrofaunal densities were lower than 50 ind m^{-2} at stations 53, 9 and 16, but at station 0 the value was greater than 3000 ind m^{-2}. The number of taxa was also low at stations 53, 9 and 16, with four species of polychaetes, one species of crustacean, and two species of sipunculans. Station 0, instead, showed a high diversification (Table 5) and was dominated by crustaceans. Molluscs, polychaetes and others showed similar contributions; only echinoderms were scarce (Fig. 3A). The SIMPER analysis showed higher dissimilarities between assemblages of the seamount flanks and of the far field area (85.7%), between the assemblages of the seamount flanks and of the summit (100%) and between the assemblages of the far field and of the summit

(94.7%). The macrofaunal biomass showed patterns similar to the density, but the taxa contribution was different: at the summit station crustaceans accounted for 94%, while at the other stations polychaetes made higher contributions (Fig. 3B).

Environmental variables

Seawater autotrophic biomass. The distribution of the autotrophic biomass in the water column (integrated values in the 0–120 m layer for the chlorophyll-a concentration) is shown in Fig. 4. The highest values in the area studied were found at station 0 (seamount summit, 0.43 µg l^{-1}) and in the SW and NW sectors (0.41 µg l^{-1} at station 53). The waters surrounding the summit had low (0.14 µg l^{-1} at station 14) and rather low (from 0.22 to 0.24 µg l^{-1} at stations 9, 16, 28 and 32) values. The SE sector also displayed rather low values (0.20 µg l^{-1} at station 41) while values higher than 0.24 µg l^{-1} were observed in the NE sector.

Sediment texture. The mean grain size of the sediment (Fig. 5) was in the range of silt & clay for half of the sampled stations, while only the shallowest station 0 had a medium-sand texture. The coarser grain sizes (gravel, >2 mm) were found only at the shallowest stations 0 and 14, while the silt & clay fraction was highly represented in the 0–2 cm layer of station 16 (40%). On average, the most represented fraction was that ranging from 0.064 to 0.212 mm (52%), followed by the fraction ranging from 0.212 to 0.5 mm (30%) and by the fraction lower than 0.064 mm (15%).

The Sed-H results (Fig. 5) highlighted the highest sediment diversity at stations 0 and 14, and the lowest at stations 25 and 32.

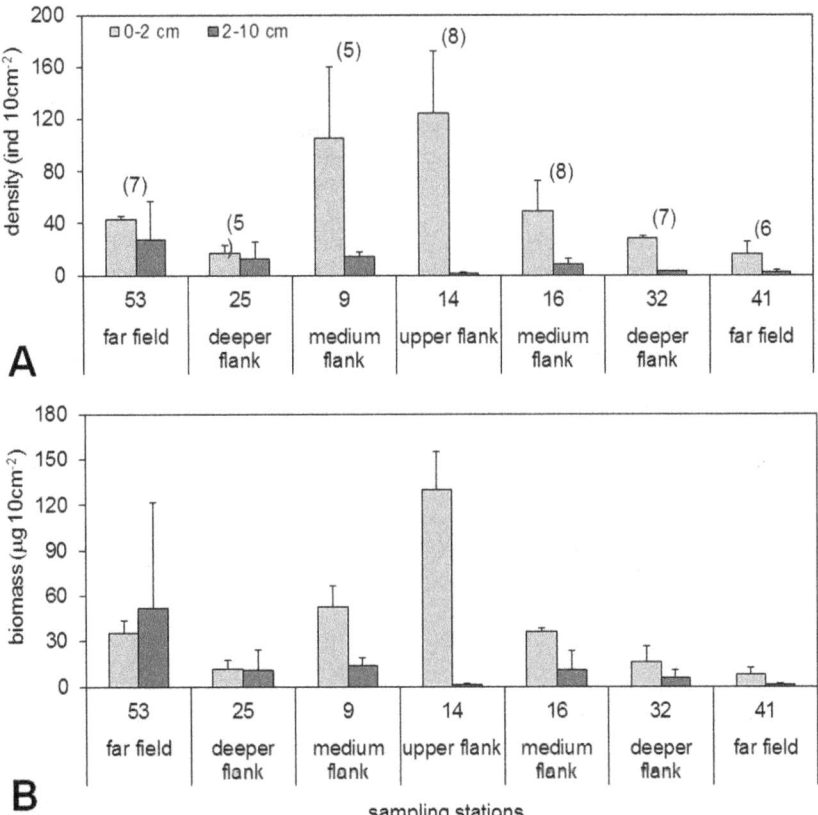

Figure 2. Preliminary meiofaunal abundance and biomass data for the two sediment layers of each station. Values averaged over replicates collected in the same deployment: bars denote standard error. A: abundance (individuals 10 cm^{-2}), in brackets the number of taxa, B: biomass (µg 10 cm^{-2}, the values have been calculated with the gravimetric method).

Table 3. Distance-based linear model (DISTLM) for meiofaunal abundance and selected environmental variables.

MARGINAL TESTS				
Variable	SS(trace)	Pseudo-F	P	Prop.
H sed	2323.2	3.788	0.052	0.43
water depth	3688.4	10.840	0.016	0.68
protein turnover time	418.6	0.421	0.597	0.08
carbohydrate turnover time	471.5	0.479	0.542	0.09
labile P turnover time	2295.3	3.709	0.107	0.43
proteins	1468.1	1.872	0.240	0.27
carbohydrates	4276.6	19.212	0.002	0.79
lipid	469.6	0.477	0.510	0.09
labile P	1154.5	1.363	0.277	0.21

SEQUENTIAL TESTS					
Variable	Adjust.R^2	SS(trace)	Pseudo-F	P	Prop.
+carbohydrates	0.7522	4276.6	19.212	0.003	0.79
+labile P	0.7561	236.6	1.080	0.364	0.04
+water depth	0.8380	439.8	3.022	0.140	0.08
+carbohydrate turnover time	0.9452	338.2	6.870	0.060	0.06
+protein turnover time	0.9857	85.6	6.660	0.172	0.02

Marginal tests: explanation of variation for each variable taken alone. Sequential tests: conditional tests of individual variables in constructing the model. Each test examines whether adding the variable contributes significantly to the explained variation. Selection procedure: step-wise, selection criterion: adjusted R^2. Prop.: % variation explained.

The sediment diversity was significantly correlated to the meiofaunal biomass (R = 0.77, n = 7, p<0.05), and also the meiofaunal density was directly proportional to the sediment diversity, although not significantly (Table 3).

Sedimentary organic matter (OM). The carbohydrate content (Fig. 6A) had the highest values of the OM components (on average 583.1±232.2 μg g^{-1} for the two layers). All the upper- and medium-flank stations of the seamount showed 0–2 cm layer values significantly higher than the 2–10 cm layer ones (one-way ANOVA and NK post-hoc test, p<0.05), with stations 9, 14 and 16 reaching the highest absolute values (from 776.9±125.6 to 1169.3±43.4 μg g^{-1} for stations 16 and 14, respectively). Considering the 0–2 cm layer, the correlation between meiofauna abundance and carbohydrates was highly significant (R = 0.95, n = 7, p<0.01). The DISTLM analysis confirmed that, within the environmental trophic features, the carbohydrate content was the variable mainly linked to the meiofaunal abundance.

Lower values were observed for the lipid content (on average 123.7±106.2 μg g^{-1} for the two layers), even lower if the upper- and medium-flank stations of the seamount were considered (on average 47.9±21.0 μg g^{-1}) (Fig. 6B). This difference was significant (one-way ANOVA and NK post-hoc test, p<0.01). On the other hand, no significant differences were observed between the two sediment layers of the upper-and medium-flank stations of the seamount, while the other stations showed a significant accumulation in the 2–10 cm layer (one-way ANOVA and NK post-hoc test, p<0.01).

The protein content values (Fig. 6C) were more homogeneous throughout area (on average 114.0±27.1 μg g^{-1}), without significant differences except for the two layers of station 41 (one-way ANOVA and NK post-hoc test, p<0.05). The labile P values (Fig. 6D) did not show significant differences between the stations. The OC and N contents (Figs. 6E and 6F) of the two layers of the upper- and medium-flank stations of the seamount,

considered together, showed significantly lower values than the other stations (one-way ANOVA and NK post-hoc test, p<0.01). Significant differences between the two layers were rare for protein, P labile, OC and N contents.

Sedimentary enzymatic activity and OM turnover. The enzymatic activities are shown in Fig. 7 and the related OM turnover times in Table 2.

The LA activity was on average 29.7±12.8 nmol g^{-1}h^{-1} for the whole area and the stations didn't show significant differences. The surface sediment layer always had higher values than the deep one, sometimes significant (one-way ANOVA and NK post-hoc test, p<0.05). The protein turnover times were, on average, 2.9±2.1 days for the whole area and the surface layer showed generally lower values than the deep one, except for stations 53 and 14. The deepest station, station 41, showed the highest value, followed by the shallowest, station 14.

The AP activity showed the highest values in the 0–2 cm layer of station 14, but station 9 also showed notable activity in its surface layer. The turnover times were significantly (one-way ANOVA and NK post-hoc test, p<0.05) lower for the upper- and medium-flank stations than for the other stations (on average 0.05±0.03 vs 0.13±0.07 days, respectively).

The BG activity was higher at stations 14 and 16, lower at the southernmost stations 32 and 41, and rather high variability was recorded for the upper- and medium-flank stations. The surface sediment layer showed significantly higher BG values than the deep layer (one-way ANOVA and NK post-hoc test, p<0.05). The carbohydrate turnover times were very high (on average 607±332 days) and the surface layer showed lower values than the deep one.

Keeping in mind that the meiofaunal data were preliminary due to the limited sampling procedure, the DISTML analysis (Table 3) indicated that the best model for the meiofaunal abundance and the environmental features included, together with the water depth and the carbohydrate sedimentary content, the turnover times for

Table 4. List of the meiofaunal taxa found in the sampling stations and position of each station with respect to the seamount morphology.

stations taxa	53 far-field	25 deeper flank	9 medium flank	14 upper flank	16 medium flank	32 deeper flank	41 far-field
Amphipoda						1.3	
Copepoda	6.1	10.0	8.2	9.2	4.2	11.3	7.4
Nauplii		2.5	0.4	1.7	1.2		
Gastrotricha					0.6		
Kinorhyncha	0.5						
Nematoda	79.7	81.2	82.6	78.1	85.5	77.5	83.3
Oligochaeta					1.2		
Ostracoda	1.0		1.7	0.6	0.6		1.8
Polychaeta	11.7	3.8	3.2	8.3	4.2	3.7	
Rotifera	0.5		3.9	1.2	1.9	2.5	1.9
Sipuncula		1.3		0.3			
Tanaidacea				0.3			
Thermosbaenacea	0.5			0.3		2.5	1.9
Turbellaria		1.3			0.6	1.2	3.7

Percentage contributions of the taxa to the meiofaunal total abundance for each station are presented.

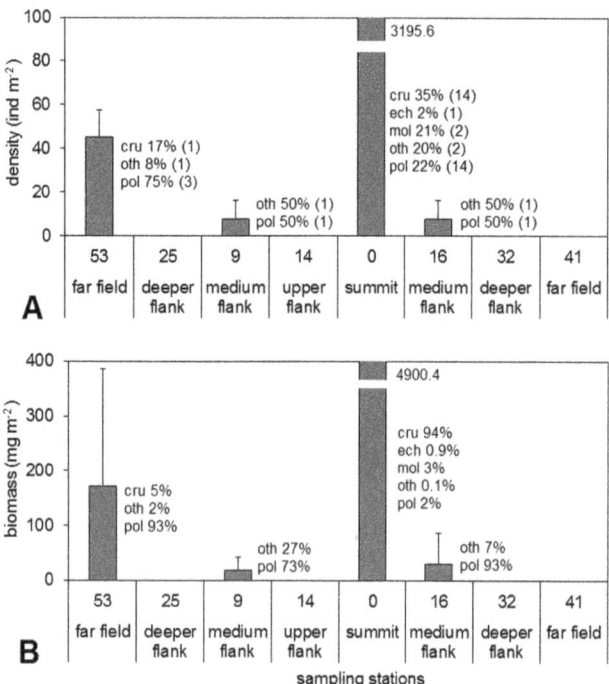

Figure 3. Average macrofaunal abundance and biomass and the contribution of each taxonomical group (cru: crustacean, pol: polychaetes, mol: molluscs, ech: echinoderms and oth: others) for each station. Values averaged over independent replicates: bars denote standard error. A: abundance (individuals m^{-2}), in brackets the number of taxa, B: biomass(μg m^{-2}, the values have been calculated with the gravimetric method).

proteins and carbohydrates. The PCA analysis (Fig. 8), focused on the indexes of OM lability, showed thatPCA1 and PCA2 explained 78.5% of the variation. The superimposed bubbles are proportional to the meiofaunal abundances and the asterisks indicate where macrofauna were found.

The statistical analyses indicated that the C/N and the protein/carbohydrate ratios failed to give a reasonable picture of the OM lability, at least for the sediments. For instance, the highest protein turnover times of station 41, indicating the poor trophic quality of the protein content, disagrees with the high protein/carbohydrate ratio values, thus indicating refractory protein accumulation rather than a large pool of trophic resources. This may be related to an analytical bias, because the protein method overestimates the concentration of proteins because of the presence of humic substances [42].

Discussion

The Tyrrhenian basin has been considered as one of the more oligotrophic of the western Mediterranean in terms of water column features [43]. In agreement with this scenario, previous studies on meiofauna [8] highlighted rather low abundances compared to extra-Mediterranean ones (in the Atlantic Ocean, for instance) [16], while benthic prokaryote abundance and biomass fell within the range of values reported in the literature for deep-sea sediments worldwide [7]. However, no exhaustive information has been provided for the deep Tyrrhenian Sea macrofaunal organisms, nor for the seamount-summit macrofauna of this area.

The summit station, 0, is a world apart for macrofauna, due firstly to the shallow depth (within the photic zone and enriched by

good trophic-quality phytopigments) and to a coarse granulometry that allows for deep oxygenation of the sediment. The heterogeneity of the substrata at the summit (rocky substrate alternating with coarse, biogenic debris) may provide different habitats for a high number of species [15]. In addition, the presence of a *Laminaria rodriguezii* meadow at the seamount summit [11]would provide the macrofauna with a surplus of food resources through the detrital food chain. In fact, as reported for the coastal *Posidonia oceanica* meadows of the Ligurian Sea [44], the macrobenthic community at the summit is dominated by carnivores (94% of the total abundance) such as Aciculata families (89% of polychaetes), Decapoda, Turbellaria, Lyssianassidae and the amphipod *Leucothoe occulta* (Krapp & Schickel, 1975) (Table 5). The good quality of the organic particulate matter [45] favours suspension feeders and mixed suspension-deposit feeders (ca. 6%) such as the echinoderm *Acrocnida brachiata* (Montagu, 1804), the bivalve *Lucinella divaricata* (Linnaeus, 1758) and cyprinid ostracods (Table 5). On the contrary, only deposit and mixed suspended-deposit feeders, such as Scolecida and Sipuncula, were found at the medium-flank stations (Table 5).

In our study we observed that, with the exception of the summit station, the benthic communities of the Vercelli Seamount area were poor in macrofauna (for instance, lower than at the Atlantic Condor Seamount [15]) and moderately rich in meiofauna (abundance and number of taxa higher than those reported for the sediments surrounding the Tyrrhenian Palinuro and Marsili seamounts[8] but lower than the abundances found at the Atlantic Condor Seamount [16]). Benthic communities are generally influenced by water depth [16,46]. In accordance with the above, a significant correlation between depth and our preliminary data of total meiofaunal abundance was found, and the macrofaunal organisms have only been found at some shallower stations.

Leduc et al. [47] showed that the sediment texture, transformed into a sediment-diversity index (Sed-H) by means of the Shannon-Wiener diversity index, deeply influences soft-bottom communities, especially those groups, such as meiofauna, that have a limited expansion rate due to their small size. In our study the preliminary data on meiofaunal abundance agreed with these observations, showing the highest abundances and biomasses where the Sed-H was the highest (station 14) and the lowest where the Sed-H was the lowest (stations 25 and 32).

Macrofauna did not directly respond to the Sed-H as the meiofauna did. Considering other environmental features that may be involved in the macrofauna distribution, we observed that three stations (9, 16 and 28, although the latter was not sampled for benthic communities) had a particular exposition degree, opposite that of the regular seamount flank slope. This implies the presence of irregularities along the flank that interfere with the incoming current, modifying the local hydrodynamic processes, allowing slowing of the current speed and increasing deposition. This would explain the fine granulometry of these stations, finer than the other flank-stations. Higher sedimentation rates on the northern flank of the seamount were also revealed by the composition and trophic strategies of the megafauna assemblages found there by Bo et al. [11]. The peculiar environment generated by these morphological features seems to favour macrofaunal development. Except for the summit station, only stations 9 and 16 within the very poor area of the Vercelli Seamount showed the presence of macrofaunal organisms. The lack of macrofauna at shallow station 14 may, then, be related to the high bottom slope and to an excessive hydrodynamic forcing, as this station lacks appropriate aprons. In addition, the high meiofaunal density and biomass at this site may have exerted more efficient competition

Table 5. List of the macrofaunal taxa found on the Vercelli Seamount summit (station 0), medium flank (stations 9 and 16) and in the far-field area (station 53).

Phylum	Class	Family	Species	stations 0	9–16	53
Annelida	**Polychaeta**					
	Scolecida	Capitellidae sp1			X	
		Capitellidae sp2		X		
		Opheliidae		X		
		Paraonidae sp1				X
		Paraonidae sp2		X		
	Aciculata	Chrysopetalidae sp1		X		
		Chrysopetalidae sp2		X		
		Exogoninaejuv.		X		
		Goniadidae		X		
		Hesionidae		X		
		Nereididae		X		
		Phyllodocidae sp1				X
		Phyllodocidae sp2		X		
		Sigalionidae		X		
		Syllidae sp1		X		X
		Syllidae sp2		X		
Artropoda	**Malacostraca**					
	Decapoda	Leucosiidae	*Ebalia sp*	X		
		Parthenopidae	*Parthenopoides massena*	X		
	Amphipoda	Ampithoidae	*Ampithoe helleri*	X		
		Leucothoidae	*Leucothoe occulta*	X		
		Lyssianassidae		X		
	Tanaidacea	Apseudidae	*Apseudopsis elisae*	X		
			Apseudes sp			X
	Isopoda sp1			X		
	Isopoda sp2			X		
	Isopoda sp3			X		
	Isopoda sp4			X		
	Maxillopoda					
	Copepoda			X		
	Ostracoda	Cypridinidae		X		
		Bythocytheridae		X		
Mollusca	**Bivalvia**	Limidae	*Limaria hians*	X		
		Lucinidae	*Lucinella divaricata*	X		
Echinodermata	**Ophiuroidea**	Amphiuridae	*Acrocnida brachiata*	X		
Sipuncula	**Sipunculidea**	Phascolionidae	*Phascolion (Phascolion) strombus*		X	
		Phascolosomatidae	*Phascolosoma (Phascolosoma) granulatum*		X	X
Platyhelminthes	**Turbellaria**			X		
Nematoda				X		

with macrofauna and/or a top-down control of macrofaunal juveniles [48].

However, water depth, sediment texture and sediment aspect are not enough to explain benthic community distribution [17,19], especially in a variable environment such as the Vercelli Seamount area. The quantitative features of the trophic supply in deep sediments have been indicated as a limiting factor for the growth

and activity of microbes [39] and small metazoan [49], and their qualitative features (bioavailability) have been invoked to explain community distribution as well [1,50].

Surprisingly, but in agreement with previous literature data [8], our preliminary data on meiofaunal abundance were related to a rather refractory fraction of the organic matter (OM), namely the carbohydrates, indicating that in this peculiar system the

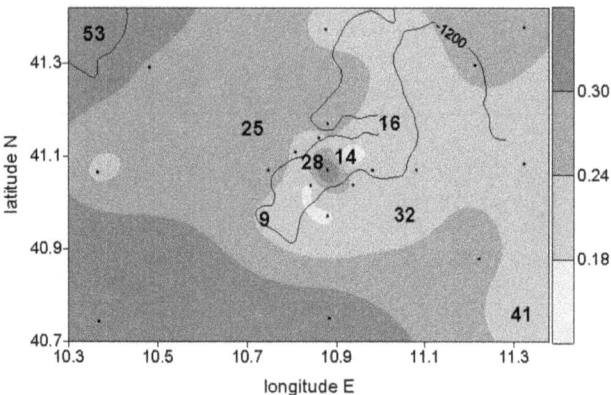

Figure 4. Seawater autotrophic pigment distribution. Chlorophyll-a concentrations (μg l^{-1}) in the water column have been averaged for the 0–120 m layer. The numbers indicate the stations also sampled for the sedimentary parameters, the dots indicate all the stations sampled for the autotrophic biomass determination in the water column. The line reports the 1200 m depth isobath.

carbohydrate trophic supply would play a role in the distribution of the meiofaunal organisms. The absence of significant relationships for the 2–10 cm layer suggested that these carbohydrates may derive from deposition via pelagic-benthic coupling.

In fact, the shallower seamount stations (9, 28, 14 and 16) showed a significant enrichment of the upper sediment layer, indicating that this carbohydrate distribution was related to the flux of OM from the surface layer. Frontal zones, as observed in the Vercelli Seamount area, are known to induce the sinking of particulate material towards the deeper layers [39]. Therefore, most of the organic bulk in these benthic areas is composed of high-molecular-weight compounds, so that the benthic zone is enriched with surface OM, which usually does not reach the seafloor. The phytoplanktonic debris, the main OM source in pelagic environments, was probably highly involved in these processes, as suggested by Speicher et al. [51]. While the more labile components of the OM flux (proteins and lipids, for instance) could be consumed and re-cycled during their sinking to the

sediment; structural carbohydrates (cellulose, for instance) were preserved and reached the sediments at a higher rate.

The complex hydrodynamic assessment of the studied area led to a carbohydrate accumulation that did not exactly match the chlorophyll-a distribution in the water column recorded during the same sampling cruise. The pelagic-benthic coupling was based on oblique fluxes rather than on vertical ones; the fluxes (and the accumulation) being higher in the shallower stations where the dilution is proportionally lower. In the southernmost section of the sampling area, the Bonifacio gyre flows from the SW [13], reaching the seamount area at station 9. The favourable conditions to phytoplanktonic biomass accumulation in the surface layer of the SW sector, that led to the very high chlorophyll-a signal recorded, could have generated a high OM flux, carrying high carbohydrate amounts to the sediments of station 9. During our cruise high chlorophyll-a values were also measured in the water column of the NE section. Sinking phytoplanktonic biomass, pushed by the gyre previously reported by Vetrano et al. [13], could have reached our station 16, increasing the carbohydrate content of the surface sediments. Station 14 was probably the only one directly subjected to the "seamount effect", indicated by the increase of phytoplanktonic biomass centred on the seamount summit that we recorded. Lateral transport of this OM could have reached shallow station 14, generating the significantly higher carbohydrate contents of the surface sediment.

Although the carbohydrate contents of the entire Vercelli Seamount area showed rather high values, the other biochemical fractions were lower than those reported for other Mediterranean and Atlantic areas [8,52], studied using the same analytical methods. The C-OM (namely the sum of the C contribution of proteins, carbohydrates and lipids) of the Portuguese, Adriatic and Cretan margins was three times the average value we found in the Vercelli Seamount area. Our results confirmed the general oligotrophy of the Tyrrhenian Sea and the tendency to ultraoligotrophy of the Vercelli area, highlighted by a very low protein content [53].

The depositional features of the studied area may have a role in these low values. Speicher et al. [51] highlighted a rather low particulate organic carbon (POC) export flux in the northern Tyrrhenian Sea (where the Vercelli Seamount is placed) during late May-early June compared to the middle and southern

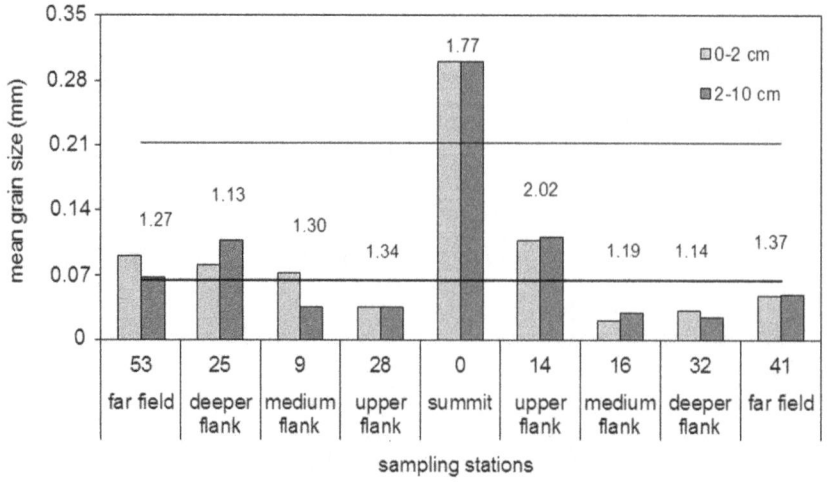

Figure 5. Sediment texture. Mean grain size (mm) for the sediment 0–2 cm layer and 2–10 cm layer. The black horizontal line highlights the upper limit of the silt & clay fraction (0.064 mm), the grey line the division between fine and medium sand (0.212 mm). The numbers above the bars report the sedimentary diversity index (Sed-H) for each station.

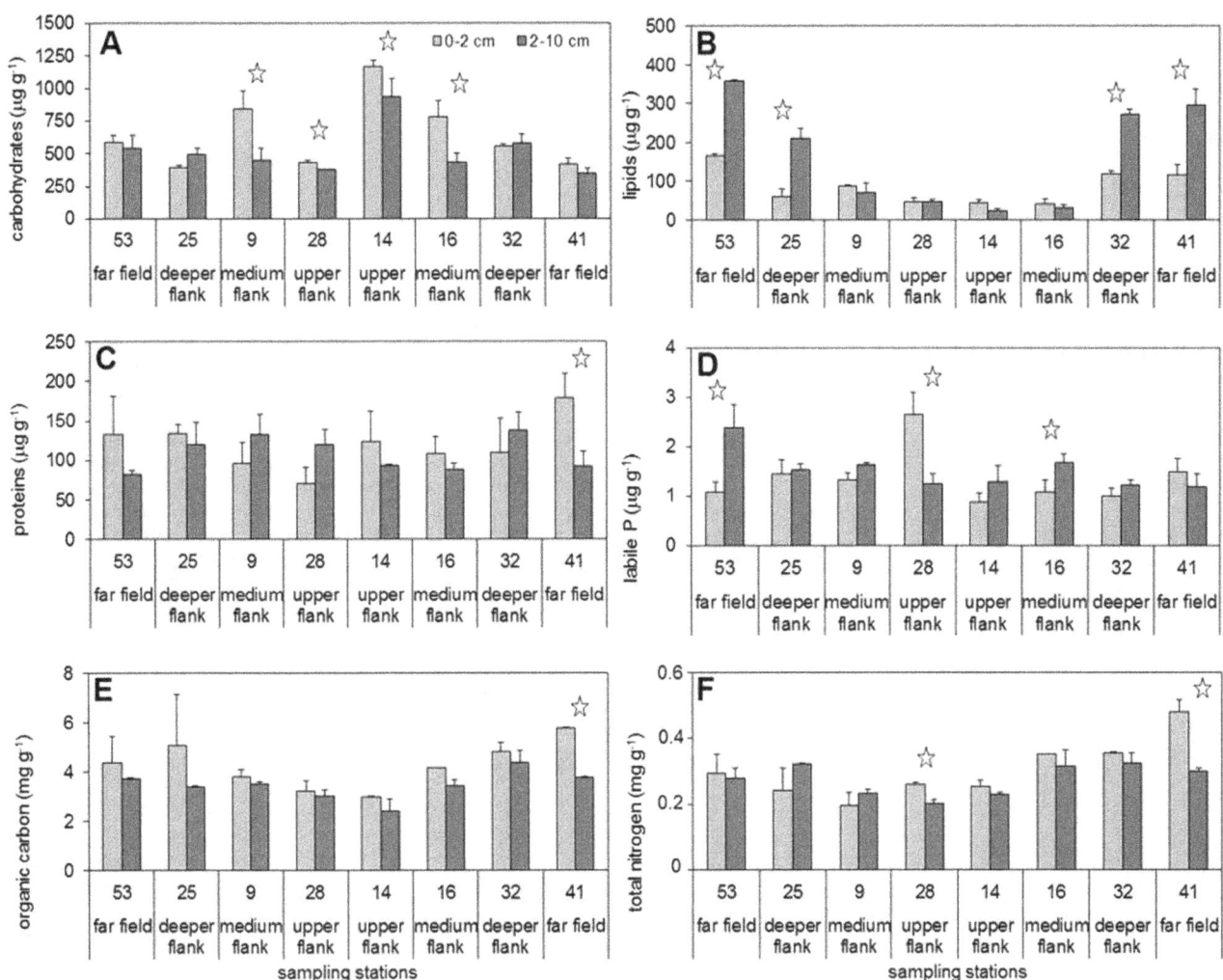

Figure 6. Organic matter contents for the two sediment layers. Bars denote standard deviations. A: carbohydrates (μg g^{-1}), B: lipids (μg g^{-1}), C: proteins (μg g^{-1}), D: labile P (μg g^{-1}), E: organic carbon (OC, mg g^{-1}), F: total nitrogen (N, mg g^{-1}). Stars denote significantly different values for the 0–2 cm and the 2–10 cm sediment layers at each station (one-way ANOVA, NK post-hoc test, $p<0.05$).

Tyrrhenian. In addition, it is well known that the sediment texture regulates the OM accumulation, the coarser sediments providing a lower surface to be coated by OM. The generally low contribution of the silt & clay fraction (grain size <63 μm) in the Vercelli area (on average 15%) compared to the 22–98% of the slope areas reported by Pusceddu et al. [8] agrees with this statement.

Together with the common quality indexes (C/N ratio and protein/carbohydrate ratio), the hydrolytic enzymatic activity may give clues on the potential lability of the OM [41], following the optimal resource allocation model of Sinsabaugh et al. [54], who stated that osmotrophic assemblages will optimize their energy expenditures by expressing high levels of particular hydrolases only if polymeric substrates are abundant and if the monomeric hydrolysates required for growth are scarce. The low values of the substrate quantity: enzyme activity ratio, namely low turnover times, highlight high trophic value OM. A major role of the OM turnover times was observed, and the highest meiofaunal abundances were found in stations where the OM was more labile (surface layer of stations 9, 14, 16) as indicated by the multivariate analysis.

The BG activity and the carbohydrate turnover time may, therefore, be useful to explain meiofaunal as well as macrofaunal

distribution. BG activity was always higher in the surface sediment layer, probably becoming too energy expensive for the trophically-limited deeper sediment layer. This matches the vertical meiofaunal distribution, with more than 80% of the total abundance found in the surface layer. Considering, instead, the macrofauna, its biomass value at station 16, higher than that of station 9, was in agreement with the higher trophic quality of the sediment, characterised by rather low turnover times for carbohydrates and proteins. In fact station 9, although showing a notable carbohydrate content, showed higher turnover times, indicating a lower lability for the OM reaching the bottom, due probably to the longer time the OM spent in the water column.

The deepest station, station 41, showed the highest protein turnover time values in both layers, indicating that the accumulated proteins did not induce LA activity as they did at the other stations of our study area. This may be because peptidase activity has been described as being negatively correlated to depth [55] but, from a trophic point of view, it indicates the lower lability of station 41 protein content. At the seamount-influenced sites, although the fluxes and/or in situ production were low, the proteins were generally more labile than those of the non-seamount station 41.

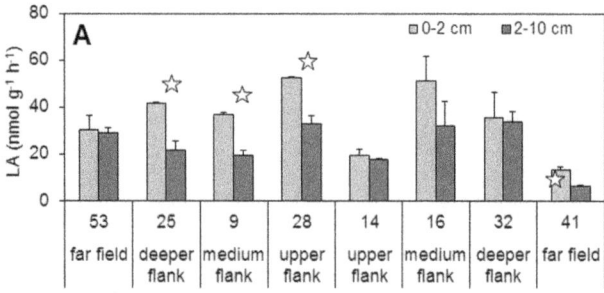

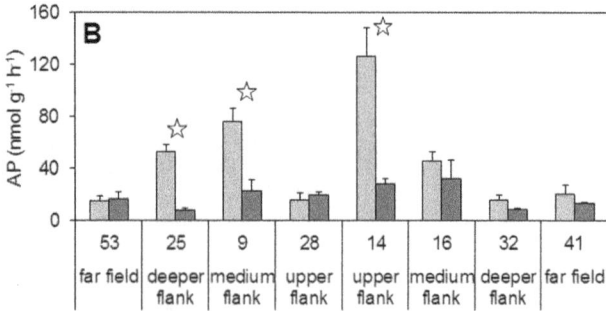

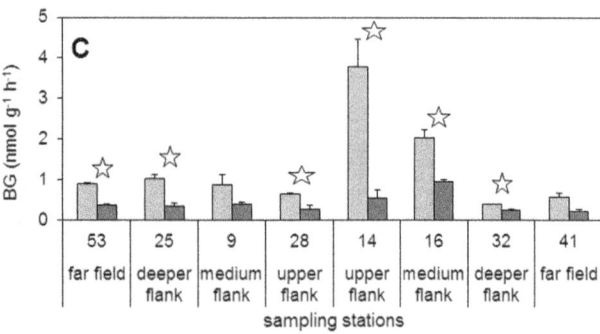

Figure 7. Enzymatic activities (nmol g⁻¹ h⁻¹) for the two sediment layers. Bars denote standard deviations. A: leucine aminopeptidase (LA), B: alkaline phosphatase (AP), C: β-glucosidase (BG). Stars denote significantly different values for the 0–2 cm and the 2–10 cm sediment layers at each station (one-way ANOVA, NK post-hoc test, p<0.05).

The relationship between benthic organisms and OM lability, expressed as a short turnover time, is complex. In fact, macrofaunal presence and also the preliminary information on meiofaunal abundance may explain the anomalous turnover times for proteins we detected for stations 53 and 14, which showed higher values in the surface sediment than in the deeper layer. If the coarser grain size of these stations allowed for higher mixing of the OM of the surface and deep sediment layers, station 53 showed high meiofaunal and rather high macrofaunal abundances, and the diffusion of OM is strongly stimulated by macrofaunal galleries [56]. At station 14, instead, the bioturbation was limited to the action of meiofauna, which showed the highest densities, due to the lack of macrofaunal organisms.

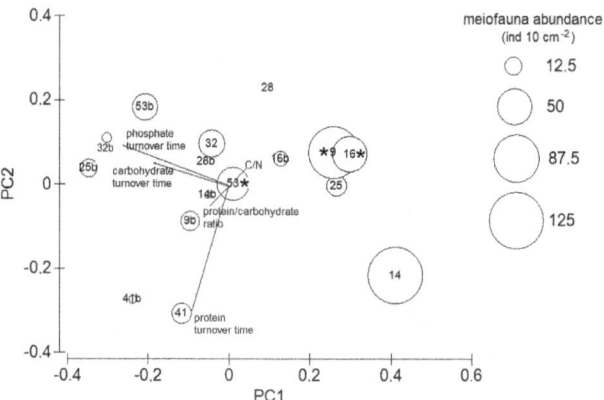

Figure 8. PCA analysis of the OM trophic-quality features (see text and Table 2 for details) for the upper flank stations 14 and 28 (at 400 and 877 m depth), the medium-flank stations 9 and 16 (at 1232 and 1166 m depth), the deeper-flank stations 25 and 32 (at 1833 and 1728 m depth), the far-field stations 53 and 41 (at 887 and 2646 m depth). C/N: (carbon/nitrogen ratio) the lower the ratio the higher the trophic quality, protein/carbohydrate: the higher the ratio the higher the trophic quality, protein-carbohydrate-labile P turnover times: the lower the time the higher the trophic quality. Meiofaunal total biomass is superimposed as bubbles and asterisks denote where macrofauna was found. "b" after the station number denotes the 2–10 cm layer of the station.

Conclusions

The Mediterranean Sea has been studied intensively, but its deep areas need more research to increase our understanding of its ecosystem functions. In particular, the studies on the deep areas of the Tyrrhenian Sea are scarce, although its morphological features (seamounts, for instance) suggest the presence of diversified habitats and ecosystems. In the present paper we observed that in the Vercelli Seamount area the peculiar environmental features (physical, morphological and trophic) differently shaped the benthic assemblages. Macrofauna showed different community composition comparing the seamount flanks and the far-field stations, and the summit station was a world apart in terms of density, diversity and biomass. Bottom inclination and aspect allowed the presence of aprons against the hydrodynamic forcing, favourable to macrofauna development. Our preliminary data on meiofaunal abundance showed, instead, a link to water depth and to trophic supply, especially to the lability of the OM. A variable pelagic-benthic coupling links the seawater with the bottom indicating that, irrespectively of being far from the coast and/or placed in the deep sea, these areas may be sensitive to global processes.

Acknowledgments

We would like to thank the captain and crew of the R/V *Urania* for their unstinting assistance during the cruise.

Author Contributions

Conceived and designed the experiments: ACH CM. Performed the experiments: ACH CM M. Bo M. Borghini MM FM AM. Analyzed the data: ACH CM FM. Contributed reagents/materials/analysis tools: ACH CM MC PP. Wrote the paper: ACH CM. Project planning: GB PP.

References

1. Danovaro R, Company JB, Corinaldesi C, D'Onghia G, Galil B, et al. (2010) Deep-Sea Biodiversity in the Mediterranean Sea: The Known, the Unknown, and the Unknowable. PLoS ONE 5(8): e11832. doi: 10.1371/journal.pone. 0011832.

2. Danovaro R, Dinet A, Duineveld G, Tselepides A (1999) Benthic response to particulate fluxes in different trophic environments: A comparison between the Gulf of Lions-Catalan Sea (Western Mediterranean) and the Cretan Sea (Eastern Mediterranean). Progr Oceanogr 44(1–3): 287–312.

3. Psarra S, Tselepides A, Ignatiades L (2000) Primary productivity in the oligotrophic Cretan Sea (NE Mediterranean): Seasonal and interannual variability. Progr Oceanogr 46: 187–204.

4. Tselepides A, Papadopoulou N, Podaras D, Plaiti W, Koutsoubas D (2000) Macrobenthic community structure over the continental margin of Crete (South Aegean Sea, NE Mediterranean). Progr Oceanogr 46(2–4): 401–428.

5. Galil BS, Goren M (1994) The deep sea Levantine fauna, new records and rare occurrences. Senckenb Marit 25 (1–3): 41–52.

6. Galil BS, Zibrowius H (1998) First benthos samples from Eratosthenes Seamount, Eastern Mediterranean. Senckenb Marit 28 (4–6): 111–121.

7. Danovaro R, Corinaldesi C, Luna GM, Magagnini M, Manini E, et al. (2009) Prokaryote diversity and viral production in deep-sea sediments and seamounts. Deep-Sea Res II 56: 738–747.

8. Pusceddu A, Gambi C, Zeppilli D, Bianchelli S, Danovaro R (2009) Organic matter composition, metazoan meiofauna and nematode biodiversity in Mediterranean deep-sea sediments. Deep-Sea Res II 56: 755–762.

9. Artale V, Astraldi M, Buffoni G, Gasparini GP (1994) Seasonal variability of gyre-scale circulation in the northern Tyrrhenian Sea. J Geophys Res 99 (C7): 14,127–14,137.

10. Nair R, Cattini E, Gasparini GP, Rossi G (1994) Circolazione ciclonica e distribuzione dei nutrienti nel Tirreno settentrionale. Proceedings of the X Symposium of the Italian Association of Limnology and Oceanology (AIOL). pp. 65–76.

11. Bo M, Bertolino M, Borghini M, Castellano M, Covazzi Harriague A, et al. (2011) Characteristics of the mesophotic megabenthic assemblages of the Vercelli Seamount (North Tyrrhenian Sea). PLoS ONE 6(2): e16357. doi: 10.1371/journal.pone.0016357.

12. Wezel FC (1985) Structural features and basin tectonics of the Tyrrhenian Sea. In: Stanley DJ, Wezel FC, editors.Geological evolution of the Mediterranean Basin.Springer-Verlag, New York. pp.153–194.

13. Vetrano A, Napolitano E, Iacono R, Schroeder K, Gasparini GP (2010) Tyrrhenian Sea circulation and water mass fluxes in spring 2004: Observations and model results. J Geophys Res. 115 (C06023), 18 pp.

14. Dower J, Freeland H, Juniper K (1992) A strong biological response to oceanic flow past Cobb Seamount. Deep-Sea Res A 39: 1139–1145.

15. Bongiorni L, Ravara A, Parretti P, Santos RS, Rodrigues CF, et al. (2013) Organic matter composition and macrofaunal diversity in sediments of the Condor Seamount (Azores, NE Atlantic). Deep-Sea Res II 98: 75–86.

16. Zeppilli D, Bongiorni L, Cattaneo A, Danovaro R, Serrão Santos R (2013) Meiofauna assemblages of the Condor Seamount (North-East Atlantic Ocean) and adjacent deep-sea sediments. Deep-Sea Res 98: 87–100.

17. Clark MR, Rowden A, Schlacher T, Williams A, Consalvey M, Stocks KI, Rogers AD, O'Hara TD, White M, Shank TM, Hall-Spencer JM (2010) The ecology of seamounts: structure, function and human impacts. Ann Rev Mar Sci 2: 253–278.

18. Gooday AJ (2002) Biological responses to seasonally varying fluxes of organic matter to the ocean floor: a review. J Oceanogr 58(2): 305–332.

19. Piepenburg D, Müller B (2004) Distribution of epibenthic communities on the Great Meteor Seamount (North-east Atlantic) mirrors pelagic processes. Arch Fish Mar Res 51: 55–70.

20. Gad G, Schminke HK (2004) How important are seamounts for the dispersal of interstitial meiofauna. Arch Fish Mar Res51: 43–54.

21. George KH (2004) Description of two new species of Bodinia, a new genus incertae sedis in Argestidae Por, 1986 (Copepoda, Harpacticoida), with reflections on argestid colonization of the Great Meteor Seamount plateau. Org Divers Evol 4 (4): 241–254.

22. Christiansen B, Wolff G (2009) The oceanography, biogeochemistry and ecology of two NE Atlantic seamounts: The OASIS project. Deep-Sea Res II 56 (25): 2579–2581.

23. Moen J (1984) Variability and mixing of the surface layer in the Tyrrhenian Sea: MILEX-80, Final Report. Saclancen Report SR-75, 128 pp.

24. Krivosheya VG (1983) Water circulation and structure in the Tyrrhenian Sea. Oceanology 23: 166–171.

25. Danovaro R (2010) Methods for the Study of Deep-sea Sediments their Functioning and Biodiversity. CRC Press, Boca Raton, FL. pp. 420.

26. Buchanan J B, Kain JM (1971) Measurement of the physical and chemical environment. In Holme NA, Mc Intyre AD, editors. Methods for the Study of Marine Benthos. Blackwell Scientific Publications, Oxford, Edinburgh. pp. 30–52.

27. Etter RJ, Grassle JF (1992) Patterns of species diversity in the deep sea as a function of sediment particle size diversity. Nature 369: 576–578.

28. Holm-Hansen O, Lorenzen CJ, Holmes RW, Strickland JDH (1965) Fluorometric determination of chlorophyll. J Cons – Cons Perm Int Explor Mer 30: 3–15.

29. Hartree EF (1972) Determination of proteins: a modification of the Lowry method that give a linear photometric response. Anal Biochem 48: 422–427.

30. Dubois M, Gilles K, Hamilton JK, Rebers PA, Smith F(1956) Colorimetric method for determination of sugars and related substances. Anal Chem 28: 350–356.

31. Bligh EG, Dyer WJ (1959) A rapid method for total lipid extraction and purification. Canadian J Biochem Physiol 37: 911–917.

32. Marsh BJ, Weinstein DB (1966) Simple charring method for determination of lipids. J Lipid Res 7: 574–576.

33. Ruttenberg KC (1992) Development of a sequential extraction method for different forms of phosphorous in marine sediments. Limnol Oceanogr 37: 1460–1482.

34. Koroleff F (1983) Determination of phosphorus. In: Grassoff, K, Ehrhardt, M, Kremling, K, editors.Methods of Seawater Analysis. Verlag Chemie, Weinheim, Haugland. pp.125–139.

35. Hansen HP, Grasshoff K (1983) Automated chemical analysis. In: Grassoff K, Ehrhardt M, Kremling K, editors.Methods of Seawater Analysis. Verlag Chemie, Weinheim, Haugland. pp. 347–379.

36. Hedges JI, Stern JH (1984) Carbon and nitrogen determination of carbonate-containing solids. Limnol Oceanogr 29: 657–663.

37. Hoppe HG (1983) Significance of exoenzymatic activities in the ecology of brackish water: measurements by means of methylumbelliferyl substrates. Mar Ecol–Prog Ser 11: 299–308.

38. Caruso G, Monticelli L, Azzaro F, Azzaro M, Decembrini F, et al. (2005) Dynamics of extracellular enzymatic activities in a shallow Mediterranean ecosystem (Tindari Ponds, Sicily). Mar Freshw Res 56: 173–188.

39. Bianchi A, Calafat A, De Wit R, Garcin J, Tholosan O, et al. (2003) Microbial activity at the deep water sediment boundary layer in two highly productive systems in the Western Mediterranean: the Almeria-Oran front and the Malaga upwelling. Oceanol Acta 25: 315–324.

40. Bhaskar PV, Bhosle NB(2008) Bacterial production, glucosidase activity and particle-associated carbohydrates in Dona Paula bay, west coast of India. Estuar Coast Shelf Sci 3: 413–424.

41. Mudryk ZJ, Skórczewski P (2004) Extracellular enzyme activity at the air-water interface of an estuarine lake. Estuar Coast Shelf Sci 59: 59–67.

42. Vakondios N, Koukouraki EE, Diamadopoulos E (2014) Effluent organic matter (EfOM) characterization by simultaneous measurement of proteins and humic matter. Water Res 63: 62–70.

43. Bosc E, Bricaud A, Antoine D (2004) Seasonal and interannual variability in algal biomass and primary production in the Mediterranean Sea, as derived from 4 years of SeaWiFS observations. Global Biogeochemical Cycles 18 (1).

44. Covazzi Harriague A, Bianchi CN, Albertelli G (2006) Soft-bottom macrobenthic community composition and biomass in a Posidonia oceanica meadow in the Ligurian Sea (NW Mediterranean). Estuar Coast Shelf Sci 70: 251–258.

45. Misic C., Bavestrello G, Bo M, Borghini M, Castellano M, et al. (2012) The "seamount effect" as revealed by organic matter dynamics around a shallow seamount in the Tyrrhenian Sea (Vercelli Seamount, western Mediterranean). Deep-Sea Res I 67: 1–11.

46. Rex MA, Etter RJ, Morris JS, Crouse J, McClain CR, et al. (2006) Global bathymetric patterns of standing stock and body size in the deep-sea benthos. Mar Ecol–Prog Ser 317: 1–8.

47. Leduc D, Rowden AA, Probert PK, Pilditch CA, Nodder SD, et al. (2012) Further evidence for the effect of particle-size diversity on deep-sea benthic biodiversity. Deep-Sea Res I 63: 164–169.

48. Zobrist EC, Coull BC (1992) Meiobenthic interactions with macrobenthic larvae and juveniles: an experimental assessment of the meiofaunal bottleneck. Mar Ecol–Prog Ser 88: 1–8.

49. Fonseca G, Soltwedel T (2009) Regional patterns of nematode assemblages in the Arctic deep seas. Polar Biol 32: 1345–1357.

50. Heinz P, Ruepp D, Hemleben C (2004) Benthic foraminifera assemblages at Great Meteor Seamount. Mar Biol 144(5): 985–998.

51. Speicher EA, Moran SB, Burd AB, Delfanti R, Kaberi H, et al. (2006) Particulate organic carbon export fluxes and size-fractionated POC/^{234}Th ratios in the Ligurian, Tyrrhenian and Aegean Seas. Deep-Sea Res I 53: 1810–1830.

52. Pusceddu A, Bianchelli S, Canals M, Sanchez-Vidal A, Durrieu De Madron X, et al. (2010) Organic matter in sediments of canyons and open slopes of the Portuguese, Catalan, Southern Adriatic and Cretan Sea margins. Deep-Sea Res I 57: 441–457.

53. Rossi S, Grémare A, Gili J-M, Amouroux J-M, Jordana E, et al. (2003) Biochemical characteristics of settling particulate organic matter at two north-western Mediterranean sites: a seasonal comparison. Estuar Coast Shelf Sci 58: 423–434.

54. Sinsabaugh RL, Findlay S, Franchini P, Fischer D (1997) Enzymatic analysis of riverine bacterioplankton production. Limnol Oceanogr 42: 29–38.

55. Tamburini C, Garcin J, Ragot M, Bianchi A (2002) Biopolymer hydrolysis and bacterial production under ambient hydrostatic pressure through a 2000 m water column in the NW Mediterranean. Deep-Sea Res II 49: 2109–2123.

56. De Wit R, Bouloubassi I (1998) Oxygen penetration depth and aerobic microbial respiration in sediments of the Western Mediterranean. Third MTP-II Workshop on the Variability of the Mediterranean Sea. Rhodos (Greece). pp. 207.

PERMISSIONS

All chapters in this book were first published in PLOS ONE, by The Public Library of Science; hereby published with permission under the Creative Commons Attribution License or equivalent. Every chapter published in this book has been scrutinized by our experts. Their significance has been extensively debated. The topics covered herein carry significant findings which will fuel the growth of the discipline. They may even be implemented as practical applications or may be referred to as a beginning point for another development.

The contributors of this book come from diverse backgrounds, making this book a truly international effort. This book will bring forth new frontiers with its revolutionizing research information and detailed analysis of the nascent developments around the world.

We would like to thank all the contributing authors for lending their expertise to make the book truly unique. They have played a crucial role in the development of this book. Without their invaluable contributions this book wouldn't have been possible. They have made vital efforts to compile up to date information on the varied aspects of this subject to make this book a valuable addition to the collection of many professionals and students.

This book was conceptualized with the vision of imparting up-to-date information and advanced data in this field. To ensure the same, a matchless editorial board was set up. Every individual on the board went through rigorous rounds of assessment to prove their worth. After which they invested a large part of their time researching and compiling the most relevant data for our readers.

The editorial board has been involved in producing this book since its inception. They have spent rigorous hours researching and exploring the diverse topics which have resulted in the successful publishing of this book. They have passed on their knowledge of decades through this book. To expedite this challenging task, the publisher supported the team at every step. A small team of assistant editors was also appointed to further simplify the editing procedure and attain best results for the readers.

Apart from the editorial board, the designing team has also invested a significant amount of their time in understanding the subject and creating the most relevant covers. They scrutinized every image to scout for the most suitable representation of the subject and create an appropriate cover for the book.

The publishing team has been an ardent support to the editorial, designing and production team. Their endless efforts to recruit the best for this project, has resulted in the accomplishment of this book. They are a veteran in the field of academics and their pool of knowledge is as vast as their experience in printing. Their expertise and guidance has proved useful at every step. Their uncompromising quality standards have made this book an exceptional effort. Their encouragement from time to time has been an inspiration for everyone.

The publisher and the editorial board hope that this book will prove to be a valuable piece of knowledge for researchers, students, practitioners and scholars across the globe.

LIST OF CONTRIBUTORS

Xiaodong Nie, Zhongwu Li, Bin Huang, Yan Zhang, Wenming Ma, Yanbiao Hu and Guangming Zeng
College of Environmental Science and Engineering, Hunan University, Changsha, PR China
Key Laboratory of Environmental Biology and Pollution Control (Hunan University), Ministry of Education, Changsha, PR China

Jinquan Huang
Department of Soil and Water Conservation, Yangtze River Scientific Research Institute, Wuhan, PR China

Sara M. Smith
Dauphin Island Sea Lab, Dauphin Island, Alabama, United States of America

Eric L. Sparks and Just Cebrian
Dauphin Island Sea Lab, Dauphin Island, Alabama, United States of America, Marine Sciences, University of South Alabama, Mobile, Alabama, United States of America

James G. Kairo
Kenya Marine and Fisheries Research Institute, Mombasa, Kenya

Joseph Kipkorir Sigi Lang'at
Kenya Marine and Fisheries Research Institute, Mombasa, Kenya
School of Life, Sport and Social Sciences, Edinburgh Napier University, Edinburgh, United Kingdom

Mark Huxham
School of Life, Sport and Social Sciences, Edinburgh Napier University, Edinburgh, United Kingdom

Maurizio Mencuccini
School of Geosciences, University of Edinburgh, Crew Building, Edinburgh, United Kingdom
Centre for Ecological Research and Forestry Applications, Universitat Autònoma de Barcelona, Barcelona, Spain

Steven Bouillon
Department of Earth and Environmental Sciences, Katholieke Universiteit
Leuven, Leuven, Belgium

Martin W. Skov
School of Ocean Sciences, Bangor University, Menai Bridge, Anglesey, United Kingdom

Susan Waldron
School of Geographical and Earth Sciences, University of Glasgow, Glasgow, South Lanarkshire, United Kingdom

Remo Freimann
Institute of Molecular Health Sciences, Professorship of Genetics, ETH Zurich, Zurich, Switzerland
Department of Aquatic Ecology, Swiss Federal Institute of Aquatic Science and Technology, Eawag, Dübendorf, Switzerland and Institute of Integrative Biology, ETH-Zurich, Zurich, Switzerland

Christopher T. Robinson
Department of Aquatic Ecology, Swiss Federal Institute of Aquatic Science and Technology, Eawag, Dübendorf, Switzerland and Institute of Integrative Biology, ETH-Zurich, Zurich, Switzerland

Helmut Bürgmann
Department of Surface Waters – Research and Management, Swiss Federal Institute of Aquatic Science and Technology, Eawag, Kastanienbaum, Switzerland

Stuart E. G. Findlay
Cary Institute of Ecosystem Studies, Millbrook, New York, United States of America

Gaurav Srivastava and R.C. Mehrotra
Birbal Sahni Institute of Palaeobotany, Lucknow, India

Mingguo Zheng
Key Laboratory of Water Cycle and Related Land Surface Processes, Institute of Geographic Sciences & Natural Resources Research, Chinese Academy of Sciences, Beijing, China

Yishan Liao
Guangdong Institute of Eco-environment and Soil Sciences, Guangzhou, China

Jijun He
Base of the State Laboratory of Urban Environmental Processes and Digital Modelling, Capital Normal University, Beijing, China

Michael Haslam
Research Laboratory for Archaeology and the History of Art, University of Oxford, Oxford, United Kingdom

Raphael Moura Cardoso
Institute of Psychology, University of São Paulo, São Paulo, Brazil

Elisabetta Visalberghi
Istituto di Scienze e Tecnologie della Cognizione, Consiglio Nazionale delle Ricerche, Roma, Italy

Dorothy Fragaszy
Department of Psychology, University of Georgia, Athens, Georgia, United States of America

Xiaoyun Huang
College of Biological Science and Technology, Fuzhou University, Fuzhou 350108, P.R. China

Guozeng Wang, Juan Lin and Xiu Yun Ye
College of Biological Science and Technology, Fuzhou University, Fuzhou 350108, P.R. China
National Engineering Laboratory for High-efficiency Enzyme Expression, Fuzhou 350002, P. R. China

Tzi Bun Ng
School of Biomedical Sciences, Faculty of Medicine, The Chinese University of Hong Kong, Shatin, New Territories, Hong Kong, China

Nicole A. Hill, Neville Barrett and Justin Hulls
Institute for Marine and Antarctic Studies, University of Tasmania, Hobart, Tasmania, Australia

Emma Lawrence
Digital Productivity Flagship, Commonwealth Scientific and industrial Research Organisation (CSIRO), Brisbane, QLD, Australia

Jeffrey M. Dambacher and Keith R. Hayes
Digital Productivity Flagship, Commonwealth Scientific and industrial Research Organisation (CSIRO), Hobart, Tasmania, Australia

Scott Nichol
Geoscience Australia, Canberra, ACT, Australia

Alan Williams
Oceans and Atmosphere Flagship, Commonwealth Scientific and industrial Research Organisation (CSIRO), Hobart, Tasmania, Australia

David B. McWethy
Department of Earth Sciences, Montana State University, Bozeman, Montana, United States of America

Jamie R. Wood and Matt S. McGlone
Landcare Research, Lincoln, New Zealand

Janet M. Wilmshurst
Landcare Research, Lincoln, New Zealand
School of Environment, University of Auckland, Auckland, New Zealand

Cathy Whitlock
Department of Earth Sciences, Montana State University, Bozeman, Montana, United States of America
Institute on Ecosystems, Montana State University, Bozeman, Montana, United States of America

Yimin Gao, Pengcheng Gao, Fen Liu, Zuoping Zhao and Yan Pang
College of Natural Resources and Environment, Northwest A&F University, Yangling, China

Xiaoying Wang and Yanan Tong
College of Natural Resources and Environment, Northwest A&F University, Yangling, China
Key Laboratory of Plant Nutrition and the Agri-environment in Northwest China, Ministry of Agriculture, Yangling, China

Willem Stock, Koen Sabbe and Marleen De Troch
Department of Biology, Ghent University, Ghent, Belgium

Kim Heylen and Anne Willems
Department of Biochemistry and Microbiology, Ghent University, Ghent, Belgium

Jessica J. Eichmiller, Przemyslaw G. Bajer and Peter W. Sorensen
Department of Fisheries, Wildlife, and Conservation Biology, Minnesota Aquatic Invasive Species Research Center, University of Minnesota, Twin Cities, St. Paul, Minnesota, United States of America

Xiao Wen Hu, Yan Pei Wu, Xing Yu Ding, Rui Zhang and Yan Rong Wang
State Key Laboratory of Grassland Agro-ecosystems, College of Pastoral Agriculture Science and Technology, Lanzhou University, Lanzhou, 730020, China

Jerry M. Baskin
Department of Biology, University of Kentucky, Lexington, Kentucky 40506-0225, United States of America

Carol C. Baskin
Department of Biology, University of Kentucky, Lexington, Kentucky 40506-0225, United States of America
Department of Plant and Soil Sciences, University of Kentucky, Lexington, Kentucky 40546-0312, United States of America

Rashmi Srivastava and Gaurav Srivastava
Cenozoic Palaeoflorist Laboratory, Birbal Sahni Institute of Palaeobotany, 53 University Road, Lucknow- 226 007, Uttar Pradesh, India

David L. Dilcher
Department of Geology, Indiana University, 1001 E. Tenth St. Bloomington- 47405, Indiana, United States of America

Maarten J. Schaafsma
B-WARE Research Centre, Radboud University Nijmegen, Mercator 3, Nijmegen, The Netherlands

Josepha M. H. van Diggelen, Gijs van Dijk and Alfons J. P. Smolders
B-WARE Research Centre, Radboud University Nijmegen, Mercator 3, Nijmegen, The Netherlands Institute for Water and Wetland Research, Department of Aquatic Ecology and Environmental Biology, Radboud University Nijmegen, Nijmegen, The Netherlands

Leon P. M. Lamers and Jan G. M. Roelofs
Institute for Water and Wetland Research, Department of Aquatic Ecology and Environmental Biology, Radboud University Nijmegen, Nijmegen, The Netherlands

Ina Severin, Eva S. Lindström and Örjan Östman
Department of Ecology and Genetics/Limnology, Uppsala University, Uppsala, Sweden

Sarah J. Ivory
Brown University, Providence, Rhode Island, United States of America

Michael M. McGlue
University of Kentucky, Lexington, Kentucky, United States of America

Geoffrey S. Ellis
U.S. Geological Survey, Denver, Colorado, United States of America

Anne-Marie Lézine
LOCEAN, CNRS, Paris, France

Andrew S. Cohen
University of Arizona, Tucson, Arizona, United States of America

Annie Vincens
CEREGE, CNRS, Aix-en-Provence, France

Lucy G. Gillis, Peter M. J. Herman and Tjeerd J. Bouma
Spatial Ecology Department, Royal Netherlands Institute for Sea Research (NIOZ), Yerseke, Zealand, The Netherlands

Alan D. Ziegler
Geography Department, National University of Singapore (NUS), Singapore, Singapore

Dick van Oevelen
Ecosystems Studies Department, Royal Netherlands Institute for Sea Research (NIOZ), Yerseke, Zealand, The Netherlands

Cecile Cathalot
Laboratoire Environnement Profond (LEP), French Research Institute for Exploitation of the Sea, Polouzane, Brittany, France

Jan W. Wolters
Department of Biology, University of Antwerp, Antwerp, Flanders, Belgium

Yi Zhou, Shaojun Zhang, Ying Liu and Hongsheng Yang
Key Laboratory of Marine Ecology and Environmental Sciences, Institute of Oceanology, Chinese Academy of Sciences, Qingdao, P. R. China

Simon P. Rout, Jessica Radford, Andrew P. Laws, Francis Sweeney, Ahmed Elmekawy, Lisa J. Gillie and Paul N. Humphreys
Department of Chemical and Biological Sciences, School of Applied Sciences, University of Huddersfield, Huddersfield, United Kingdom

John D. Hedley
Environmental Computer Science Ltd., Tiverton, Devon, United Kingdom

Kathryn McMahon
School of Natural Sciences and Centre for Marine Ecosystems Research, Edith Cowan University, Joondalup, Western Australia

Peter Fearns
Department of Imaging and Applied Physics, Curtin University of Technology, Perth, Western Australia

John Gichimu Mbaka, Celia Somlai, Denis Köpfer, Andreas Maeck, Andreas Lorke and Ralf B. Schäfer
Institute for Environmental Sciences, University of Koblenz-Landau, Landau, Rhineland-Palatinate State, Germany

Anabella Covazzi Harriague, Giorgio Bavestrello, Marzia Bo, Michela Castellano, Margherita Majorana, Francesco Massa, Alessandro Montella, Paolo Povero, Cristina Misic
Dipartimento di Scienze della Terra, dell'Ambiente e della Vita - DiSTAV – University of Genoa, Italy

Mireno Borghini
CNR-ISMAR, Institute of Marine Sciences, National Research Council, Section of La Spezia, Pozzuolo di Lerici, Italy

Index